Metal-Organic Frameworks for Anticorrosive Coatings

This book provides a comprehensive analysis of the application of metal-organic frameworks (MOFs) in corrosion protection, consolidating current knowledge and offering insights into potential future advancements in the field. It explores the methods used to fabricate MOF coatings, discussing their advantages, disadvantages, and suitability for various harsh environments. The book also includes industrial case studies, highlighting real-world applications and providing guidance on the future potential of MOFs as anticorrosive coatings.

Features:

- Explores the function of MOFs as anticorrosive coatings, addressing their characteristics, methods of manufacture, and real-world uses.
- Discovers how MOF coatings offer superior durability and longevity compared to traditional coatings.
- Offers step-by-step guidance on choosing the right protective measures for different materials and environments.
- Demonstrates binding effectiveness of MOF coatings with different metallic surfaces using experimental and computational approaches.
- Reviews the broad applicability of MOF coatings across various industries.

This book is aimed at researchers, graduate students in materials science, corrosion science, and related fields, as well as industry professionals seeking to understand and apply the latest developments in corrosion protection.

Emerging Materials and Technologies

Series Editor: Boris I. Kharissov

The *Emerging Materials and Technologies* series is devoted to highlighting publications centered on emerging advanced materials and novel technologies. Attention is paid to those newly discovered or applied materials with potential to solve pressing societal problems and improve quality of life, corresponding to environmental protection, medicine, communications, energy, transportation, advanced manufacturing, and related areas.

The series takes into account that, under present strong demands for energy, material, and cost savings, as well as heavy contamination problems and worldwide pandemic conditions, the area of emerging materials and related scalable technologies is a highly interdisciplinary field, with the need for researchers, professionals, and academics across the spectrum of engineering and technological disciplines. The main objective of this book series is to attract more attention to these materials and technologies and invite conversation among the international R&D community.

Boron Nitride Nanostructures
Applications in Energy and Healthcare
Indranil Lahiri, Amrita De Adhikari, Souvik Ghosh, Rita Joshi, Satish Jaiswal, and Debrupa Lahiri

Smart Polymers
An Emerging Vista for Nanotheranostics
Dipankar Chattopadhyay, Adrija Ghosh, Jonathan Tersur Orasugh and Suprakas Sinha Ray

Lightweight Materials for Electric Vehicles
Sustainable Materials, Production Process, and Modeling Techniques
Edited by R. Suresh, C. Durga Prasad, Satish Kumar, Bharath K.N. and Ajith G. Joshi

Graphitic Carbon Nitrides
Michio Inagaki, Masahiro Toyoda and Yasushi Soneda

The Science and Applications of 3D Bioprinting
Edited by Thiago Domingues Stocco and Anderson de Oliveira Lobo

Metal-Organic Frameworks for Anticorrosive Coatings
Edited by Ashish Kumar, Ambrish Singh and Humira Assad

For more information about this series, please visit: www.routledge.com/Emerging-Materials-and-Technologies/book-series/CRCEMT

Metal-Organic Frameworks for Anticorrosive Coatings

Edited by
Ashish Kumar, Ambrish Singh
and Humira Assad

CRC Press
Taylor & Francis Group
Boca Raton London New York

CRC Press is an imprint of the
Taylor & Francis Group, an **informa** business

First edition published 2026
by CRC Press
2385 NW Executive Center Drive, Suite 320, Boca Raton, FL 33431

and by CRC Press
4 Park Square, Milton Park, Abingdon, Oxon, OX14 4RN

CRC Press is an imprint of Taylor & Francis Group, LLC

ISBN: 9781032936246 (hbk)
ISBN: 9781032945200 (pbk)
ISBN: 9781003571162 (ebk)

DOI: 10.1201/ 9781003571162

Typeset in Times
by Newgen Publishing UK

Contents

About the Editors

Ashish Kumar is currently working as a Professor (Chemistry), Dean Academics and Head, Applied Science and Humanities at Nalanda College of Engineering, Bihar Engineering University under the Department of Science, Technology and Technical Education, Government of Bihar, India. He has published more than 150 research papers in high-impact quality journals. He has also authored and reviewed several books and published diverse chapters (100+) in referred books with international publishers. He is a guest editor and an Editorial Board member of several international journals, as well as a keynote speaker, session chair, and a member of technical committee in various national and international conferences. His area of research includes materials chemistry, corrosion, nanomaterials, environmental science, surface chemistry, and solution thermodynamics. He is listed in Stanford University's list of the World's Top 2% researchers and Scopus 2023 and is the recipient of several research awards, including Asia's Prominent Scientists Award 2023. His current h-index is 36.

Ambrish Singh is working as a Professor in the Department of Chemistry, Nagaland University, Lumami, Nagaland, India. He has worked as a professor in the School of New Energy and Materials, Southwest Petroleum University, China, for a decade before coming back to Nagaland. He is included as a Fellow in the Royal Society of Chemistry (FRSC). His research interests are mainly focused on corrosion, electrochemistry, green chemistry, quantum chemistry, smart coatings, polymers, nanomaterials, composites, and petroleum engineering. He has received the prestigious Sichuan 1000 Talent Award from the Sichuan government, China, for his outstanding research contributions as a faculty. He has also received the President's Award for exceptional post-doctoral research work. Moreover, he has received the Young Scientist Award from UPCST, Lucknow, India. He has published more than 175 SCI peer-reviewed research papers in high-impact journals. He is acting as a reviewer for more than 53 high-impact journals and an editor for a few journals. Dr. Singh has been invited to present his work at several national and international conferences, seminars, webinars, and workshops. He is the author as well as editor of several book chapters and books. He has drafted 50 patents based on his new and innovative findings and awarded with one in China, one in India, two in South Africa, and published 21 in India. He has completed several state and provincial projects in China and India. He was a consultant to KFUPM, Saudi Arabia, for an international project. He is a life member of ISCA, NESA, CRSI, ITS, AMPP, SPE, and ACS. His research paper citations are over 9000, with an h-Index of 52 and i10-index of 133. He is listed in the top 2% scientists of the world for the last three years according to the list of Stanford and Elsevier. He is a visiting professor at Al-Farabi National Kazakh University (QS Rank-150), Almaty, Kazakhstan.

Humira Assad is currently a Postdoctoral Fellow in the Department of Metallurgical Engineering and Materials Science at the Indian Institute of Technology Bombay (IIT Bombay). With nearly five years of dedicated research experience, Dr. Assad specializes in a diverse range of fields, including electrochemistry, corrosion, nanomaterials, green chemistry, polymer composite coatings, surface chemistry, and computational chemistry. Dr. Assad has made significant contributions to the scientific community, having authored more than 40 highly regarded research papers and 50 book chapters published by leading global publishers. Her work has earned her recognition as a potential reviewer for prominent journals such as ACS, RSC, Elsevier, etc. With an h-index of 24 and an i10-index of 32, Dr. Assad's research has been cited approximately 1969 times, underscoring her impact in her areas of expertise. Her excellence in research has been acknowledged through various accolades, including the Best Oral Presentation Award at the 1st National Conference on Contemporary Innovations in Science and Technology in 2019. Dr. Assad has also been recognized as one of the top 2% of scientists globally, according to the Stanford University and Scopus 2025 data. Among her many honors, she is the recipient of the Eurodec Paints Research Excellence Award 2024 and the Best Alumni Award 2025 for her outstanding contributions to research.

Contributors

A. Mushira Banu
Assistant Professor, PG and Research Department of Chemistry
Jamal Mohamed College, Affiliated to Bharathidasan University
Tiruchirappalli, Tamilnadu, India

Muhamad Rizal Addin
Department of Metallurgical Engineering,
Faculty of Engineering, University of Sultan Ageng Tirtayasa,
Indonesia

F.M. Mashood Ahamed
Assistant Professor, PG and Research Department of Chemistry
Jamal Mohamed College, Affiliated to Bharathidasan University
Tiruchirappalli, Tamilnadu, India

Farhat A Ansari
Department of Chemistry
Integral University
Lucknow, Uttar Pradesh, India

Humira Assad
Department of Metallurgical Engineering and Materials Science
Indian Institute of Technology Bombay
Mumbai, India

B. Arifa Farzana
Assistant Professor, PG and Research Department of Chemistry
Jamal Mohamed College, Affiliated to Bharathidasan University
Tiruchirappalli, Tamil Nadu, India

Youssef Belaoufi
Laboratory of Organic, Bio-organic and Environmental Chemistry
Faculty of Sciences
Chouaib Doukkali University of El Jadida, El Jadida, Morocco

Elyor Berdimurodov
Chemical & Materials Engineering
New Uzbekistan University,
Tashkent, Uzbekistan
and
University of Tashkent for Applied Sciences
Uzbekistan Faculty of Chemistry
National University of Uzbekistan
Tashkent, Uzbekistan

Khasan Berdimuradov
Faculty of Industrial Viticulture and Food Production Technology
Shahrisabz Branch of the Tashkent Institute of Chemical Technology
Shahrisabz, Uzbekistan

Ayoub Cherrat
LS3MN2E, CERNE2D, Faculty of Sciences
Mohammed V University
Rabat, Morocco

Mohamed Dalimi
LS3MN2E, CERNE2D, ENSAM
Mohammed V University
Rabat, Morocco

Houda Damour
Faculty of Sciences and Technologies of Mohammedia
University Hassan II of Casablanca
Morocco

Hicham Elmsellem
Higher Institute of Nursing Professions and Health Techniques (ISPITSO)
Oujda, Morocco
Laboratory of Applied Chemistry and Environment (LCAE)
Department of Chemistry, Faculty of Sciences,
University Mohamed I
Oujda, Morocco

Samia Essadki
Higher Institute of Nursing Professions and Health Techniques (ISPITSO)
Oujda, Morocco

Neeta Gupta
Department of Chemistry
Govt. E. Raghavendra Rao P.G. Science College
Bilaspur, Chhattisgarh, India

Sangeeta Gupta
Department of Chemistry
Govt. Shivnath Science College
Rajnandgaon, Chhattisgarh, India

Souad El Hajjaji
Faculty of Sciences and Technologies of Mohammedia
University Hassan II of Casablanca
Morocco

Fadly Daffa Hardiansyah
Department Metallurgical Engineering
Faculty of Engineering, University of Sultan Ageng Tirtayasa
Indonesia

Arockia Selvi J
Department of Chemistry,
SRM Institute of Science and Technology,
Kattankulathur, India

Bhawana Jain
Department of Chemical Sciences, Siddhachalam Laboratory
Raipur, India

Bening Nurul Hidayah Kambuna
Department Metallurgical Engineering
Faculty of Engineering, University of Sultan Ageng Tirtayasa
Indonesia

Anvar Khamidov
Department of Natural-Mathematical Sciences
Turin Polytechnic University in Tashkent
Tashkent, Uzbekistan

İlayda Bersu Kul
Department of Molecular Biology and Genetics
Science Faculty, Sivas Cumhuriyet University
Sivas, Turkey

Ashish Kumar
Nalanda College of Engineering
Bihar Engineering University, Science, Technology and Technical Education Department, Government of Bihar, India

Houda Labjar
Faculty of Sciences and Technologies of Mohammedia
University Hassan II of Casablanca
Morocco

Najoua Labjar
LS3MN2E, CERNE2D, ENSAM, Mohammed V University
Rabat, Morocco

Nabil Lahrache
LS3MN2E, CERNE2D, ENSAM, Mohammed V University
Rabat, Morocco

Limasenla Longkumer
Department of Chemistry
Nagaland University
Zunheboto, Nagaland, India

M. Arthanareeswari
Department of Chemistry
SRM Institute of Science and Technology
Kattankulathur, Tamilnadu, India

Utkir Madatov
Department of Physical Chemistry
National University of Uzbekistan
Tashkent, Uzbekistan

Sourav Maity
Centre for the Environment
Indian Institute of Technology Guwahati
Guwahati, Assam, India

Jaykhun Mamatov
Department of Physical Chemistry
National University of Uzbekistan
Tashkent, Uzbekistan

Ali Mehri
Institute for Convergence Science & Technology
Center for Bioscience & Technology
Sharif University of Technology
Tehran, Iran

Ammar Fahmy Almasarie Nahdi
Department Metallurgical Engineering
Faculty of Engineering, University of Sultan Ageng Tirtayasa
Indonesia

Hamid Nasrellah
ESEF, University Chouaib Doukkali University of El Jadida
El Jadida, Morocco

Muhammad Revanza Naufal Novianto
Department Metallurgical Engineering
Faculty of Engincering, University of Sultan Ageng Tirtayasa
Indonesia

Muhammad Rizky Nurzaman
Department Metallurgical Engineering
Faculty of Engineering, University of Sultan Ageng Tirtayasa
Indonesia

Fransiska Maura Meilan Kristianti Pasaribu
Department Metallurgical Engineering
Faculty of Engineering, University of Sultan Ageng Tirtayasa
Indonesia

Muhamad Eko Prasetio
Department Metallurgical Engineering
Faculty of Engineering, University of Sultan Ageng Tirtayasa
Indonesia

Siska Prifiharni
Research Center for Metallurgy
National Research and Innovation Agency
Republic of Indonesia

Bharathi Kannan Raja
Department of Chemistry
SRM Institute of Science and Technology
Kattankulathur, Tamilnadu, India

Yusufboy Rajabov
Department of Physical Chemistry
National University of Uzbekistan
Tashkent, Uzbekistan

Samariddin Rakhimov
Department of Physical Chemistry
National University of Uzbekistan
Tashkent, Uzbekistan

Reena Negi Rawat
Department of Chemical Sciences
Siddhachalam Laboratory
Raipur, India

Vilabeilie Rutsa
Department of Chemistry
Nagaland University
Zunheboto, Nagaland, India

Ahmad Ramazani Saadatabadi
Chemical and Petroleum Engineering Department
Sharif University of Technology
Tehran, Iran

Rajesh Kumar Sahu
Department of Chemistry
Vishwavidyalaya Engineering College
Ambikapur, Chhattisgarh, India

Therola Sangtam
Department of Chemistry
Nagaland University
Zunheboto, Nagaland, India

Goncagül Serdaroğlu
Sivas Cumhuriyet University, Department of Math. and Sci. Edu.,
Sivas, Turkey

Praveen Kumar Sharma
Department of Chemistry
School of Chemical Engineering and Physical Sciences
Lovely Professional University
Punjab, India

Ambrish Singh
Department of Chemistry
Nagaland University
Zunheboto, Nagaland, India

Fotima Sobirova
Department of Pharmacy and Chemistry
Alfraganus University
Tashkent, Uzbekistan

Hanae Steli
Mechanical & Energy Laboratory, Faculty of Sciences
Mohammed Premier University
Oujda, Morocco

Burak Tüzün
Plant and Animal Production Department
Technical Sciences Vocational School of Sivas
Sivas Cumhuriyet University
Sivas, Turkey

Gamze Tüzün
Department of Chemistry, Faculty of Science
Cumhuriyet University
Sivas, Turkey

Dilara Ülger Özbek
Advanced Technology and Research Centre
Sivas Cumhuriyet University
Sivas, Turkey

Vetezo Venuh
Department of Chemistry
Nagaland University
Zunheboto, Nagaland, India

Alok Kumar Yadav
Department of Biotechnology
Parul University
Vadodara, Gujarat, India

Akhiu K. Yimchunger
Department of Chemistry
Nagaland University
Zunheboto, Nagaland, India

1 Introduction to Corrosion

Fundamentals and Implication

Bening Nurul Hidayah Kambuna, Siska Prifiharni, Fadly Daffa Hardiansyah, Muhamad Rizal Addin, Muhammad Revanza Naufal Novianto, Muhammad Rizky Nurzaman, Fransiska Maura Meilan Kristianti Pasaribu, Muhamad Eko Prasetio, and Ammar Fahmy Almasarie Nahdi

1.1 INTRODUCTION

Corrosion refers to the degradation of metal and alloy surfaces in particular environments. Certain metals have superior corrosion resistance compared to others, attributable to conditions such as ambient humidity, abrasive metal polishing, oils, acids, alkaline salts, liquid chemicals, and gaseous pollutants including sulfur, ammonia, and compounds with acidic pH levels. Corrosion is the spontaneous degradation of materials that impacts various industrial sectors worldwide, leading to significant economic challenges, including the costly replacement of equipment. Consequently, it is essential to investigate methods for its control, including manufacturing, chemical processing, and the oil sector, among others that utilize metals. Issues such as chemical leaks, failures in oil pipelines, and even fire hazards often arise from the corrosion of materials when exposed to electrical components. The primary types of corrosion include pitting, uniform crack, galvanic, intergranular corrosion, and erosion. The interaction of cathodic and anodic reactions involving metals and environment can reduce the risk of corrosion [1–3].

With the widespread consequences of corrosion, the development and implementation of corrosion protection coatings have become essential across various industries. Coating technology, a key component of surface engineering, has demonstrated significant promise for future industrial applications. As coating technologies continue to evolve and advance, they have become essential in industrial manufacturing, exhibiting a range of superior properties, including superhydrophobicity and self-cleaning capabilities, enhanced antibacterial characteristics, and improved resistance to corrosion. These coatings function as a barrier between materials and their environments, effectively preventing or mitigating corrosive processes. The significance of corrosion protection coatings is particularly evident in sectors such as oil

DOI: 10.1201/9781003571162-1

and gas industry, where pipelines and storage tanks face severe conditions and corrosive agents, rendering them vulnerable to deterioration. The application of protective coatings not only shields these assets from corrosive environments but also prolongs their service life, reduces maintenance expenditures, and enhances operational safety. Thus, corrosion protection coatings represent a vital defense strategy, safeguarding materials, infrastructure, and ecosystems from the detrimental impacts of corrosion [4, 5].

1.2 FACTOR CAUSING CORROSION

Metal corrosion is a widespread phenomenon that occurs whenever an active corrosive environment is present. This section delves into the multifaceted factors that influence metallic corrosion. These factors encompass metal purity, the nature of surface films, the characteristics of corrosion products, temperature, air humidity, and pH levels. A thorough investigation of these variables will facilitate a more in-depth comprehension of the corrosion mechanisms in metals [6]. The presence of impurities within a metal generally accelerates its corrosion rate. This phenomenon arises from the formation of localized electrochemical cells at impurity sites, rendering these regions more susceptible to anodic dissolution. For instance, zinc containing impurities such as iron (Fe) or lead (Pb) exhibit an increased rate of corrosion. In an oxygen-rich environment, all metals naturally develop a thin oxide layer on their surface. The influence of this protective film is significantly impacted by the "specific volume ratio," a parameter that quantifies the volumetric relationship between the metal oxide and the underlying metal. A higher specific volume ratio typically correlates with a slower oxidation rate. Illustratively, nickel, cobalt, and tungsten exhibit specific volume ratios of 1.6, 2.0, and 3.6, respectively. Consequently, tungsten demonstrates the lowest oxidation rate, even under elevated temperature conditions. Furthermore, corrosion processes are significantly exacerbated when the corrosion products exhibit solubility within the corrosive medium. Moreover, if these products possess volatile properties, their subsequent evaporation exposes the underlying metal surface to renewed corrosive attack, thereby accelerating the overall corrosion rate [7]. Corrosion rates exhibit an inverse relationship with pH, decreasing as pH increases. In contrast, corrosion rates generally increase with rising temperatures, often following an exponential trend. However, the temperature–corrosion relationship can be more intricate, as temperature fluctuations can also influence the effect of other contributing factors [6]. Experiments conducted on aluminum and copper at temperatures of 25°C and 50°C, across pH values of 4.8, 7, and 8.2, revealed that corrosion rates generally decrease with increasing pH. Conversely, a rise in temperature consistently led to an increase in corrosion rates. Although an exponential relationship between temperature and corrosion rate is often observed, the influence of temperature can be more intricate, as it may also modulate the impact of other contributing factors [6].

Conversely, relative humidity exerts a significant influence on corrosion rates, with a sharp increase observed beyond a critical humidity threshold. This heightened corrosion activity at elevated humidity levels is attributed to the oxide film's propensity to absorb moisture, thereby facilitating accelerated electrochemical corrosion. Furthermore, atmospheric moisture provides the necessary water for electrolyte formation, a critical component in the establishment of an electrochemical cell [7]. Furthermore, pH is a crucial determinant of corrosion rates, with lower pH values generally associated with higher corrosion activity. Acidic environments (pH < 7) exhibit greater corrosivity compared to alkaline or neutral media. Although noble metals such as gold and platinum exhibit exceptionally low corrosion rates, largely unaffected by solution pH, their high cost restricts their widespread application [6]. In contrast, for metals like aluminum, zinc, and lead, corrosion rates significantly increase in both acidic and alkaline environments relative to neutral solutions. This phenomenon is attributed to the enhanced solubility of the oxides of these metals in both acidic and alkaline conditions [6].

1.3 CORROSION MECHANISM

The gradual degradation of metal structures occurs consequent to chemical reactions with environmental factors or electrochemical mechanisms, a phenomenon known as metal corrosion. This process occurs naturally, as metals revert to their original state during these reactions. As a result, they transform into various metallic compounds, including oxides, sulfates, and hydroxides. Just as the element compositions of various materials differ, so does the corrosion mechanism as well [7, 8].

The occurrence of corrosion requires two electrochemical processes. In electrochemical corrosion, the anode reaction refers to the oxidation process, while the cathode reaction describes the reduction process. In the anode reaction, metal is lost. However, to maintain charge neutrality, the cathode reaction must consume the electrons released by the anode. Therefore, electrons are transferred from the anode to the cathode as part of the corrosion process within a conductive solution. Corrosion of the anodic metal occurs because it has a more negative reduction potential value, in contrast to the cathodic metal, which has a more positive reduction potential value. Typically, when exposed to an aqueous electrolyte, a metal surface has sites for both anodic and cathodic reactions. A corrosion cell is composed of these locations. The anodic reaction illustrated in Figure 1.1 involves the separation of metal, resulting in the production of either an insoluble metallic compound, commonly an oxide, or a soluble ionic product. In the cathodic reaction depicted in Figure 1.1b, hydrogen gas can be produced by reducing water or oxygen gas. An electrochemical cell is created through the concurrent reactions of anodic and cathodic processes [9–11].

In water completely devoid of oxygen, the conversion of hydrogen gas from hydrogen ions occurs during the cathodic reaction, as illustrated in Equation 1.1:

$$2H^+ + 2e^- \rightarrow H_{2(g)} \tag{1.1}$$

The cathodic reaction occurring in the system, when considerable quantities of oxygen are present, is illustrated in Equation 1.2:

$$2H^+ + \frac{1}{2}O_2 + 2e^- \rightarrow H_2O \tag{1.2}$$

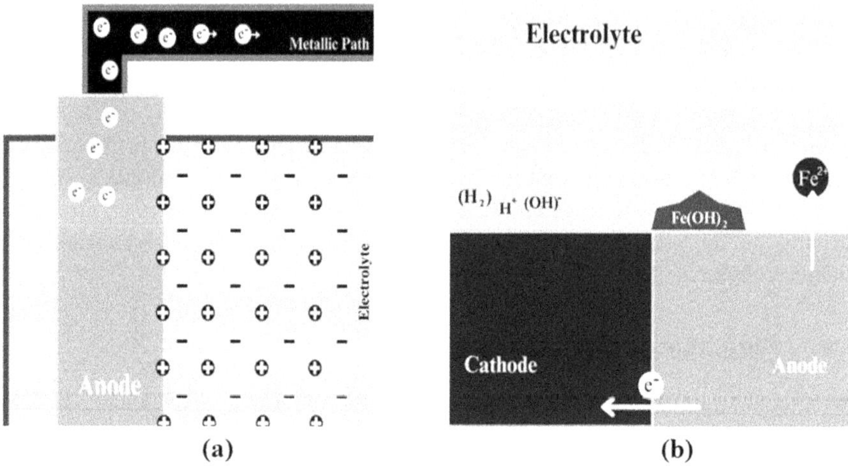

FIGURE 1.1 Electrochemical reactions. (a) Anodic process; (b) cathodic process.

The ubiquitous dissolution of water into hydroxyl ions leads to the existence of the hydrogen ion in water, as indicated in Equation 1.3:

$$2H_2O \rightarrow 2H^+ + 2(OH)^- \tag{1.3}$$

Ferrous ions are produced at the anode through the dissociation of iron, as illustrated in Equation 1.4:

$$Fe \rightarrow Fe^{2+} + 2e^- \tag{1.4}$$

Insoluble ferrous hydroxide will be formed through the interaction of the ferrous ion with the hydroxyl ion, as shown in Equation 1.5:

$$Fe^{2+} + 2(OH)^- \rightarrow Fe(OH)_2 \tag{1.5}$$

In neutral and alkaline conditions, the reactions at the anode and cathode are illustrated in Figure 1.2.

For corrosion to occur, specific conditions must be met, and distinct regions (cathodic and anodic) must be established that allow current to flow through a conducting medium. Oxidation of the metal occurs at the anode, resulting in a depletion of the metal surface as it reacts with oxygen, which ensures that corrosion consistently takes place at this electrode. Reduction takes place at the cathode, producing ions such as OH^- or O_2^-, when dissolved oxygen accepts electrons from the anode.

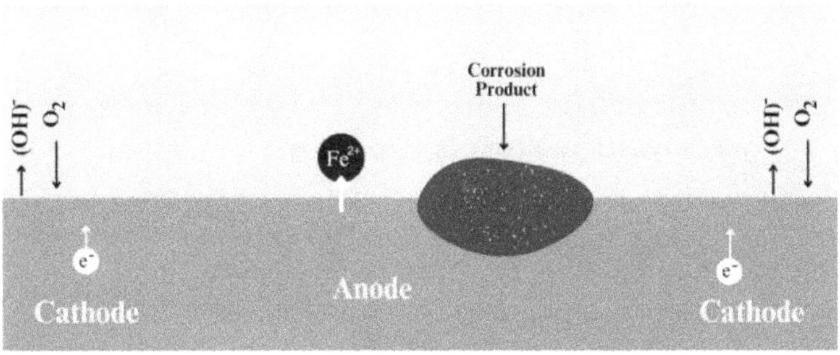

FIGURE 1.2 Corrosion processes in neutral and alkaline conditions.

The merging of ions released at the various electrodes leads to the creation of a product of corrosion in the space dividing them [10].

1.4 TYPES OF CORROSION

Corrosion can occur in a variety of ways, the form and characteristics of which are influenced by various factors, such as the surrounding environment, the nature of the materials present, the specific conditions in the vicinity (aqueous environment, open space, or high temperature environment), and the electrochemical mechanisms that occur during the corrosive process [12]. Corrosion damage can occur through various mechanisms, such as cracking of the material, reduction in structural strength, and other forms of degradation. We can classify corrosion based on the type of corrosive or medium in which it occurs, as well as its form. There are two primary types of corrosion: localized corrosion and uniform or widespread corrosion [13]. A thorough examination of the various forms of corrosion in metals, including both general and more specialized forms, is provided here.

1.4.1 Uniform Corrosion

A form of corrosion damage known as uniform corrosion is defined by a comparatively even degradation of the metal surface [14]. The impacted area experiences this corrosive process at the same pace. The primary catalyst in the corrosion process of common materials like steel and cast iron that are susceptible to uniform corrosion is oxygen [6]. Those materials will take on the look of rust when exposed to a humid environment. Steel buildings that have been left to rust are frequently affected by uniform corrosion, which is the corrosion assault that results in the most loss of metal weight. In actuality, the vivid hues produced by the corrosion of certain metals have been used in visually appealing external applications, such as weathering steel in buildings and sculptures and using copper as a long-lasting roofing material. The uniform distribution of corrosive agents on metal surfaces due to their identical chemical composition and microstructure can result in uniform corrosion. Temperature parameters have a significant impact on the uniform corrosion rate; high temperatures

FIGURE 1.3 Phenomena of uniform corrosion on piping, pumps, and control panel of a water system.

will decrease the amount of oxygen present in the aqueous solution, which will then alter the rate of corrosion. Uniform corrosion is the most prevalent type of corrosion that causes a metal surface to lose thickness overall when exposed to a corrosive environment (Figure 1.3). The final surface's smoothness or roughness is often determined by the corrosion product's level of solubility [15]. This type of corrosion is most common in atmospheric corrosion, acids, etc. [16]. Unless the corroded material is hidden from direct view, like corrosion on the inside of pipes, hidden components, or immersed and submerged structures, uniform corrosion is regarded as less problematic than other forms of corrosion from the perspective of corrosion inspection because it is relatively detectable and its effects are predictable [17]. The metal surface will become rough if surface corrosion persists, which can then result in the formation of more hazardous forms of corrosion; therefore, it is important to closely monitor uniform corrosion [18].

1.4.2 Localized Corrosion

Localized corrosion is a type of corrosion characterized by severe attacks on specific areas of the metal surface, occurring in small regions that are more susceptible to corrosion compared to the entire surface in a corrosive environment. This phenomenon is caused by the heterogeneous state of the substance or environment, as well as the influence of certain concentrations of biodiversity or reactive systems. The consequences of this phenomenon are far more severe than those of uniform corrosion due to the sudden nature of the damage. Areas with a full oxygen supply become cathodes, while regions with a limited oxygen supply act as anodes. These regions exhibit a variety of corrosion types, including fretting, stress cracking, galvanic, erosion, pitting, fatigue, and crevice corrosion. These types of corrosion are typically caused by local environmental variations or the metal's surface characteristics [3, 14].

Wormhole corrosion is characterized as a localized one-dimensional (1D) process, unlike the more commonly recognized forms of corrosion that are assessed in two or three dimensions [16]. A study provides an example of a corrosion scenario involving a Ni-20Cr alloy attacked using molten fluoride salts, which the researchers described as 1D wormhole corrosion. The study shows that this type of corrosion can create pathways for mass flow, different from traditional diffusion pathways, resulting in a faster and more aggressive attack on the material [19].

Pitting corrosion is the initial type of localized corrosion, which affects metals that are protected by a film, either from spontaneous environmental reactions or from surface treatments like coatings [20]. Figure 1.4 shows the diagram process of pitting corrosion. Pitting begins at mechanical flaws in this protective layer, and as these pits expand, they can perforate the heat transfer surfaces. Environmental factors, polarization effects, metallurgical and structural features, and the presence of corrosion products are some of the factors that affect the onset of pitting. High temperatures, low pH levels, and high chloride levels all encourage pitting corrosion in stainless steel [21]. Aluminum, nickel, titanium, and their alloys are among the metals frequently employed in heat exchanger construction that may suffer from localized deterioration of their protective coatings [22]. The Pitting Resistance Equivalent Number (PREN), which is determined by the proportions of chromium (Cr), molybdenum (Mo), tungsten (W), and nitrogen (N) in the alloy, can be used to forecast the susceptibility of stainless steel to pitting [23]. Type 304 stainless steel, for example, has a PREN of 19, although more highly alloyed varieties can have a PREN of over 50. When using saltwater at 20°C, a PREN of at least 32 is advised [24]. The middle tubes of a shell-and-tube heat exchanger used in a geothermal plant had deep pits, while the upper and lower tubes were unharmed, according to the failure study. Due to the condensing steam and brine mixture that these middle tubes received, the water's concentration of acidic gases such as CO_2 and H_2S rose, hastening the formation of pits. Furthermore, the pits' elemental arsenic may have worsened corrosion by having galvanic effects

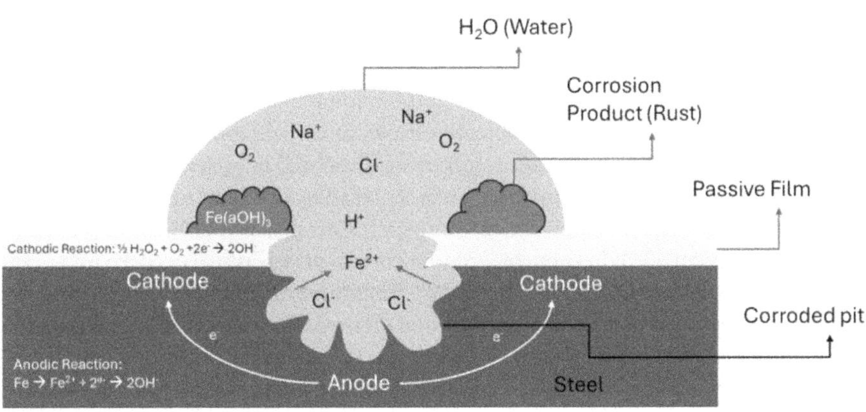

FIGURE 1.4 Pitting corrosion diagram process.

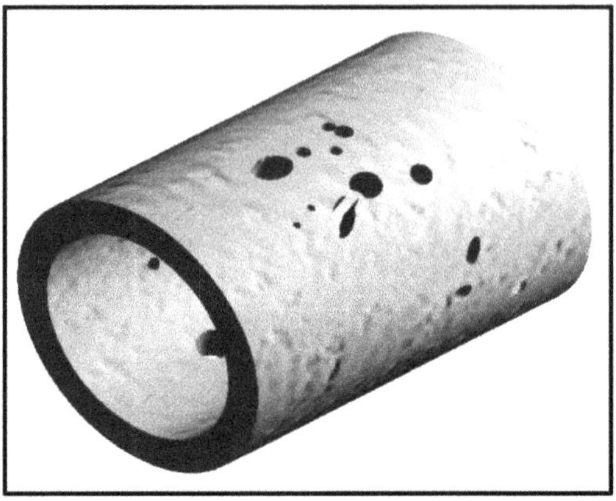

FIGURE 1.5 Pits in the galvanized water pipe that contributed to the erosion–corrosion of the gas line and the subsequent leak in a sewer main that caused the Guadalajara 1992 explosion.

[25] (Figure 1.5). Another study examined a plate heat exchanger made from type 316L stainless steel (PREN ≈ 24.2) used in a textile mill's cooling loop. After 2000 hours of operation, several plates showed perforation due to pitting caused by high chloride levels in the cooling water [26]. Pitting corrosion manifests as small cavities on metal surfaces and is often challenging to detect during design stages because it develops gradually over time and can become obscured by corrosion products [27]. Despite their small size, these pits can lead to significant structural failures in pipelines. This type of corrosion is particularly prevalent in both sour and sweet field pipelines [28]. Another research has identified a novel form of pitting corrosion termed "moundless pitting corrosion," discovered in Noboribetsu City, Hokkaido, Japan. This study by Masahiro Sakai involved field surveys and laboratory simulations to analyze this unique corrosion type, characterized by open pits without accumulation of corrosion products, very small diameters (<1 mm), extreme depths, and surrounding verdigris glass. The findings indicated that water quality significantly influences pit formation, particularly with high silica content and low bicarbonate ion concentration [29]. Pitting corrosion is also prevalent in aircraft environments due to chloride ions present in electrolytic conditions. It typically results in vertical growth into the material and may initiate fatigue cracks that compromise structural integrity. Although various factors contribute to pitting initiation, such as localized damage to protective oxide films and adverse water chemistry, its complex nature makes it difficult to predict or mitigate effectively [30]. In summary, pitting corrosion is a critical concern for engineers due to its potential for severe damage despite minimal overall metal loss. Its detection often requires meticulous inspection methods due to its stealthy progression and can lead to catastrophic failures if not properly managed [31].

A severe and dangerous type of localized corrosion that results from a localized environment that is different from the surrounding bulk environment is called "crevice corrosion" (Figure 1.6). This type of corrosion usually occurs in protected places

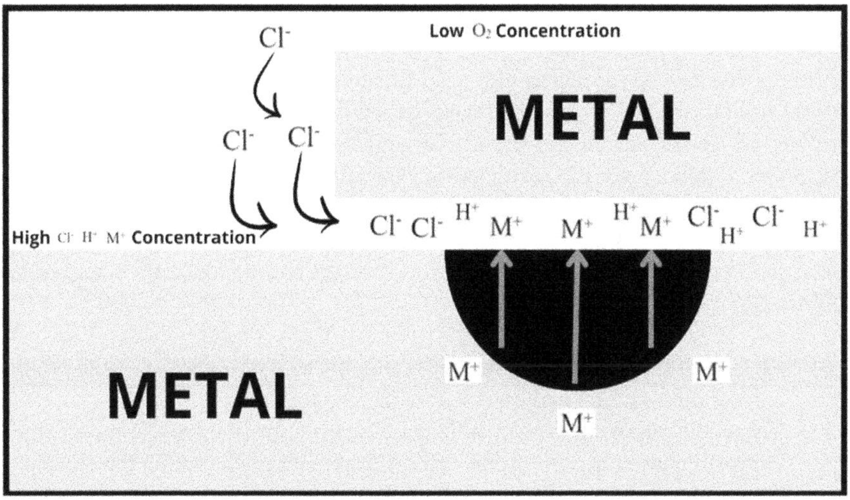

FIGURE 1.6 Crevice corrosion diagram process.

or cracks where standing liquid might cause oxygen to be depleted. This change in oxygen content affects when crevice corrosion starts, and the presence of chlorides can make the issue worse [32]. As the crevice narrows, the rate of corrosion tends to increase due to a larger cathodic area [23]. Crevices are frequently seen in plate heat exchangers at contact sites between plates, beneath deposits, and at gaskets—where metal meets non-metal [32]. The difficulties in avoiding crevice corrosion in plate heat exchangers used in marine applications are highlighted by Refs. [33] and [34] in their studies. Through laboratory research and exposure to actual saltwater with regulated chlorine levels to reduce marine development, they tested a variety of metals at high temperatures (up to 70°C). Only more recent nickel–chromium–molybdenum alloys, including alloy 686 (UNS N06686), were resistant to this corrosive environment; the majority of tested metals, including titanium, displayed evidence of crevice corrosion under high circumstances. In plates composed of alloy 31, severe cases of crevice corrosion have been observed at gasket locations (UNS N08031) [33, 34]. Crucial locations vulnerable to crevice corrosion in shell-and-tube heat exchangers are bolt holes, unprotected welds, and spaces between rolled tubes and tube sheets. Sharp edges and intricate geometric patterns are especially vulnerable [22]. Additional studies provide further examples of failures attributed to crevice corrosion found between tubes and tube sheets, as well as in gaps between tubes and baffles [35, 36]. Crevice corrosion typically occurs when two metal surfaces come into close contact, creating a confined space that restricts the flow of corrosive agents and oxygen. This results in more severe attacks on the metal surface within the crevice compared to surrounding areas [37]. It often develops around protected areas such as gaskets, insulation materials, joints, washers, clamps, and threads [21]. The localized nature of this corrosion makes it particularly damaging in environments with poor fluid circulation [12]. In aircraft environments, crevice corrosion is commonly found in gaps or crevices where condensates stagnate and electrolyte circulation is hindered. Even materials known for their excellent corrosion resistance can suffer significant damage due to crevice corrosion, which may lead to sudden failures [30]. Three steps can be used to describe the process that leads to the formation of severe crevices: first, dissolved oxygen is consumed by reactions with the metal inside the crevice, causing differential aeration; second, restricted oxygen diffusion creates a differential aeration cell between the external surface and the microenvironment inside the crevice; and third, this imbalance produces extremely corrosive conditions that facilitate additional metal dissolution. Crevice corrosion severity is influenced by a number of elements [17]:

- Metal ions produced during anodic reactions can hydrolyze, releasing protons and forming acidic conditions within the crevice.
- The acidification can significantly increase the corrosion rate for most metals.
- Corrosion products can further seal off the environment within the crevice.
- The accumulated positive charges draw caustic negative ions such as sulfates and chlorides.

In conclusion, crevice corrosion is a type of localized assault that occurs in small gaps between metal surfaces or between metals and non-metals. It is characterized by differential aeration that creates anodic conditions conducive to rapid corrosion within these restricted areas. This type of corrosion poses significant challenges in various applications, particularly in marine environments where stainless steel is commonly used.

When a flowing corrosive fluid interacts with a metal surface, the combined effects of mechanical erosion and chemical corrosion are referred to as erosion corrosion. This synergistic effect is more severe than either erosion or corrosion alone, particularly under turbulent flow conditions. The fluid's motion removes protective films from metals like stainless steel and aluminum, exposing them to rapid corrosion. A critical velocity threshold often exists, beyond which the protective layer cannot reform, leading to accelerated degradation [38]. In seawater, high chloride concentrations prevent stainless steel from achieving passivity, increasing corrosion rates as flow velocity rises [20]. Guedes-Soares et al. [39] first discovered a relationship between seawater flow velocity and carbon steel corrosion rates, which they then refined into an exponential function. Systems like heat exchangers frequently experience erosion corrosion, particularly at inlets. In plate heat exchangers, abrasive solids such as sand exacerbate the damage [38]. Inlet-tube corrosion is a similar phenomenon that shell-and-tube heat exchangers experience on the tube side [40]. Additional forms include cavitation erosion, where collapsing vapor bubbles impact surfaces, commonly in evaporators, and fretting corrosion, caused by repeated rubbing at joints or contact points. For example, Kuźnicka observed erosion corrosion pits on copper tubes in a heat exchanger, aligned with turbulence regions [41]. Klein et al. described fretting corrosion in a wastewater heat exchanger, where debris caused tube oscillation and contact damage [42]. Erosion corrosion also affects pipelines, valves, and pumps in offshore operations involving slurry transport, leading to significant annual repair costs [43, 44]. Factors influencing this process include protective film integrity, material properties, environmental conditions, and flow velocity. Solids in the fluid contribute further by causing abrasion or attrition, depending on their hardness relative to the metal [45, 46]. Effective mitigation strategies include selecting harder materials, optimizing flow geometry, and reducing velocities. Wang et al. proposed design adjustments, such as helically shaped pipes, increased elbow radii, and gradual fluid movement changes [47]. Controlling pH levels and minimizing particle size can also reduce erosion corrosion impacts. Attention to these factors is critical for maintaining system efficiency and reducing maintenance costs [48, 49].

Intergranular corrosion is a form of metal degradation that occurs specifically along the boundaries between metal crystals or grains, targeting narrow paths between the grain structures while leaving the central grain areas relatively untouched [50, 51]. Inappropriate heat treatment, welding procedures, high temperatures, and the presence of specific pollutants can all produce this kind of corrosion, which can result in structural or chemical variations at the grain boundaries [22, 42]. The consequences can be very severe, including a significant reduction in material strength, decreased ductility, and the potential for total structural failure [52]. Materials commonly affected include stainless steel, aluminum alloys, and high-strength metal alloys, with corrosion often resulting from the depletion of corrosion-resistant elements or changes in electrical potential between grain boundaries and the main grain structure [53].

A severe type of localized corrosion known as stress corrosion cracking (SCC) occurs when a metal is exposed to both mechanical stress and a corrosive environment (Figure 1.7). This type of corrosion is particularly troublesome because it can occur at relatively low stress levels, much below the yield strength of the material, making it challenging to predict and control. SCC involves the interaction of static tensile stress, a corrosive medium, and metallurgical factors such as chemical composition and grain structure. The process typically starts with the nucleation of cracks at microscopic sites, followed by slow propagation over time, sometimes taking months or years to develop fully [23, 38, 54]. Key factors contributing to SCC include [55, 56]:

1. *Presence of Susceptible Materials:* Specific alloys are more prone to SCC depending on their chemical makeup and crystallographic arrangement.
2. *Corrosive Environment:* Exposure to mild corrosive agents accelerates the cracking process.
3. *Static Tensile Stresses:* Internal or external stresses, especially those generated during manufacturing processes like welding or cold forming, enhance the likelihood of SCC.

Residual stresses from processing steps like quenching or machining can also induce SCC. The corrosion products accumulated in narrow spaces can exert high local pressures, further increasing the risk of cracking [40]. To combat SCC, strategies focus on managing both the material properties and the environmental conditions. Techniques include selecting materials with high resistance to cracking, controlling hardness and stress levels, and minimizing exposure to corrosive substances [57, 58].

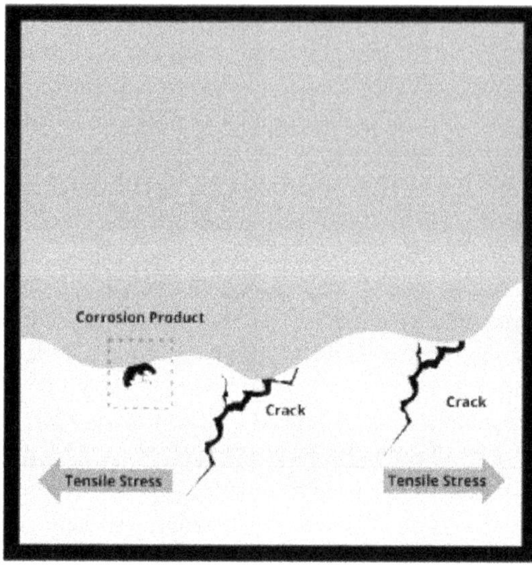

FIGURE 1.7 SCC diagram process.

Understanding the temporal dynamics of SCC is crucial, as the rate of crack propagation plays a pivotal role in predicting component lifespan [55].

1.5 CORROSION THERMODYNAMICS AND KINETICS

Corrosion is a natural process, whether chemical or electrochemical, which causes the deterioration of metals, ceramics, and alloys, rendering them unsuitable for their intended purposes [59, 60]. All metals, except for gold, exhibit thermodynamic instability concerning their oxides. Metal corrosion indicates that pure metals and their alloys generally revert to thermodynamically stable compounds, including oxides, sulfides, carbonates, hydroxides, and chlorides. This results from their electrochemical properties in the environment [61].

Thermodynamics and kinetics are two essential areas of knowledge that are used in a variety of scientific and technical professions. They are also quite vital in a whole lot of distinct sectors. As regards chemical processes, thermodynamics is often used to provide predictions and suggestions about the directions, energy, products, and equilibrium conditions [62]. Kinetics, on the other hand, may be seen as a technique that can be used to investigate the pace of chemical reactions and to gain knowledge of how various processes are impacted [63]. The Pourbaix diagram and Tafel equation are fundamental principles regarding the thermodynamics and kinetics associated with electrochemical corrosion, which significantly impact global corrosion research and development [64]. On the other hand, the link between kinetics and thermodynamics of electrochemical corrosion is not very strong for the most part. For instance, the Pourbaix diagram may demonstrate passive behavior for Fe and Cr when it is applied to the appropriate pH zone. On the other hand, the findings of the kinetic analysis indicate that Cr demonstrates passive behavior, but Fe does not demonstrate passive behavior [65, 66].

The electrode potential and the concentration of hydrogen ions (pH) are two factors that may be used to define the thermodynamics of electrochemical corrosion. This type of diagram is usually known as a potential-pH diagram, which is also referred to as the Pourbaix diagram [67]. The Pourbaix diagram illustrates the electrochemical stability of different elements in water, influenced by the solution's pH and electrode potential. It provides the boundary conditions for the separation of specific elements found in water. The relationship between the pH and the redox potential (E_h) and their respective values with regard to the standard hydrogen electrode (SHE) may be used to establish it. In the fields of electrochemical and corrosion research, Pourbaix diagrams have been used extensively for predicting the stages of speciation in water at a certain pH and potential concentration [68, 69].

The corrosion rate (C_R) is a term that is used in a general sense to refer to the amount of metal that is lost due to corrosion over a specified period of time [67]. There are a few different methods that may be used to compute the kinetics of corrosion. However, the most common way is using the Tafel extrapolation [70]. Tafel extrapolation represents a widely utilized polarization technique in the measurement of corrosion rates, offering a more efficient experimental alternative to the traditional weight-loss technique. Corrosion rates determined through Tafel extrapolation of polarization curves typically differ from those measured by weight loss. The various reactions involved in corrosion can be articulated through the Tafel equation, allowing for the representation of mixed potential theory. This method facilitates the prediction rates of corrosion and potentials through the analysis of the kinetics and thermodynamics of reactions occurring at the electrode surface. Tafel kinetics provide a precise framework for understanding corrosion kinetics in scenarios where mass transport limitations are negligible, as corrosion conditions typically deviate from the reversible potentials of all reactions involved [71, 72]. These approaches have limited applicability for on-site corrosion testing due to their reliance on lengthy and slow polarization experiments, which are time-consuming and may induce permanent alterations in the reinforcement. In addition, these tests may be time-consuming [73].

1.6 IMPACT OF CORROSION

Corrosion-related failures result in significant environmental contamination. Corroded pipelines, exemplified by the Prudhoe Bay event, result in oil leaks, while the dissolution of heavy metals through water systems endangers ecosystems and the health of society. Corrosion in maritime habitats intensifies biodiversity loss by releasing pollutants [74]. Corrosion endangers cultural heritage, as historic structures such as the Statue of Liberty necessitate significant restoration owing to galvanic corrosion. Appropriate mitigation measures encompass the selection of anti-corrosion materials, the application of protective coatings, and the implementation of cathodic protection systems. These preventative approaches are crucial to mitigate the widespread effects of corrosion, assuring sustainability and safety across several sectors [75].

Corrosion has a significant economic impact, costing industries around the globe billions of dollars each year in repairs, maintenance, and replacements. Sectors such as infrastructure, transportation, and energy are especially at risk, as corrosion can result in structural failures and periods of inactivity. The Berlin Congress Hall in 1957 was celebrated for its innovative design and construction, featuring a distinctive curved canopy made of post-tensioned concrete panels. This system involved placing a metal or plastic duct in the formwork, where a high-strength steel rod or strand was inserted and anchored at one end. After the concrete cured, the free end of the steel was pulled to create tension, and the duct was filled with grout to facilitate load transfer between the concrete and the steel. The alkaline environment of the concrete and grout protected the steel from corrosion by forming a passive film. However, in 1980, the building collapsed, resulting in one death. Investigations revealed that the ducts in some panels were incorrectly installed, leaving the post-tensioned steel inadequately surrounded by concrete and grout, which compromised its protection. Fractography indicated corrosion along grain boundaries and numerous internal cracks, with the failure ultimately linked to hydrogen embrittlement [75].

The ferry *Princess Ashika* tragically sank just a month after it was launched in 2009, resulting in the loss of all 74 lives on board. An investigation by a Royal Commission uncovered significant safety issues that contributed to the disaster. A marine engineer who inspected the vessel prior to its departure presented photos showing that many doors and entrances could not close properly due to corrosion. He had raised concerns about the ship's seaworthiness and advised against its sailing. This incident highlights the critical importance of proper maintenance and safety measures in preventing such tragedies [75].

There are numerous instances of injuries and fatalities (for both humans and animals) related to corrosion. By far, the most severe corrosion-related disaster in terms of human deaths, injuries, and subsequent health issues was the explosion at a chemical plant in Bhopal, India, in 1984. A series of design issues and management failures led to a devastating explosion. Reports indicate that when steel pipes corroded, water leaked into tanks holding methylisocyanate. The resulting corrosion created a catalyst that triggered a reaction, causing the plant to explode. This disaster released methylisocyanate and other harmful gases into the air, resulting in the tragic loss of over 8000 lives [76].

1.7 FINANCIAL BURDEN OF CORROSION

The cost associated with corrosion can be categorized into two: direct and indirect costs. Direct costs include replacement of damaged component, maintenance and inspection, corrosion protection, and environmental control. Indirect costs are often challenging to quantify but can sometimes far exceed direct costs. Indirect cost include production loss, product failure, environmental contamination, and many others. When combined with direct costs, the total financial impact of corrosion reaches \$276 billion, contributing over 3% to the U.S. Gross Domestic Product (GDP), surpassing the economic output of both the mining and agriculture sectors. A significant portion of this cost is allocated to corrosion prevention, amounting to \$121.4 billion, which represents nearly 90% of the total direct costs of corrosion. Of this expenditure, approximately 90% is dedicated to protective coatings. In the case of organic coatings, material costs account for only 4–20% of the total, depending on project complexity, while the remaining portion is attributed to application expenses.

Among metallic coatings, zinc is the most widely used material, primarily applied through hot-dip galvanization of steel, as well as in combination with aluminum through thermal spraying or electroplating. Financially, corrosion represents a significant burden, with its total cost exceeding 3% of the U.S. GDP. The expenditure on prevention alone constitutes almost 90% of the direct corrosion costs, with protective coatings making up a similar proportion of this preventative spending. Within organic coatings, material expenses range from only 4% to 20% of the total, highlighting that application costs dominate the overall expenditure [76]. Corrosion has a significant economic impact across various industrial sectors. Therefore, effective corrosion control strategies must be implemented to minimize economic losses and enhance industrial sustainability.

1.8 CHAPTER SUMMARY

Corrosion refers to the degradation of metal and alloy surfaces in particular environments. The factors that can cause corrosion include metal purity, the nature of surface films, the characteristics of corrosion products, temperature, air humidity, and pH levels. The corrosion mechanism takes place when a metal that has a higher reduction potential acts as an anode, while another metal with a lower potential acts as a cathode, so that under certain conditions the anode metal will corrode. There are two main types of corrosion: uniform or widespread corrosion and localized corrosion, including pitting corrosion, crevice corrosion, wormhole corrosion, erosion corrosion, SCC, and intergranular corrosion. Corrosion is influenced by thermodynamics, which predicts reaction directions, and kinetics, which examines reaction rates. The Pourbaix diagram and Tafel equation are fundamental principles regarding the thermodynamics and kinetics associated with electrochemical corrosion, which significantly impact global corrosion research and development. Corrosion causes severe environmental, economic, and safety impacts, leading to oil leaks, heavy metal contamination, and biodiversity loss, while also endangering cultural heritage and costing industries billions annually. To prolong the life of metal structures and lower the risks to safety and the economy, preventative measures like cathodic protection and protective coatings are crucial. Creating more corrosion-resistant materials, enhancing early detection techniques, and safeguarding infrastructure in harsh settings like the deep sea and space are some of the challenges of the future. Furthermore, the industry's survival depends on the development of eco-friendly technologies that reduce the negative effects of corrosion on ecosystems.

REFERENCES

[1] O. S. I. Fayomi, I. G. Akande, and S. Odigie, "Economic impact of corrosion in oil sectors and prevention: An overview," *Journal of Physics: Conference Series*, vol. 1378, 022037, Dec. 2019. doi: 10.1088/1742-6596/1378/2/022037.

[2] M. F. P. Carlos, G. K. P. Barboza, and A. Echevarria, "Anticorrosive effect of halogenated Aniline enaminoesters on carbon steel in HCl," *International Journal of Corrosion*, vol. 2022, 2022, doi: 10.1155/2022/7218063.

[3] A. Popoola, O. Olorunniwo, and O. Ige, "Corrosion resistance through the application of anti-corrosion coatings," in M. Aliofkhazraei (ed.), *Developments in Corrosion Protection*, InTech, 2014. doi: 10.5772/57420.

[4] H. S. Aljibori, A. Alamiery, and A. A. H. Kadhum, "Advances in corrosion protection coatings: A comprehensive review," *International Journal of Corrosion and Scale Inhibition*, vol. 12, pp. 1476–1520, 2023. doi: 10.17675/2305-6894-2023-12-4-6.

[5] X. Chen, B. Zheng, S. Zhou, C. Shi, Y. Liang, and L. Hu, "Development and application of intelligent coating technology: A review," *Coatings*, vol. 14, no. 10, p. 597, May 1, 2024. doi: 10.3390/coatings14050597.

[6] S. Harsimran, K. Santosh, and K. Rakesh, "Overview of corrosion and its control: A critical review," *Proceedings on Engineering Sciences*, vol. 3, no. 1, pp. 13–24, Mar. 2021, doi: 10.24874/PES03.01.002.

[7] M. Czaban, "Aircraft corrosion – Review of corrosion processes and its effects in selected cases," *Fatigue of Aircraft Structures*, vol. 2018, no. 10, pp. 5–20, Dec. 2018, doi: 10.2478/fas-2018-0001.

[8] S. Chandra Paul, "A review on reinforcement corrosion mechanism and measurement methods in concrete," *Civil Engineering Research Journal*, vol. 5, no. 3, Jun. 2018, doi: 10.19080/cerj.2018.05.555661.

[9] Z. Huang, C. Feng, J. Wang, W. Liu, and Y. Xie, "Research on metal corrosion mechanism and inhibition measures of transmission and distribution equipment in Hunan," in *IOP Conference Series: Earth and Environmental Science*, IOP Publishing, Jan. 2021. doi: 10.1088/1755-1315/621/1/012035.

[10] S. Harsimran, K. Santosh, and K. Rakesh, "Overview of corrosion and its control: A critical review," Proceedings on Engineering Sciences, vol. 3, pp. 13–24, 2021. doi: 10.24874/PES03.01.002.

[11] C. I. Ossai, "Advances in asset management techniques: An overview of corrosion mechanisms and mitigation strategies for oil and gas pipelines," *ISRN Corrosion*, vol. 2012, pp. 1–10, Nov. 2012, doi: 10.5402/2012/570143.

[12] S. Al-Baghdadia and A. A. Alamierya, "A critical review of current corrosion research," *Journal of Materials Science*, vol. 3, no. 2, pp. 164–185, 2024. [Online]. Available: www.aspur.rs/jme/archive/v3/n2/2.pdf

[13] R. Aslam *et al.*, "Corrosion inhibition of steel using different families of organic compounds: Past and present progress," *Journal of Molecular Liquids*, vol. 348, 118373, Feb. 2022, doi: 10.1016/j.molliq.2021.118373.

[14] J. K. Balangao, "Corrosion of metals: Factors, types and prevention strategies," *Journal of Chemical Health Risks*, vol. 14, pp. 79–87, 2024. [Online]. Available: https://hal.science/hal-04372559/

[15] A. K. Lahiri, *Applied Metallurgy and Corrosion Control: A Handbook for the Petrochemical Industry Indian Institute of Metals Series*. Springer, 2017.

[16] S. E. Apeh, S. I. Kuye, E. A. Ogwu, and B. Martins, "A review of corrosion types, management strategies and its consequences in oil and gas as well as construction sectors in Nigeria," *Nigerian Research Journal of Engineering and Environmental Sciences*, vol. 9, Jun. 2024, doi: 10.5281/zenodo.12599422.

[17] P. R. Roberge and P. Eng, *Corrosion Engineering: Principle and Practice.* International Book, 2008.

[18] A. I. Ayodeji, O. S. I. Fayomi, A. A. Daniyan, K. O. Babaremu, P. O. Abioye, and O. Agboola, "Corrosion phenomena and the occurrences; A comment," *IOP Conference*

Series: Materials Science and Engineering, vol. 1107, no. 1, 012101, Apr. 2021, doi: 10.1088/1757-899X/1107/1/012101.

[19] Y. Yang *et al.*, "One dimensional wormhole corrosion in metals," *Nature Communications*, vol. 14, no. 1, p. 988, Feb. 2023, doi: 10.1038/s41467-023-36588-9.

[20] R. W. Revie and H. H. Uhlig, *Corrosion and Corrosion Control: An Introduction to Corrosion Science and Engineering*. Wiley Publication, 2008.

[21] A. Qasim, M. S. Khan, B. Lal, and A. M. Shariff, "A perspective on dual purpose gas hydrate and corrosion inhibitors for flow assurance," *Journal of Petroleum Science and Engineering*, vol. 183, 106418, Dec. 2019, doi: 10.1016/j.petrol.2019.106418.

[22] K. Thulukkanam, *Heat Exchanger Design Handbook*. Taylorfrancis.com, 2000, doi: 10.1201/9781420026870.

[23] M. Schütze, "*Corrosion Books: Handbook of Corrosion Engineering*. By Pierre R. Roberge – Materials and Corrosion 4/2002," *Materials and Corrosion*, vol. 53, pp. 284–284, https://doi.org/10.1002/1521-4176(200204)53:4<284::AID-MACO1111284>3.0.CO;2-8.

[24] J. Veldkamp, T. Goldberg, P. Bressers, and F. Wilschut, "TNO report: corrosion in Dutch geothermal systems," *Energy/Geological Survey of the Netherlands*, 2016.

[25] R. Julian, K. Lichti, E. Mroczek, and B. Mountain, "Heavy metal scaling and corrosion in a geothermal heat exchanger," in *Proceedings World Geothermal Congress*, IGA, 2015.

[26] K. M. Deen, M. A. Virk, C. I. Haque, R. Ahmad, and I. H. Khan, "Failure investigation of heat exchanger plates due to pitting corrosion," *Engineering Failure Analysis*, vol. 17, pp. 886–893, 2010. [Online]. Available: www.sciencedirect.com/science/article/pii/S1350630709002647

[27] K. L. Heppner and R. W. Evitts, "Modelling of the effect of hydrogen ion reduction on the crevice corrosion of titanium," in S. A. Shipilov, R. H. Jones, J.-M. Olive, and R. B. Rebak (eds.), *Environment-Induced Cracking of Materials*, Elsevier, 2008, pp. 95–104, doi: 10.1016/B978-008044635-6.50011-X.

[28] A. Pardo, M. C. Merino, A. E. Coy, R. Arrabal, F. Viejo, and E. Matykina, "Corrosion behaviour of magnesium/aluminium alloys in 3.5wt.% NaCl," *Corros Sci*, vol. 50, no. 3, pp. 823–834, Mar. 2008, doi: 10.1016/j.corsci.2007.11.005.

[29] M. Sakai, "A new type of silica-induced 'moundless' pitting corrosion in copper observed in Japan," *Heliyon*, vol. 8, no. 8, e10110, Aug. 2022, doi: 10.1016/j.heliyon.2022.e10110.

[30] M. Czaban, "Aircraft corrosion – Review of corrosion processes and its effects in selected cases," *Fatigue of Aircraft Structures*, vol. 2018, no. 10, pp. 5–20, Dec. 2018, doi: 10.2478/fas-2018-0001.

[31] J. M. Malo, V. Salinas, and J. Uruchurtu, "Stray current corrosion causes gasoline pipeline failure," *Materials and Performance*, vol. 33, 1994. [Online]. Available: https://pascal-francis.inist.fr/vibad/index.php?action=getRecordDetail&idt=4273800

[32] W. Faes *et al.*, "Corrosion and corrosion prevention in heat exchangers," *Corrosion Reviews*, vol. 37, no. 2, pp. 131–155, Mar. 2019, doi: 10.1515/corrrev-2018-0054.

[33] J. R. Crum, P. Hazeldine, L. E. Shoemaker, and J. Peschel, "Evaluation of materials for seawater plate heat exchanger applications," in *Corrosion 2007*, Association for Materials Protection and Performance, 2007. [Online]. Available: https://onepetro.org/NACECORR/proceedings-abstract/CORR07/All-CORR07/118547

[34] T. Cassagne, P. Houlle, D. Zuili, P. Bluzat, J. M. Corrieu, and N. Larché, "Replacing titanium in sea water plate heat exchangers," in *Corrosion 2007*, Association for Materials Protection and Performance, 2010. [Online]. Available: https://onepetro.org/NACECORR/proceedings-abstract/CORR10/All-CORR10/127013

[35] R. Kaul, N. G. Muralidharan, T. Jayakumar, K. V. Kasiviswanathan, B. Raj, R. K. Dayal, V. R. Raju, J. B. Gnanamoorthy, and S. Pattu, "Failure analysis of carbonate reboiler

heat exchangers," *Engineering Failure Analysis*, vol. 2, pp. 165–174, 1995. [Online]. Available: www.sciencedirect.com/science/article/pii/135063079500017K

[36] S. R. Allahkaram, P. Zakersafaee, and S. A. Haghgoo, "Failure analysis of heat exchanger tubes of four gas coolers," *Engineering Failure Analysis*, vol. 18, pp. 1108–1114, 2011. [Online]. Available: www.sciencedirect.com/science/article/pii/S13506 3071000230X

[37] G. F. Kennell and R. W. Evitts, "Crevice corrosion cathodic reactions and crevice scaling laws," *Electrochim Acta*, vol. 54, no. 20, pp. 4696–4703, Aug. 2009, doi: 10.1016/j.electacta.2009.03.080.

[38] B. Sundén and R. M. Manglik, *Plate Heat Exchangers: Design, Applications and Performance*. Books.google.com, 2007. [Online]. Available: https://books.google.com/books?hl=en&lr=&id=P3gTR8YHLHgC&oi=fnd&pg=PA1&dq=plate+heat+exchangers+design&ots=5gDOF9fF2q&sig=5_YnNwPTkuHnyWSennKk7Z2tYuc

[39] C. Guedes-Soares, Y. Garbatov, A. Zayed, "Effect of environmental factors on steel plate corrosion under marine immersion conditions," *Corrosion Engineering, Science and Technology*, vol. 46, no. 4, 524–541, 2011.

[40] G. Walker, "Industrial heat exchangers: A basic guide," (*No Title*), 1982. [Online]. Available: https://cir.nii.ac.jp/crid/1130000795034614528

[41] B. Kuźnicka, "Erosion–corrosion of heat exchanger tubes," *Engineering Failure Analysis*, vol. 16, pp. 2382–2387, 2009. [Online]. Available: www.sciencedirect.com/science/article/pii/S1350630709000776

[42] U. Klein, A. Zunkel, and A. Eberle, "Breakdown of heat exchangers due to erosion corrosion and fretting caused by inappropriate operating conditions," *Engineering Failure Analysis*, vol. 43, pp. 271–280, 2014. [Online]. Available: www.sciencedirect.com/science/article/pii/S1350630714001046

[43] S. S. Rajahram, T. J. Harvey, and R. J. K. Wood, "Erosion–corrosion resistance of engineering materials in various test conditions," *Wear*, vol. 267, no. 1–4, pp. 244–254, Jun. 2009, doi: 10.1016/j.wear.2009.01.052.

[44] F. L. Wang, *Polydimethylsiloxane Modification of Segmented Thermoplastic Polyurethanes and Polyureas*. Search.proquest.com, 1998. [Online]. Available: https://search.proquest.com/openview/10a43ebb835a4c0aa38a6a9705a03598/1?pq-origsite=gscholar&cbl=18750&diss=y

[45] R. Kumar, R. Singh, and S. Kumar, "Erosion and hot corrosion phenomena in thermal power plant and their preventive methods: A study," *Asian Review of Mechanical Engineering*, vol. 7, pp. 38–45, 2018. [Online]. Available: https://armejournal.org/index.php/arme/article/view/2436

[46] R. Kumar, R. Kumar, and S. Kumar, "Erosion corrosion study of HVOF sprayed thermal sprayed coating on boiler tubes: A review," *IJSMS*, vol. 1, pp. 1–6, 2018. [Online]. Available: www.ijsmsjournal.org/2018/volume-1%20issue-3/ijsms-v1i3p101.pdf

[47] J. Wang, V. Giridharan, V. Shanov, Z. Xu, B. Collins, L. White, Y. Jang, J. Sankar, N. Huang, and Y. Yun, "Flow-induced corrosion behavior of absorbable magnesium-based stents," *Acta Biomaterials*, vol. 10, pp. 5213–5223, 2014, [Online]. Available: www.sciencedirect.com/science/article/pii/S174270611400381X

[48] A. Bahadori, *Corrosion and Materials Selection: A Guide for the Chemical and Petroleum Industries*. Wiley, 2014. [Online]. Available: https://books.google.co.id/books?id=mVrwAwAAQBAJ

[49] N. Andrews, L. Giourntas, A. M. Galloway, and A. Pearson, "Effect of impact angle on the slurry erosion-corrosion of Stellite 6 and SS316," *Wear*, vol. 320, pp. 143–151, Dec. 2014, doi: 10.1016/j.wear.2014.08.006.

[50] J. M. Kolotyrkin, "Pitting corrosion of metals," *Corrosion*, vol. 19, no. 8, pp. 261t–268t, Aug. 1963, doi: 10.5006/0010-9312-19.8.261.
[51] J. S. Fernandes and F. Montemor, "Corrosion," in M. Clara Gonçalves and Fernanda Margarido (eds.), *Materials for Construction and Civil Engineering*, Springer International Publishing, 2015, pp. 679–716, doi: 10.1007/978-3-319-08236-3_15.
[52] O. T. Joshua, O. S. I. Fayomi, and E. T. Omeje, "Development of an in-situ composite doped coating for corrosion protection and mechanical properties enhancements in process engineering," *Journal of Minerals and Materials Characterization and Engineering*, vol. 07, no. 04, pp. 171–179, 2019, doi: 10.4236/jmmce.2019.74013.
[53] W. Zhang and G. S. Frankel, "Transitions between pitting and intergranular corrosion in AA2024," *Electrochim Acta*, vol. 48, no. 9, pp. 1193–1210, Apr. 2003, doi: 10.1016/S0013-4686(02)00828-9.
[54] G. Palanisamy, "Corrosion inhibitors," in Ambrish Singh (ed.), *Corrosion Inhibitors*, IntechOpen, 2019, doi: 10.5772/intechopen.80542.
[55] G. A. Martinchek and M. R. Yaffe, "Detection of pitting corrosion," US Patent 6,015,484, 2000. [Online]. Available: https://patents.google.com/patent/US6015484A/en
[56] Y. Li, L. Song, G. Sun,, "Key factors of stress corrosion cracking of X70 pipeline steel in simulated deep-sea environment: Role of localized strain and stress," *International Journal of Electrochemical Science*, vol. 13, no. 11, 2018, https://doi.org/10.20964/2018.11.53.
[57] T. E. Perez, "Corrosion in the oil and gas industry: An increasing challenge for materials," *JOM*, vol. 65, no. 8, pp. 1033–1042, Aug. 2013, doi: 10.1007/s11837-013-0675-3.
[58] C. I. Ekeocha, B. I. Onyeachu, I.-I. N. Etim, and E. E. Oguzie, "Review of forms of corrosion and mitigation techniques: A visual guide," *African Scientific Reports*, 117, Oct. 2023, doi: 10.46481/asr.2023.2.3.117.
[59] F. Nwosu and M. Muzakir, "Thermodynamic and adsorption studies of corrosion inhibition of mild steel using lignin from Siam weed (Chromolaena odorata) in acid medium," *Journal of Materials and Environmental Science*, vol. 7, pp. 1663–1673, 2016.
[60] K. H. Hariom, H. P. Kumar, and A. Bala, "Corrosion inhibition of mild steel by using hexylamine as corrosion inhibitor in acidic medium," Der *Pharma Chemica*, vol. 8, pp. 268–278, 2016. [Online]. Available: https://api.semanticscholar.org/CorpusID:54991441
[61] A. L. Cummings, S. Waite, and D. K. Nelsen, "Technical article corrosion and corrosion enhancers in amine systems presented at the Brimstone sulfur conference," in *The Brimstone Sulfur Conference*, Brimstone STS, 2005, pp. 1–19.
[62] Y. Wang, J. He, C. Liu, W. H. Chong, and H. Chen, "Thermodynamics versus kinetics in nanosynthesis," *Angewandte Chemie International Edition*, vol. 54, no. 7, pp. 2022–2051, Feb. 2015, doi: 10.1002/anie.201402986.
[63] S. H. Fogler, *Elements of Chemical Reaction Engineering*, 5th ed., Pearson, 2016.
[64] Y. Zhao, J. Xie, G. Zeng, T. Zhang, D. Xu, and F. Wang, "Pourbaix diagram for HP-13Cr stainless steel in the aggressive oilfield environment characterized by high temperature, high CO2 partial pressure and high salinity," *Electrochim Acta*, vol. 293, pp. 116–127, Jan. 2019, doi: 10.1016/j.electacta.2018.08.156.
[65] W. G. Cook and R. P. Olive, "Pourbaix diagrams for the iron–water system extended to high-subcritical and low-supercritical conditions," *Corrosion Science*, vol. 55, pp. 326–331, Feb. 2012, doi: 10.1016/j.corsci.2011.10.034.
[66] Y. Zhao, T. Zhang, H. Xiong, and F. Wang, "Bridge for the thermodynamics and kinetics of electrochemical corrosion: Modeling on dissolution, ionization, diffusion and deposition in metal/solution interface," *Corrosion Science*, vol. 191, 109763, Oct. 2021, doi: 10.1016/j.corsci.2021.109763.

[67] L. O. Olasunkanmi, "Corrosion: Favoured, Yet Undesirable – Its Kinetics and Thermodynamics," in Fahmina Zafar, Anujit Ghosal, and Eram Sharmin (eds.), *Corrosion – Fundamentals and Protection Mechanisms*, IntechOpen, 2022. doi: 10.5772/intechopen.98545.

[68] N. D. Kalisa, T. Muhizi, J. J. Y. Niyotwizera, J. B. Barutwanayo, and J. B. Nkuranga, "Kinetics and thermodynamics investigations on corrosion inhibiting properties of coffee husks extract on mild steel in acidic Medium," *Rwanda Journal of Engineering, Science, Technology and Environment*, vol. 3, no. 1, 2020, doi: 10.4314/rjeste.v3i1.10.

[69] M. Layegh, P. Yan, and J. W. Bennett, "The formation and stability of 3D and 2D materials," *Progress in Crystal Growth and Characterization of Materials*, vol. 70, no. 1, 100615, Feb. 2024, doi: 10.1016/j.pcrysgrow.2023.100615.

[70] S. Papavinasam, "Electrochemical polarization techniques for corrosion monitoring," in Lietai Yang (ed.), *Techniques for Corrosion Monitoring*, Elsevier, 2021, pp. 45–77. doi: 10.1016/B978-0-08-103003-5.00003-5.

[71] G. S. Frankel, "Fundamentals of Corrosion Kinetics," in Anthony E. Hughes, Johannes M.C. Mol, Mikhail L. Zheludkevich, and Rudolph G. Buchheit (eds.), *Active Protective Coatings*, Springer Nature, 2016, pp. 17–32. doi: 10.1007/978-94-017-7540-3_2.

[72] A. Atrens, G. Song, M. Liu, Z. Shi, F. Cao, and M. S. Dargusch, "Review of recent developments in the field of magnesium corrosion," *Advanced Engineering Materials*, vol. 17, no. 4, pp. 400–453, Apr. 2015, doi: 10.1002/adem.201400434.

[73] Z.-T. Chang, B. Cherry, and M. Marosszeky, "Polarisation behaviour of steel bar samples in concrete in seawater. Part 1: Experimental measurement of polarisation curves of steel in concrete," *Corrosion Science*, vol. 50, no. 2, pp. 357–364, Feb. 2008, doi: 10.1016/j.corsci.2007.08.009.

[74] H. Wu *et al.*, "Impact of corrosion on the degradation of the mechanical properties of 2195 and 2297 Al alloys in the marine environment," *Metals (Basel)*, vol. 12, no. 8, Aug. 2022, doi: 10.3390/met12081371.

[75] C. M. Hansson, "The impact of corrosion on society," *Metallurgy and Materials Transactions A Physical Metallurgy and Materials Science*, vol. 42, no. 10, pp. 2952–2962, Oct. 2011, doi: 10.1007/s11661-011-0703-2.

[76] C. M. Hansson, "The impact of corrosion on society," *Metall Mater Trans A Phys Metall Mater Sci*, vol. 42, no. 10, pp. 2952–2962, Oct. 2011, doi: 10.1007/s11661-011-0703-2.

2 Corrosion Prevention
Current and Historical Advancements and Future Directions

Vilabeilie Rutsa, Limasenla Longkumer, Vetezo Venuh, Akhiu K. Yimchunger, Therola Sangtam, and Ambrish Singh

2.1 INTRODUCTION

Material deterioration brought on by chemical reactions with their environment is known as corrosion [1, 2]. It occurs in various forms, like uniform [3], pitting [4], crevice [5], filiform [6], erosion, [7] and galvanic corrosion [8]. Corrosion is mostly caused by water, humidity in the air, metal polishing, acids, oils, particulates, base salts, fluid chemicals, sulfur, ammonia, and gases with an acidic pH [9, 10]. Corrosion failures have become increasingly costly, leading to increased awareness. According to published research, a lack of corrosion protection led to the high cost of corrosion-related accidents, which included maintenance, repairs, and replacing damaged buildings [11]. If corruption caused by corrosion goes unchecked, it can irreversibly harm a nation's economy, making recovery impossible. The cost of corrosion was estimated to be 2.5 trillion US dollars worldwide in 2013, or 3–4% of global GDP [12]. This statistic emphasizes the crucial need for comprehensive corrosion management procedures, which may potentially save between $375 billion and $875 billion per year globally by using accessible corrosion prevention methods [13]. The cost of corrosion can be further classified as follows [14–16]:

Direct Costs: Those that can be quantified. They include replacement costs, prevention costs, research, and development, as well as corrosion inhibition.

Indirect Costs: One cannot quantify these. Examples include product contamination, income loss from disruptions, product loss from spills and fires, inefficient machinery, environmental pollution, and overdesigning to account for metal damage.

Sector-Specific Cost: Industries like oil and gas have extremely significant corrosion costs, with studies showing that the overall cost of corrosion in this industry alone approaches billions of dollars each year owing to both direct and indirect effects.

DOI: 10.1201/9781003571162-2

Preventing corrosion is essential to maintaining the longevity and integrity of metal machinery and buildings across many industries. The economic, operational, and safety consequences associated with corrosion highlight the significance of developing effective preventative techniques [17, 18]. Corrosion failures pose safety risks and disrupt plant operations, prompting researchers and experts to focus on preventing and controlling the process [19, 20]. It is almost impossible to eradicate corrosion; however, it can be mitigated to some extent [21]. Some of the methods [22–24] that are used to control corrosion can be broadly categorized as follows:

1. *Removal of Oxidizing Agent:* Reduced/eliminated exposure to oxidizing substances such as oxygen and moisture aids in slowing corrosion processes.
2. *Prevention of Surface Reactions:*
    a. *Cathodic Protection (CP):* This approach includes using the metal surface as the cathode in an electrochemical cell, avoiding oxidation. Sacrificial anodes or impressed current devices can help achieve this.
    b. *Anodic Protection:* This method lowers corrosion rates by shifting the potential of a metal surface by supplying a shielding current.
3. *Surface Reaction Inhibition:* This involves employing chemical inhibitors and controlling process/system variables like pH.
4. *Protective Coatings:* A protective barrier is created between the metal and corrosive substances through the application of coatings like paints, galvanization, electroplating, and anodizing. This approach is commonly utilized because of its efficiency in avoiding direct interaction with corrosive substances.
5. *Metal Modification:* Incorporating additional metals into steel (for example, stainless steel) improves its mechanical attributes and oxidation resistance.
6. *Surface Condition Modification:* Regularly removing corrosive chemicals and constructing buildings to eliminate cracks or avoid reactive metal mixtures may considerably reduce corrosion hazards.

These approaches can be used alone or in combination, depending on the environmental conditions, material kinds, and economic factors. Effective corrosion control not only extends the life of materials but also ensures safety and dependability in a variety of applications. The most economical approach to corrosion control has been identified as corrosion inhibition. When used in low concentrations, corrosion inhibitors are chemicals that reduce corrosion without interfering with the system [25]. Although chemical inhibitors have proven to be an efficient means of preventing corrosion, their application is restricted due to their toxicity, cost, and availability [26, 27]. Understanding corrosion barrier coatings and their mechanism is essential for successful corrosion control [28]. Synthetic organic and traditional inorganic corrosion inhibitors (CIs) are two well-known conventional inhibitors that have been the subject of much research. The bulk of traditional CIs are organic chemicals, which harm the environment and are a crucial consideration when evaluating CIs [29]. Hydrophobic components and polar functional groups with heteroatoms (O, S, or N) and π electrons are features of organic inhibitors that help keep corrosive aqueous species away from the material surface. Polymers, fatty amides, pyridines, 1,3-azoles, and imidazolines are all components of **surfactant-organic corrosion inhibitors** (SOCIs) [30–34]. In comparison to organic CIs, inorganic CIs are less expensive and can operate at higher temperatures for longer periods. However, when it comes to acids stronger than 17% HCl, they are ineffective [35]. Some studied inorganic inhibitors include chromates [36, [37], phosphates [38, 39], nitrates [40], silicates [41], and arsenates [42, 43].

The primary focus of this chapter is based on assessing the different methodologies applied in corrosion prevention and protection starting from its initial stage of development, analyzing current techniques, and determining future scopes, further discussing the developing trends in corrosion inhibition.

2.2 HISTORICAL BACKGROUND OF CORROSION INHIBITION

Corrosion study has grown from ancient civilizations, beginning with Michael Faraday's 1834 [44] work on electrochemical mechanisms. Modernization in the 19th century resulted in more rigorous research, with electrochemical notions such as Evans' mixed-potential theory [45] providing further insight. Advances in corrosion management, such as stainless-steel materials and protective coatings, have increased the longevity of metal structures while reducing corrosion risk [46, 47]. In 1824, Sir Humphrey Davy [48] established the theory and practice of CP, which was successfully applied to iron anodes to shelter copper sheathing from seawater corrosion. The origins of CP can be traced back to the early 19th century when pioneering electrochemical research by Volta, Davy, and Faraday laid the groundwork for corrosion control [49]. Only in the early 20th century did scientists notice that iron/steel's CP differed from that of other metals, such as copper. This observation helped clarify the various methods by which copper and iron/steel may prevent corrosion [50, 51]. A common method for stopping corrosion in metal structures exposed to severe, conductive-rich ion environments, like soils and seas, is CP [52]. Galvanizing is another method that has been used extensively in the early century (Figure 2.1). Another common process of applying a zinc layer on steel is called hot-dip galvanization, which increases the lifetime and durability of the metallic structures [53].

The development of rust-resistant alloys, such as titanium and stainless steel, has improved corrosion resistance even more, reducing the need for extra safety precautions in harsh and challenging conditions [54]. Along with the development of coatings and linings, epoxy and polyurethane have enhanced performance and durability in various applications from underground pipelines to maritime vessels [55]. Additionally, the development of corrosion inhibitors has considerably reduced the huge expenses by offering efficient chemical barriers or forming a thin protective layer which slows the rate of corrosion in a variety of industrial sectors [56]. These corrosion inhibitors are vital both scientifically and economically because of their use in protecting metallic surfaces and it has come a long way in preventing corrosion. In the 19th century, chromium is introduced as a corrosion inhibitor [57], and in the 1930s, the classification of corrosion inhibitors is done according to their mode of action, such as cathodic and anodic inhibitors [58]. The first inhibitors of

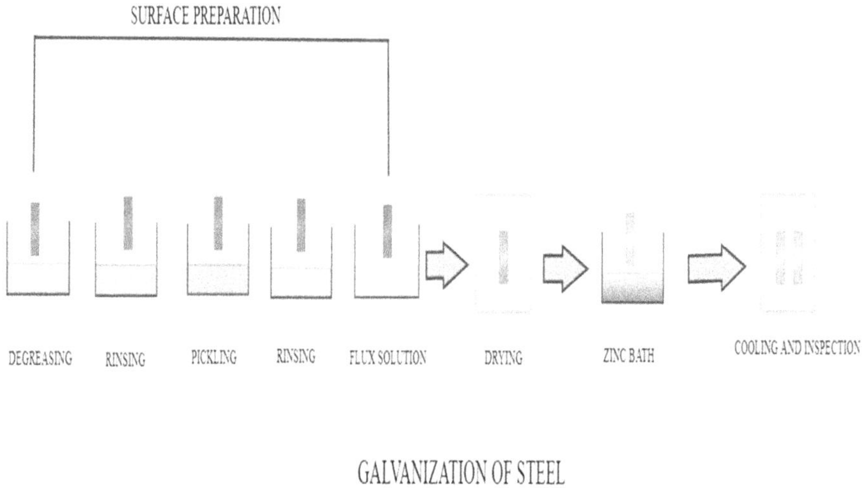

FIGURE 2.1 A basic process of galvanization of steel.

the adsorption type were created in the 1950s. Nonetheless, since 1970, research on ecologically friendly inhibitors has grown in popularity. It is crucial to remember that inorganic inhibitors have several drawbacks, including toxicity and environmental risk, despite their extensive use (Figure 2.2). Their makeup includes lead and chrome, which are not biodegradable and have a propensity to build up in the environment, endangering both human health and the environment [59, 60].

The development of protective coatings for metals, particularly for structures like bridges and buildings, began to gain traction. These coatings, often based on oil-based paints, helped to shield metal from moisture and corrosive elements [61]. The use of CP systems has become more common, particularly for pipelines and marine applications. By introducing a direct current into the metal framework, this method considerably lowers the rate of corrosion [62]. Introducing corrosion inhibitors into paints and coatings provided an additional layer of protection (Table 2.1). These inhibitors can slow down the electrochemical reactions that lead to corrosion [63]. Active corrosion protection is the capacity of a coating to shield a scratch or other flaw by releasing a soluble inhibitor. This characteristic, which is most frequently linked to coatings that include chromate, makes corrosion-resistant coatings very alluring. Strong inhibitory pigments include chromate pigments like calcium, strontium, zinc, barium, and lead [64]. When dissolved hexavalent chromate is reduced to its insoluble trivalent oxidation state, it functions as a potent oxidant on the metal surface, preventing anodic and cathodic reactions (Figure 2.3). There are several different solubilities in the chromate inhibitor family [65]. It is commonly known that galvanized steel provides CP. Either zinc or an alloy of zinc and aluminum makes up the galvanized coating. If fractured, the polarization of the galvanized coating protects the steel substrate underneath from corrosion. The galvanized coating's rate of corrosion is accelerated in return. Sacrificial CP is the term used to describe this type of CP as the galvanized layer corrodes to protect the substrate underneath.

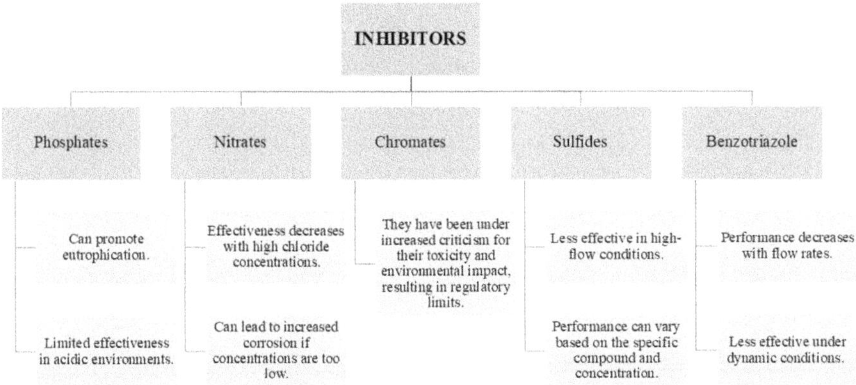

FIGURE 2.2 Some examples of toxic inhibitors.

TABLE 2.1
Development of Synthetic Corrosion Inhibitors

Year	Creation of Synthetic Corrosion Inhibitor	Impact	Ref.
1928	Sulfonation of anthracene or its oil	Resulting in the production of "Antra"	[69]
1936	Recognition of inhibitors on metal surfaces	Aliphatic amines hinder mild steel's ability to absorb nitrogen in sulfuric acid.	[70]
1958	The impact of molecular structure on the prevention of corrosion	In 1 M HCl, iron corrosion is inhibited by aniline derivatives and certain alkylamines.	[71]
1963	Tetraphenylphosphonium chloride, diphenylphosphinodithioic acid, Benzenephosphonic acid	The link between hydrogenation processes and Raney nickel toxicity caused by catalysis	[72]
1970	Use of ethylene glycol-bis (trialkyl phosphonium acetates) or "phosphonium compounds"	Iron, steel, and ferrous alloys are protected from corrosion in corrosive conditions with oil in pipelines and equipment.	[72]
1984	Phosphonium salt corrosion inhibitors	Phosphorus salts are used in deep, high-pressure wells where high temperatures and zinc halides create a very caustic combination.	[69]

Sacrificial CP is a unique corrosion prevention technique that makes zinc-rich paints (ZRPs) stand out [66]. ZRPs are made up of zinc or zinc oxide pigments mixed with either inorganic or organic binding carriers. Zinc oxide (ZnO) and metallic zinc particles [67] or a mixture of spherical metallic particles with a diameter of 2–8 μm [68] are the constituents of pigments.

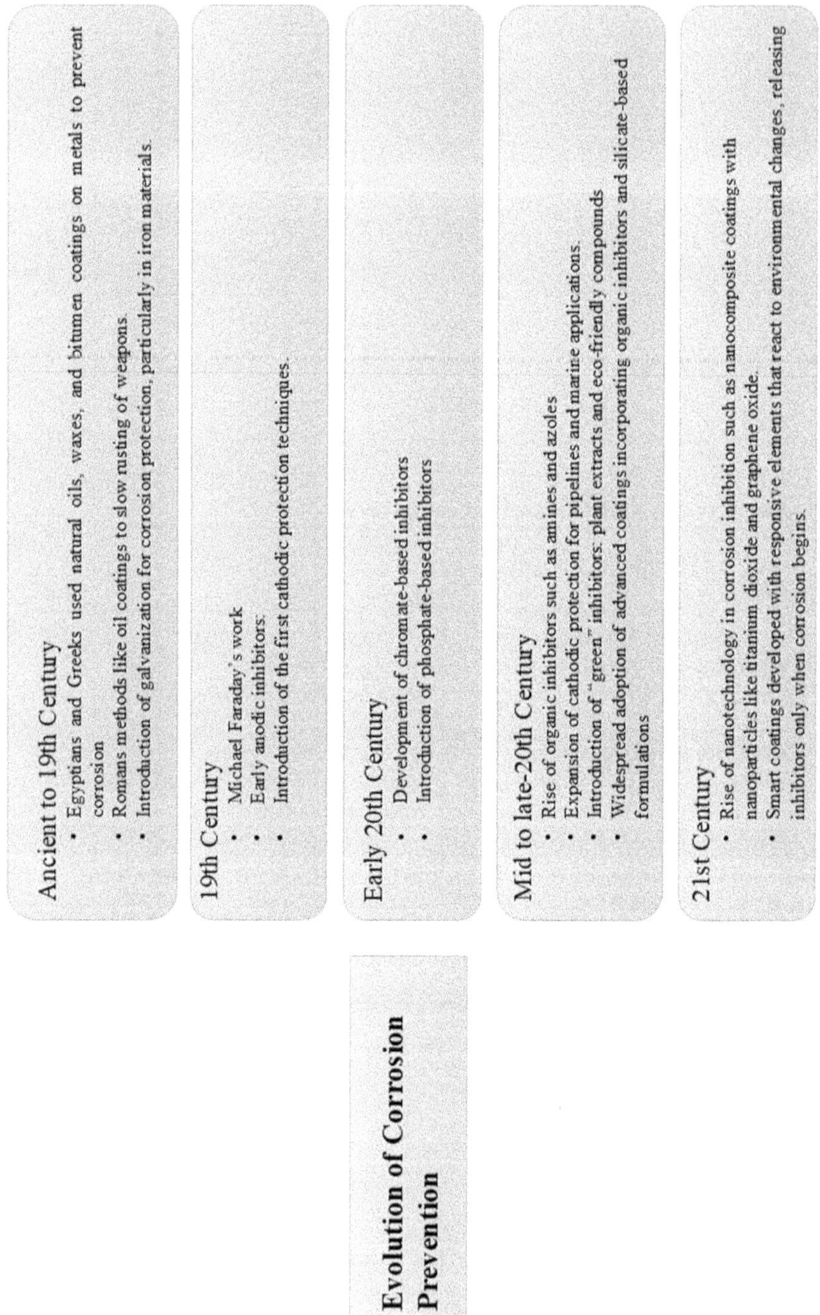

FIGURE 2.3 Evolution of corrosion prevention.

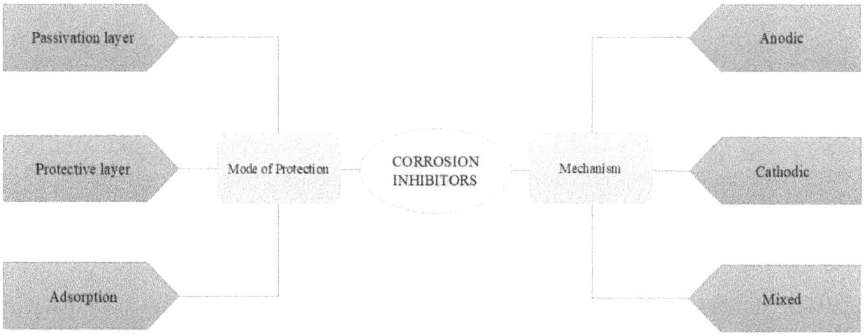

FIGURE 2.4 Scheme for corrosion inhibitor role.

2.3 CURRENT ADVANCEMENT IN CORROSION INHIBITION

Corrosion protection films are crucial for reducing the negative effects of corrosion on materials, especially metals. By establishing a barrier between materials and their environment, coatings provide protection [73]. In addition to providing a sufficient physical barrier to exclude harmful species from reaching the metallic contact, coatings designed to prevent corrosion must also be able to reduce the rate of corrosion even if the protective layer is broken [61, 74]. A variety of alternatives have been investigated, including recent advances in organic coatings, infusion of nanomaterials, metallic coatings, etc. The three types of corrosion inhibitors—anodic, cathodic, and mixed inhibitors—are classified based on how they work. The process of anodic metal breakdown is impeded by anodic inhibitors. A crucial step in corrosion processes, the reduction reaction, is stopped by cathodic inhibitors. By reducing both anodic and cathodic processes, mixed inhibitors provide more thorough corrosion inhibition. A higher overall inhibitory efficiency may arise from the anodic and cathodic mechanisms working in concert. Depending on the type of protection, inhibitors can form a protective layer (Figure 2.4), a passive layer, or adsorb on the metal surface [75–77].

2.4 ORGANIC COATINGS

Organic coatings have great resistivity to ionic conductivity, making them effective barriers that prevent the flow of chemicals like water, ions, and oxygen to and from metal surfaces [78]. Organic coatings' permeability to water and oxygen limits their barrier characteristics and it has been well acknowledged that more water can permeate organic coatings of a suitable density than is necessary for the corrosion process [79]. With a focus on novel formulations, self-healing properties, and the use of nanotechnology for enhanced corrosion resistance, this section explores current advancements in organic coatings. The adoption of graphene-based coatings has shown significant improvement in their capacity to endure various corrosive challenges, while also displaying increased resistance to the damaging impacts of ultraviolet (UV) radiation and mechanical wear [80–82]. Experts investigated how graphene coatings protect steel,

Al, and Cu against corrosion. They discovered that graphene's resistance and ability to block moisture and ion transport reduce degradation rates in corrosive settings [83–85]. One prominent approach is the use of microcapsules or nanocontainers packed with specific agents, such as corrosion inhibitors or polymer precursors that can be embedded in the coating matrix [86]. In situ polymerization [87, 88] and interfacial [89] polymerization are two common microencapsulation procedures; however, alternative ways include in situ condensation, solvent evaporation, dialysis, and supercritical fluid expansion, or by employing classical polymerization [90]. These containers offer the ability to heal themselves by releasing encapsulated chemicals that can patch cracks or damage in the covering. This procedure automatically restores the integrity of the coating, extending the lifespan of the base metal and lowering the cost of upkeep [91, 92]. When the covering is broken by outside pressures, microcapsules may burst, releasing therapeutic compounds into the injured area. According to an experiment by Suiyanarayana et al. [93], a coating flaw allowed encapsulated linseed oil to escape; the synthetic coating then filled up the break in the coating matrix. A solid layer finally formed inside the crevice because of the oxygen in the surrounding atmosphere oxidizing the linseed oil. Aluminum substrates are vulnerable to corrosion in acidic conditions. Self-healing films use incorporated microcapsules which discharge alkaline chemicals when damaged, neutralizing corrosive agents and enabling coating restoration [94]. Some examples of studies highlight the potential of encapsulating functional chemicals for corrosion prevention, with claims of self-healing activated by external or internal stimuli which is presented in Table 2.2.

Additionally, nanotechnology's ability to improve coating performance and safeguard metals has also become a critical consideration [101]. Incorporating nanoparticles (NPs) into organic coatings improves their durability, barrier properties, and resistance to UV rays [102]. Nanocoatings such as Halloysite ($Al_2Si_2O_5(OH)_4 \cdot nH_2O$) polyelectrolyte [103], SiO_2-imidazoline [104], hollow mesoporous silica NPs [104], carbon nanowire [105], graphene nanoflakes [106], nano-titanium dioxide,

TABLE 2.2
Examples of Self-Healing Methods That Use Active Species Mediators to Create Smart Coatings

Metal	Coating	Reservoir	Healing Agents	Trigger	Ref.
Al	CS coating	CS	Cerium ions	pH	[95]
Al	Epoxy	$Ce_2Mo_2O_9$ capsules	Silyl ester	Moisture	[96]
Al and galvanneal	Epoxy	Polyelectrolyte nanocapsules	Mercaptobenzo-thiazole	pH	[97]
Zinc-coated steel	Epoxy	LDH, $Ce_2Mo_2O_9$	Mercaptobenzo-thiazole	pH, Cl^-	[98]
Glass	Silane	Sodium montmorillonite	Sodium montmorillonite	Moisture caused expansion	[99]
Steel	Epoxy polymer	$C_2H_6N_2O_2$ microcapsules	Siloxane	Crack propagation	[100]

nano-silicon dioxide pigments [107], 3-aminopropyltrimethoxysilane [108], and polyaniline-ZnO nanocomposites [109] have been investigated by researchers in the recent time. With the implementation of nanotechnology and the usage of NPs, it has been made possible to improve the mechanical characteristics, barrier efficiency, and other special functions of organic coatings; such developments have great potential to create corrosion inhibitors with long-term efficiency.

2.5 INORGANIC COATINGS

On metal surfaces, inorganic inhibitors such as chromates, phosphates, molybdates, and silicates apply a protective layer to stop corrosion. They are frequently used in boilers, pipelines, and cooling water systems, among other industrial applications. [110, 111]. Some recent inorganic inhibitors that have shown great inhibiting efficiency in preventing corrosion are highlighted in Table 2.3 (Figure 2.5).

TABLE 2.3
Some Examples of Inorganic Inhibitors

Metal Substrate	Inhibitor	Corrosive Media	Inhibitor Type	Efficiency (%)	Ref.
Carbon steel	Zn^{2+}	Seawater	Cathodic inhibitor	83.3	[112]
Carbon steel	Sodium nitrite	3% NaCl	Anodic inhibitor	94.1	[113]
Mild steel	Sodium tungstate + $ZnSO_4$	Natural seawater	Mixed inhibitor	84.81	[114]
Iron-based alloys	Sodium tungstate	10% H_2SO_4	Anodic inhibitor	97	[115]
Carbon steel	Sodium molybdate	Simulated cooling water	Anodic inhibitor	98.5	[116]
Carbon steel	Sodium nitrite	Chlorine containing solution	Anodic inhibitor	65	[117]
Carbon steel	Sodium nitrite	5 mM chloride	Anodic inhibitor	90	[118]
N-80 steel	Sodium vanadate	Conc. $K_4P_2O_7$ solution	Anodic inhibitor	99.85	[119]
Carbon steel	Sodium molybdate	0.01 M NaCl + 0.1 M $NaHCO_3$	Anodic inhibitor	97.9	[120]
Carbon steel	Sodium phosphate	Simulated cooling water	Anodic inhibitor	90	[118]
Carbon steel	Sodium nitrite	250 NaCl	Anodic inhibitor	99.06	[121]
Carbon steel	Sodium molybdate	Saturated $Ca(OH)_2$ + 0.5 M NaCl	Anodic inhibitor	97	[122]
Carbon steel	Sodium silicate	3.5% NaCl	Anodic inhibitor	92	[123]

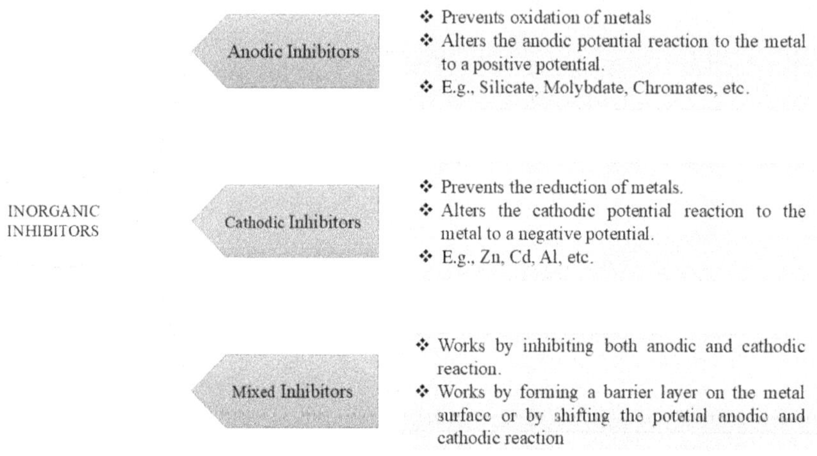

FIGURE 2.5 Mechanism of inorganic inhibitors.

2.6 BIOPOLYMER INHIBITORS

Biopolymers are natural polymers derived from sources that are renewable animals, plants, and microorganisms (proteins, polysaccharides, bacteria, and nucleic acids) [124]. Polymers are now a vital part of both industrial and research operations. The application of polymers in recent decades in nanotechnology research serves as an instance [125]. Numerous biopolymers have been evaluated as possible inhibitors of corrosion; for instance, the use of carbohydrates as inhibitors offers a compelling substitute because of their minimal environmental effect, renewable nature, and availability (Table 2.4). There are many ways to get carbohydrates, such as through bacterial fermentation processes, marine crustacean shells (such as chitosan (CS) and alginate), and plant components (such as starch and cellulose) [126, 127]. A range of characteristics support the use of CS as a corrosion inhibitor, including its polymer nature, ability to completely adsorb onto and cover a variety of metallic surfaces, and effective corrosion protection [128, 129]. A study involving CS as an inhibitor reported 96% and 93% corrosion inhibition efficacy at 60°C and 70°C, respectively, when used for MS protection [130]. Another abundant biopolymer produced as a result of the paper bulb process is lignin, which normally is considered as a waste product. Surfaces of ferrous metals exposed to acidic environments were treated with lignin as an inhibitor which reported an efficiency of 80.79% [131]. Even though lignin has been shown to have corrosion-inhibiting qualities, research is still being carried out to determine how to make it even more effective [132, 133]. Alginate, a polysaccharide found in brown seaweed, has shown potential as an ecologically friendly inhibitor. Alginate's eco-friendliness and unique features have made it a popular alternative to standard and harmful corrosion inhibitors among industries and experts [134]. Sodium alginate's (SA) inhibition on mild steel in an acidic medium exhibited a good efficiency of 90.9%, highlighting its potential as a robust inhibitor in an acidic medium [135]. 2-Acrylamido-2-methylpropanesulfonic acid synthesized

TABLE 2.4
Some Biopolymers Evaluated as Corrosion Inhibitors

Metal Substrate	Inhibitor	Corrosive Media	Inhibitor Type	Efficiency (%)	Ref.
J55 steel	Salicylaldehyde-chitosan Schiff base (SCBS)	3.5% NaCl	Mixed-type inhibitor	95.4	[129]
Cold-rolled steel	Cassava starch-graft copolymer	0.3 mol/L glycolic acid	Mixed-type inhibitor	92.61	[140]
Mg alloy	Starch	3% NaCl	Mixed-type inhibitor	86	[141]
Carbon steel	*Mangifera indica* gum tree (MA gum)	1 M HCl	Mixed-type inhibitor	98	[142]
Carbon steel	CS	1 M HCl	—	93.2	[143]
Carbon steel	Chitosan-cinnamaldehyde Schiff base	3.5 NaCl	Mixed-type inhibitor	92.67	[144]
Mild steel	Guar gum	1 M H_2SO_4	Mixed-type inhibitor	89.37	[145]
Mild steel	4-(Dimethylamino)-benzaldehyde chitosan	1 M HCl	—	90.65	[146]
Mild steel	Gum acacia	1 M HCl	Mixed-type inhibitor	97	[147]
Mild steel	Amylose-acetate-blended carboxymethyl chitosan (AA-CMCh)	1 M H_2SO_4	—	91.09	[148]
Zinc	Polyacrylamide grafted starch (St-g-PAM)	1 M HCl	Mixed-type inhibitor	89.67	[149]

from Guar gum has also exhibited excellent corrosion inhibition of 95% at 600 mg/L in a 3.5% NaCl environment [136]. Researchers employed nanotechnology with biopolymers to improve the inhibition of corrosion in metals, either as coatings or as inhibitors [137], and focused on biopolymers as a medium for nanocomposites, such as CS acetate for Fe_2O_3-TiO_2 and Urchin-like CdS-Fe_3O_4 nanostructures [138, 139].

2.7 GREEN PLANT-BASED INHIBITORS

Plant-based materials provide biodegradable and sustainable substitutes. Because plant extracts are abundant in phytochemicals, they are useful in both industrial and biological processes. Stabilizing agents derived from plants can also serve as environmentally benign corrosion inhibitors in metals, alloys, and electrolytic systems [150]. Preventing corrosion in metals including zinc, copper, steel, and aluminum was the focus of the majority of the research in the literature. Silver nanoparticles (AgNPs) derived from *Eichhornia crassipes* [151] (Figure 2.6) leaf extract effectively

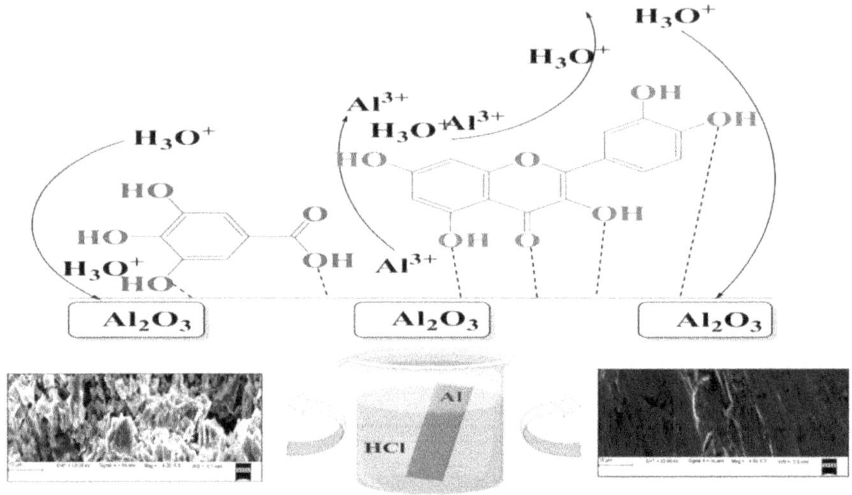

FIGURE 2.6 Schematic indication of the adsorption mechanism on the mild steel surface of Al corrosion in 1 M HCl solution (0.4 mg/L of electrochemical solution).

inhibited aluminum corrosion in an acidic setting at room temperature, with an efficiency of 99.6%. Using *Moringa oleifera* [152] leaf extract to synthesize CuO NPs resulted in a corrosion inhibition efficiency of 95.06%. Zinc corrosion behavior has been extensively studied in hydrochloric [153], sulfuric, phosphoric, and sulfamic acids. Pais et al. [154] synthesized Glycogen NPs (GLY-Np) employing a microwave-mediated nanoprecipitation technique. The NPs act as a sustainable barrier for zinc corrosion in sulfamic acid. The inhibitory effectiveness reached 92% at a GLY-Np concentration of 0.02 g/L. To ensure durability, compatibility, and adhesion, green NP mixtures and application techniques must be customized to fit different metal substrates and coatings. Assessing interactions between sustainable NPs and existing inhibitors of corrosion or surface coatings is crucial for ensuring optimal performance and avoiding compatibility issues [155]. Some of the plants that are investigated for corrosion inhibition are listed in Table 2.5.

2.8 CARBON DOTS AS INHIBITORS OF CORROSION

Carbon dots (CDs) are nanomaterials that range in size from 2 to 10 nm. They are used as corrosion inhibitors due to their environmental friendliness, low cost, stiffness, variety of applications, and low toxicity (Figure 2.7). CDs are carbon-based nanostructures with unique characteristics that limit corrosion and serve many additional functions like catalysis, biomedicine, sensors, energy storage, optoelectronics, and anticounterfeiting [167]. As anti-corrosion agents, a variety of CD-based materials are frequently used, such as ionic liquids, nitrogen-doped CDs [168], polymer-based CDs [169], and SN-co-doped CDs [170]. It was investigated how well new N and S co-doped CD (N, S CDs) inhibited 5052 aluminum alloys in an acidic environment (0.1 M HCl). According to the data collected, the N and S-CDs were effective corrosion

TABLE 2.5
Green Synthesized Nanoparticles Used to Prevent Corrosion

Plant	Nanoparticle	Corrosive Media	Metal	Efficiency (%)	Ref.
Tobacco	Ag	HCl	Carbon steel	98	[156]
Lonchocarpus laxiflorus	Ag	1.0 M HCl	Mild steel	96.44	[157]
Kola nut	Ag–Au	1.0 M HCl	Mild steel	96.11	[158]
Pomegranate	ZnO	De-ionized H_2O	Stainless steel	99.9	[159]
Citrus reticulata	Ni	1 M HCl	X80 steel	87.3	[160]
Acacia nilotica	NiO	HCl	Carbon steel	93.6%	[161]
Lemon peel waste	Fe_3O_4 nanoparticles	Deionized H_2O	ASTM A-662 steel	85	[162]
Chromolaena odorata	Titanium nanocomposite	1 M HCl	Mild steel	92.3	[163]
Solanum tuberosum	Copper	0.5 H_2SO_4	Stainless steel	98.8	[164]
Musa sapientum	Iron oxide nanoparticles	HCl	Mild steel	94.4	[165]
Parkia biglandulosa	Zirconia nanoparticles (ZrO_2 NPs)	NaCl	Mild steel	99.26	[166]

inhibitors for Al-metal, exhibiting an 85.9% efficacy at low concentrations in acidic circumstances [171, 172]. N-doped CDs were created by microwave synthesis from cost-effective substances such as urotropine, citric acid, and diethylenetriamine. This method created affordable N-CDs that might be used as corrosion inhibitors in the oil and gas sectors. The study also noted that N-CDs have water solubility, biocompatibility, minimal environmental impact, and are good optical agents. They are also effective anti-corrosion resistance compounds for steel materials in severely acidic and corrosive circumstances. According to the results, N-CDs have an 81.2% inhibitory efficiency at 60°C, effectively suppressing corrosion. Many processes, including as laser ablation, hydrothermal reactions, ultrasonication (US), microwave (MW) irradiation, electrochemical oxidation, and hydrothermal reactions, can produce CDs. Materials are characterized by spectral analyses including transmission electron microscopy, Fourier-transform infrared spectroscopy, X-ray diffraction, energy-dispersive X-ray spectroscopy, X-ray photoelectron spectroscopy [173].

Some CDs studied for corrosion inhibition are listed in Table 2.6.

2.9 FUTURE DIRECTIONS

Technological developments, ecologically friendly processes, and novel materials are all necessary for the future of corrosion inhibitors. As industrial organizations seek

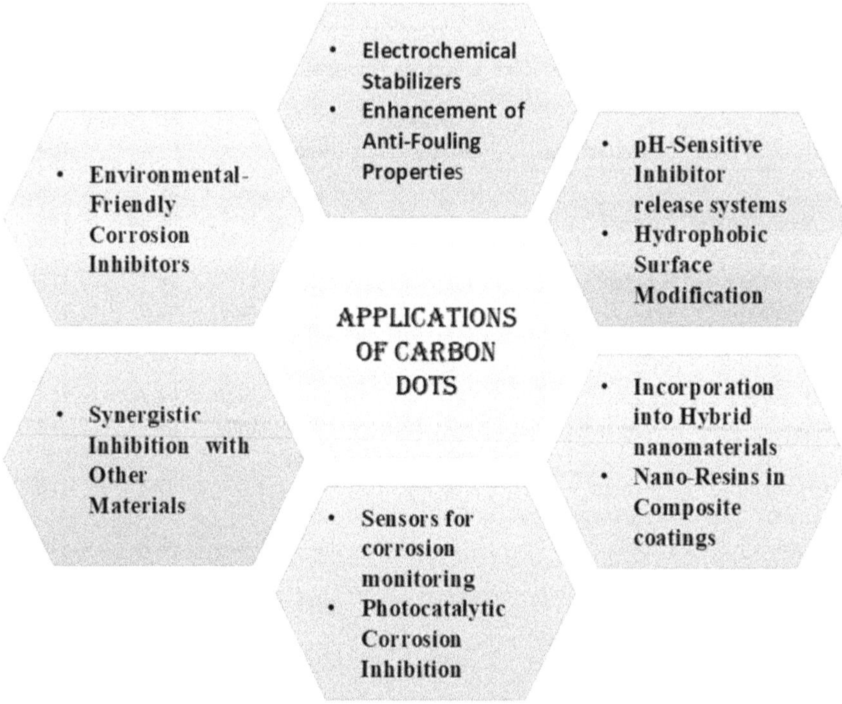

FIGURE 2.7 Various applications of CDs.

more effective and environmentally friendly solutions, more research is expected to yield innovative formulations and strategies for tackling both commercial and environmental challenges in corrosion control. There is also increasing curiosity in multifunctional coatings that combine corrosion resistance with other features like anti-fouling, antimicrobial, or electrically conductive abilities. Such multifunctional coatings are especially useful in areas such as maritime, medical, and electronics, where metals are susceptible to an extensive range of corrosive hazards. Coatings having anti-fouling and antibacterial qualities, for example, provide dual protection against both bio-corrosion and conventional electrochemical corrosion, increasing the life of metallic surfaces in aquatic or moist conditions. This approach emphasizes the trend toward creating corrosion inhibitors that not only protect but also provide more attributes customized to specific applications. Another approach could be the use of coatings integrated with sensors that monitor electrochemical characteristics that can detect the initial stages of corrosion, allowing researchers to react before major damage occurs and it could potentially provide real-time insights into structural integrity, resulting in a more refined and dependable approach to corrosion management. Advanced computational approaches are also increasingly being used for estimating corrosion inhibitor activity at the molecular level. Quantum mechanical simulations can give insights into adsorption processes and thermodynamics, assisting in the creation of more potent inhibitors. This novel technique to corrosion

TABLE 2.6
CDs as CIs

CDs/Source	Metal	Corrosive Media	Inhibitor Type	Efficiency (%)	Ref.
N-doped CDs/*citric and urea*	Carbon steel	0.5 M H_2SO_4	Mixed-type inhibitor	97.8	[168]
N-doped CDs/*citric acid monohydrate and ethanolamine*	Q235 carbon steel	0.1 M HCl	Mixed-type inhibitor	89.98	[174]
N and S co-doped CD/ —	5052 aluminum alloy	0.1 M HCl	—	85.9	[171]
N-doped CDs/ *ammonium bicarbonate electrolyte*	Cu	1.0 mol/L H_2SO_4	Mixed-type inhibitor	96.32	[175]
Citric acid-based CDs/ *imidazole*	Q235 steel	3.5% NaCl	—	85.7	[176]
N and S co-doped CDs/*aminosalicylic acid and thiourea*	Carbon steel	CO_2 saturated 3% NaCl	Mixed-type inhibitor	93	[177]
N-doped CDs/ *dopamine*	Q235 carbon steel	1 M HCl	Mixed-type inhibitor	96.1	[178]
N- and S-doped CDs/ *O-phenylenediamine and thiourea*	Cu	0.5% H_2SO_4	Mixed-type inhibitor	99.88	[179]
N-doped CDs/ *ammonium citrate*	Q235 carbon steel	1 M HCl	Mixed-type inhibitor	96.96	[180]
N-doped CDs/ *diethylenetriamine and urotropine*	N80 steel	1 M HCl	Mixed-type inhibitor	84.93	[172]

protection might significantly increase the lifespan of metal infrastructure while reducing negative environmental effects, leading to a more resilient and sustainable corrosion prevention strategy.

2.10 CONCLUSION

To preserve the longevity and integrity of metal structures and machinery across a variety of industries, the study highlights how important it is to prevent corrosion. Safety hazards and operational interruptions brought on by corrosion failures can be reduced with efficient corrosion control. It draws attention to corrosion's enormous financial cost, which amounts to billions of dollars every year. The necessity of thorough corrosion management plans is emphasized as they have the potential to save significant sums of money by lowering maintenance and replacement expenses. Among the corrosion inhibitors covered in the study are biopolymers such as lignin and CS, which have demonstrated encouraging effectiveness in preventing metal corrosion. These developments point to a move toward more environmentally friendly and efficient corrosion control techniques.

The use of nanotechnology, such as graphene-based coatings, and the creation of self-healing coatings have greatly increased corrosion resistance. These developments offer active defense against corrosive substances, which can prolong the life of metal substrates. To create corrosion prevention techniques that are more economical, sustainable, and effective, additional research is required. To improve inhibitor design, this entails investigating novel materials, real-time monitoring tools, and computational techniques. To reduce the financial and safety hazards connected with corrosion, our findings highlight the continuous need for innovation in corrosion prevention techniques. Globally, corrosion is still a major problem with serious structural, economic, and environmental repercussions. Despite its unavoidability, scientific and technological advancements have made it possible to create practical plans for lessening its effects. From conventional methods like CP and galvanization to more recent approaches utilizing nanotechnology, biopolymers, and plant-based inhibitors, this chapter has examined the development of corrosion prevention. Protective coatings have become essential for preventing corrosion. Strong barriers against environmental variables are provided by organic coatings, particularly those reinforced with nanomaterials and graphene, increasing resilience and durability. A major advancement in self-healing coatings is the use of microcapsules or nanocontainers, which enable materials to self-heal damage and retain their protective effectiveness. Comparably, inorganic coatings—like those made of chromates, phosphates, and silicates—have proven to be highly effective in a variety of industrial applications, but their negative effects on the environment make it necessary to look for alternatives. Interest in corrosion inhibitors derived from plants and biopolymers has increased as sustainability has become more important. Because of their natural abundance, renewability, and environmental friendliness, biopolymers such as CS, lignin, and alginate have demonstrated strong corrosion prevention efficacy. Green technology's promise is further highlighted by plant-derived NPs, such as those made from iron oxide, zinc oxide, and silver. In addition to reducing environmental hazards, these inhibitors operate on par with or better than traditional inhibitors, especially in severe environments. Integrating multifunctional systems is the key to corrosion avoidance in the future. Redefining material protection will be possible with coatings and inhibitors that combine anti-corrosion qualities with other features including antimicrobial, anti-fouling, and real-time monitoring capabilities. The development of highly effective, customized inhibitors is expected to be accelerated by computational methods such as molecular simulations and quantum modelling. Despite tremendous advancements, there are still obstacles to overcome, especially when it comes to addressing long-term durability and scaling sustainable solutions for industrial usage. Driving innovation will require sustained multidisciplinary research and cooperation between academic institutions, business, and policymakers. Society may lessen the negative effects of corrosion on the economy and environment by using cutting-edge, environmentally friendly technology, guaranteeing the durability and integrity of vital infrastructure for future generations.

REFERENCES

[1] B. S. Mahdi *et al.*, "Corrosion inhibition of mild steel in hydrochloric acid environment using terephthaldehyde based on Schiff base: gravimetric, thermodynamic, and computational studies," *Molecules*, vol. 27, no. 15, p. 4857, 2022.

[2] A. A. Al-Amiery, W. N. R. W. Isahak, and W. K. Al-Azzawi, "Corrosion inhibitors: natural and synthetic organic inhibitors," *Lubricants*, vol. 11, no. 4, p. 174, 2023.

[3] P. Geethamani and P. K. Kasthuri, "Adsorption and corrosion inhibition of mild steel in acidic media by expired pharmaceutical drug," *Cogent Chem*, vol. 1, no. 1, p. 1091558, 2015.

[4] R. Singh, *Corrosion control for offshore structures: cathodic protection and high-efficiency coating*. Gulf Professional Publishing, 2014.

[5] N. Rashidi, S. Alavi-Soltani, and R. Asmatulu, "Crevice corrosion theory, mechanisms and prevention methods," in *Proceedings of the 3rd Annual GRASP Symposium*. Wichita State University, 2007.

[6] J. M. C. Mol, B. R. W. Hinton, D. H. Van Der Weijde, J. H. W. De Wit, and S. Van Der Zwaag, "A filiform corrosion and potentiodynamic polarisation study of some aluminium alloys," *J Mater Sci*, vol. 35, pp. 1629–1639, 2000.

[7] P. Pedeferri and M. Ormellese, *Corrosion science and engineering*, vol. 720. Springer, 2018.

[8] A. Kadhim *et al.*, "A mini review on corrosion, inhibitors and mechanism types of mild steel inhibition in an acidic environment," *Int J Corros Scale Inhib*, vol. 10, no. 3, pp. 861–884, 2021.

[9] D. Musik, K. Wójcik, M. Sekuła-Wybańska, M. Konopacki, and R. Rakoczy, "Analysis of the corrosion process with the application of the novel type of coupon installation," *Processes*, vol. 10, no. 12, p. 2468, 2022.

[10] B. E. A. Rani and B. B. J. Basu, "Green inhibitors for corrosion protection of metals and alloys: an overview," *IntJ Corros*, vol. 2012, no. 1, p. 380217, 2012.

[11] S. Z. Salleh *et al.*, "Plant extracts as green corrosion inhibitor for ferrous metal alloys: a review," *J Clean Prod*, vol. 304, p. 127030, 2021.

[12] A. H. Al-Moubaraki and I. B. Obot, "Corrosion challenges in petroleum refinery operations: sources, mechanisms, mitigation, and future outlook," *J Saudi Chem Soc*, vol. 25, no. 12, p. 101370, 2021.

[13] C. Verma *et al.*, "Principles and theories of green chemistry for corrosion science and engineering: design and application," *Green Chem*, vol. 26, no. 8, pp. 4270–4357, 2024.

[14] C. Okoroafor, "Cathodic protection as a means of saving national asset," *J Corr Sci Tech*, vol. 1, no. 1, pp. 1–6, 2004.

[15] G. H. Koch, M. P. H. Brongers, N. G. Thompson, Y. P. Virmani, and J. H. Payer, "*Corrosion cost and preventive strategies in the United States*," United States. Federal Highway Administration, 2002.

[16] M. A. J. Mazumder, "Global impact of corrosion: occurrence, cost and mitigation," *Glob J Eng Sci*, vol. 5, no. 4, pp. 1–5, 2020.

[17] H. Aljibori, A. Al-Amiery, and W. N. R. Isahak, "Advancements in corrosion prevention techniques," *J Bio Tribo-Corros*, vol. 10, no. 4, p. 78, 2024.

[18] H. S. Aljibori, A. Alamiery, and A. A. H. Kadhum, "Advances in corrosion protection coatings: a comprehensive review," *Int J Corros Scale Inhib*, vol. 12, no. 4, pp. 1476–1520, 2023.
[19] L. L. Shreir, *Corrosion: corrosion control*. Newnes, 2013.
[20] K. Elayaperumal and V. S. Raja, *Corrosion failures: theory, case studies, and solutions*. Wiley, 2015.
[21] L. T. Popoola, A. S. Grema, G. K. Latinwo, B. Gutti, and A. S. Balogun, "Corrosion problems during oil and gas production and its mitigation," *Int J Ind Chem*, vol. 4, pp. 1–15, 2013.
[22] R. Zuo, "Biofilms: strategies for metal corrosion inhibition employing microorganisms," *Appl Microbiol Biotechnol*, vol. 76, no. 6, pp. 1245–1253, 2007.
[23] O. S. I. Fayomi, I. G. Akande, and S. Odigie, "Economic impact of corrosion in oil sectors and prevention: an overview," in *Journal of Physics: Conference Series*, IOP Publishing, 2019, p. 022037.
[24] O. N. Mohanty and G. Jha, "Corrosion control through surface modification," R&D, TATA STEEL: http://eprints. nmlindia. org/4100/1/O_N_Mohanty-1. PDF. 1998.
[25] E. Bardal, *Corrosion and protection*, vol. 68. Springer-Verlag, 2004.
[26] M. Dargahi, A. L. J. Olsson, N. Tufenkji, and R. Gaudreault, "Green technology: tannin-based corrosion inhibitor for protection of mild steel," *Corrosion*, vol. 71, no. 11, pp. 1321–1329, 2015.
[27] I. Milošev, "Corrosion inhibition of aluminium alloys by molybdate ions: a critical review of the chemistry, mechanisms and applications," *Corros Sci*, p. 111854, 2024.
[28] L. M. Muresan, "Nanocomposite coatings for anti-corrosion properties of metallic substrates," *Materials*, vol. 16, no. 14, p. 5092, 2023.
[29] C. Verma, E. E. Ebenso, I. Bahadur, and M. A. Quraishi, "An overview on plant extracts as environmental sustainable and green corrosion inhibitors for metals and alloys in aggressive corrosive media," *J Mol Liq*, vol. 266, pp. 577–590, 2018.
[30] K. R. Ansari, M. A. Quraishi, and A. Singh, "Pyridine derivatives as corrosion inhibitors for N80 steel in 15% HCl: electrochemical, surface and quantum chemical studies," *Measurement*, vol. 76, pp. 136–147, 2015.
[31] S. A. Abd El-Maksoud and A. S. Fouda, "Some pyridine derivatives as corrosion inhibitors for carbon steel in acidic medium," *Mater Chem Phys*, vol. 93, no. 1, pp. 84–90, 2005.
[32] O. Olivares-Xometl, N. V. Likhanova, M. A. Domínguez-Aguilar, E. Arce, H. Dorantes, and P. Arellanes-Lozada, "Synthesis and corrosion inhibition of α-amino acids alkylamides for mild steel in acidic environment," *Mater Chem Phys*, vol. 110, no. 2–3, pp. 344–351, 2008.
[33] M. M. Antonijević, S. M. Milić, and M. B. Petrović, "Films formed on copper surface in chloride media in the presence of azoles," *Corros Sci*, vol. 51, no. 6, pp. 1228–1237, 2009.
[34] M. N. Rahuma, "Corrosion in oil and gas industry: a perspective on corrosion inhibitors," *J Mater Sci Eng*, vol. 3, no. 03, p. 1, 2014.
[35] K. Tamalmani and H. Husin, "Review on corrosion inhibitors for oil and gas corrosion issues," *Appl Sci*, vol. 10, no. 10, pp. 3389, 2020.
[36] M. W. Kendig and R. G. Buchheit, "Corrosion inhibition of aluminum and aluminum alloys by soluble chromates, chromate coatings, and chromate-free coatings," *Corrosion*, vol. 59, no. 5, pp. 379–400, 2003.
[37] S. Zehra, M. Mobin, and J. Aslam, "Chromates as corrosion inhibitors," in Chandrabhan Verma, Jeenat Aslam and Chaudhery Mustansar Hussain (eds.), *Inorganic anticorrosive materials*, Elsevier, 2022, pp. 251–268.

[38] D. M. Bastidas, M. Criado, V. M. La Iglesia, S. Fajardo, A. La Iglesia, and J. M. Bastidas, "Comparative study of three sodium phosphates as corrosion inhibitors for steel reinforcements," *Cem Concr Compos*, vol. 43, pp. 31–38, 2013.
[39] L. Yohai, M. Vázquez, and M. B. Valcarce, "Phosphate ions as corrosion inhibitors for reinforcement steel in chloride-rich environments," *Electrochim Acta*, vol. 102, pp. 88–96, 2013.
[40] M. Murmu, S. K. Saha, N. C. Murmu, and P. Banerjee, "Nitrate as corrosion inhibitor," in Chandrabhan Verma, Jeenat Aslam and Chaudhery Mustansar Hussain (eds.) *Inorganic anticorrosive materials*, Elsevier, 2022, pp. 269–296.
[41] H. Gao, Q. Li, F. N. Chen, Y. Dai, F. Luo, and L. Q. Li, "Study of the corrosion inhibition effect of sodium silicate on AZ91D magnesium alloy," *Corros Sci*, vol. 53, no. 4, pp. 1401–1407, 2011.
[42] N. Betti, A. A. Al-Amiery, W. K. Al-Azzawi, and W. N. R. W. Isahak, "Corrosion inhibition properties of Schiff base derivative against mild steel in HCl environment complemented with DFT investigations," *Sci Rep*, vol. 13, no. 1, pp. 8979, 2023.
[43] R. Hsissou *et al.*, "Insight into the corrosion inhibition of novel macromolecular epoxy resin as highly efficient inhibitor for carbon steel in acidic mediums: synthesis, characterization, electrochemical techniques, AFM/UV–Visible and computational investigations," *J Mol Liq*, vol. 337, pp. 116492, 2021.
[44] K. A. Shukla and T. P. Kumar, "Pillars of modern electrochemistry: a brief history," 2008. http://electrochem.cwru.edu/ed/encycl/art-p05-pillars-of-ec.htm.
[45] M. G. Fontana and N. D. Greene, *Corrosion Engineering*. McGraw-hill, 1967.
[46] A. Goyal, H. S. Pouya, E. Ganjian, and P. Claisse, "A review of corrosion and protection of steel in concrete," *Arab J Sci Eng*, vol. 43, pp. 5035–5055, 2018.
[47] M. Taghavikish, N. K. Dutta, and N. Roy Choudhury, "Emerging corrosion inhibitors for interfacial coating," *Coatings*, vol. 7, no. 12, p. 217, 2017.
[48] J. E. McCoy, "Corrosion control by cathodic protection-theoretical and design concepts for marine applications," *Transactions*, vol. 82, no. 1, 1970.
[49] F. A. J. L. James, *Davy in the Dockyard: Humphry Davy, The Royal Society and the electrochemical protection of the copper sheeting of his majesty's ship in the mid 1820s*. 1992.
[50] R. Kuhn, "Galvanic currents on cast-iron pipe," in *The first corrosion conference*, US Bureau of Standards, 1928.
[51] W. van Wullen-Scholten, "Der Schutz des Eisens durch elektrischen Strom," *Gwf*, vol. 73, p. 403, 1930.
[52] A. Nicoleta and A. Ovidiu, "Cathodic protection by sacrificial anodes or impressed current, comparative analysis," *Annals (Constanța Maritime University)*, vol. 14, no. 19, 2013.
[53] C. Ahner, P. Maaß, and P. Peißker, *Handbook of hot-dip galvanization*. Wiley, 2011.
[54] K. Perflikowski and M. Wińcza, "Finishing works after welding of structures made of rust-resistant, high-alloy steels," *Weld Int*, vol. 26, no. 10, pp. 747–751, 2012.
[55] K. B. Tator, "Introduction to coatings and linings," in Stephen D. Cramer, Bernard S. Covino, Jr. (eds.), Corrosion: Fundamentals, Testing, and Protection. 2003.
[56] E.-S. M. Sherif, "The role of corrosion inhibitors in protecting metallic structures against corrosion in harsh environments," in Monzer Fanun (ed.), *The role of colloidal systems in environmental protection*, Elsevier. pp. 509–526, 2014.
[57] A. E. Hughes *et al.*, "Corrosion inhibition, inhibitor environments, and the role of machine learning," *Corros Mater Degrad*, vol. 3, no. 4, pp. 672–693, 2022.
[58] C. G. Dariva and A. F. Galio, "Corrosion inhibitors–principles, mechanisms and applications," *Devel Corros Protect*, vol. 16, pp. 365–378, 2014.

[59] J. Telegdi, "History of phosphorus-containing corrosion inhibitors: from the beginning till the present time," in Zahid Amjad and Konstantinos D. Demadis (eds.), *Water-formed deposits*, Elsevier, 2022, pp. 49–68.
[60] K. Aramaki, "Historical and Future Perspectives on Corrosion Inhibitor Studies," *Zairyo-to-Kankyo*, vol. 53, no. 7, pp. 348–351, 2004.
[61] A. A. Nazeer and M. Madkour, "Potential use of smart coatings for corrosion protection of metals and alloys: a review," *J Mol Liq*, vol. 253, pp. 11–22, 2018.
[62] P. E. Francis, "Cathodic Protection," National Physical Laboratory, p. 1, 2007.
[63] M. Kendig and D. J. Mills, "An historical perspective on the corrosion protection by paints," *Prog Org Coat*, vol. 102, pp. 53–59, 2017.
[64] A. J. Eickhoff and W. S. Hebre, "corrosionl Nl Hl bitive pigments and GGGG GGGGGG LLLLLLGGGGGG".
[65] O. Gharbi, S. Thomas, C. Smith, and N. Birbilis, "Chromate replacement: what does the future hold?," *Npj Mater Degrad*, vol. 2, no. 1, pp. 12, 2018.
[66] R. G. Buchheit, "Corrosion resistant coatings and paints," in Myer Kutz (ed.), *Handbook of environmental degradation of materials*, Elsevier, 2005, pp. 367–385.
[67] H. F. Payne, "Organic coating technology," 1954.
[68] J. R. Vilche, E. C. Bucharsky, and C. A. Giudice, "Application of EIS and SEM to evaluate the influence of pigment shape and content in ZRP formulations on the corrosion prevention of naval steel," *Corros Sci*, vol. 44, no. 6, pp. 1287–1309, 2002.
[69] T. J. Harvey, F. C. Walsh, and A. H. Nahlé, "A review of inhibitors for the corrosion of transition metals in aqueous acids," *J Mol Liq*, vol. 266, pp. 160–175, 2018.
[70] C. A. Mann, B. E. Lauer, and C. T. Hultin, "Organic inhibitors of corrosion, aliphatic amines," *Ind Eng Chem*, vol. 28, no. 2, pp. 159–163, 1936.
[71] H. Kaesche and N. Hackerman, "Corrosion inhibition by organic amines," *J Electrochem Soc*, vol. 105, no. 4, p. 191, 1958.
[72] L. Horner and F. Roettger, "Corrosion inhibitors. I. Relations between chemical structure and corrosion protection with old and new inhibitors of iron corrosion in acid solution," *Korrosion*, vol. 16, pp. 57–71, 1963.
[73] M. F. Montemor, "Functional and smart coatings for corrosion protection: a review of recent advances," *Surf Coat Technol*, vol. 258, pp. 17–37, 2014.
[74] M. Fenker, M. Balzer, and H. Kappl, "Corrosion protection with hard coatings on steel: past approaches and current research efforts," *Surf Coat Technol*, vol. 257, pp. 182–205, 2014.
[75] I. A. W. Ma, S. Ammar, S. S. A. Kumar, K. Ramesh, and S. Ramesh, "A concise review on corrosion inhibitors: types, mechanisms and electrochemical evaluation studies," *J Coat Technol Res*, vol. 19, pp. 241–268, 2022.
[76] H. Assad and A. Kumar, "Understanding functional group effect on corrosion inhibition efficiency of selected organic compounds," *J Mol Liq*, vol. 344, p. 117755, 2021.
[77] K. M. Shwetha, B. M. Praveen, and B. K. Devendra, "A review on corrosion inhibitors: types, mechanisms, electrochemical analysis, corrosion rate and efficiency of corrosion inhibitors on mild steel in an acidic environment," *Results Surf Interfaces*, vol. 73, p. 100258, 2024.
[78] R.-G. Hu, S. Zhang, J.-F. Bu, C.-J. Lin, and G.-L. Song, "Recent progress in corrosion protection of magnesium alloys by organic coatings," *Prog Org Coat*, vol. 73, no. 2–3, pp. 129–141, 2012.
[79] G. Grundmeier, W. Schmidt, and M. Stratmann, "Corrosion protection by organic coatings: electrochemical mechanism and novel methods of investigation," *Electrochim Acta*, vol. 45, no. 15–16, pp. 2515–2533, 2000.

[80] N. H. Othman, M. C. Ismail, M. Mustapha, N. Sallih, K. E. Kee, and R. A. Jaal, "Graphene-based polymer nanocomposites as barrier coatings for corrosion protection," *Prog Org Coat*, vol. 135, pp. 82–99, 2019.

[81] S. S. A. Kumar, S. Bashir, K. Ramesh, and S. Ramesh, "New perspectives on graphene/graphene oxide based polymer nanocomposites for corrosion applications: the relevance of the graphene/polymer barrier coatings," *Prog Org Coat*, vol. 154, p. 106215, 2021.

[82] S. S. A. Kumar, S. Bashir, K. Ramesh, and S. Ramesh, "A comprehensive review: super hydrophobic graphene nanocomposite coatings for underwater and wet applications to enhance corrosion resistance," *FlatChem*, vol. 31, p. 100326, 2022.

[83] A. R. Heiba, F. A. Taher, R. M. Abou Shahba, and N. A. Abdel Ghany, "Corrosion mitigation of carbon steel in acidic and salty solutions using electrophoretically deposited graphene coatings," *J Coat Technol Res*, vol. 18, no. 2, pp. 501–510, 2021.

[84] Q. Wang *et al.*, "The multiple faces of graphene on anticorrosion: advances and prospects," *Corrosion Engineering, Science and Technology*, vol. 54, pp. 417–434, 2024.

[85] A. Thakur and A. Kumar, *Sustainability, safety, and applications of nanomaterials-based corrosion inhibitors*. IGI Global, 2024.

[86] A. M. Fadl, M. I. Abdou, M. A. Hamza, and S. A. Sadeek, "Corrosion-inhibiting, self-healing, mechanical-resistant, chemically and UV stable PDMAS/TiO_2 epoxy hybrid nanocomposite coating for steel petroleum tanker trucks," *Prog Org Coat*, vol. 146, p. 105715, 2020.

[87] S. Lang and Q. Zhou, "Synthesis and characterization of poly (urea-formaldehyde) microcapsules containing linseed oil for self-healing coating development," *Prog Org Coat*, vol. 105, pp. 99–110, 2017.

[88] H. Wang and Q. Zhou, "Evaluation and failure analysis of linseed oil encapsulated self-healing anticorrosive coating," *Prog Org Coat*, vol. 118, pp. 108–115, 2018.

[89] P. Vijayan and M. A. AlMaadeed, "'Containers' for self-healing epoxy composites and coating: trends and advances," *Express Polym Lett*, vol. 10, no. 6, 2016.

[90] X. Xu, Y. Guo, W. Li, and L. Zhu, "Electrochemical behavior of different shelled microcapsule composite copper coatings," *Int J Min Metall Mater*, vol. 18, no. 3, pp. 377–384, 2011.

[91] H. Wei *et al.*, "Advanced micro/nanocapsules for self-healing smart anticorrosion coatings," *J Mater Chem A Mater*, vol. 3, no. 2, pp. 469–480, 2015.

[92] K. Lee, C. G. Cavazos, J. Rouse, X. Wei, M. Li, and S. Wei, "Advanced micro/nanocapsules for self-healing smart anticorrosion coatings: a review of recent developments," in Hatem M.A. Amin, Ahmed Galal (eds.), *Corrosion protection of metals and alloys using graphene and biopolymer based nanocomposites*, pp. 169–190, 2021. CRC Press.

[93] C. Suryanarayana, K. C. Rao, and D. Kumar, "Preparation and characterization of microcapsules containing linseed oil and its use in self-healing coatings," *Prog Org Coat*, vol. 63, no. 1, pp. 72–78, 2008.

[94] U. Eduok, E. Ohaeri, and J. Szpunar, "Self-healing composite coatings with protective and anticorrosion potentials: classification by healing mechanism," in *Self-healing composite materials*, Elsevier, 2020, pp. 123–162.

[95] J. Carneiro *et al.*, "Chitosan-based self-healing protective coatings doped with cerium nitrate for corrosion protection of aluminum alloy 2024," *Prog Org Coat*, vol. 75, no. 1–2, pp. 8–13, 2012.

[96] S. J. García *et al.*, "Self-healing anticorrosive organic coating based on an encapsulated water reactive silyl ester: synthesis and proof of concept," *Prog Org Coat*, vol. 70, no. 2–3, pp. 142–149, 2011.

[97] M. Plawecka, D. Snihirova, B. Martins, K. Szczepanowicz, P. Warszynski, and M. F. Montemor, "Self healing ability of inhibitor-containing nanocapsules loaded in epoxy coatings applied on aluminium 5083 and galvanneal substrates," *Electrochim Acta*, vol. 140, pp. 282–293, 2014.

[98] M. F. Montemor, D. V. Snihirova, M. G. Taryba, S. V. Lamaka, and I. A. Kartsonakis, A. C. Balaskas, G. C. Kordas, J. Tedim, A. Kuznetsova, M. L. Zheludkevich, M. G. S. Ferreira, "Evaluation of self-healing ability in protective coatings modified with 37 combinations of layered double hydroxides and cerium molibdate nanocontainers filled 38 with corrosion inhibitors," *Electrochim Acta*, vol. 60, pp. 31–40, 2012.

[99] F. Micciché, H. Fischer, R. Varley, and S. Van der Zwaag, "Moisture induced crack filling in barrier coatings containing montmorillonite as an expandable phase," *Surf Coat Technol*, vol. 202, no. 14, pp. 3346 3353, 2008.

[100] S. H. Cho, S. R. White, and P. V. Braun, "Self-healing polymer coatings," *Adv Mater*, vol. 21, no. 6, pp. 645–649, 2009.

[101] S. E. Elugoke, A. S. Adekunle, O. E. Fayemi, B. B. Mamba, and E. E. Ebenso, "Nanocomposites as steel corrosion inhibitors–the journey so far," in Chandrabhan Verma, Chaudhery Mustansar Hussain, Eno Ebenso (eds.), Anticorrosive Nanomaterials: Future perspectives, pp. 207–267, 2022.

[102] M. H. Nazari *et al.*, "Nanocomposite organic coatings for corrosion protection of metals: a review of recent advances," *Prog Org Coat*, vol. 162, p. 106573, 2022.

[103] Y. C. Feng and Y. F. Cheng, "Fabrication of Halloysite nanocontainers and their compatibility with epoxy coating for anti-corrosion performance," *Corrosion Engineering, Science and Technology*, vol. 51, no. 7, pp. 489–497, 2016.

[104] Y. Liang *et al.*, "Facile synthesis of smart nanocontainers as key components for construction of self-healing coating with superhydrophobic surfaces," *Nanoscale Res Lett*, vol. 11, pp. 1–12, 2016.

[105] V. Valencia-Goujon, T. Hawkins, E. Maya-Visuet, and H. Castaneda, "Electrochemical characterization in CO_2 saturated environment of Zn-rich epoxy nanocoatings on API X52 pipeline grade steel substrate under flow conditions," in *NACE CORROSION*, NACE, 2015, p. NACE–2015.

[106] T. Monetta, A. Acquesta, A. Carangelo, and F. Bellucci, "The effect of graphene on the protective properties of water-based epoxy coatings on Al2024-T3," *Int J Corros*, vol. 2017, no. 1, p. 1541267, 2017.

[107] A. M. Kamalan Kirubaharan, M. Selvaraj, K. Maruthan, and D. Jeyakumar, "Synthesis and characterization of nanosized titanium dioxide and silicon dioxide for corrosion resistance applications," *J Coat Technol Res*, vol. 9, pp. 163–170, 2012.

[108] M. Behzadnasab, S. M. Mirabedini, and M. Esfandeh, "Corrosion protection of steel by epoxy nanocomposite coatings containing various combinations of clay and nanoparticulate zirconia," *Corros Sci*, vol. 75, pp. 134–141, 2013.

[109] A. Mostafaei and F. Nasirpouri, "Epoxy/polyaniline–ZnO nanorods hybrid nanocomposite coatings: synthesis, characterization and corrosion protection performance of conducting paints," *Prog Org Coat*, vol. 77, no. 1, pp. 146–159, 2014.

[110] A. Alamiery, "Short report of mild steel corrosion in 0.5 m H2SO4 by 4-ethyl-1-(4-oxo-4-phenylbutanoyl) thiosemicarbazide," *J Tribol*, vol. 30, no. 8, pp. 90–99, 2021.

[111] A. A. Alamiery, W. N. R. Wan Isahak, and M. S. Takriff, "Inhibition of mild steel corrosion by 4-benzyl-1-(4-oxo-4-phenylbutanoyl) thiosemicarbazide: gravimetrical, adsorption and theoretical studies," *Lubricants*, vol. 9, no. 9, p. 93, 2021.

[112] R. Sabzi and R. Arefinia, "Investigation of zinc as a scale and corrosion inhibitor of carbon steel in artificial seawater," *Corros Sci*, vol. 153, pp. 292–300, 2019.
[113] S. Kareem Mohammed, "Effect use of two chemical compounds sodium nitrate and sodium silicate as corrosion inhibitor to steel reinforcement," *Al-Khwarizmi Eng J*, vol. 14, no. 2, pp. 123–128, 2018.
[114] J. Zhang, L. Zhang, G. Tao, and F. Cui, "Synergistic inhibition effect of sodium tungstate and zinc sulphate on mild steel corrosion in seawater," *Int J Electrochem Sci*, vol. 13, no. 7, pp. 6522–6536, 2018.
[115] G. A. Gaber, H. A. Aly, and L. Z. Mohamed, "Effect of sodium tungstate on the corrosion behavior of Fe-base alloy in H2SO4 solution," *Int J Electrochem Sci*, vol. 15, no. 8, pp. 8229–8240, 2020.
[116] K. Yee, K. Kee, S. Hassas, and M. Ismail, "Corrosion inhibition of molybdate and nitrite for carbon steel corrosion in process cooling water," *J Engi Sci Technol*, vol. 14, no. 4, pp. 2431–2444, 2019.
[117] M. Shaglouf, H. F. Emrayed, M. Milad, A. Areef, G. Idris, and M. Yousif, "Influence of nitrite and molybdate blend on carbon steel inhibition in chloride containing solutions," *Sirte Uni Sci J*, vol. 12, no. 1, pp. 108–121, 2022.
[118] H. Khani and R. Arefinia, "Inhibition mechanism of nitrite on the corrosion of carbon steel in simulated cooling water systems," *Mater Corrosi*, vol. 69, no. 3, pp. 337–347, 2018.
[119] J. Zhao, Q. Yang, C. Zhang, and Y. Wang, "Corrosion of N80 steel in a concentrated tetrapotassium pyrophosphate solution and corrosion control by vanadates," *Int J Electrochem Sci*, vol. 14, no. 7, pp. 6209–6222, 2019.
[120] H. Bensabra, A. Franczak, O. Aaboubi, N. Azzouz, and J.-P. Chopart, "Inhibitive effect of molybdate ions on the electrochemical behavior of steel rebar in simulated concrete pore solution," *Metall Mater Transac A*, vol. 48, pp. 412–424, 2017.
[121] M. Hayyan, S. A. Sameh, A. Hayyan, and I. M. AlNashef, "Utilizing of sodium nitrite as inhibitor for protection of carbon steel in salt solution," *Int J Electrochem Sci*, vol. 7, no. 8, pp. 6941–6950, 2012.
[122] Y. T. Tan, S. L. Wijesinghe, and D. J. Blackwood, "Effect of molybdate on the passivation of carbon steel in alkaline solutions under open-circuit conditions," *J Electrochem Soc*, vol. 163, no. 10, pp. C649, 2016.
[123] F. B. Mainier, A. A. M. Figueiredo, A. E. R. de Freitas, and A. A. M. de Alencar Junior, "The use of sodium silicate as a corrosion inhibitor in a saline drilling fluid: a nonaggressive option to the environment," *J Environ Prot (Irvine, Calif)*, vol. 7, no. 13, p. 2025, 2016.
[124] G. Machado *et al.*, "Biopolymers from lignocellulosic biomass: feedstocks, production processes, and applications," in Avinash P. Ingle, Anuj Kumar Chandel, Silvio Silvério da Silva (eds.), *Lignocellulosic biorefining technologies*, pp. 125–158, 2020. Wiley.
[125] M. Yousefi, F. Gholamian, D. Ghanbari, and M. Salavati-Niasari, "Polymeric nanocomposite materials: preparation and characterization of star-shaped PbS nanocrystals and their influence on the thermal stability of acrylonitrile–butadiene–styrene (ABS) copolymer," *Polyhedron*, vol. 30, no. 6, pp. 1055–1060, 2011.
[126] M. Moradi, Z. Song, and X. Tao, "Introducing a novel bacterium, Vibrio neocaledonicus sp., with the highest corrosion inhibition efficiency," *Electrochem Commun*, vol. 51, pp. 64–68, 2015.
[127] H. Ashassi-Sorkhabi and A. Kazempour, "Thermodynamic and kinetic insights into the role of amino acids in improving the adhesion of iota-carrageenan as a natural corrosion inhibitor to the aluminum surface," *J Adhes Sci Technol*, vol. 34, no. 9, pp. 961–975, 2020.

[128] C. Verma, M. A. Quraishi, A. Alfantazi, and K. Y. Rhee, "Corrosion inhibition potential of chitosan based Schiff bases: design, performance and applications," *Int J Biol Macromol*, vol. 184, pp. 135–143, 2021.
[129] K. R. Ansari, D. S. Chauhan, M. A. Quraishi, M. A. J. Mazumder, and A. Singh, "Chitosan Schiff base: an environmentally benign biological macromolecule as a new corrosion inhibitor for oil & gas industries," *Int J Biol Macromol*, vol. 144, pp. 305–315, 2020.
[130] S. A. Umoren, M. J. Banera, T. Alonso-Garcia, C. A. Gervasi, and M. V. Mirífico, "Inhibition of mild steel corrosion in HCl solution using chitosan," *Cellulose*, vol. 20, pp. 2529–2545, 2013.
[131] P. P. Rahayu, C. D. D. Sundari, and I. Farida, "Corrosion inhibition using lignin of sugarcane bagasse," in *IOP Conference Series: Materials Science and Engineering*, IOP Publishing, 2018, p. 012087.
[132] N. Aissiou, M. Bounoughaz, and A. Djeddi, "Lignin-phenylhydrazone as a corrosion inhibitor of API X52 carbon steel in 3.5% NaCl and 0.1 mol/L HCl medium," *Chem Res Chin Univ*, vol. 37, pp. 718–728, 2021.
[133] J. Wang, F. Seidi, Y. Huang, and H. Xiao, "Smart lignin-based polyurethane conjugated with corrosion inhibitor as bio-based anticorrosive sublayer coating," *Ind Crops Prod*, vol. 188, p. 115719, 2022.
[134] L. Muthulakshmi, S. Mohan, N. Hariharan, and J. Aslam, "Sodium alginate: grafted alginates as sustainable corrosion inhibitors," in Jeenat Aslam, Chandrabhan Verma, Ruby Aslam (eds.), *Grafted biopolymers as corrosion inhibitors: safety, sustainability, and efficiency*, pp. 365–382, 2023. Wiley.
[135] I. Nadi *et al.*, "Sargassum muticum extract based on alginate biopolymer as a new efficient biological corrosion inhibitor for carbon steel in hydrochloric acid pickling environment: gravimetric, electrochemical and surface studies," *Int J Biol Macromol*, vol. 141, pp. 137–149, 2019.
[136] A. Singh, M. Liu, E. Ituen, and Y. Lin, "Anti-corrosive properties of an effective guar gum grafted 2-acrylamido-2-methylpropanesulfonic acid (GG-AMPS) coating on copper in a 3.5% NaCl solution," *Coatings*, vol. 10, no. 3, p. 241, 2020.
[137] T. H. Mekonnen, T. Haile, and M. Ly, "Hydrophobic functionalization of cellulose nanocrystals for enhanced corrosion resistance of polyurethane nanocomposite coatings," *Appl Surf Sci*, vol. 540, p. 148299, 2021.
[138] A. Abbasi, D. Ghanbari, M. Salavati-Niasari, and M. Hamadanian, "Photo-degradation of methylene blue: photocatalyst and magnetic investigation of Fe 2 O 3–TiO 2 nanoparticles and nanocomposites," *J Mater Sci Mater Electron*, vol. 27, pp. 4800–4809, 2016.
[139] D. Ghanbari and M. Salavati-Niasari, "Synthesis of urchin-like CdS-Fe3O4 nanocomposite and its application in flame retardancy of magnetic cellulose acetate," *J Ind Eng Chem*, vol. 24, pp. 284–292, 2015.
[140] L. X. Li XiangHong, D. S. Deng ShuDuan, L. Y. Li YunXian, and L. J. Liu JianXiang, "Corrosion inhibition of steel in glycolic acid solution by cassava starch grafted copolymer," *Chem Ind For Prod*, vol. 37, pp. 101–108, 2017.
[141] S. Babu, A. Sivarajan, R. Sudhakaran, T. Kasilingam, K. Raja, and T. Asokan, "Corrosion inhibition on magnesium alloy in 3% NaCl solution by polysaccharide," *Ann Rom Soc Cell Biol*, vol. 25, no. 4, pp. 19569–19578, 2021.
[142] H. F. Chahul, E. Maji, and T. B. Danat, "Adsorptive, inhibitive and thermodynamics studies on the corrosion of mild steel in the presence of Mangifera indica gums," *Ovidius Univ Ann Chem*, vol. 30, no. 2, pp. 75–80, 2019.

[143] A. E. Okoronkwo, S. J. Olusegun, and O. O. Oluwasina, "The inhibitive action of chitosan extracted from Archachatina marginata shells on the corrosion of plain carbon steel in acid media," *Anti-Corros Methods Mater*, vol. 62, no. 1, pp. 13–18, 2015.

[144] D. S. Chauhan, M. A. J. Mazumder, M. A. Quraishi, and K. R. Ansari, "Chitosan-cinnamaldehyde Schiff base: a bioinspired macromolecule as corrosion inhibitor for oil and gas industry," *Int J Biol Macromol*, vol. 158, pp. 127–138, 2020.

[145] R. Kushwah and R. K. Pathak, "Open circuit potential, polarization and thermometric study of guar gum as corrosion inhibitor on mild steel by in acidic media," *Asian J Chem Sci*, vol. 8, no. 2, pp. 55–60, 2020.

[146] J. Haque, V. Srivastava, D. S. Chauhan, H. Lgaz, and M. A. Quraishi, "Microwave-induced synthesis of chitosan Schiff bases and their application as novel and green corrosion inhibitors: experimental and theoretical approach," *ACS Omega*, vol. 3, no. 5, pp. 5654–5668, 2018.

[147] K. Azzaoui *et al.*, "Eco friendly green inhibitor Gum Arabic (GA) for the corrosion control of mild steel in hydrochloric acid medium," *Corros Sci*, vol. 129, pp. 70–81, 2017.

[148] M. Erna, H. Herdini, and D. Futra, "Corrosion inhibition mechanism of mild steel by Amylose-Acetate/Carboxymethyl Chitosan composites in acidic media," *Int J Chem Eng*, vol. 2019, no. 1, pp. 8514132, 2019.

[149] W. Yating *et al.*, "Inhibition for Zn corrosion by starch grafted copolymer," *J Chinese Soc Corros Protect*, vol. 41, no. 1, pp. 131–138, 2021.

[150] N. Hossain, M. Asaduzzaman Chowdhury, and M. Kchaou, "An overview of green corrosion inhibitors for sustainable and environment friendly industrial development," *J Adhes Sci Technol*, vol. 35, no. 7, pp. 673–690, 2021.

[151] L. V. Hublikar, S. V. Ganachari, N. Raghavendra, V. B. Patil, and N. R. Banapurmath, "Green synthesis silver nanoparticles via Eichhornia Crassipes leaves extract and their applications," *Curr Res Green Sustain Chem*, vol. 4, p. 100212, 2021.

[152] A. M. Hyba, H. M. El Refay, S. Shahen, and G. A. Gaber, "Comparison fabrication, identification and avoidance of corrosion potential of M-CuO NPs/S-CuO NPs to suppress corrosion on steel in an acidic solution," *Chem Papers*, vol. 77, no. 9, pp. 5395–5407, 2023.

[153] C. X. Sun *et al.*, "Research on the corrosion inhibitors of zinc in hydrochloric acid," in *IOP Conference Series: Materials Science and Engineering*, IOP Publishing, 2017, p. 012043.

[154] M. Pais, S. D. George, and P. Rao, "Glycogen nanoparticles as a potential corrosion inhibitor," *Int J Biol Macromol*, vol. 182, pp. 2117–2129, 2021.

[155] A. Thakur, A. Kumar, S. Kaya, R. Marzouki, F. Zhang, and L. Guo, "Recent advancements in surface modification, characterization and functionalization for enhancing the biocompatibility and corrosion resistance of biomedical implants," *Coatings*, vol. 12, no. 10, p. 1459, 2022.

[156] S. R. Al-Mhyawi, "Green synthesis of silver nanoparticles and their inhibitory efficacy on corrosion of carbon steel in hydrochloric acid solution," *Int J Electrochem Sci*, vol. 18, no. 9, p. 100210, 2023.

[157] G. A. Ijuo, N. Surma, and J. O. Igoli, "Ag-nanoparticles mediated by Lonchocarpus laxiflorus stem bark extract as anticorrosion additive for mild steel in 1.0 M HCl solution," *Progres Chem Biochem Res*, vol. 5, no. 2, pp. 133–146, 2022.

[158] J. K. Odusote *et al.*, "Silver–Gold Synthesized Nanoparticle as Corrosion Inhibitor on Mild Steel in 1.0 M HCl Environment," *Chem Afr*, vol. 7, no. 1, pp. 243–256, 2024.

[159] E. A. Ibadi, H. Al-Hetty, M. A. W. Alwardi, R. S. Alazragi, H. A. Almashhadani, and M. M. Kadhim, "Anticorrosion, antimicrobial and antioxidant study of ZnO nanoparticles synthesized from Punica granatum (Pomegranate) extract," *Int J Corros Scale Inhib*, vol. 11, no. 4, pp. 1569–1582, 2022.
[160] E. Ituen, A. Singh, and L. Yuanhua, "Synthesis of bio-based nickel nanoparticles composite, characterization and corrosion inhibition in simulated oilfield microbial and acidizing environments," *J Adhes Sci Technol*, vol. 35, no. 1, pp. 15–34, 2021.
[161] H. M. K. Sheit, M. S. Mubarak, M. V. Mohamed, M. Jayaprasadh, K. S. Mohan, and K. V. Gunavathy, "Enhanced anti-corrosion efficiency and antimicrobial properties of green synthesised nickel oxide (NiO) nanoparticles," 2023.
[162] N. Elizondo-Villarreal *et al.*, "Green synthesis of magnetic nanoparticles of iron oxide using aqueous extracts of lemon peel waste and its application in anti-corrosive coatings," *Materials*, vol. 15, no. 23, p. 8328, 2022.
[163] E. F. Olasehinde, B. E. Agbaffa, M. A. Adebayo, and J. Enis, "Corrosion protection of mild steel in acidic medium by titanium-based nanocomposite of Chromolaena odorata leaf extract," *Mater Chem Phys*, vol. 281, p. 125856, 2022.
[164] O. O. Adenike, O. Oluwafemi, and O. O. Johnson, "Investigation of Eco-friendly plant nanoparticles inhibitor on the corrosion behaviour of duplex stainless steel 2101 and 2205 in 0.5 MH2so4". Journal of Multidisciplinary Engineering Science Studies, Vol. 5, Issue 2, 2019.
[165] E. Ituen, V. Mkpenie, and E. Ekemini, "Adsorptive Fe-nanoparticles mediated by Musa sapientum peels extract as anticorrosion additive for aqueous oilfield descaling solution," *Sci Afr*, vol. 3, p. e00075, 2019.
[166] N. Muthulakshmi, A. Kathirvel, M. Senthil, and R. Subramanian, "Green synthesis of zirconia nanoparticles and their characterization, anticancer activity and corrosion inhibition properties," *Journal of the Indian Chemical Society*, vol. 100, no. 10, p. 101076, 2023.
[167] W. Wang, L. Cheng, and W. Liu, "Biological applications of carbon dots," *Sci China Chem*, vol. 57, pp. 522–539, 2014.
[168] S. Cao *et al.*, "Nitrogen-doped carbon dots as high-effective inhibitors for carbon steel in acidic medium," *Colloids Surf A Physicochem Eng Asp*, vol. 616, p. 126280, 2021.
[169] H. D. Zhang, A. Y. Chen, B. Gan, H. Jiang, and L. J. Gu, "Corrosion protection investigations of carbon dots and polydopamine composite coating on magnesium alloy," *J Magnes Alloys*, vol. 10, no. 5, pp. 1358–1367, 2022.
[170] V. Saraswat and M. Yadav, "Carbon dots as green corrosion inhibitor for mild steel in HCl solution," *ChemistrySelect*, vol. 5, no. 25, pp. 7347–7357, 2020.
[171] H. Cen, X. Zhang, L. Zhao, Z. Chen, and X. Guo, "Carbon dots as effective corrosion inhibitor for 5052 aluminium alloy in 0.1 M HCl solution," *Corros Sci*, vol. 161, p. 108197, 2019.
[172] J. Lv, L. Fu, B. Zeng, M. Tang, and J. Li, "Synthesis and acidizing corrosion inhibition performance of N-doped carbon quantum dots," *Russ J Appl Chem*, vol. 92, pp. 848–856, 2019.
[173] W. Ghann *et al.*, "The synthesis and characterization of carbon dots and their application in dye sensitized solar cell," *Int J Hydrogen Energy*, vol. 44, no. 29, pp. 14580–14587, 2019.
[174] M. Cui, Y. Qiang, W. Wang, H. Zhao, and S. Ren, "Microwave synthesis of eco-friendly nitrogen doped carbon dots for the corrosion inhibition of Q235 carbon steel in 0.1 M HCl," *Int J Electrochem Sci*, vol. 16, no. 1, p. 151019, 2021.
[175] Q. Zhou *et al.*, "Large-scale electrochemical fabrication of nitrogen-doped carbon quantum dots and their application as corrosion inhibitor for copper," *J Mater Sci*, vol. 56, pp. 12909–12919, 2021.

[176] D. Yang, Y. Ye, Y. Su, S. Liu, D. Gong, and H. Zhao, "Functionalization of citric acid-based carbon dots by imidazole toward novel green corrosion inhibitor for carbon steel," *J Clean Prod*, vol. 229, pp. 180–192, 2019.
[177] H. Cen, Z. Chen, and X. Guo, "N, S co-doped carbon dots as effective corrosion inhibitor for carbon steel in CO_2-saturated 3.5% NaCl solution," *J Taiwan Inst Chem Eng*, vol. 99, pp. 224–238, 2019.
[178] M. Cui, Y. Yu, and Y. Zheng, "Effective corrosion inhibition of carbon steel in hydrochloric acid by dopamine-produced carbon dots," *Polymers (Basel)*, vol. 13, no. 12, p. 1923, 2021.
[179] Y. Zhang, B. Tan, X. Zhang, L. Guo, and S. Zhang, "Synthesized carbon dots with high N and S content as excellent corrosion inhibitors for copper in sulfuric acid solution," *J Mol Liq*, vol. 338, p. 116702, 2021.
[180] Y. Ye *et al.*, "A high-efficiency corrosion inhibitor of N-doped citric acid-based carbon dots for mild steel in hydrochloric acid environment," *J Hazard Mater*, vol. 381, p. 121019, 2020.

3 Role of Coatings in Corrosion Protection

Neeta Gupta, Sangeeta Gupta, Rajesh Kumar Sahu, Reena Negi Rawat, and Bhawana Jain

3.1 INTRODUCTION

Corrosion, a natural process that affects materials in virtually every environment, has been a persistent challenge for human civilization since the dawn of metallurgy. This electrochemical phenomenon, primarily associated with metals, involves the gradual degradation of materials due to their interaction with the surrounding environment. While often perceived as a slow, inevitable process, corrosion's impact on modern society is both profound and far-reaching, affecting industries, infrastructure, and daily life in ways that are often underestimated. The rate and nature of corrosion are influenced by a multitude of factors, including the properties of the metal itself, environmental conditions, and the presence of various chemical species. For instance, the pH of the surrounding medium can significantly affect corrosion rates, with acidic environments generally accelerating the process. Temperature also plays a crucial role, with higher temperatures typically increasing the rate of chemical reactions and thus corrosion (Wang et al., 2023).

Moreover, the products of corrosion themselves can have significant impacts. Rust, the familiar reddish-brown substance formed by the corrosion of iron, is not merely a cosmetic issue. The formation of rust can lead to expansion and flaking of the metal surface, compromising structural integrity and potentially leading to catastrophic failures in critical infrastructure. In industrial settings, corrosion can lead to the contamination of products, particularly in industries such as food processing and pharmaceuticals where purity is paramount. The release of metal ions from corroding equipment can introduce impurities into production lines, necessitating stringent quality control measures and potentially leading to costly product recalls. The study of corrosion, known as corrosion science, has evolved significantly over the past century, driven by the need to understand, predict, and mitigate its effects. As our reliance on metals and other susceptible materials has grown with technological advancement, so too has the urgency to address the challenges posed by corrosion. From the steel reinforcements in our buildings and bridges to the intricate components of advanced machinery, corrosion threatens the integrity, safety, and longevity of countless structures and systems (Pant, 2023).

Recent advancements in materials science and nanotechnology have opened new avenues for corrosion research. For instance, the development of self-healing

 DOI: 10.1201/9781003571162-3

materials that can automatically repair corrosion damage represents a promising frontier in corrosion prevention. Similarly, the use of advanced computational models to predict corrosion behavior in complex systems is enabling more targeted and effective corrosion management strategies.

This comprehensive analysis aims to explore the multifaceted nature of corrosion, its underlying mechanisms, and its wide-ranging impacts on various sectors of society. By examining current research and real-world examples, we will delve into the economic, environmental, and safety implications of corrosion, as well as the strategies and technologies being developed to combat this pervasive issue. Understanding the chemical basis of corrosion and its far-reaching impacts is essential for developing effective solutions to this ongoing challenge, ensuring the sustainability and safety of our infrastructure and technologies for future generations.

3.1.1 Understanding Corrosion: Mechanisms and Types

At its most fundamental level, corrosion is rooted in the principles of electrochemistry and thermodynamics. The chemical basis of corrosion lies in the tendency of metals to return to their more stable, oxidized state. In nature, most metals exist as ores, which are typically oxides, sulfides, or other compounds. The process of extracting and refining metals requires energy input, leaving the purified metals in a higher energy state. Corrosion, therefore, represents the spontaneous return of these metals to their lower energy, more thermodynamically stable state. The basic electrochemical reaction involved in corrosion can be represented by the general equation:

$$M \rightarrow M^{n+} + n^{e-}$$

where M represents the metal, M^{n+} is the metal ion, and n is the number of electrons (e^-) released in the oxidation process. This anodic reaction is coupled with a corresponding cathodic reaction, which in many cases involves the reduction of oxygen:

$$O_2 + 2H_2O + 4e^- \rightarrow 4OH^-$$

These reactions form the basis of what is known as an electrochemical cell, with the metal acting as the anode (where oxidation occurs) and the surrounding environment serving as the cathode (where reduction takes place). The presence of an electrolyte, often water or moisture in the air, facilitates the flow of ions and completes the circuit necessary for corrosion to proceed.

Research by Revie and Uhlig (2008) in their seminal work "Corrosion and Corrosion Control" highlights several key types of corrosion (Figure 3.1), each with its own characteristics and challenges:

1. *Uniform Corrosion:* This is the most common form of corrosion, characterized by a relatively even loss of material across the surface. While often predictable, uniform corrosion can lead to significant thinning of materials over time, compromising structural integrity.

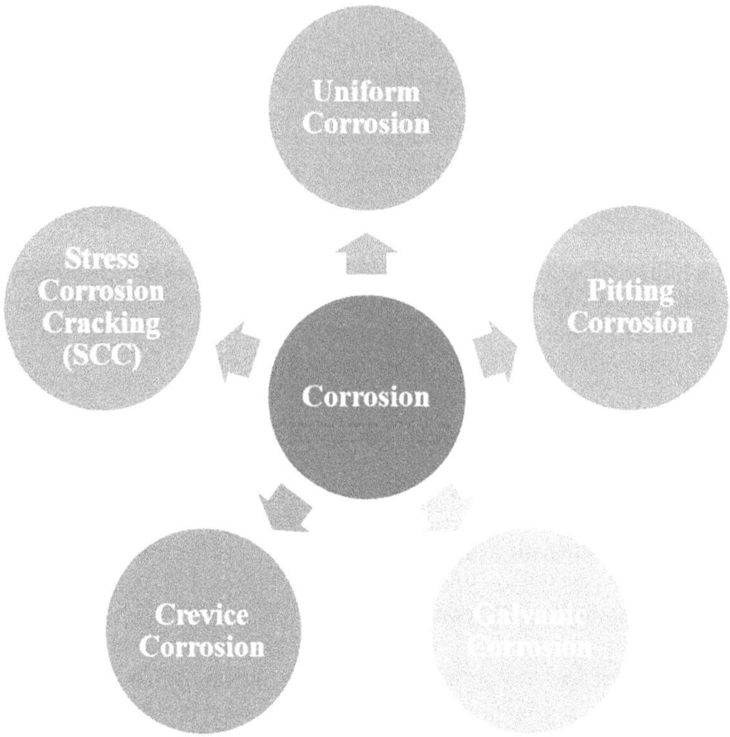

FIGURE 3.1 Types of corrosion.

2. *Pitting Corrosion:* A localized form of corrosion that results in small holes or "pits" in the material. Pitting is particularly insidious as it can cause rapid penetration with minimal overall material loss, leading to unexpected failures.
3. *Crevice Corrosion:* This occurs in narrow gaps or crevices where a small volume of stagnant solution is trapped. The confined space leads to a depletion of oxygen and a buildup of aggressive ions, accelerating corrosion in these areas.
4. *Galvanic Corrosion:* When two dissimilar metals are in electrical contact in the presence of an electrolyte, the more active metal corrodes preferentially. This principle is sometimes used beneficially in cathodic protection systems but can also lead to unexpected and severe corrosion in mixed-metal systems.
5. *Stress Corrosion Cracking (SCC):* A complex phenomenon involving the combined action of tensile stress and a corrosive environment. SCC can lead to sudden, catastrophic failures in materials that otherwise appear undamaged.

Understanding these various types of corrosion is crucial for developing effective prevention and mitigation strategies. As noted by Marcus and Oudar (1995) in their comprehensive review "Corrosion Mechanisms in Theory and Practice," the specific mechanisms of corrosion can vary widely depending on the material, environment, and operating conditions. This complexity underscores the need for ongoing research and tailored approaches to corrosion management across different industries and applications.

3.2 FUNDAMENTALS OF PROTECTIVE COATINGS

Protective coatings play a crucial role in safeguarding various materials and structures against environmental degradation, particularly corrosion. These specialized layers serve as a barrier between the substrate and its surrounding environment, effectively prolonging the lifespan of the protected surface and maintaining its integrity. The science and application of protective coatings have evolved significantly over the years, driven by the need for more durable, efficient, and environmentally friendly solutions across a wide range of industries.

At its core, a protective coating system consists of one or more layers of coating materials applied to a substrate. The primary function of these coatings is to prevent or slow down the degradation of the underlying material by creating a physical barrier against aggressive environmental factors such as moisture, oxygen, and chemical contaminants. However, the effectiveness of a protective coating goes beyond mere physical separation. Modern coating systems often incorporate active corrosion inhibition mechanisms, sacrificial protection, and even self-healing properties to provide comprehensive protection (Hughes et al., 2010).

The selection and design of protective coating systems are governed by a complex interplay of factors, including the nature of the substrate, the environmental conditions it will be exposed to, and the specific performance requirements of the application. As noted by Samimi and Zarinabadi (2011), the success of a coating system depends on three critical aspects: proper surface preparation, appropriate coating selection, and correct application methods. Each of these factors plays a vital role in ensuring the long-term effectiveness of the protective coating.

Surface preparation is often considered the most critical step in the coating process. It involves cleaning and modifying the substrate surface to ensure optimal adhesion of the coating. Techniques such as abrasive blasting, chemical cleaning, and heat treatment are commonly employed to remove contaminants, create a suitable surface profile, and activate the surface for better coating adhesion. Research by Momber has shown that the quality of surface preparation can significantly impact the performance and longevity of the coating system, with poorly prepared surfaces often leading to premature coating failure (Momber, 2016).

The selection of appropriate coating materials is another crucial aspect of protective coating systems. Coatings can be broadly categorized into organic and inorganic types, each with its own set of properties and applications. Organic coatings, which include epoxies, polyurethanes, and acrylics, are widely used due to their versatility and ease of application. These coatings form a continuous film that acts as a barrier against environmental factors. On the other hand, inorganic coatings such as zinc-rich primers and ceramic coatings offer excellent corrosion resistance and can withstand high temperatures (Schweitzer, 2006).

Recent advancements in nanotechnology have led to the development of nanocomposite coatings, which incorporate nanoparticles to enhance various properties of the coating. For instance, the addition of nano-silica particles to epoxy coatings has been shown to improve their scratch resistance and thermal stability. Similarly, the incorporation of carbon nanotubes in polymer coatings has been found to enhance their mechanical strength and electrical conductivity, opening up new possibilities for multifunctional protective coatings (Deyab, 2014).

The performance evaluation and quality control of protective coatings are essential aspects of their application. Standard test methods, such as salt spray testing, electrochemical impedance spectroscopy (EIS), and accelerated weathering tests, are commonly used to assess the corrosion resistance and durability of coating systems. However, it is important to note that these accelerated tests may not always accurately predict long-term performance in real-world conditions. As such, there is ongoing research into developing more reliable and representative testing methodologies (D'Alessandro et al., 2019).

3.2.1 Types of Protective Coatings

Protective coatings represent a diverse and evolving field of materials science, offering a wide array of solutions for safeguarding surfaces against corrosion, wear, and environmental degradation (Figure 3.2). The selection of an appropriate coating type is

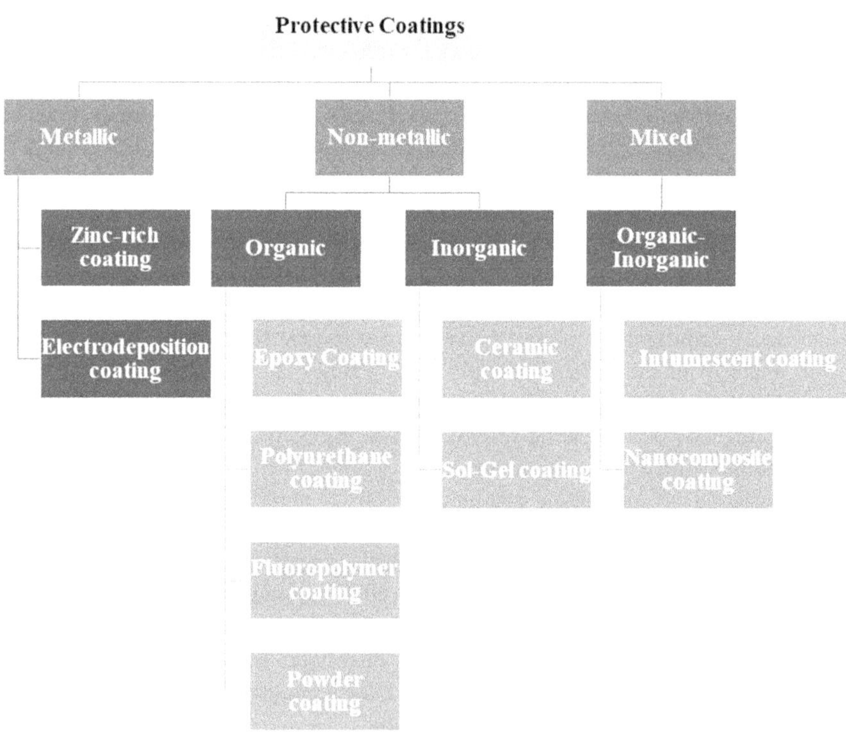

FIGURE 3.2 Types of protective coatings.

crucial for ensuring optimal protection in various applications, ranging from industrial machinery to architectural structures. This comprehensive overview explores the main types of protective coatings, their compositions, properties, and specific applications, drawing insights from recent research and industry developments.

3.2.1.1 Epoxy

Epoxy coatings stand out as one of the most versatile and widely used types of protective coatings. These coatings are formed through the polymerization of epoxy resins with various hardeners, resulting in a cross-linked structure that provides excellent adhesion, chemical resistance, and durability. Zhang et al. (2019) highlight the superior barrier properties of epoxy coatings, noting their effectiveness in protecting metal substrates from corrosive environments. The versatility of epoxy coatings is further enhanced by the ability to modify their properties through the incorporation of different additives and fillers. For instance, the addition of ceramic particles can significantly improve the wear resistance of epoxy coatings, making them suitable for applications in abrasive environments (Zhang et al., 2019).

3.2.1.2 Polyurethane

Polyurethane coatings represent another major category, known for their exceptional flexibility, abrasion resistance, and weatherability. These coatings are formed by the reaction between isocyanates and polyols, allowing for a wide range of formulations to meet specific performance requirements. Chattopadhyay and Raju (2007) discuss the superior ultraviolet (UV) resistance of aliphatic polyurethane coatings, making them particularly suitable for exterior applications where color and gloss retention are crucial. Recent advancements in polyurethane coating technology have focused on improving their environmental profile, with the development of water-based and high-solids formulations that reduce volatile organic compound (VOC) emissions while maintaining performance characteristics (Israel et al., 2013).

3.2.1.2 Zinc-Rich Coating

Zinc-rich coatings offer a unique approach to corrosion protection, particularly for steel substrates. These coatings contain high levels of zinc particles, typically 65–95% by weight in the dry film, which provide cathodic protection to the underlying steel. Schweitzer (1989) explains that when the coating is damaged, the zinc particles preferentially corrode, protecting the steel substrate. This sacrificial protection mechanism makes zinc-rich coatings highly effective in severe environments, such as marine and industrial settings. Recent research has explored the use of nano-sized zinc particles to enhance the protective properties and reduce the overall zinc content required for effective protection (Shirehjini et al., 2016).

3.2.1.3 Ceramic Coatings

Ceramic coatings represent a class of high-performance protective solutions, offering exceptional resistance to high temperatures, wear, and chemical attack. These coatings can be applied through various methods, including thermal spraying, chemical vapor deposition, and sol–gel processes. Bose (2017) discusses the application of ceramic coatings in aerospace and automotive industries, where their ability to withstand extreme conditions is particularly valuable. Advanced ceramic coatings, such as those based on yttria-stabilized zirconia, play a crucial role in protecting turbine blades in jet engines, significantly extending their operational life under high-temperature conditions (Wu et al., 2002).

3.2.1.4 Fluoropolymer Coatings

Fluoropolymer coatings, including polytetrafluoroethylene (PTFE) and fluoroethylene vinyl ether (FEVE)-based systems, offer exceptional chemical resistance and non-stick properties. These coatings are characterized by their low surface energy, which imparts excellent release properties and resistance to fouling. McKeen (2015) highlights the use of fluoropolymer coatings in chemical processing equipment, where their inertness to a wide range of chemicals is particularly advantageous. Recent developments in fluoropolymer coating technology have focused on improving their durability and expanding their application range. For instance, the development of fluorourethane coatings combines the chemical resistance of fluoropolymers with the toughness and flexibility of polyurethanes, offering enhanced performance in demanding environments (Gardiner, 2015).

3.2.1.5 Intumescent Coatings

Intumescent coatings represent a specialized category of protective coatings designed for fire protection. These coatings expand when exposed to high temperatures, forming an insulating char layer that protects the substrate from fire damage. Bourbigot and Duquesne (2007) discuss the mechanisms of intumescent coatings and their effectiveness in prolonging the fire resistance of steel structures. Recent research has focused on enhancing the efficiency of intumescent coatings through the incorporation of nanoparticles and novel intumescent agents, leading to improved char formation and better insulation properties (Rao et al., 2019).

3.2.1.6 Sol–Gel Coatings

Sol–gel coatings have gained increasing attention in recent years due to their ability to form thin, highly adherent layers with excellent barrier properties. These coatings are produced through the hydrolysis and condensation of metal alkoxides, typically silicon or titanium-based precursors. Wang and Bierwagen (2009) review the applications of sol–gel coatings in corrosion protection, highlighting their potential for creating tailored surface properties through the incorporation of various functional groups. The versatility of the sol–gel process allows for the development of hybrid organic–inorganic coatings that combine the durability of ceramic materials with the flexibility of organic polymers (Figueira et al., 2015).

3.2.1.7 Powder Coatings

Powder coatings represent an environmentally friendly alternative to traditional liquid coatings, offering zero VOC emissions and high transfer efficiency. These coatings are applied as dry powder and then cured under heat to form a continuous film. Talbert (2008) discusses the advantages of powder coatings, including their durability, uniform coverage, and ability to produce thick coatings in a single application. Recent developments in powder coating technology have focused on expanding their application range to heat-sensitive substrates through the development of UV-curable and low-temperature curing powder coatings (Czachor-Jadacka & Pilch-Pitera, 2021).

3.2.1.8 Nanocomposite Coatings

Nanocomposite coatings represent the cutting edge of protective coating technology, incorporating nanoparticles or nanostructures to enhance performance characteristics. These coatings can offer improvements in hardness, scratch resistance, and barrier properties compared to conventional coatings. Zheludkevich et al. (2012) review the use of nanocontainers in self-healing coatings, where encapsulated corrosion inhibitors are released in response to coating damage, providing active corrosion protection. The incorporation of graphene and carbon nanotubes in protective coatings has also shown promise in enhancing their mechanical and barrier properties (Wang et al., 2020).

3.2.1.9 Electrodeposited Coatings

Electrodeposited coatings, including electroplating and electroless plating, offer unique advantages in terms of uniformity and the ability to coat complex geometries. These coatings are widely used in the automotive and electronics industries for both decorative and functional purposes. Walsh (2014) discusses recent advancements in electrodeposition technology, including the development of pulse plating techniques for enhanced coating properties. The use of ionic liquids as electrolytes in electrodeposition has opened up new possibilities for depositing metals and alloys that are difficult to plate using conventional aqueous systems (Abbott & McKenzie, 2006).

3.2.2 Coating Components and Their Functions

Protective coatings are complex formulations designed to shield surfaces from environmental degradation, corrosion, and wear (Figure 3.3). The effectiveness of these coatings stems from the careful selection and combination of various components, each serving specific functions that contribute to the overall performance of the coating system At the heart of any coating formulation lies the binder, also known as the resin or film-former. The binder serves as the primary component that forms the continuous film adhering to the substrate and binding together all other components of the coating (Sørensen et al., 2009) describe binders as the backbone of the coating system, responsible for many of the coating's physical and chemical properties. Common types of binders include epoxies, polyurethanes, acrylics, and alkyds, each offering distinct characteristics suitable for various applications. For instance, epoxy resins are renowned for their excellent adhesion and chemical resistance, making them ideal for protective coatings in harsh industrial environments. The choice of binder significantly influences the coating's durability, flexibility, and resistance to environmental factors. Recent advancements in binder technology have focused on developing hybrid systems that combine the benefits of different resin types. For example, Hua et al. (2015) report on the development of epoxy–acrylic hybrid

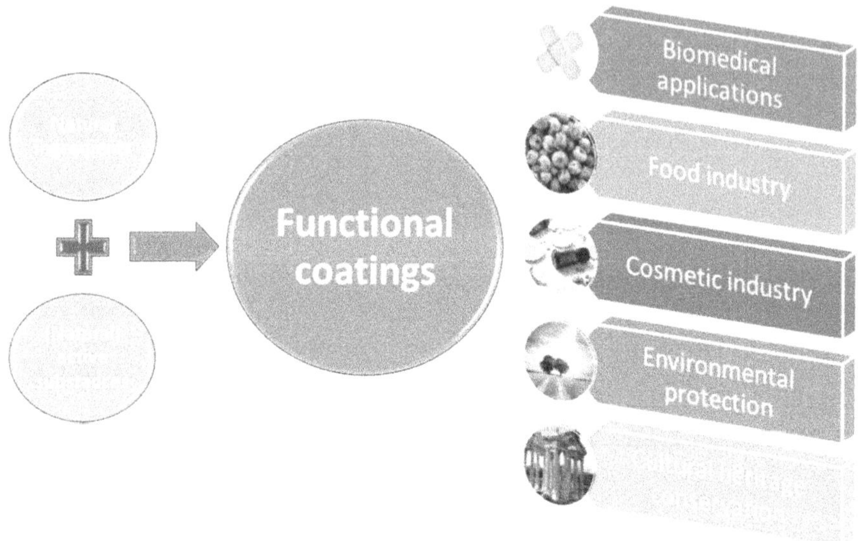

FIGURE 3.3 Coating components and their functions. (Fierascu et al., 2021.)

binders that offer improved weatherability and chemical resistance compared to traditional epoxy systems.

Pigments constitute another crucial component of protective coatings, serving multiple functions beyond mere color provision. These finely divided solid particles are dispersed throughout the binder and play a vital role in determining the coating's optical properties, durability, and protective capabilities. Talbert (2008) categorizes pigments into prime pigments, extender pigments, and functional pigments, each contributing uniquely to the coating's performance. Prime pigments, such as titanium dioxide, provide color and opacity, essential for both aesthetic appeal and UV protection of the underlying layers (Ewulonu et al., 2016). Extender pigments, such as calcium carbonate or talc, modify the coating's physical properties and can reduce cost. Functional pigments, including zinc dust or iron oxide, offer specific protective properties such as corrosion inhibition or UV absorption. The selection and concentration of pigments can dramatically influence a coating's barrier properties and overall effectiveness. For instance, Kalendová et al. (2015) demonstrate that the proper selection and dispersion of anticorrosive pigments can significantly enhance the corrosion resistance of protective coatings.

Solvents play a critical role in coating formulations by dissolving or dispersing the binder and other components, facilitating application, and promoting film formation. These volatile liquids adjust the coating's viscosity, allowing for proper application and flow characteristics. Goldschmidt and Streitberger (2007) discuss the importance of solvent selection in determining the coating's drying time, flow behavior, and final film properties. While organic solvents have traditionally been widely used, environmental concerns have driven a shift toward water-based and high-solids coating systems (Faccini et al., 2021).

Additives, though used in small quantities, play a disproportionately large role in modifying and enhancing coating properties. These specialized chemicals are incorporated to impart specific characteristics or to address particular performance issues. Florio and Miller (2004) provide a comprehensive overview of coating additives, categorizing them based on their functions, which include improving flow and leveling, enhancing adhesion, preventing foaming, and providing UV stability. For example, rheology modifiers control the flow behavior of the coating during application, ensuring proper thickness and preventing sagging on vertical surfaces. Biocides are crucial in preventing microbial growth in the wet coating and on the dried film, particularly important in humid environments. Recent research has focused on developing multifunctional additives that can address multiple performance aspects simultaneously. For instance, Manoj et al. (2020) report on the development of nanostructured additives that combine corrosion inhibition with self-healing properties, potentially revolutionizing long-term corrosion protection strategies.

Corrosion inhibitors represent a specialized class of additives particularly important in protective coatings for metal substrates. These compounds work by interfering with the electrochemical processes that lead to corrosion. Revie and Uhlig (2008) classify corrosion inhibitors into anodic, cathodic, and mixed inhibitors based on their mechanism of action. Anodic inhibitors, such as chromates, have been widely used due to their effectiveness but face regulatory restrictions due to environmental concerns. This has spurred research into environmentally friendly alternatives. Abdel-Gaber et al. (2021) investigate the use of plant extracts as green corrosion inhibitors, demonstrating their potential as sustainable alternatives to traditional synthetic inhibitors.

The concept of functional fillers has gained prominence in recent years, blurring the line between traditional pigments and additives. These materials are incorporated into coatings to impart specific properties such as improved abrasion resistance, thermal conductivity, or even self-cleaning capabilities. For example, Yao et al. (2018) demonstrate the use of nano-silica particles to enhance the scratch resistance and durability of polymeric coatings. Similarly, the incorporation of phase-change materials as functional fillers can impart thermal management properties to coatings, as shown by Ling and Poon (2013) in their study on temperature-regulating building coatings.

Cross-linking agents play a crucial role in many coating systems, particularly in two-component formulations such as epoxies and polyurethanes. These reactive compounds initiate the curing process, forming a three-dimensional network that significantly enhances the coating's mechanical properties, chemical resistance, and durability. Jin et al. (2015) discuss the importance of cross-linking density in determining coating performance, highlighting the balance required between hardness and flexibility. Recent advancements in cross-linking technology have focused on developing systems that can cure under milder conditions or provide additional functionalities. For instance, Yang et al. (2015) report on the development of self-healing coatings utilizing dynamic covalent bonds, which can reform after being broken, providing a mechanism for autonomous repair of coating damage.

Plasticizers are another important component in many coating formulations, used to impart flexibility and improve the coating's ability to withstand mechanical stress and temperature fluctuations. These compounds work by increasing the free volume between polymer chains, lowering the glass transition temperature of the coating.

TABLE 3.1
Coating Components and Their Function

S. No.	Coating Component	Function
1.	Binder	Binding together all other components of the coating.
2.	Pigments	Optical properties, durability, and protective capabilities.
3.	Solvents	Role in coating formulations, promoting film formation.
4.	Additives	Modifying and enhancing coating properties.
5.	Corrosion inhibitors	Interfering with the electrochemical processes that lead to corrosion.
6.	Fillers	Impart specific properties such as improved abrasion resistance, thermal conductivity, or even self-cleaning capabilities.
7.	Cross-linking agents	Enhances the coating's mechanical properties, chemical resistance, and durability.
8.	Plasticizers	Impart flexibility and improve the coating's ability to withstand mechanical stress and temperature fluctuations.

Wypych (2017) provides a comprehensive review of plasticizers, discussing their mechanisms of action and the factors influencing their selection. Although traditional phthalate-based plasticizers have been widely used, concerns over their environmental impact have led to the development of alternative, bio-based plasticizers. Carbonell Verdu et al. (2016) investigate the use of epoxidized vegetable oils as environmentally friendly plasticizers, demonstrating their potential as sustainable alternatives in coating formulations.

The development of smart or stimuli-responsive components represents the cutting edge of coating technology. These advanced materials can change their properties in response to environmental stimuli such as temperature, pH, or mechanical stress. Zhao et al. (2017) review the development of smart coatings for corrosion protection, discussing technologies such as pH-sensitive microcapsules that release corrosion inhibitors in response to local pH changes caused by the onset of corrosion. Similarly, Zhang and Zhai (2019) demonstrate the use of thermochromic pigments in coatings for visual indication of temperature changes, which can be useful for safety applications or energy management in buildings.

3.3 MECHANISMS OF CORROSION PROTECTION BY COATINGS

Corrosion protection through coatings is essential in various industries, especially where metal durability is crucial (Figure 3.4). The mechanisms by which coatings protect substrates from corrosion can be broadly categorized into barrier, sacrificial, and self-healing mechanisms. Barrier coatings function primarily by preventing the corrosive environment from coming into direct contact with the metal surface. These

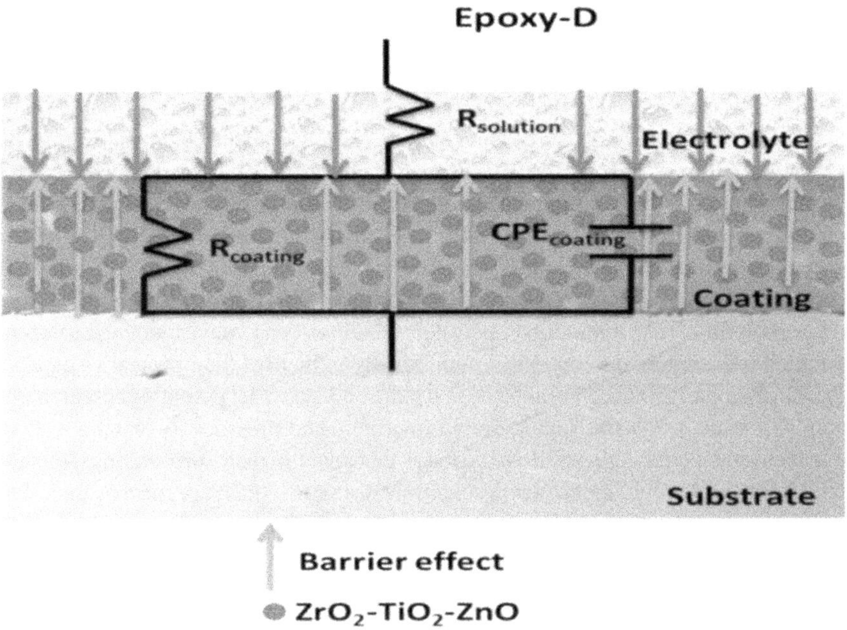

FIGURE 3.4 Mechanisms of corrosion protection by coatings. (Ferrel-Alvarez et al., 2022.)

coatings act as a partition between the environment and the surface, minimizing moisture and corrosive agents' penetration. Layers such as paint, varnishes, or synthetic polymers are typically created to deliver this protective function. Research indicates that coatings can significantly enhance the lifespan of structures by ensuring that localized corrosion does not compromise performance, as highlighted by the results from recent advancements in polymer coatings that improve barrier properties.

Conversely, sacrificial coatings, often made from metals such as zinc, corrode preferentially to the underlying substrate. This method of protection is critical in galvanizing processes, where the sacrificial anode effectively diverts corrosion away from important structures, ensuring integrity over time. The electrochemical mechanism of sacrificial anodes relies on the principle of galvanic corrosion, where the less noble metal (e.g., zinc) oxidizes before the more noble substrate (e.g., steel) can. This technique is widely recognized for its efficiency, particularly in harsh environments where moisture and salts can accelerate corrosion rates.

Another novel area of research focuses on self-healing coatings, which actively repair corrosion damage without human intervention. The self-healing mechanism often incorporates microcapsules filled with healing agents that release upon mechanical damage, healing cracks and restoring the barrier properties of the coating. These coatings are designed to extend the lifespan of the materials they protect further and reduce maintenance costs (Kim et al., 2017). Self-healing technology represents a significant advancement in corrosion protection, combining barrier and healing properties to ensure the longevity of coated surfaces.

The electrochemical behavior of coatings is also fundamental to their effectiveness in corrosion protection. Modern investigations into polymeric coatings have revealed that their efficacy can be enhanced by modifying the coating's microstructure or composition. For example, incorporating nanoparticles into coatings can improve their mechanical properties and corrosion resistance. Nanocomposite coatings leverage the unique properties of nanoparticles, such as increased surface area and enhanced barrier performance, which significantly mitigate corrosion processes. Researchers have noted that coating durability and performance can be accessed through EIS, allowing for real-time monitoring of corrosion processes and the effectiveness of protective coatings.

Furthermore, the interaction between coatings and environmental factors, including temperature and humidity, significantly influences their protective capabilities. Coating performance diminishes in harsh conditions, necessitating research into coatings that can withstand fluctuations in environmental stressors. Recent innovations have focused on creating responsive coatings that adapt to their surroundings, thereby maintaining protection under various exposure scenarios. Such approaches underline the importance of understanding the fundamental mechanisms of corrosion and the development of smart coatings capable of providing lasting protection against the aggressive effects of corrosion (Blaiszik et al., 2010).

3.3.1 Barrier Protection

Barrier protection is a pivotal strategy in corrosion mitigation, designed to prevent corrosive agents from coming into contact with metal substrates. This technique involves the application of protective coatings that form a physical barrier against moisture, salt, and other corrosive materials. The effectiveness of these barriers significantly depends on their adhesion properties, chemical stability, and resistance to environmental stresses. Among various types of coatings, epoxy resins are highly regarded for their exceptional adhesion and durability, making them suitable for metal surfaces in diverse environments. Studies indicate that epoxy-based coatings can effectively reduce the likelihood of localized corrosion by offering robust protective layers (Hernández-Maya et al., 2023).

Moreover, surface preparation prior to the application of these coatings is crucial for maximizing their protective efficacy. Contaminants such as dust, oil, or rust can impair adhesion, leading to premature coating failure. Therefore, proper surface cleaning and profiling must accompany the coating application to ensure optimal performance. Overall, barrier protection remains a fundamental component of corrosion control strategies, integrating traditional and advanced materials to safeguard infrastructure and enhance the longevity of metal components in various industries.

3.3.2 Inhibitive Protection

Inhibitive protection is a crucial method utilized to mitigate corrosion in metals and alloys, significantly enhancing their durability and service life. This process revolves around the use of chemical inhibitors that interrupt the electrochemical reactions responsible for corrosion. By creating a protective layer on the surface of metals, inhibitors can effectively slow down or prevent the corrosion process, thus preserving the structural integrity of materials exposed to aggressive environments.

The effectiveness of corrosion inhibitors depends on their ability to absorb onto metal surfaces, forming a barrier against corrosive agents such as moisture, oxygen, and salts. Various types of inhibitors are employed, including anodic, cathodic, and mixed inhibitors, each tailored to interrupt specific electrochemical reactions. For example, anodic inhibitors work by passivating the anode of a corrosion cell, while cathodic inhibitors reduce the cathodic reaction, effectively controlling the corrosion rate. Recent studies confirm that the use of organic inhibitors, such as those derived from azoles and thiol compounds, can provide significant protection, particularly in acidic environments where metals are more susceptible to corrosion.

Despite their advantages, the use of corrosion inhibitors is not without challenges. The long-term effectiveness of certain inhibitors can be compromised due to leaching, degradation, or changes in environmental conditions. Thus, ongoing research is essential to develop more robust formulations that maintain their protective qualities over extended periods. Additionally, understanding the mechanism of action of various inhibitors through advanced computational methods provides valuable insights that guide the development of more effective and targeted corrosion protection strategies.

3.3.3 Galvanic (Sacrificial) Protection

Galvanic or sacrificial protection is an effective corrosion mitigation technique extensively utilized in various industrial applications, particularly for submerged and exposed metal structures (Figure 3.5). This method involves attaching a more anodic metal, known as a sacrificial anode, to a cathodic metal structure, which reduces the corrosion rate of the cathodic material. The sacrificial anode, usually made from zinc,

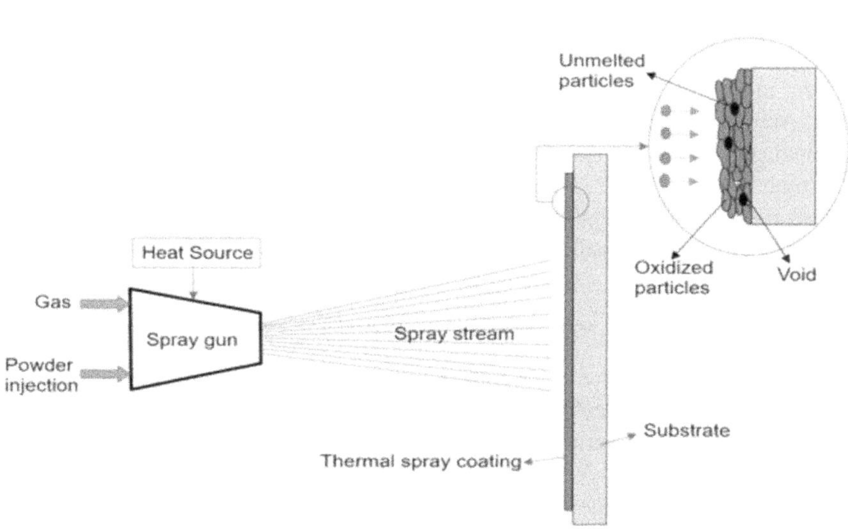

FIGURE 3.5 Mechanisms for corrosion protection.

magnesium, or aluminum, preferentially corrodes, thereby protecting the underlying metal structure from electrochemical degradation. The principle of galvanic protection is grounded in the electrochemical series, where the more anodic metal provides electrons to the cathodic metal, creating a corrosion cell that inhibits the corrosion process.

Recent advancements in the design of sacrificial anodes have improved their performance and longevity. For instance, the integration of rare earth metals has increased the efficiency of magnesium anodes for use in marine applications, providing enhanced protection even in challenging environments (Zhang et al., 2023). Moreover, ongoing research into the combination of sacrificial anodes with other protective measures, such as coating systems, is promising for developing robust strategies against corrosion. Thus, galvanic protection remains a practical and cost-effective solution for enhancing the durability of structural components in various sectors.

3.4 ADVANCED COATING TECHNOLOGIES

Advanced coating technologies are critical in enhancing the corrosion resistance of metals and alloys in various industrial applications. These innovative coatings are designed to create barriers that prevent corrosive substances from reaching the substrate, thereby extending the service life of components. Among the emerging materials, graphene-based coatings have gained significant attention due to their superior mechanical and thermal properties, which contribute to enhanced barrier performance and durability (Ollik & Lieder, 2020). Likewise, the incorporation of nanoparticles into epoxy and polymer coatings has shown promising results in improving protective capabilities, leading to coatings that are both effective and multifunctional.

Additionally, super hydrophobic coatings, which prevent water and contaminants from adhering to surfaces, have been developed to provide further protection in corrosive environments. These coatings impart self-cleaning properties while minimizing moisture retention, significantly reducing the likelihood of corrosion initiation. Another notable advancement is the advent of self-healing coatings that can automatically repair themselves upon damage, maintaining their protective function over time (Ma et al., 2022). Such developments represent a proactive approach to corrosion management, highlighting how advanced coating technologies can effectively combat corrosion challenges across diverse sectors.

3.4.1 Nanocoatings for Corrosion Resistance

Nanocoatings have emerged as a significant advancement in corrosion resistance technologies, offering enhanced protection for various metallic substrates. These coatings, formulated with nanomaterials, improve the barrier properties significantly compared to traditional coatings by providing a defect-free and dense layer that hinders the penetration of corrosive agents. The unique characteristics of nanocoatings stem from their nanoscale thickness, typically less than 100 nm, which allows them to exhibit superior adhesion, mechanical strength, and chemical resistance (Farag, 2020).

Research indicates that incorporating nanoparticles such as silica, titanium dioxide, and zinc oxide into coatings can enhance their performance, particularly in aggressive environments where conventional coatings have limitations. These nanoparticles not only increase the durability of the coatings but also improve their self-cleaning and anti-fouling properties, further contributing to long-term corrosion protection (Huang et al., 2021). Additionally, advancements in layered and hybrid nanocoatings have shown promise in achieving multifunctionality, where protective, antibacterial, and hydrophobic features coexist. Thus, the utilization of nanocoatings represents a transformative approach in the field of corrosion protection, ensuring the longevity and integrity of vital infrastructure and components across various industries.

3.4.2 Smart and Self-Healing Coatings

Smart and self-healing coatings represent a significant advancement in materials science, particularly in the field of corrosion protection. These innovative coatings are designed to respond autonomously to damage and, in some cases, can restore their integrity without external intervention. Self-healing mechanisms are largely based on the incorporation of microcapsules or nanocontainers that release healing agents upon damage to the coating. For instance, coatings may contain urea-formaldehyde microcapsules filled with a liquid healing agent, which are triggered to release their contents in response to cracks or mechanical stress, thereby repairing the damage and enhancing durability (Lang & Zhou, 2017).

Recent research has shown that these smart coatings can employ various materials to improve performance, such as halloysite nanotubes, which serve as nanocarriers for corrosion inhibitors, offering controlled release and enhanced protection against corrosive environments. Furthermore, advancements in multifunctional polymeric smart coatings have demonstrated improved self-healing capabilities through the integration of reactive materials that can chemically bond with the exposed surfaces, thereby sealing cracks more effectively (Udoh et al., 2022). Overall, the development of smart and self-healing coatings offers a promising pathway to improve the longevity and reliability of coated surfaces used in diverse applications, from infrastructure to automotive industries.

3.5 SURFACE PREPARATION AND COATING APPLICATION TECHNIQUES

Surface preparation and coating application techniques play a crucial role in ensuring effective corrosion protection for various materials and structures (Figure 3.6). The integrity and longevity of protective coatings are heavily dependent on the quality of surface preparation, which serves as the foundation for successful coating adhesion and performance (Jamali & Mills, 2014). Proper surface preparation involves the removal of contaminants, such as rust, scale, oil, and debris, which can compromise the coating's ability to adhere to the substrate and provide long-term protection.

One of the most widely used methods for surface preparation is abrasive blasting, which employs high-pressure propulsion of abrasive materials to clean and profile the surface. This technique not only removes contaminants but also creates a roughened

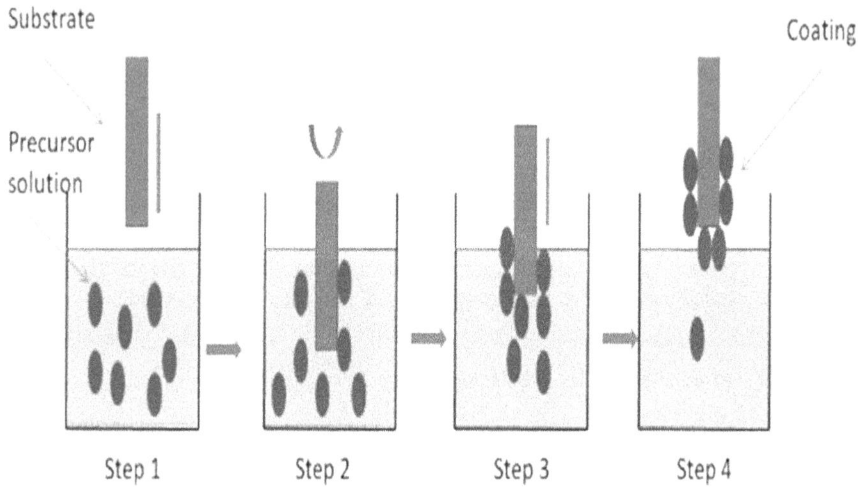

FIGURE 3.6 Surface preparation and coating. (Ramezani et al., 2023.)

surface profile that enhances coating adhesion. Chemical cleaning methods, including acid etching and alkaline degreasing, are also employed to remove specific types of contaminants and prepare surfaces for coating application (Zaki et al., 2017).

Once the surface is properly prepared, the selection and application of appropriate coating systems are paramount to achieving optimal corrosion protection. Modern coating technologies encompass a wide range of formulations, including epoxy-based coatings, polyurethanes, and zinc-rich primers, each designed to provide specific protective properties. The application technique chosen can significantly impact the coating's effectiveness, with methods such as spray application, brush coating, and electrostatic deposition being employed based on factors like substrate geometry, coating properties, and environmental conditions (Thomas et al., 2022).

Advanced coating application techniques, such as thermal spraying and electrodeposition, have gained prominence in recent years due to their ability to produce highly uniform and adherent coatings with enhanced corrosion resistance (Galedari et al., 2019). These methods allow for the deposition of metallic, ceramic, and composite coatings that can provide superior protection in aggressive environments.

The importance of quality control throughout the surface preparation and coating application process cannot be overstated. Proper inspection and monitoring of environmental conditions, such as temperature and humidity, are essential to ensure optimal coating performance (Bettridge & Ubank, 1986). Furthermore, the development of innovative coating formulations and application techniques continues to drive advancements in corrosion protection strategies, emphasizing the need for ongoing research and development in this critical field.

3.5.1 Importance of Surface Preparation

Surface preparation is a critical factor in ensuring effective corrosion protection for various materials and structures. The quality of surface preparation significantly influences the adhesion, performance, and longevity of protective coatings, ultimately determining their ability to resist corrosive environments (Jamali & Mills, 2014). Proper surface preparation involves the removal of contaminants such as rust, scale, oil, and debris, which can compromise the coating's ability to bond with the substrate and provide long-term protection.

Research has shown that inadequate surface preparation is a leading cause of premature coating failure and subsequent corrosion. The removal of surface contaminants and the creation of an appropriate surface profile are essential for maximizing coating adhesion and minimizing the risk of underfilm corrosion. Various methods, including abrasive blasting, chemical cleaning, and power tool cleaning, are employed to achieve the desired surface condition, with each technique offering specific advantages depending on the substrate material and environmental conditions.

The importance of surface preparation extends beyond initial coating application, as it also plays a crucial role in maintenance and repair processes. Proper surface preparation during maintenance activities can significantly extend the service life of protective coatings and reduce the overall cost of corrosion protection (Jamali & Mills, 2014). As such, ongoing research in surface preparation techniques and their impact on coating performance remains essential for advancing corrosion protection strategies across various industries.

3.5.2 Modern Coating Application Methods

Modern coating application methods have significantly advanced the field of corrosion protection, offering improved performance, efficiency, and versatility. These innovative techniques have been developed to address the limitations of traditional methods and to meet the increasing demands for more durable and effective protective coatings across various industries (Aljibori et al., 2023).

One of the most prominent advancements in coating application is the use of thermal spray technologies. These methods, including high-velocity oxy-fuel (HVOF) and plasma spraying, enable the deposition of metallic, ceramic, and composite coatings with exceptional adhesion and density. Thermal spray coatings have demonstrated superior corrosion resistance in aggressive environments, making them particularly valuable for applications in marine and chemical processing industries (Sidhu et al., 2005).

Electrodeposition techniques, such as electroplating and electroless plating, have also gained traction in corrosion protection applications. These methods offer precise control over coating thickness and composition, allowing for the creation of highly uniform and adherent protective layers. Additionally, the emergence of nanostructured coatings applied through methods like sol–gel processing has opened new avenues for corrosion protection, offering enhanced barrier properties and self-healing capabilities (Voevodin et al., 2003).

Advanced spray technologies, including airless and electrostatic spraying, have revolutionized the application of liquid and powder coatings. These methods provide improved transfer efficiency, reduced overspray, and more uniform coating distribution, resulting in enhanced corrosion protection and reduced material waste. Furthermore, the integration of robotics and automation in coating application processes has led to increased precision and reproducibility, ensuring consistent quality in large-scale industrial applications (Arrais et al., 2021).

3.6 PERFORMANCE EVALUATION OF PROTECTIVE COATINGS

Performance evaluation of protective coatings is a critical aspect of corrosion protection research and development, providing essential insights into coating effectiveness, durability, and longevity. Comprehensive assessment methodologies encompass a wide range of laboratory and field testing techniques designed to simulate various environmental conditions and assess coating performance under diverse corrosive scenarios (G. Bierwagen et al., 2003).

EIS has emerged as a powerful non-destructive technique for evaluating coating performance and degradation over time. This method allows researchers to monitor changes in coating resistance and capacitance, providing valuable information about barrier properties and the onset of corrosion processes. Additionally, accelerated weathering tests, such as salt spray and cyclic corrosion testing, are widely employed to assess coating durability under aggressive environmental conditions (Hinderliter & Sapper, 2015).

Advanced surface analysis techniques, including scanning electron microscopy (SEM) and X-ray photoelectron spectroscopy (XPS), play a crucial role in characterizing coating morphology, composition, and interfacial properties. These methods offer insights into degradation mechanisms and failure modes, enabling researchers to optimize coating formulations and application processes. Furthermore, the development of in-situ monitoring techniques, such as embedded sensors and electrochemical noise analysis, has facilitated real-time assessment of coating performance in operational environments (Sacher et al., 2019).

The integration of artificial intelligence and machine learning algorithms in coating performance evaluation has gained traction in recent years. These advanced data analysis techniques enable researchers to identify complex relationships between coating properties, environmental factors, and corrosion protection performance, leading to more accurate predictive models and optimized coating systems (Liu et al., 2018).

3.6.1 Accelerated Corrosion Testing

Accelerated corrosion testing is a crucial methodology in evaluating the performance of protective coatings and materials under simulated aggressive environments. These tests aim to reproduce long-term corrosion effects in a compressed timeframe, enabling researchers and industries to assess corrosion resistance efficiently. Common techniques include salt spray testing, cyclic corrosion testing, and electrochemical methods such as potentiodynamic polarization. While these accelerated tests provide valuable insights, it is essential to recognize their limitations in fully replicating real-world conditions. Researchers continue to refine testing protocols and develop new methods to improve the correlation between accelerated tests and field performance, enhancing the predictive capabilities of these techniques (Prosek et al., 2016).

3.6.2 Electrochemical Impedance Spectroscopy

EIS is a powerful non-destructive technique widely employed in corrosion science for evaluating the performance of protective coatings and studying electrochemical processes at interfaces. This method involves applying a small amplitude sinusoidal potential perturbation to an electrochemical system and measuring the resultant current response over a range of frequencies (G. Bierwagen et al., 2003). EIS provides valuable insights into coating degradation, barrier properties, and the kinetics of corrosion reactions by analyzing the complex impedance data through equivalent circuit modeling. The technique's sensitivity to subtle changes in electrochemical systems makes it particularly useful for monitoring coating performance over time and detecting the onset of corrosion processes beneath protective layers (Filotás et al., 2022).

3.7 ENVIRONMENTAL CONSIDERATIONS IN COATING SELECTION

Environmental considerations play a crucial role in the selection of protective coatings for corrosion protection, driven by increasing regulatory pressures and growing awareness of ecological impacts. The shift toward more sustainable coating systems has led to the development of eco-friendly alternatives to traditional solvent-based coatings, such as water-based, high-solids, and powder coatings. These environmentally conscious options aim to reduce VOC emissions and minimize the release of hazardous air pollutants during application and curing processes.

Furthermore, the incorporation of bio-based materials and renewable resources in coating formulations has gained traction, offering potential environmental benefits without compromising corrosion protection performance. The selection of coating systems also considers end-of-life scenarios, with emphasis on recyclability and ease of removal to facilitate maintenance and reduce waste generation. Additionally, the development of self-healing and smart coatings presents opportunities for extending service life and reducing the frequency of reapplication, thereby minimizing the overall environmental footprint of corrosion protection strategies.

3.7.1 VOC Regulations and Compliance

VOC regulations have significantly impacted the corrosion protection industry, driving innovation in coating formulations and application techniques. These regulations, aimed at reducing air pollution and improving environmental and human health, have necessitated a shift away from traditional solvent-based coatings towards more environmentally friendly alternatives. Compliance with VOC regulations has led to the development and adoption of water-based coatings, high-solids formulations, and powder coatings, which offer reduced emissions without compromising corrosion protection performance (Xia et al., 2023).

The implementation of VOC regulations varies globally, with regions such as the European Union and California enforcing particularly stringent standards. This regulatory landscape has spurred research into novel low-VOC and zero-VOC coating technologies, including UV-curable systems and reactive diluents. Additionally, the adoption of advanced application methods, such as electrostatic spraying and airless systems, has further contributed to VOC reduction efforts by improving transfer efficiency and minimizing overspray (Wang et al., 2022). As regulations continue to evolve, the corrosion protection industry faces ongoing challenges in balancing environmental compliance with coating performance and cost-effectiveness.

3.7.2 Eco-Friendly Coating Options

The development of eco-friendly coating options for corrosion protection has gained significant momentum in recent years, driven by increasing environmental concerns and stricter regulations. These sustainable alternatives aim to minimize environmental impact while maintaining effective corrosion resistance. Water-based coatings have emerged as a prominent eco-friendly option, offering reduced VOC emissions and improved worker safety. Additionally, high-solids coatings provide enhanced film build with lower solvent content, contributing to VOC reduction efforts (Park et al., 2008).

Powder coatings represent another environmentally conscious solution, eliminating the need for solvents entirely and offering excellent durability and corrosion resistance (Yang et al., 2022). Bio-based coatings derived from renewable resources, such as plant oils and natural polymers, have also shown promise in providing sustainable corrosion protection. Furthermore, the development of self-healing coatings incorporating environmentally benign components offers the potential for extended service life and reduced maintenance requirements. As research in this field progresses, the integration of nanotechnology and smart materials is expected to further enhance the performance and sustainability of eco-friendly corrosion protection systems (Akbarzadeh et al., 2020).

3.8 COATING FAILURES: CAUSES AND PREVENTION

Coating failures in corrosion protection systems can lead to significant economic and safety consequences, necessitating a comprehensive understanding of their causes and prevention strategies. Common failure mechanisms include adhesion loss, blistering, cracking, and undercutting, often resulting from inadequate surface preparation, improper application techniques, or environmental stressors (G. P. Bierwagen et al., 2003). Poor substrate cleanliness and incompatible coating systems are frequent contributors to premature failure, highlighting the critical importance of proper surface preparation and coating selection (Hanák et al., 2015).

Environmental factors, such as moisture, temperature fluctuations, and chemical exposure, can accelerate coating degradation, emphasizing the need for thorough environmental assessment in coating design (Tiong & Clark, 2011). Preventive measures include rigorous quality control during surface preparation and application, as well as the implementation of multi-layer coating systems to provide enhanced protection against corrosive elements (Hu et al., 2022). Additionally, the development of self-healing coatings and the incorporation of corrosion inhibitors offer promising approaches to mitigate coating failures and extend service life (Boumezgane et al., 2022). Regular inspection and maintenance programs, coupled with advanced monitoring techniques, play crucial roles in early detection and prevention of coating failures in corrosion protection systems.

3.8.1 Common Coating Defects

Common coating defects pose significant challenges in corrosion protection systems, potentially compromising their effectiveness and longevity. These defects can manifest in various forms, including blistering, cracking, delamination, and pinholing, each with distinct causes and implications for coating performance. Blistering, characterized by localized swelling of the coating film, often results from osmotic pressure buildup or trapped solvents, leading to reduced barrier properties (Effendy et al., 2021). Cracking, which can occur due to internal stresses or environmental factors, creates pathways for corrosive agents to reach the substrate (Bjerkén & Ortiz, 2010).

Delamination, the separation of the coating from the substrate, frequently stems from inadequate surface preparation or incompatible coating systems, severely compromising corrosion protection (Leng et al., 1998). Pinholing, the formation of small holes in the coating film, can arise from improper application techniques or solvent entrapment, providing direct access for corrosive elements to the underlying material (Ahn et al., 2004). Understanding these common defects and their root causes is crucial for developing effective prevention strategies and improving the overall performance of corrosion protection coatings in various industrial applications.

3.8.2 Strategies for Extending Coating Lifespan

Extending the lifespan of protective coatings is crucial for enhancing long-term corrosion protection and reducing maintenance costs. Several strategies have been developed to achieve this goal, focusing on both coating formulation and application techniques. One approach involves the incorporation of nanosized particles, such as silica or zinc oxide, into coating systems to improve barrier properties and mechanical strength (Zhang et al., 2020). Self-healing coatings, which can autonomously repair minor damage, have shown promise in prolonging coating life by preventing the propagation of defects (Manoj et al., 2020).

The use of multilayer coating systems, combining primers, intermediate coats, and topcoats, provides enhanced protection against environmental stressors and mechanical damage (Fotovvati et al., 2019). Additionally, the development of smart coatings with stimuli-responsive properties offers potential for adapting to changing environmental conditions and extending service life (Ulaeto et al., 2017). Proper surface preparation and application techniques remain critical factors in maximizing coating longevity, with advances in surface treatment technologies contributing to improved adhesion and durability (Momber, 2008).

Regular maintenance and inspection protocols, coupled with the use of non-destructive evaluation techniques, enable early detection of coating degradation and timely intervention (Kušter Marić et al., n.d.). Furthermore, the integration of corrosion inhibitors and sacrificial elements in coating formulations provides additional protection mechanisms, further extending the effective lifespan of corrosion protection systems (Presuel-Moreno et al., 2008).

3.9 INDUSTRY-SPECIFIC COATING SOLUTIONS

Industry-specific coating solutions for corrosion protection are tailored to address the unique challenges and environmental conditions encountered in various sectors. In the marine industry, for instance, coatings must withstand constant exposure to saltwater, UV radiation, and biofouling. In the aerospace industry, coatings must provide corrosion protection while meeting strict weight and performance requirements. Advanced antifouling coatings incorporating biocides or utilizing foul-release technologies have been developed to combat marine growth and maintain vessel efficiency (Selim et al., 2017). The oil and gas sector demands coatings capable of resisting extreme temperatures, chemical exposure, and abrasion. High-performance epoxy and fluoropolymer coatings are often employed in these harsh environments to protect pipeline infrastructure and offshore structures (Popoola et al., 2013).

The development of these industry-specific solutions often involves collaborative efforts between coating manufacturers, end-users, and research institutions to optimize performance and meet evolving regulatory requirements (Figure 3.7). As industries continue to face new challenges, such as the transition to renewable energy sources or the adoption of advanced materials, coating technologies must adapt to provide effective corrosion protection in these emerging applications (Rani & Basu, 2012).

3.9.1 Marine Coatings

Marine coatings play a crucial role in protecting naval vessels, offshore structures, and marine infrastructure from the harsh and corrosive seawater environment. These specialized coatings must contend with unique challenges, including constant exposure to saltwater, UV radiation, temperature fluctuations, and biofouling. The

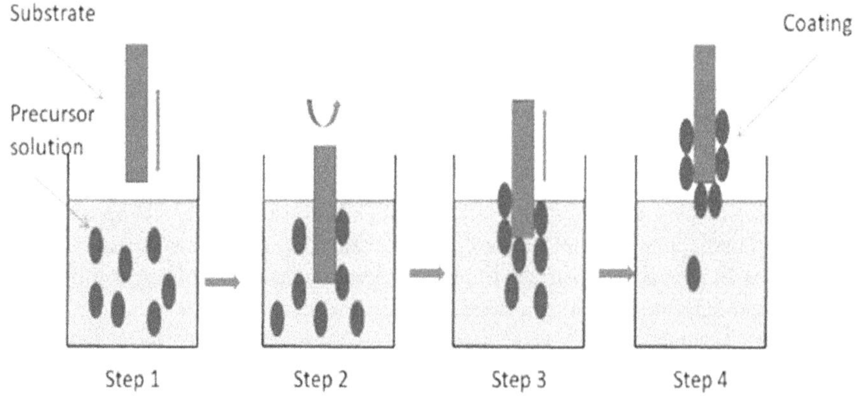

FIGURE 3.7 Industry-specific coating solutions.

development of advanced marine coatings focuses on enhancing corrosion resistance, reducing friction, and preventing the attachment of marine organisms to submerged surfaces.

Antifouling coatings represent a significant advancement in marine protection, utilizing various mechanisms to deter the settlement of barnacles, algae, and other marine life. Self-polishing copolymer (SPC) coatings, which gradually release biocides through a controlled erosion process, have been widely adopted for their long-term effectiveness. However, environmental concerns have led to the development of more eco-friendly alternatives, such as foul-release coatings based on silicone or fluoropolymer technologies, which create a low surface energy that inhibits organism adhesion.

Corrosion protection in marine environments often employs multi-layer coating systems, typically consisting of zinc-rich primers, epoxy barrier coats, and durable topcoats. The incorporation of sacrificial anodes and corrosion inhibitors further enhances the protective capabilities of these systems (Cao et al., 2019). Recent advancements in nanotechnology have led to the development of nanocomposite coatings that offer improved barrier properties and mechanical strength. Additionally, the integration of self-healing mechanisms in marine coatings shows promise in extending service life and reducing maintenance requirements in challenging offshore environments (Cao et al., 2022).

3.9.2 Aerospace Coatings

Aerospace coatings for corrosion protection are engineered to meet the stringent requirements of the aviation industry, balancing durability, weight considerations, and environmental compliance. These specialized coatings must withstand extreme temperature fluctuations, UV radiation, and chemical exposure while maintaining structural integrity. The development of chromate-free primers has been a significant focus in recent years, driven by environmental and health concerns associated with hexavalent chromium compounds (Vaghefinazari et al., 2022).

Advanced aerospace coating systems typically employ multi-layer configurations, including primers, intermediate coats, and topcoats, to provide comprehensive protection against various forms of corrosion. The incorporation of nanoparticles and self-healing mechanisms has shown promise in enhancing coating performance and extending service life. Additionally, the use of electrodeposition techniques for primer application has gained traction due to its ability to provide uniform coverage on complex geometries. As the aerospace industry continues to evolve, with increasing use of composite materials and more electric aircraft designs, coating technologies must adapt to address new challenges in corrosion protection and electromagnetic shielding (Piscitelli & Volpe, 2023).

3.9.2 Pipeline Coatings

Pipeline coatings are crucial for preventing corrosion, which significantly impacts the integrity and longevity of pipeline systems. Corrosion, often driven by electrochemical reactions and environmental factors, can lead to severe financial and ecological consequences. Effective coatings act as a barrier between the metal and corrosive elements, thereby extending the lifespan of pipelines and reducing maintenance costs. Various types of coatings, including fusion-bonded epoxy and polyethylene, offer differing levels of protection depending on the specific environmental conditions wherein they are employed. Moreover, advancements in coating technologies continue to evolve, providing enhanced adhesion and resistance to impact and chemicals, thereby addressing the challenges posed by pipeline corrosion more effectively.

3.10 FUTURE TRENDS IN CORROSION-RESISTANT COATINGS

Future trends in corrosion-resistant coatings are set to be influenced by significant advancements in materials science and technology. Emerging methods are increasingly focusing on sustainable and eco-friendly solutions, such as biodegradable corrosion inhibitors and coatings made from renewable resources, to meet environmental regulations. Additionally, the integration of nanotechnology into coating formulations is enhancing the barrier properties and longevity of these protective coatings, which can withstand harsh environmental conditions and provide superior corrosion. Innovations such as self-healing coatings are also gaining traction, where embedded microcapsules release healing agents upon damage, thereby prolonging the lifespan of coated surfaces. Furthermore, the adoption of smart corrosion monitoring systems utilizing advanced sensors is anticipated to revolutionize maintenance protocols, allowing for real-time assessment of corrosion-related risks (Wright et al., 2019).

3.10.1 Emerging Materials and Technologies

Emerging materials and technologies for corrosion protection are shaping the future landscape of anticorrosion strategies, contributing significantly to material longevity and sustainability. Nanotechnology plays a pivotal role in enhancing the barrier properties of coatings, with nanomaterials such as graphene and carbon nanotubes providing superior strength and resistance to degradation. Additionally, smart corrosion monitoring systems are being developed, allowing for real-time tracking of corrosion processes and facilitating timely maintenance interventions. Biodegradable corrosion inhibitors are also gaining attention, as they offer effective protection while minimizing environmental impact, aligning with the principles of sustainable engineering. Furthermore, layered double hydroxides and transition metal dichalcogenides are being explored for their unique properties, enabling innovative application in protective coatings (Sukanya et al., 2022).

3.10.2 Integration with Internet of Things for Corrosion Monitoring

The integration of the Internet of Things (IoT) into corrosion monitoring is transforming traditional corrosion protection strategies, allowing for real-time data collection and analysis. This technological advancement facilitates the continuous monitoring of corrosion-prone structures such as pipelines and storage tanks, thereby enhancing the ability to detect corrosion at early stages and implement timely maintenance strategies. IoT-enabled sensors can capture critical conditions such as temperature, humidity, and the presence of corrosive agents, relaying this information to centralized systems for analysis. Such capabilities promote proactive rather than reactive management, optimizing operational efficiency and reducing maintenance costs. Furthermore, machine learning algorithms can analyze the data collected to predict corrosion trends, allowing for tailored maintenance schedules that enhance asset lifespan while minimizing risks (Chou et al., 2017).

3.11 ECONOMIC ASPECTS OF CORROSION PROTECTION COATINGS

Corrosion protection coatings play a crucial role in safeguarding metal infrastructures, significantly impacting India's economy. The cost-effectiveness of these coatings can be realized through their ability to extend the lifespan of structures, thus reducing maintenance and replacement costs. The adoption of advanced corrosion protection technologies not only enhances the durability of assets but also contributes to sustainability by minimizing resource wastage associated with corroded materials. Moreover, as industries such as construction and manufacturing continue to grow, the demand for reliable protective coatings is estimated to increase, driving economic growth in related sectors (Habib et al., 2021). Increased investment in research and development of innovative coatings can also lead to competitive advantages for Indian manufacturers in the global market. Hence, a strategic focus on corrosion protection coatings is essential for enhancing infrastructure resilience and fostering economic development in India.

3.11.1 Cost–Benefit Analysis of Coating Systems

The economic aspects of corrosion protection coatings in India are critical, as corrosion incurs significant financial burdens on various industries. Reports estimate that the direct cost of corrosion in India reaches approximately \$26.1 billion annually, representing about 2.4% of the nation's Gross Domestic Product (GDP). This substantial economic impact necessitates investment in effective corrosion protection measures to mitigate losses and enhance infrastructure longevity. Corrosion protection coatings, which are essential in preserving the integrity of assets, can significantly reduce maintenance costs and extend the service life of structures, thereby offering considerable long-term savings (Montemor, 2018). Moreover, the Indian government's focus on infrastructure development, particularly in sectors like construction and transportation, further highlights the demand for high-performance coatings that can withstand environmental challenges and ensure economic sustainability.

3.11.2 Life-Cycle Cost Considerations

Life-cycle cost considerations for corrosion protection are essential for optimizing both economic and structural longevity. By assessing the total costs associated with corrosion management including initial acquisition, installation, maintenance, and rehabilitation stakeholders can make informed decisions that enhance the sustainability and performance of infrastructure. Effective life-cycle cost analysis (LCCA) provides a comprehensive framework to evaluate various corrosion protection strategies, thereby allowing for a comparative assessment of alternatives based on long-term costs rather than initial expenses alone. For instance, while high-quality coatings may involve higher upfront costs, their durability and reduced maintenance needs often result in significant savings over time (Fotovvati et al., 2019). Additionally, LCCA can account for factors such as inflation rates, service life, and discount rates, further refining the decision-making process.

3.12 CONCLUSION

The evolving role of coatings in corrosion management represents a dynamic and critical frontier in materials science and engineering. As this comprehensive study has demonstrated, protective coatings have become an indispensable tool in the fight against corrosion across numerous industries and environmental conditions. The advancements in coating technologies, from traditional barrier methods to cutting-edge nanocoatings and self-healing systems, have significantly enhanced our ability to protect materials from corrosive attacks. The integration of smart materials and IoT technologies into coating systems promises a future where corrosion protection becomes more proactive and efficient, with real-time monitoring and predictive maintenance capabilities.

Environmental considerations have driven the development of eco-friendly coating options, addressing the pressing need for sustainable solutions without compromising on performance. The economic impact of effective corrosion management through coatings cannot be overstated, with LCCAs consistently demonstrating the long-term benefits of investing in high-quality protective systems. As industries continue to push the boundaries of material performance in increasingly challenging environments, the role of coatings will undoubtedly expand, requiring ongoing research and innovation.

The future of corrosion management lies in the synergy between advanced coating formulations, precise application techniques, and sophisticated monitoring systems. This holistic approach will not only extend the lifespan of materials and structures but also contribute to resource conservation and environmental protection. As we move forward, the collaboration between materials scientists, chemical engineers, and industry practitioners will be crucial in developing the next generation of corrosion-resistant coatings. These advancements will not only address current challenges but also anticipate future needs, ensuring that protective coatings remain at the forefront of corrosion management strategies. In conclusion, the role of coatings in corrosion protection continues to evolve, promising more effective, sustainable, and intelligent solutions that will shape the future of materials preservation across all sectors.

REFERENCES

Abbott, A. P., & McKenzie, K. J. (2006). Application of ionic liquids to the electrodeposition of metals. *Physical Chemistry Chemical Physics*, *8*(37), 4265. https://doi.org/10.1039/b607329h

Abdel-Gaber, A. M., Rahal, H. T., & El-Rifai, M. S. (2021). Green approach towards corrosion inhibition in hydrochloric acid solutions. *Biointerface Research in Applied Chemistry*, *11*(6), 14185–14195.

Ahn, S. H., Lee, J. H., Kim, J. G., & Han, J. G. (2004). Localized corrosion mechanisms of the multilayered coatings related to growth defects. *Surface and Coatings Technology*, *177–178*, 638–644. https://doi.org/10.1016/S0257-8972(03)00939-3

Akbarzadeh, S., Ramezanzadeh, M., Ramezanzadeh, B., & Bahlakeh, G. (2020). A green assisted route for the fabrication of a high-efficiency self-healing anti-corrosion coating through graphene oxide nanoplatform reduction by Tamarindus indiaca extract. *Journal of Hazardous Materials*, *390*, 122147. https://doi.org/10.1016/j.jhazmat.2020.122147

Aljibori, H. S., Alamiery, A., & Kadhum, A. A. H. (2023). Advances in corrosion protection coatings: A comprehensive review. *International Journal of Corrosion and Scale Inhibition*, *12*(4), 1476–1520.

Arrais, R., Costa, C. M., Ribeiro, P., Rocha, L. F., Silva, M., & Veiga, G. (2021). On the development of a collaborative robotic system for industrial coating cells. *The International Journal of Advanced Manufacturing Technology*, *115*(3), 853–871. https://doi.org/10.1007/s00170-020-06167-z

Bettridge, D. F., & Ubank, R. G. (1986). Quality control of high-temperature protective Coatings. *Materials Science and Technology*, *2*(3), 232–242. https://doi.org/10.1179/mst.1986.2.3.232

Bierwagen, G., Tallman, D., Li, J., He, L., & Jeffcoate, C. (2003). EIS studies of coated metals in accelerated exposure. *Progress in Organic Coatings*, *46*(2), 149–158. https://doi.org/10.1016/S0300-9440(02)00222-9

Bierwagen, G. P., Wang, X., & Tallman, D. E. (2003). In situ study of coatings using embedded electrodes for ENM measurements. *Progress in Organic Coatings*, *46*(3), 163–175. https://doi.org/10.1016/S0300-9440(02)00186-8

Bjerkén, C., & Ortiz, M. (2010). Evolution of anodic stress corrosion cracking in a coated material. *International Journal of Fracture*, *165*(2), 211–221. https://doi.org/10.1007/s10704-010-9514-5

Blaiszik, B. J., Kramer, S. L. B., Olugebefola, S. C., Moore, J. S., Sottos, N. R., & White, S. R. (2010). Self-Healing Polymers and Composites. *Annual Review of Materials Research*, *40*(1), 179–211. https://doi.org/10.1146/annurev-matsci-070909-104532

Bourbigot, S. (2021). Intumescence-Based Flame Retardant. Non-Halogenated Flame Retardant Handbook.

Boumezgane, O., Suriano, R., Fedel, M., Tonelli, C., Deflorian, F., & Turri, S. (2022). Self-healing epoxy coatings with microencapsulated ionic PDMS oligomers for corrosion protection based on supramolecular acid-base interactions. *Progress in Organic Coatings*, *162*, 106558. https://doi.org/10.1016/j.porgcoat.2021.106558

Bose, S. (2017). High Temperature Coatings. Butterworth-Heinemann.

Chattopadhyay, D.K., Raju, K.V. (2007). Structural engineering of polyurethane coatings for high performance applications. *Progress in Polymer Science*, *32*(3), 352–418.

Cao, L., Wang, Q., Wang, W., Li, Q., & Chen, S. (2022). Synthesis of smart nanofiber coatings with autonomous self-warning and self-healing functions. *ACS Applied Materials & Interfaces*, *14*(23), 27168–27176. https://doi.org/10.1021/acsami.2c05048

Cao, X., Huang, F., Huang, C., Liu, J., & Cheng, Y. F. (2019). Preparation of graphene nanoplate added zinc-rich epoxy coatings for enhanced sacrificial anode-based corrosion protection. *Corrosion Science*, *159*, 108120. https://doi.org/10.1016/j.corsci.2019.108120

Carbonell-Verdu, A., Garcia-Sanoguera, D., Jordá-Vilaplana, A., Sanchez-Nacher, L., & Balart, R. (2016). A new biobased plasticizer for poly(vinyl chloride) based on epoxidized cottonseed oil. *Journal of Applied Polymer Science*, *133*(27), app.43642. https://doi.org/10.1002/app.43642

Chou, J.S., Ngo, N.T., & Chong, W.K. (2017). The use of artificial intelligence combiners for modeling steel pitting risk and corrosion rate. *Engineering Applications of Artificial Intelligence*, 65, 471–483.

Czachor-Jadacka, D., & Pilch-Pitera, B. (2021). Progress in development of UV curable powder coatings. *Progress in Organic Coatings*, *158*, 106355. https://doi.org/10.1016/j.porgcoat.2021.106355

D'Alessandro, O., Selmi, G. J., Di Sarli, A. R., Romagnoli, R., & Deyá, C. (2019). Accelerated tests, a necessary complement of electrochemical assays to evaluate anti-corrosive coatings. *Journal of Applied Electrochemistry*, *49*(8), 811–822. https://doi.org/10.1007/s10800-019-01324-z

Deyab, M. A. (2014). Corrosion protection of aluminum bipolar plates with polyaniline coating containing carbon nanotubes in acidic medium inside the polymer electrolyte membrane fuel cell. *Journal of Power Sources*, *268*, 50–55. https://doi.org/10.1016/j.jpowsour.2014.06.021

Effendy, S., Zhou, T., Eichman, H., Petr, M., & Bazant, M. Z. (2021). Blistering failure of elastic coatings with applications to corrosion resistance. *Soft Matter*, *17*(41), 9480–9498. https://doi.org/10.1039/D1SM00986A

Ewulonu, C. M., Igwe, I. O., & Onyeagoro, G. N. (2016). Performance of local Clay-Titanium Dioxide Core-Shell extender pigments in Alkyd paints. *Advances in Nanoparticles*, *05*(01), 90–102. https://doi.org/10.4236/anp.2016.51011

Faccini, M., Bautista, L., Soldi, L., Escobar, A. M., Altavilla, M., Calvet, M., Domènech, A., & Domínguez, E. (2021). Environmentally friendly anticorrosive polymeric coatings. *Applied Sciences*, *11*(8), 3446. https://doi.org/10.3390/app11083446

Farag, A. A. (2020). Applications of nanomaterials in corrosion protection coatings and inhibitors. *Corrosion Reviews*, *38*(1), 67–86. https://doi.org/10.1515/corrrev-2019-0011

Ferrel-Alvarez, A., Galicia, B. B., Ashok, A., Regmi, G., Velumani, S., & Castaneda, H. (2022). Synthesis and characterization of Epoxy-Rich TMOs deposited on stainless steel for corrosion applications. *Coatings*, *12*(3), Article 3. https://doi.org/10.3390/coatings12030387

Fierascu, R. C., Fierascu, I., & Chican, I. E. (2021). Natural ingredients in functional coatings—Recent advances and future challenges. *Coatings*, *11*(4), Article 4. https://doi.org/10.3390/coatings11040429

Figovsky, O., Shapovalov, L., Leykin, A., Birukova, O., & Potashnikova, R. (2013). Advances in the field of nonisocyanate polyurethanes based on cyclic carbonates. *Chemistry & Chemical Technology*, *7*(1), 79–87. https://doi.org/10.23939/chcht07.01.079

Figueira, R. B., Silva, C. J. R., & Pereira, E. V. (2015). Organic–inorganic hybrid sol–gel coatings for metal corrosion protection: A review of recent progress. *Journal of Coatings Technology and Research*, *12*(1), 1–35. https://doi.org/10.1007/s11998-014-9595-6

Filotás, D., Izquierdo, J., Fernández-Pérez, B. M., Nagy, L., Nagy, G., & Souto, R. M. (2022). Contributions of microelectrochemical scanning techniques for the efficient detection of localized corrosion processes at the cut edges of polymer-coated galvanized steel. *Molecules*, *27*(7), 2167. https://doi.org/10.3390/molecules27072167

Florio, J.J., Miller, D.J. (2004). Handbook of Coating Additives. CRC Press, 2004.

Fotovvati, B., Namdari, N., & Dehghanghadikolaei, A. (2019). On coating techniques for surface protection: A review. *Journal of Manufacturing and Materials Processing*, *3*(1), 28. https://doi.org/10.3390/jmmp3010028

Galedari, S. A., Mahdavi, A., Azarmi, F., Huang, Y., & McDonald, A. (2019). A comprehensive review of corrosion resistance of thermally-sprayed and thermally-diffused protective coatings on steel structures. *Journal of Thermal Spray Technology*, *28*(4), 645–677. https://doi.org/10.1007/s11666-019-00855-3

Gardiner, J. (2015). Fluoropolymers: Origin, production, and industrial and commercial applications. *Australian Journal of Chemistry*, *68*(1), 13. https://doi.org/10.1071/CH14165

Habib, S., Shakoor, R. A., & Kahraman, R. (2021). A focused review on smart carriers tailored for corrosion protection: Developments, applications, and challenges. *Progress in Organic Coatings*, *154*, 106218. https://doi.org/10.1016/j.porgcoat.2021.106218

Hao, T., Yao, W., Zhipeng, L., Jie, L., Liangang, S., Wenchao, W., Jixian, L., & Tang, J. (2021). Emerging applications of silica nanoparticles as multifunctional modifiers for high performance polyester composites. *Nanomaterials*, 11, 2810.

Hanák, M., Podjuklová, J., Siostrzonek, R., Pavelková, D., & Koldová, L. (2015). The influence of substrate surface cleanliness on the corrosion resistance of coating systems. *Materials Science Forum*, *818*, 167–170. https://doi.org/10.4028/www.scientific.net/MSF.818.167

Hernández-Maya, L. G., Zapata-Peñasco, I., Olivares-Luna, M., Campos-Silva, I., Martínez-Trinidad, J., Díaz-Barriga, L. G., Rivera-Olvera, J. N., & Garibay-Febles, V. (2023). The ZrC and TiNi nanostructures in epoxy coatings: An anticorrosion and tribological study. *Surface and Coatings Technology*, *469*, 129816. https://doi.org/10.1016/j.surfcoat.2023.129816

Hinderliter, B. R., & Sapper, E. D. (2015). Water concentration distribution in coatings during accelerated weathering protocols. *Journal of Coatings Technology and Research*, *12*(3), 477–487. https://doi.org/10.1007/s11998-015-9661-8

Hu, J. Y., Zhang, S. S., Chen, E., & Li, W. G. (2022). A review on corrosion detection and protection of existing reinforced concrete (RC) structures. *Construction and Building Materials*, *325*, 126718. https://doi.org/10.1016/j.conbuildmat.2022.126718

Hua, Z., Shu, H., Xu, A., Zheng, B., Yang, A., Fu, Z., Heji, A., Eichman, H., & Baikerikar, K. (2015). Novel acrylic epoxy hybrid system design for metal protection application. *JCT Coatingstech*, *12*(4), 58–65.

Huang, X., Wang, L.-T., Song, Y.-F., Ge, F., Zhang, Y., Meng, X.-J., Ge, H.-H., & Zhao, Y.-Z. (2021). Effect of Al2O3 nanoparticles on the corrosion behavior of aluminum alloy in simulated vehicle coolant. *Journal of Alloys and Compounds*, *874*, 159807. https://doi.org/10.1016/j.jallcom.2021.159807

Hughes, A. E., Cole, I. S., Muster, T. H., & Varley, R. J. (2010). Designing green, self-healing coatings for metal protection. *NPG Asia Materials*, *2*(4), 143–151. https://doi.org/10.1038/asiamat.2010.136

Jamali, S. S., & Mills, D. J. (2014). Steel surface preparation prior to painting and its impact on protective performance of organic coating. *Progress in Organic Coatings*, *77*(12), 2091–2099. https://doi.org/10.1016/j.porgcoat.2014.08.001

Jin, F.L., Li, X., & Park, S.J. (2015). Synthesis and application of epoxy resins: A review. *Journal of Industrial and Engineering Chemistry*, *29*, 1–1.

Kalendová, A., Veselý, D., Kohl, M., & Stejskal, J. (2015). Anticorrosion efficiency of zinc-filled epoxy coatings containing conducting polymers and pigments. *Progress in Organic Coatings*, *78*, 1–20. https://doi.org/10.1016/j.porgcoat.2014.10.009

Kim, D.-M., Cho, Y.-J., Choi, J.-Y., Kim, B.-J., Jin, S.-W., & Chung, C.-M. (2017). Low-temperature self-healing of a Microcapsule-type protective coating. *Materials*, *10*(9), 1079. https://doi.org/10.3390/ma10091079

Kušter Marić, M., Ivanković, A. M., Vlašić, A., Bleiziffer, J., Srbić, M., & Skokandić, D. (n.d.). Assessment of reinforcement corrosion and concrete damage on bridges using non-destructive testing. *Građevinar*, *71*(10), 843–862. Retrieved 15 September 2024, from http://casopis-gradjevinar.hr/archive/article/2724

Lang, S., & Zhou, Q. (2017). Synthesis and characterization of poly(urea-formaldehyde) microcapsules containing linseed oil for self-healing coating development. *Progress in Organic Coatings*, *105*, 99–110. https://doi.org/10.1016/j.porgcoat.2016.11.015

Leng, A., Streckel, H., & Stratmann, M. (1998). The delamination of polymeric coatings from steel. Part 1: Calibration of the Kelvinprobe and basic delamination mechanism. *Corrosion Science*, *41*(3), 547–578. https://doi.org/10.1016/S0010-938X(98)00166-8

Ling, T.-C., & Poon, C.-S. (2013). Use of phase change materials for thermal energy storage in concrete: An overview. *Construction and Building Materials*, *46*, 55–62. https://doi.org/10.1016/j.conbuildmat.2013.04.031

Liu, L., Tan, E., Zhen, Y., Yin, X. J., & Cai, Z. Q. (2018). AI-facilitated coating corrosion assessment system for productivity enhancement. *2018 13th IEEE Conference on Industrial Electronics and Applications (ICIEA)*, 606–610. https://doi.org/10.1109/ICIEA.2018.8397787

Ma, Y., Wang, H., Li, S., Chen, Y., Zhang, L., & Sheng, X. (2022). An eco-friendly and pH response cerium-based melamine phytate (PM) nanosheets with active and passive anticorrosion ability in water-borne epoxy coating. *Applied Surface Science*, *597*, 153726. https://doi.org/10.1016/j.apsusc.2022.153726

McKeen, L.W. (2015). Fluorinated coatings and finishes handbook: The definitive user's guide.

Marcus, P., Oudar, J., editors. (2002). Corrosion mechanisms in theory and practice. New York: Marcel Dekker.

Manoj, A., Ramachandran, R., & Menezes, P. L. (2020). Self-healing and superhydrophobic coatings for corrosion inhibition and protection. *The International Journal of Advanced Manufacturing Technology*, *106*(5–6), 2119–2131. https://doi.org/10.1007/s00170-019-04758-z

Momber A. (2007). Blast cleaning technology. Springer Science & Business Media.

Momber, A. W. (2016). Quantitative performance assessment of corrosion protection systems for offshore wind power transmission platforms. *Renewable Energy*, *94*, 314–327. https://doi.org/10.1016/j.renene.2016.03.059

Montemor, F. (2018). Coatings for corrosion protection: What do we need and where are we. *Qatar Foundation Annual Research Conference Proceedings Volume 2018 Issue 1*. Qatar National Convention Center (QNCC), Doha, Qatar.. https://doi.org/10.5339/qfarc.2018.EEPP542

Ollik, K., & Lieder, M. (2020). Review of the application of graphene-based coatings as anticorrosion layers. *Coatings*, *10*(9), Article 9. https://doi.org/10.3390/coatings10090883

Pant, P. (2023). A report on investigation on corrosion and rusting on iron. https://doi.org/10.13140/RG.2.2.23991.70564

Park, C.-N., Lee, W.-K., & Jang, S.-H. (2008). Development of eco-friendly high-solids paints for automotive coatings. *Journal of Environmental Science International*, *17*(8), 925–932. https://doi.org/10.5322/JES.2008.17.8.925

Piscitelli, F., & Volpe, A. (2023). Superhydrophobic coatings for corrosion protection of stainless steel. *Aerospace*, *11*(1), 3. https://doi.org/10.3390/aerospace11010003

Popoola, L., Grema, A., Latinwo, G., Gutti, B., & Balogun, A. (2013). Corrosion problems during oil and gas production and its mitigation. *International Journal of Industrial Chemistry*, *4*(1), 35. https://doi.org/10.1186/2228-5547-4-35

Presuel-Moreno, F., Jakab, M. A., Tailleart, N., Goldman, M., & Scully, J. R. (2008). Corrosion-resistant metallic coatings. *Materials Today*, *11*(10), 14–23. https://doi.org/10.1016/S1369-7021(08)70203-7

Prosek, T., Nazarov, A., Goodwin, F., Šerák, J., & Thierry, D. (2016). Improving corrosion stability of ZnAlMg by alloying for protection of car bodies. *Surface and Coatings Technology*, *306*, 439–447. https://doi.org/10.1016/j.surfcoat.2016.03.062

Ramezani, M., Mohd Ripin, Z., Pasang, T., & Jiang, C.-P. (2023). Surface engineering of metals: Techniques, characterizations and applications. *Metals*, *13*(7), Article 7. https://doi.org/10.3390/met13071299

Rani, B. E. A., & Basu, B. B. J. (2012). Green inhibitors for corrosion protection of metals and alloys: An overview. *International Journal of Corrosion*, *2012*, 1–15. https://doi.org/10.1155/2012/380217

Rao, T. N., Hussain, I., Lee, J. E., Kumar, A., & Koo, B. H. (2019). Enhanced thermal properties of zirconia nanoparticles and Chitosan-based intumescent flame retardant coatings. *Applied Sciences*, *9*(17), 3464. https://doi.org/10.3390/app9173464

Revie, R.W., Uhlig, H.H. (2025). Corrosion and corrosion control. John Wiley & Sons.

Sacher, S., Wahl, P., Weißensteiner, M., Wolfgang, M., Pokhilchuk, Y., Looser, B., Thies, J., Raffa, A., & Khinast, J. G. (2019). Shedding light on coatings: Real-time monitoring of coating quality at industrial scale. *International Journal of Pharmaceutics*, *566*, 57–66. https://doi.org/10.1016/j.ijpharm.2019.05.048

Samimi, A., & Zarinabadi, S. (2011). An analysis of polyethylene coating corrosion in oil and gas pipelines. *Journal of American Science, USA*. https://citeseerx.ist.psu.edu/document?repid=rep1&type=pdf&doi=29d178d8cdce7c86dfe28a66d31fde80ca11e372

Schweitzer, P.A. (1989). *Corrosion and corrosion protection handbook* (Vol. 1). CRC Press.

Schweitzer, P. E. (2006). *Corrosion of linings & coatings: Cathodic and inhibitor protection and corrosion monitoring*. CRC press. https://books.google.com/books?hl=en&lr=&id=ZvLUxwb-A3MC&oi=fnd&pg=PP1&dq=On+the+other+hand,+inorganic+coatings+such+as+zinc-rich+primers+and+ceramic+coatings+offer+excellent+corrosion+resistance+and+can+withstand+high+temperatures+(Schweitzer,+2006).&ots=wF8lkgMfLg&sig=SSCRHm8uSMLDUp00FMqXW6NMYdg

Selim, M. S., Shenashen, M. A., El-Safty, S. A., Higazy, S. A., Selim, M. M., Isago, H., & Elmarakbi, A. (2017). Recent progress in marine foul-release polymeric nanocomposite coatings. *Progress in Materials Science*, *87*, 1–32. https://doi.org/10.1016/j.pmatsci.2017.02.001

Shirehjini, F. T., Danaee, I., Eskandari, H., & Zarei, D. (2016). Effect of nano clay on corrosion protection of Zinc-rich epoxy coatings on steel 37. *Journal of Materials Science & Technology*, *32*(11), 1152–1160. https://doi.org/10.1016/j.jmst.2016.08.017

Sidhu, T. S., Prakash, S., & Agrawal, R. D. (2005). Studies on the properties of high-velocity oxy-fuel thermal spray coatings for higher temperature applications. *Materials Science*, *41*(6), 805–823. https://doi.org/10.1007/s11003-006-0047-z

Sørensen, P. A., Kiil, S., Dam-Johansen, K., & Weinell, C. E. (2009). Anticorrosive coatings: A review. *Journal of Coatings Technology and Research*, *6*(2), 135–176. https://doi.org/10.1007/s11998-008-9144-2

Streitberger, H.J., Goldschmidt, A. (2018). BASF Handbook Basics of Coating Technology. European Coatings.

Sukanya, R., Barwa, T. N., Luo, Y., Dempsey, E., & Breslin, C. B. (2022). Emerging layered materials and their applications in the corrosion protection of metals and alloys. *Sustainability*, *14*(7), 4079. https://doi.org/10.3390/su14074079

Talbert, R. (2009). CGGC Staff Interview with Rodger Talbert, Powder Coating Institute Technical Director. Accessed 25 September 2009.

Thomas, D., Reshmy, R., Philip, E., Sindhu, R., Ulaeto, S. B., Pugazhendhi, A., & Awasthi, M. K. (2022). Developments in smart organic coatings for anticorrosion applications: A review. *Biomass Conversion and Biorefinery*, *12*(10), 4683–4699. https://doi.org/10.1007/s13399-022-02363-x

Tiong, U. H., & Clark, G. (2011). Impact of mechanical strain environment on aircraft protective coatings and corrosion protection. *Journal of Aircraft*, *48*(4), 1315–1330. https://doi.org/10.2514/1.C031270

Udoh, I. I., Shi, H., Daniel, E. F., Li, J., Gu, S., Liu, F., & Han, E.-H. (2022). Active anticorrosion and self-healing coatings: A review with focus on multi-action smart coating strategies. *Journal of Materials Science & Technology*, *116*, 224–237. https://doi.org/10.1016/j.jmst.2021.11.042

Ulaeto, S. B., Rajan, R., Pancrecious, J. K., Rajan, T. P. D., & Pai, B. C. (2017). Developments in smart anticorrosive coatings with multifunctional characteristics. *Progress in Organic Coatings*, *111*, 294–314. https://doi.org/10.1016/j.porgcoat.2017.06.013

Vaghcfinazari, B., Wicrzbicka, E., Visscr, P., Posncr, R., Arrabal, R., Matykina, E., Mohedano, M., Blawert, C., Zheludkevich, M. L., & Lamaka, S. V. (2022). Chromate-free corrosion protection strategies for Magnesium alloys—A review: Part III—Corrosion inhibitors and combining them with other protection strategies. *Materials*, *15*(23), Article 23. https://doi.org/10.3390/ma15238489

Voevodin, N. N., Balbyshev, V. N., Khobaib, M., & Donley, M. S. (2003). Nanostructured coatings approach for corrosion protection. *Progress in Organic Coatings*, *47*(3–4), 416–423. https://doi.org/10.1016/S0300-9440(03)00131-0

Walsh, F.C., Wang, S., Zhou, N. (2020). The electrodeposition of composite coatings: diversity, applications and challenges. *Current Opinion in Electrochemistry*, *20*, 8–19.

Wang, D., Bierwagen, G.P. (2009). Sol–gel coatings on metals for corrosion protection. *Progress in Organic Coatings*. *64*(4), 327–338.

Wang, L., Li, S., & Fu, J. (2023). Self-healing anti-corrosion coatings based on micron-nano containers with different structural morphologies. *Progress in Organic Coatings*, *175*, 107381. https://doi.org/10.1016/j.porgcoat.2022.107381

Wang, X., Tang, F., Cao, Q., Qi, X., Pearson, M., Li, M., Pan, H., Zhang, Z., & Lin, Z. (2020). Comparative study of three carbon additives: Carbon nanotubes, graphene, and Fullerene-C60, for synthesizing enhanced polymer nanocomposites. *Nanomaterials*, *10*(5), 838. https://doi.org/10.3390/nano10050838

Wang, Z., Shi, X., Qi, B., Xu, Z., & Fang, S. (2022). Application and research of electrostatic painting process in the field of charge system equipment. In Z. Zhang & J. Zhu (Eds.), *International Conference on Advanced Sensing and Smart Manufacturing (ASSM 2022)* (p. 28). SPIE. https://doi.org/10.1117/12.2652083

Wright, R. F., Lu, P., Devkota, J., Lu, F., Ziomek-Moroz, M., & Ohodnicki, P. R. (2019). Corrosion sensors for structural health monitoring of oil and natural gas infrastructure: A review. *Sensors*, *19*(18), 3964. https://doi.org/10.3390/s19183964

Wu, J., Padture, N. P., Klemens, P. G., Gell, M., García, E., Miranzo, P., & Osendi, M. I. (2002). Thermal conductivity of ceramics in the ZrO_2-$GdO_{1.5}$ system. *Journal of Materials Research*, *17*(12), 3193–3200. https://doi.org/10.1557/JMR.2002.0462

Wypych, A. (2017). Databook of plasticizers. Elsevier.

Xia, S., Zhao, Q., Ji, J., Wu, R., Chen, L., Yin, Y., & Liu, Q. (2023). Impact of water-based coating substitution on VOCs emission characteristics for the surface-coating industries and policy effectiveness: A case study in Jiangsu Province, China. *Atmosphere*, *14*(4), 662. https://doi.org/10.3390/atmos14040662

Yang, M. S., Huang, J., Noël, J. J., Chen, J., Barker, I., Henderson, J. D., Zhang, H., Zhang, H., & Zhu, J. (2022). A mechanistic study on the anti-corrosive performance of zinc-rich polyester/TGIC powder coatings. *Processes*, *10*(9), 1853. https://doi.org/10.3390/pr10091853

Yang, W. J., Tao, X., Zhao, T., Weng, L., Kang, E.-T., & Wang, L. (2015). Antifouling and antibacterial hydrogel coatings with self-healing properties based on a dynamic disulfide exchange reaction. *Polymer Chemistry*, *6*(39), 7027–7035. https://doi.org/10.1039/C5PY00936G

Zaki, M. H. M., Mohd, Y., & Isa, N. N. C. (2017). Surface pre-treatment of aluminium by cleaning, chemical etching and conversion coating. In *AIP Conference Proceedings*, vol. 1901, no. 1, p. 120006. AIP Publishing LLC, 2017120006. https://doi.org/10.1063/1.5010556

Zhang, J., Zhang, W., Wei, L., Pu, L., Liu, J., Liu, H., Li, Y., Fan, J., Ding, T., & Guo, Z. (2019). Alternating multilayer structural epoxy composite coating for corrosion protection of steel. *Macromolecular Materials and Engineering*, *304*(12), 1900374. https://doi.org/10.1002/mame.201900374

Zhang, M., Wang, H., Nie, T., Bai, J., Zhao, F., & Ma, S. (2020). Enhancement of barrier and anti-corrosive performance of zinc-rich epoxy coatings using nano-silica/graphene oxide hybrid. *Corrosion Reviews*, *38*(6), 497–513. https://doi.org/10.1515/corrrev-2020-0034

Zhang, Y., Liu, Y., Zheng, R., Zheng, Y., & Chen, L. (2023). Research progress on corrosion behaviors and biocompatibility of rare-earth magnesium alloys *in vivo* and *in vitro*. *Journal of Rare Earths*, *41*(12), 1827–1842. https://doi.org/10.1016/j.jre.2023.03.005

Zhang, Y., & Zhai, X. (2019). Preparation and testing of thermochromic coatings for buildings. *Solar Energy*, *191*, 540–548. https://doi.org/10.1016/j.solener.2019.09.042

Zhao, D., Liu, D., & Hu, Z. (2017). A smart anticorrosion coating based on hollow silica nanocapsules with inorganic salt in shells. *Journal of Coatings Technology and Research*, *14*(1), 85–94. https://doi.org/10.1007/s11998-016-9830-4

Zheludkevich ML, Tedim J, Ferreira MG. "Smart" coatings for active corrosion protection based on multi-functional micro and nanocontainers. Electrochimica Acta. 2012 Nov 1;82:314-23.

4 Potential Use of Smart Coatings for Corrosion Protection

Nabil Lahrache, Houda Labjar, Ayoub Cherrat, Mohamed Dalimi, Hamid Nasrellah, Youssef Belaoufi, Houda Damour, Najoua Labjar, and Souad El Hajjaji

4.1 INTRODUCTION

Corrosion is etching, the destruction of metals when they come into contact with corrosive media, which may be environmental chemical reagents. According to ISO, corrosion is the set of physicochemical steps using the metal and its environment as reagents and producing changes in the metal's properties, such as the degradation of metal functions [1]. As the performance of materials can be affected by the action of certain external corrosive factors, it is necessary to protect them against corrosion. The use of coatings is the best alternative for protecting metal surfaces. Good barriers can be useful in preventing the movement of corrosive substances. These corrosion-inhibiting coatings must form a good physical barrier, avoiding the interaction of products with the metal surface [2]. Corrosion coatings are formed in one of the following steps: cathodic, anode, electrolyte, and corrosion active inhibitions.

Metal protection using coatings is an excellent means of inhibition. These products create environmental problems and harmful effects on human health, which has limited the use of such dangerous products as volatile organic products, toxic and carcinogenic chromium-based products used in coatings [3].

Intelligent coatings protect the external environment from or their reactive or intelligent applications. These coatings are produced while controlling their molecular compositions and nano morphologies used in microbial treatment, conduction, anti-fouling, and self-cleaning [4].

The nanohybrids compounds are the class of materials with dimensions that fall within the nanometric scale, generally between 1 and 100 nm. At this tiny scale, materials have unique properties that distinguish them from atoms. The properties of materials change dramatically at the nanometric scale. Nanoparticles can have a large surface area and high reactivity. For example, we find these products in sun creams and stain-resistant clothing, as well as drug delivery systems Nanomaterials can be used to develop new corrosion-resistant coatings or improve existing ones [5].

These products are used in many fields, including life sciences [6], the nature [7,8], energy [9], electronics industry [9,10], chemicals [3], and polymers [10]. They have

DOI: 10.1201/9781003571162-4

a number of applications such as bioinspired materials, medical, biomimetics, renewable sources of energy, metal protection, and sanitation [11–15]. Research into these molecules has enabled the synthesis of heterostructure nanomaterials such as layered nanohybrids. These products have been used in a number of areas, such as corrosion inhibition (due to their specific physicochemical and mechanical properties), which have not been achieved by other similar nanohybrids [16–18]. These products are prepared by intercalation reactions [19,20], by reversibly inserting inorganic, organic, or ion molecules, in a product while retaining its structural characteristics [21–26].

Unfortunately, these products deteriorated the environment and human beings, which has necessitated the creation of new coatings that perform better, last longer, and are less harmful to the environment as smart coatings [26].

The sections below will give us the different types of deterioration, the various forms of corrosion, and which intelligent smart coatings can inhibit this corrosion and potential use of their products.

4.2 DIFFERENT FORMS OF CORROSION

- *Spike Corrosion*
  Corrosion or pitting is an attack by aggressive media such as acids, and involves a local rupture of the film. This corrosion affects many materials, including nickel, steel, copper, and aluminum. This pitting germinates and the presence of a local current forms products, causing the oxide to dissolve and a salt film to form [27].
- *Galvanic Corrosion*
  This is the phenomenon representing interaction products between two metals in a conductive solvent. These galvanic corrosions are classified according to the potentials of these reactions. The contact between two different metals with two different potentials forms an electric current. This potential differential across the two dissimilar metals defines whether the corrosion is anodic or cathodic. Physical properties influence metal properties. Zinc oxide, the product coating formed by corrosion, becomes more resistant than iron in warm water. Car design companies use aluminum or magnesium instead of iron as a raw material to reduce the weight of vehicles, so these parts corrode in aggressive environments, forming galvanic corrosion oxides. The transport industry requires automated processes such as the closing and opening of bus and metro doors, etc. Because of these properties, aluminum is used as the riveting component that joins the doors to the frame. This is why these industries are also affected by this corrosion [28].
- *Corrosion by Concentration Pile*
  To achieve corrosion by concentration pile, we need to eliminate the defect on part of the same metal. In this way, the two dissimilar metals do not mate, avoiding galvanic abrasion. This deterioration occurs on same metal in the presence of aggressive media [29].
- *Corrosion by Differential Aeration*
  This is the product of the material coming into contact with two oxidizing media (e.g. a pile driven into mud-filled soil, where the near-surface part is in

contact with more oxygen than the driven part). Differential voltage between two parts triggers corrosion of same metals. The part of the surface immersed in water is closer to oxygen and aggressive elements than the part not immersed, which is not in contact with these elements. Corrosion thus affects the waterline. This type of corrosion can be found in parts of ships immersed in the mother ship of goods and passenger transport companies. It can also be found in non-isothermal water tanks or pipes in different companies, where there are hot zones with little oxygen and cold zones with little oxygen. This temperature gradient is responsible for creating this corrosion [30].

- *Intergranular Corrosion*
  This is the result of partial alteration of grain boundaries. The latter contain defects causing deposition of intermetallic phases for purification. This type of corrosion can be observed in aluminum alloys such as $CuAl_2$ or Al_3Mg_2. It is also formed in the nickel alloy slots of steam generators and pressurized water reactors in nuclear power stations. It is also found in stainless steel tanks and pipes used by fuel reprocessing companies that use nitric acid in their work procedures [30].
- *Cavernous Corrosion*
  It occurs when small volumes of fluid get trapped in the interstices, as in the case of sand, which can block these droplets. It affects the steels and alloys forming the tank and desalter pipes of the oil and gas purification stations of an oil and gas refinery company. It also concerns transport companies that wash their vehicles but do not dry them. Consequently, the metals come into contact with water droplets and corrode [31].
- *Corrosion Erosion Cavitation*
  This is the alteration of the surface due to droplets with a surface defect. It concerns the metals used in nuclear power stations, because they are subjected to high temperatures. It mainly concerns flow control valves, because the collapse of gas bubbles causes mechanical shocks that wear away the surfaces [32].
- *Stress Corrosion*
  This is caused by mechanical stress and corrosive reagents, causing inter- or transgranular cracks. Steel rockbolts used in the underground mining industry are corroded in the form of stresses and are therefore a bête noire for this industry. Similarly, the anchor bolts used in the underground coal industry and in the civil engineering sector are also strung in the same shape. Metals such as steel, used in nuclear power plants, corrode under stress due to the high temperatures and radiation they contain. Flanged connections on column head pipes in the oil industry corrode in the form of stress due to the presence of sulfides [33].
- *Fatigue corrosion*
  This erosion is the product of a corrosive environment and cyclones. This corrosion is observed when parts are subjected to low-frequency stresses (oligocyclic fatigue), as in the case of heating–cooling structures [34].
- *Fretting Corrosion*
  This is the product of contact between two surfaces. This corrosion affects metals and alloys used in the aeronautical, automotive, and biomedical

sectors. It also concerns the electrical connectors in the circuits of the various industries. Contact between the surfaces of crankshafts and bearings used in the textile industry show corrosion on their friction surfaces [35].

- *Biocorrosion*
  Biocorrosion is the result of the action of bacteria, whose metabolism accelerates a process by creating favorable conditions for its establishment, that is, when colonies of potentially corrosive microorganisms grow on the material and form a biofilm. In this case, no new form is formed, but the kinetics of existing corrosion are modified. Metal structures used in ports, transport systems, storage tanks, the energy sector, water distribution and sewage systems, and the oil and gas industry use metals in their operations, on which microorganisms supply metal-aggressive products for their metabolism [36].
- *Hot Corrosion (Fluxing)*
  This is produced by salt melt and oxides such as vanadium pentoxide V_2O_5 [37].
- *Corrosion by Molten Metals*
  Industries use molten metals in primary metallurgy for the manufacture of materials used in agri-food and foundries. These metals can be corroded due to the aggressive environment around them [38].

4.3 METAL COATING PROTECTION

These coatings are made up of inhibitors which are molecules that block or stop a redox reaction and therefore reduce the attack on the corrosive environment by aggressive agents.

They are used to prevent corrosion. The effectiveness of these inhibitors depends on [39]:

- its desolution in the environment
- corrosion damage, which costs more than inhibitors.
  At low concentrations, this protector reduces or prevents corrosion [40]. Several research projects have been carried out to find anticorrosives that can be used in different corrosive environments [41].
- *Synthetic Inhibitors*
  These inhibitors are artificial, produced by chemical interactions, and are synthetic substances such as:
  - Benzyl chloride and diethylsulfate quaternary,
  - Amine ethoxylates,
  - Organic boron products [42].
- *Green Inhibitors*
  These inhibitors are natural compounds that can be adsorbed and do no harm to the community or its environment. This 'green protector' is an 'environmental protector', as natural product extract, 2-hydroxy-napthoquinone resin, and so on [43]. These anticorrosion products interact with the free surface electrons, producing the formation of a film which inhibits reactions between offensive products and the surface [44].

Coating Types

a. The inhibitor creates electrostatic forces with the free surface electrons, producing a film used to prevent the passage of offensive elements,
b. The protein molecule inhibits the formation of anodic and cathodic transformation products, which can take place at the metal/protector layer interface [45].

These coatings formed have the following defects:

- Cracks may form due to the presence of air bubbles in these shades.
- These layers, which are made up of organic protective molecules, have poor resistance, which encourages cracking, limiting their lifespan [46,47].
- Nanohybrid is used in smart coatings for corrosion protection

The nanometric scale offers great potential for a range of fields other than corrosion, from medicine (making it easier to administer drugs) to electronics, where they facilitate the development of ultra-miniaturized devices [26]. Nanomaterials are very diverse and encompass a wide range of materials with different classification systems based on their different characteristics [26]. Table 4.1 shows a number of examples of this type with their shapes.

Besides, several examples of applications for nanomaterials in corrosion prevention are listed in Table 4.2.

4.3.1 Inhibition of Steel Corrosion by Nanohybrids Used in Smart Coatings

Because of its physical and chemical properties, steel is used in industry. Zinc as a sacrificial coating protects steel, but recently nanographene sheets have been used and have given a good reduction in the corrosion of this metal [52,53]. Lara Velasco Davoise and colleagues [54] were able to achieve good corrosion reduction in steel using graphene-coated nickel and copper. Iman Khosravi et al [55] found that the use of graphene nano-oxide inhibited the Cui et al. [56]. Researchers demonstrated that zinc coating with a high concentration of GO increased the corrosion resistance of steel [57]. Others [58] used GO sheets as sediment and phosphates, which captured metal ions and helped inhibit the corrosion of steel. Researchers [59] have reported that GO-modified ZnO species have antibacterial properties, so these pruits reduce metal deterioration. Other researchers have used a nanohybrid ZnO-graphene film and achieved high steel inhibition [60]. In another study, three nanohybrid silver nanoparticle products (heterojunction photodetectors (HPDS), 2,3-dihydroxybiphenyl 1,2-dioxygenase (HPDO)and High Density Poly Ethylene(HPDE)) were used and showed high inhibition of steel corrosion in acidic media (99.9% inhibition) [61] (Figure 4.1).

4.3.2 Inhibition of Copper Corrosion by Nanohybrids

Transition metals (MXenes) are widely used, but they oxidize in an oxygenated environment to produce metal oxides (such as TiO_2), which limits their practical

TABLE 4.1
Different Types of Nanohybrids Based on Their Dimensions

Type	Designation	Structures
Zero-dimensional (0D) nanomaterials	Clusters, atomic granules, metal-based nanoparticles, graphene quantum dots and fullerenes	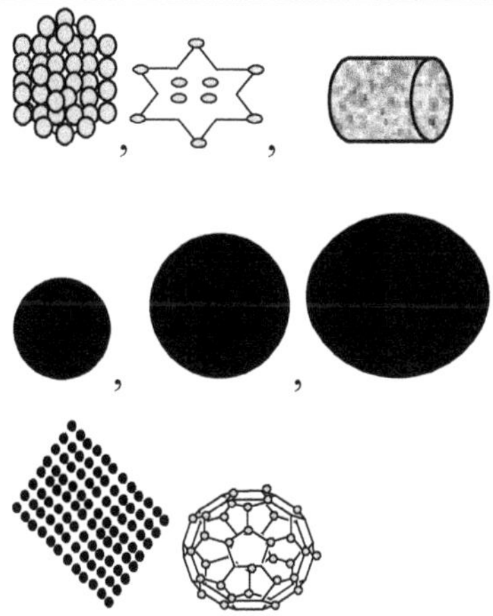
One-dimensional (1D) nanomaterials	carbon-based nanowires, nanotubes and nanoribbons.	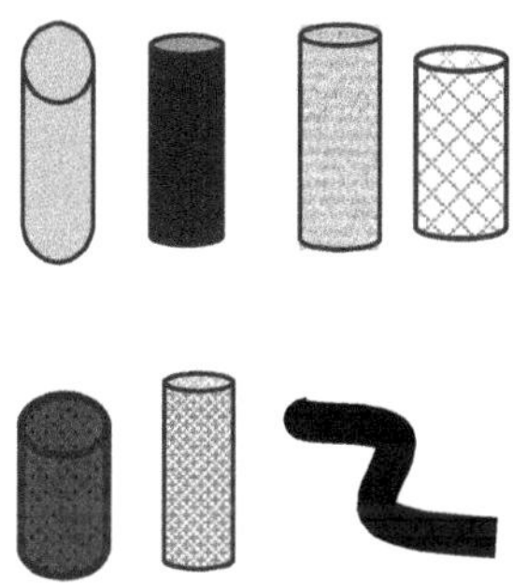
Two-dimensional (2D) nanomaterials	Nanofilms, nanolayers, graphene and its oxide,	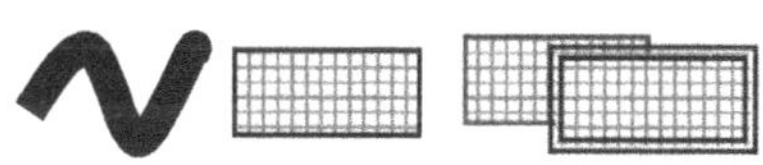
Three dimensional (3D) nanomaterials:	Graphite, polycrystals, diamond, pilared graphen, and aerogels.	

applications [62]. Figure 4.2 illustrates the layer produced on the copper surface by means of electrostatic interaction.

Surface modification in water, which stabilizes these metals by covering oxidation-sensitive defects on the nanofoil, alters their intrinsic conductivity and dispersivity [63]. Copper corrodes in alkaline environments, which affects the efficiency of devices made of copper. Adding corrosion inhibitors to electrolytes prevents copper from corroding while maintaining its high conductivity. The

TABLE 4.2
Different Types of Nanohybrids Based on Colors [48–51]

Types	Domain	Examples
Red nanohybrid	These red-layered nanohybrids are materials with biomedical applications. Red-layered nanohybrids are nano composites based on layered materials. Biotechnology used in drug production and administration techniques of antibiotics, healing or genetic uses, such as nanodrugs for medicals use, tissue engineering artificial joints and nanomedicines used in tissue engineering or artificial joints and sensors or electrodes	Drug–inorganic nanohybrids used in cellular uptake, controlled drug release, and polymer-layered nanohybrids for red application
White nanohybrids	For energy and environmental application. Nanohybrids based on stratified inorganic solids have energy and environmental applications. Recently, stratified inorganic solids and stratified metal oxides have been used as photocatalysts or as electrodes in supercapacitors and lithium-ion batteries, while nanohybrids based on stratified inorganic solids have energy and environmental applications	Components of supercapacitors, rechargeable lithium batteries, photocatalyst-based nanohybrids with polymer or carbon
Green nanohybrids	Agricultural and food sectors	Super-absorbent polymer nanohybrids to prevent desertification and improve the bioavailability
Blue nanohybrids	Aquatic and marine sectors	Remediation of radionuclides in water and the ocean

addition of anticorrosion additives to electronic devices has become a necessity to guarantee their long-term operation. To remedy this problem, researchers have used the most extensively studied transition metal Ti_3C_2Tx (T indicates terminations) as a conductive substrate. The optimal Ti_3C_2Tx@PEI (M@nP, n being the PEI/MXene mass ratio) obtained was used as a nanohybrid inhibitor for copper, and gave high copper protection [64].

4.3.3 Inhibition of Metal Corrosion by Nanohybrid Coatings

Below are cited some examples of nanohybrids used to inhibit the corrosion of metals against aggressive elements, such as:

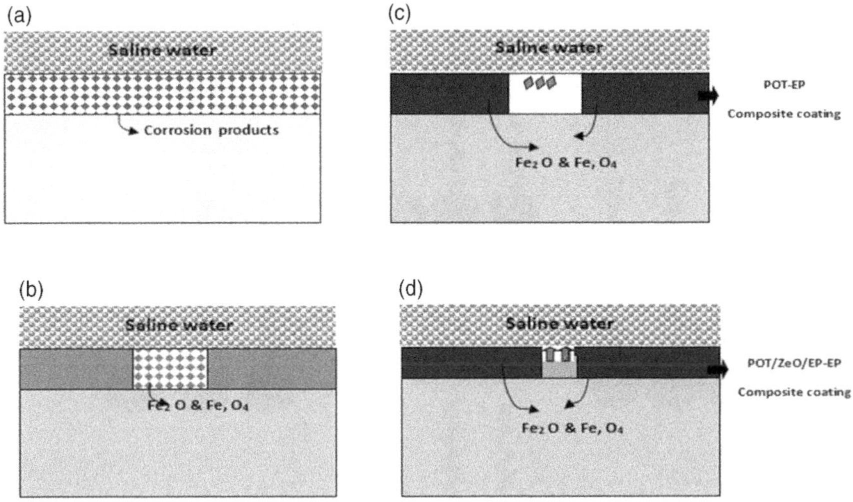

FIGURE 4.1 Nanohybrid inhibition of iron corrosion.

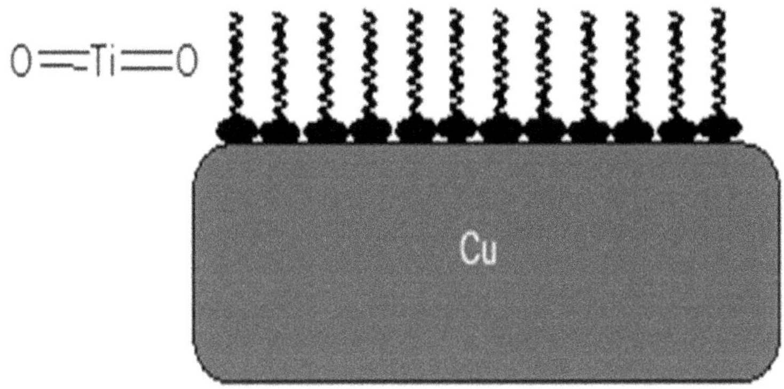

FIGURE 4.2 Fixation of nanohybrid TiO_2 on the copper surface by electrostatic force.

- *Nano-silica (SiO_2)*
  Anticorrosion SiO_2 nanofillers added to coating polymers form a corrosion-resistant nanohybrid [65]. They also serve to improve the host's mechanical strength and hydrophobicity [66] as shown in Figure 4.3. This molecule blocks the entry of corrosive ions by occupying the gaps in the polymer [67]. Researchers have shown that the use of an epoxy resin with SiO_2 fillers inhibited the corrosion of DSS 2205 stainless steel [68]. Other researchers used poly (vinylpyrrolidone)/acrylate/silane + SiO_2 and obtained a corrosion resistance of 83.4% [69].
- *Nano-zinc Oxide (ZnO)*
  ZnO nanoparticles added to polymers increase their hardness, decrease their refractive index, increase their hydrophobicity, and enhance their corrosion

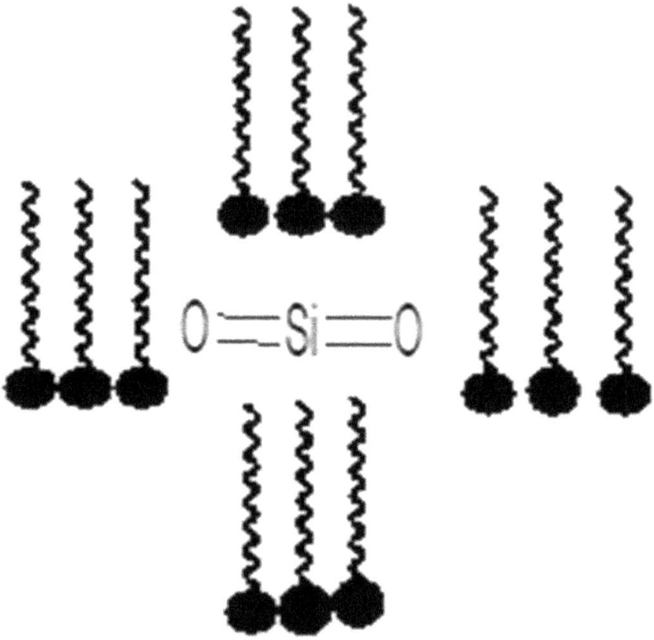

FIGURE 4.3 Product obtained by addition of a nanohybrid to a polymer.

inhibition of metals [70], while one study of zinc nanoxide to metal increased their corrosion protection [71]. Another study interspersed ZnO with epoxypolyamide polymer, resulting in a good metal corrosion-inhibiting nanohybrid [72].

- *Cerium Oxides (CeO_2)*
  CeO_2 is used as a catalyst [73], UV absorber [74], antioxidant [75], and as a coating additive for corrosion protection [76]. One study added CeO_2 to a polymer to form an anticorrosive nanohybrid on various metal substrates [77,78]. Another study used the polyurethane (PU) nanohybrid WCNT-CeO_2 and achieved good corrosion reduction [79].
  Another team studied nanohybrid epoxy resin with CeO_2 and obtained a corrosion inhibition rate of 75% [80].
- *Nano-zirconia (ZrO_2)*
  Research has added ZrO_2 to hydroxyapatite (HA) and TiO_2 to form a nanohybrid with coating performance [81]. The mixture of ZrO_2 and polymer (PMC) forms a nanohybrid resistant to metal corrosion [82], with: high resistance to cracking, wear, and high hardness [82]. A study employing a coating formed from polymeric epoxy resin with ZrO_2 formed a nanohybrid resistant to metal corrosion [82].
- *Nano-iron Oxide (FeO, Fe_2O_3, Fe_3O_4)*
  Iron nano-oxide FeO, Fe_2O_3, Fe_3O_4 has exceptional physical and thermal properties [83]. One study incorporated Fe_3O_4 into a polyaniline polymer

to form a good metal corrosion inhibitor [84]. Another study developed a nanohybrid using an epoxy-acrylate resin and Fe_3O_4, which inhibits corrosion by 98.21% [85].

- *Nano-alumina (Al_2O_3)*
  Nano-alumina (Al_2O_3) has high strength, excellent electrical insulation, and resistance to wear and corrosion [86]. The addition of Al_2O_3 to polymers forms nanocomposites with good properties [87]. Researchers have used the coating obtained from an epoxy resin polymer with an Al_2O_3 filler and achieved good corrosion reduction [88].
- *Nano-titania (TiO_2)*
  TiO_2 has good electrical conductivity and UV adsorption, and therefore it serves as an additive to a coating, increasing its UV resistance, serving as an anticorrosion, anti-microbial, self-cleaning and self-smoothing agent [89]. Studies have shown that the addition of TiO_2 to polymers increases its corrosion resistance [89]. A study on the nanohybrid formed by an epoxy polymer and TiO_2 gave good corrosion inhibition [89].
- *Silicate-based Polymer Nanocomposite Coating*
  Nanohybrid coatings based on silicate polymers, the substitution of a polymer with silicates, reduces water absorption and therefore protects metals against corrosion [90], increases nanohybrid metal and ensures the stacking pattern of the coating, improves the alignment of polyaniline chain [90], and ensures biocompatibility and adhesion [90].

  One study used a silicate-based polymeric nanohybrid coating as an anticorrosion coating and achieved a high inhibition rate [90]. Another study worked with a modified epoxy resin-based coating and silicates and demonstrated increased metal protection [90].

4.4 SYNTHESIZING METHODS FOR SMART COATINGS

Recently synthesized point materials used in different applications such as modern industries must guarantee a high level of protection against corrosion in different corrosive environments and guarantee the protection of the environment and human beings, which has necessitated the creation of new coatings that perform better, last longer, and are less harmful to the environment.

These coatings protect the different metal surfaces, preventing corrosive agents from attacking the material and forming oxides that form a passive layer. As they must be environmentally friendly, they are intelligent, possessing different properties such as self-extinguishing, corrosion resistance, water resistance, abrasion resistance, wear resistance, water resistance, etc. [91]. They must be able to resist various corrosive aggressions. The various techniques and methodologies for synthesizing these intelligent coatings to inhibit various metal surfaces are presented below (Figure 4.4).

- *Chemical Conversion Coatings*
  These chemical transformation films are the products resulting from the conversion of metal layer in its passive stage by production of a mixed film

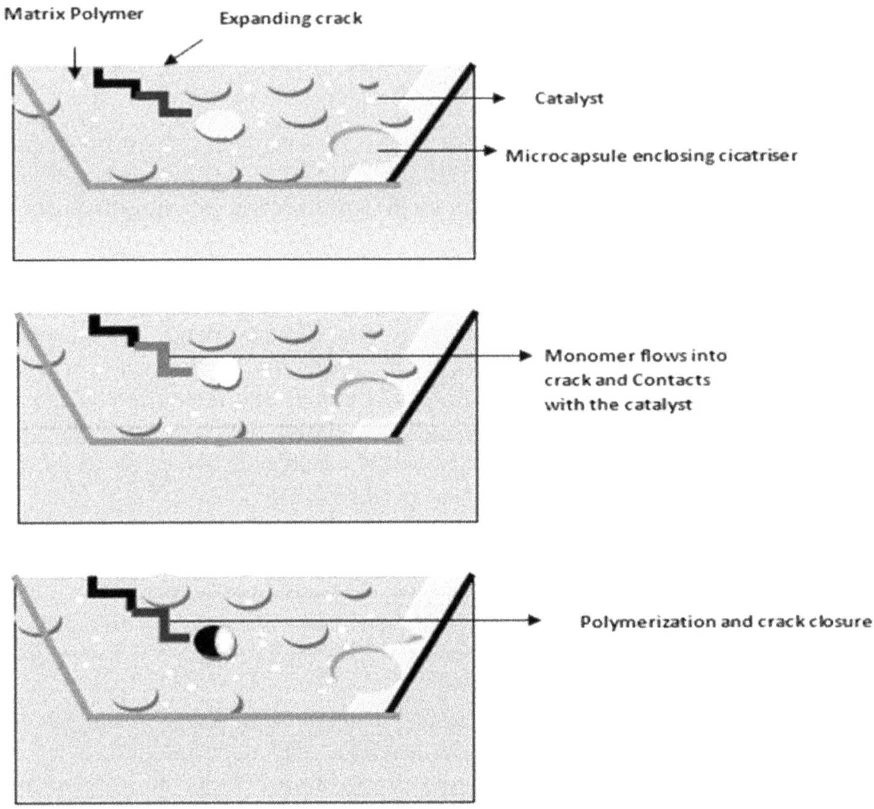

FIGURE 4.4 Diagram represents the self-healing technique based on incorporated microcapsules using embedded microcapsule [91].

improving physicochemical properties by enhancing inhibition, robustness, and longevity. It also helps to develop coloration of the product, making it more marketable, functional, and suited to coating or painting. The characteristics of a high-quality protective conversion coating include high adhesive strength, low cost and high performance, strong mechanical characteristics, environmental protection, industrial application, and uniform distribution on different metal surfaces. An example of the chemical conversion coatings in commercial use is chromate conversion coatings, designed to provide long inhibition for the metals, thereby solving the problem of damage to polymer coatings causing them to corrode. These coatings are considered pioneering intelligent coatings by virtue of their self-healing behavior, simple to install, low cost, high efficiency, and pronounced electrical conductivity, giving them the ability to restore corroded surfaces release chromate from the coating applied in solution. However, because of their carcinogenic effects, these chromates are harmful to the environment, which has restricted the use of their compounds in various applications. As a result, less toxic, less costly, and environmentally friendly coatings have been discovered that are self-healing and capable of protecting the metal surfaces used in various applications.

Other chromate gels have been developed to avoid chromate toxicity, with a capacity for self-healing and capable of protecting the metal surfaces used in various applications [92]. The way in which these salts are preserved is linked to the ability of these coatings to resist aggressive, corrosive agents, by producing protective oxides capable of healing the damage caused to the surface, because of their self-healing function. The precipitation of these soluble forms in a high oxidation state on defective sites on the metal surface inhibits further corrosion. These coatings are non-hazardous to the environment with high protective powers when deposited on the organic topcoat. These coatings inhibit magnesium and aluminum because they have good self-repair properties [93].

- *Nano/microcapsules Formed from Polymeric Compounds*
  Restoring visible damage to coatings can be achieved using a number of complicated, costly, and time-consuming techniques, making them unusable on an industrial scale. Therefore, the production of new intelligent self-healing compounds is necessary to address these invisible cracks. These compounds have a high self-healing and healing property of damage or cracks reacting to changes occurring in the surrounding environment that affect active metal surfaces. Micro/nanocontainers are designed to contain active substances (protectors or polymerizable products) and are then dispersed in the coating layer. Their ability to eliminate the appearance of cracks in coatings is obtained by diffusing the active substances, made up of micro/nanoparticles, from the coating matrix. These are used in many areas, especially in the manufacture of electronic components, molecular biology, medical imaging, photonic materials, coatings, and capsules [94]. The various techniques of micro/nanocapsule synthesis are processed by several compounds as organic coatings composed of polymers, mesoporous silica nanoparticles, etc. Yang et al. carried out research into self-healing coatings based on microencapsulated epoxy resin that gave high corrosion protection and efficiency on steels by adding microcapsules to paints. The rough shape of their envelope strengthens the bond between the epoxy resin and the microcapsule. The disadvantage of this method is the considerable time required to produce capsules filled after with catalyst and epoxy [95]. To manufacture intelligent coatings with self-repairing properties, polyuria formaldehyde microcontainers filled with polymerizable compounds are used, as demonstrated by a research project [96], as shown in Figure 4.5, in which the self-healing mechanisms are stimulated by cracking of incorporated capsules. poly (formaldehyde-urea) (PUF) is considered to be an important polymer microcontainer in the coatings with intelligent properties. These microcapsules are prepared in one or two stages [97]. In the first stage of this process, formaldehyde and urea are polymerized and stabilized in the base product in the form of an emulsion in which urea, resorcinol and NH_4^+ have been in solution by means of a surfactant. Once the emulsion stabilizes, formaldehyde is added until polymerization is complete.

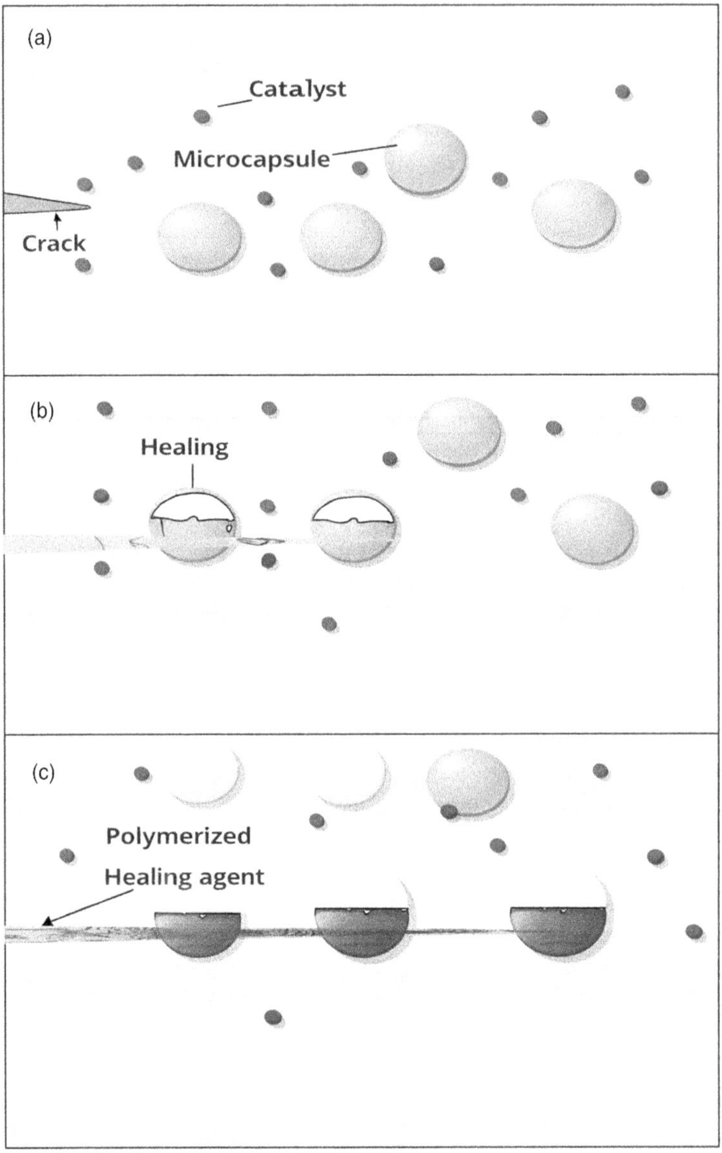

FIGURE 4.5 Process map of autonomic healing concept. (a) Cracks; (b) microcapsules and the healing product; (c) healing agent contacts the catalyst [96].

> The second stage is a pre-polymer which is prepared using one catalyst, mixed with a stabilized emulsion of the active ingredient.

The medium is then acidified using heat, and polycondensation begins rapidly. The two-stage process has also been used to synthesize various nanocontainers.

Initially, filled polystyrene nanospheres were produced by emulsion polymerization followed by spreading of polystyrene gels. This method was used to produce nanocontainers of Fe/Ti oxides, Ce/Ti oxides, Ce/Mo oxides, Si/Ca oxides, polyanilines, and polypyrroles [98]. On the other hand, this technique has its drawbacks: the uneven depth of the manufactured coating and the high heat make this process difficult to apply in industrial manufacturing. The main media for preparing these materials are polymer micro/nanocontainers, which are employed successfully in various inhibition techniques. It is possible to prepare these polymer micro/nanocontainers using several techniques, such as sol–gel, polymerization, etc.

The main technique is in situ oil-water emulsion polymerization, serving for encapsulating dynamic substances, which is characterized by its ease of preparation [99]. It involves dissolving the hydrophilic monomer envelope product in the aqueous sequence and adding the activated agents employed for producing the envelope under the effect of heat. Polymerization is then activated on the surface of the droplets of base product to produce a solid polymeric envelope. Researchers looked at another intelligent inhibitor coating founded on compounds contained in polymeric micronanocontainers obtained in emulsion by interfacial polymerization, made up of polyurea-based micro/nanocontainers containing the inhibitors 8-hydroxyquinoline and 2-methyl-benzothiazol [100]. Studies have also been carried out on PU as a polymer by interfacial polymerization after dissolving products containing several isocyanates in an emulsion so that they interact with the polyalcohols soluble in the immiscible part so that encapsulating the main product takes place in situ simultaneously as formation of the polymer containers. Numerous PU-based microcapsules containing water-based hexamethylene diisocyanate (HDI) have been studied [101] and are intended for self-repair without catalyst. Most corrosion control methods are based on the inhibition of metal surfaces and the formation of a new physical barrier with separating properties using sealing agents or corrosion inhibitors in the crack zone. The synergy between coating passivation effects formation and its water-repellent properties has made it possible to protect metal surfaces [102]. By exposing the coating to a very humid environment, it is damaged and releases the alkoxysilanes which establish electrostatic bonds with the hydroxyls of the metal's outer layer, rendering it electrochemically inactive, as indicated in Figure 4.6. Samples composed of PU/PUF bilayers were also studied as microcontainers for self-healing inhibitor coatings. These products were found to have excellent properties compared with a single layer of PUF or PU. The external PUF coating employed in this scheme provides the desired robustness and the internal PU coating delivers the flexibility needed to function correctly.

Another polymer container is polystyrene filled with benzotriazole enveloped during polymerization used as a protector. A layer of polyethyleneimine was also added during polymerization to control the release of Benzotriazole (BTA) and prevent leakage. This method offers several advantages such as low cost, ease of fabrication, and industrial application, but its disadvantage is its poor compatibility with inorganic films [103].

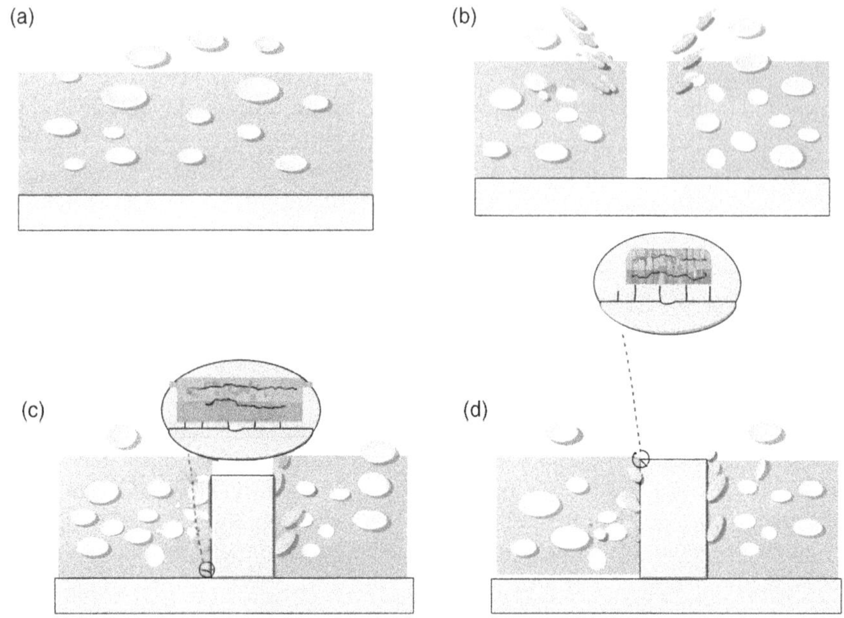

FIGURE 4.6 Self-healing mechanism scheme: (a) self-healing coating consisting of microcontainers product, (b) deterioration of the film; encapsulated substance on the product surface, (c) interaction alkoxysilanes/H_2O, (d) damaged surface is passive and production of hydrophobic layer of silanes [103].

4.5 NANOCONTAINERS WITH LAYER-BY-LAYER ASSEMBLY

The main method used for layer-by-layer (LBL) production is to produce nanocontainers that release the corrosion inhibitor they contain in a controlled manner.

This mechanism relies on the deployment of electrostatic coats composed of nanoparticles on the face of the material to produce a film that controls the spread and penetration of protector depending on environmental agents and has various functions, Electrostatic nanoparticle coating involves charging particles to create a protective, often self-healing, film on a substrate, allowing for controlled release of corrosion inhibitors or functional agents. This mechanism uses pH-responsive, smart nanocarriers—frequently applied via electrophoretic deposition—to release inhibitors in response to localized corrosion, managing both anodic and cathodicElectrostatic nanoparticle coating involves charging particles to create a protective, often self-healing, film on a substrate, allowing for controlled release of corrosion inhibitors or functional agents. This mechanism uses pH-responsive, smart nanocarriers—frequently applied via electrophoretic deposition—to release inhibitors in response to localized corrosion, managing both anodic and cathodic. When pH modification is used to control protector release in the anodic and cathodic parts. Research has revealed a new method for producing a multilayer inhibitor film, involving pretreatment of the aluminum surface by sonication, followed by pouring in the inhibitor

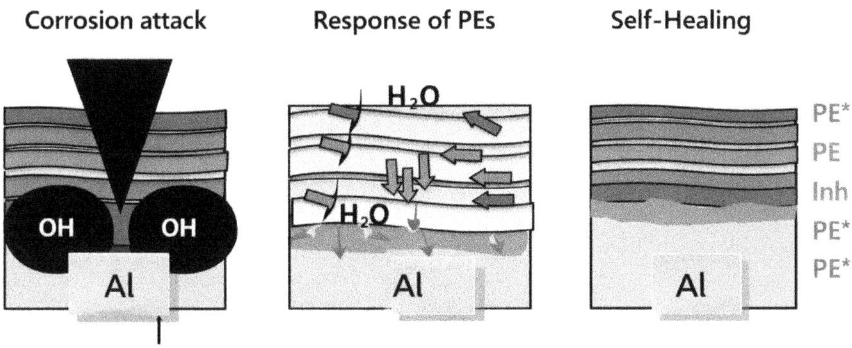

FIGURE 4.7 Diagram of the 'intelligent' polyelectrolyte layer [105].

substances and polyelectrolytes [104]. The coatings produced have shown good adhesion and wettability due to the high electrostatic forces established between the layers and the metal, forming a homogeneous polymer spread over the aluminum surface. This coating provided high corrosion resistance for a period of 3 weeks. The advantages of this coating are that it is economical, easy to prepare, and not harmful to the environment. It releases the inhibitors by varying the pH in a controllable way due to the reactions undergone by the polyelectrolyte molecules. Unfortunately, this technique forms shells with low mechanical stability and requires complex preparation methods for its application on an industrial scale. Figure 4.7 shows the mechanism by which an anticorrosion protective film composed of 'intelligent' polyelectrolytes self-heals [105]. Capsules are formed by depositing LBL on colloidal materials and then decomposing the catalytic center, to create hollow structures of specific shape and size similar to those of the models represented in Figure 4.8.

4.6 SHAPE RECOVERY POLYMER COATING

Memorization of structure polymers is one of the new approaches to wound healing. These polymers can take on a new shape after being deformed and return to their original form under the effect of an external risk factor, such as light or heat. They are employed in the textile sector, in the manufacture of electronic devices and packaging, and in the medical field. In their protective mechanism, these techniques rely on a chemical transformation likely to produce undesirable reversible transformations that reduce healing capacity [106].

They are thermally induced by the shape recovery process, founded on the thermal diffusion from the zones corresponding to the solid to liquid phase change temperature of the segments [107]. When the medium temperature is above T_{trans}, the switching zones become flexible and the polymer network becomes elastic above T_{trans}. If the film is damaged, physical healing takes place by raising the temperature, causing the switching segment of the coating to relax 107]. Lutz et al. [108] looked at shape-recovery polymer coatings with self-healing properties. They looked at two new self-healing UV-cured PU acrylate-polycaprolactone coatings. These products revealed

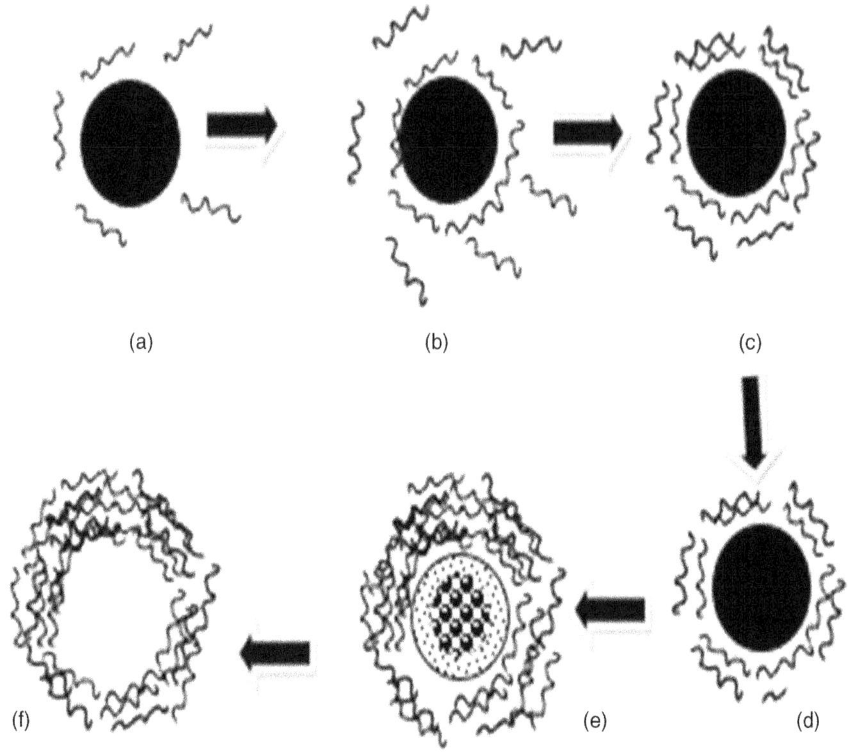

FIGURE 4.8 Schematic diagram of capsule formation. (a–d) Assembly process; (e,f) breakdown of core forming hollow capsules.

excellent mechanical properties and a high Young's ratio. Following thermal activation, 97% of the deformed films regained their form. Because of the ability of PCL cross-linking system to regain its shape, it was possible to close the scratch before the linear PCL chains repaired themselves on the surface crack, which is known as shape–memory-assisted self-repair [109]. Together with Diels–Alder chemistry, this technique has made possible the reversible transposition of thermally stable shape-memorizing PUs [110]. In line with research by D'Hollander et al., polymer segments based on PCL (AB)n have excellent shape-memory characteristics [111,112].

4.7 CARBON AND CLAY NANOTUBE FILMS

Researchers have shown that paint or carbon nanotube (CNT) pitting can give rise to intelligent multilayer self-healing coatings, because of a p-–n junction that reacts electrically to corrosion damage [113]. A team of researchers investigated the application of CNTs for healing as reservoirs by conducting a molecular fluid dynamics study that simulated the secretion of a curative product from the damaged CNT. It concluded that crack propagation was slowed by a CNT nanocontainer as soon as the crack began to penetrate the matrix, causing the liberation of the restorative product outside the face of the damaged CNT and the final closure of the product. According to these molecular simulations, CNTs are used as consolidation and self-repair containers to repair cracks by themselves, restrict the damage caused by the matrix, and consolidate the film before and after repair. The toxicity and high cost of CNTs restrict their use [114].

Halloysite nanotubes (HNTs) are also widely used in the preparation of two-layer aluminosilicate clay minerals because they are cheaper and more abundant, with lengths from 1 to 15 μm and diameters from 10 to 150 nm [115]. They are characterized by their high dissolution in epoxy without the need for surface treatment, and they also serve as epoxy toughness modifiers [115]. Abdullayev et al. have succeeded in obtaining these products encapsulated with inhibitors and biocides whose contents are released in a controlled manner and they can play the role of nanoscale containers for coatings [116].

Other researchers incorporated benzotriazole into these nanotubes to prevent corrosion, then introduced these nanotubes into a ZrO_2–SiO_2 sol–gel coating and used the aluminum alloy AA2024, obtaining excellent inhibition efficiency and great self-healing power [117]. In one investigation, it was deduced that the CNT nanocontainer slows crack propagation when the crack initiates through the matrix by diffusion of the healing compound outside the face of the damaged CNT, leading to the sealing of the CNT, as shown in Figure 4.9 [118].

Finally, it can be said that halloysite clay nanotubes have various advantages for use as a self- coating, including their bass cost and mechanical robustness. On the other hand, there are various drawbacks to their industrial application as a protective coating, such as the difficulty of controlling the formation of plugs on the surface

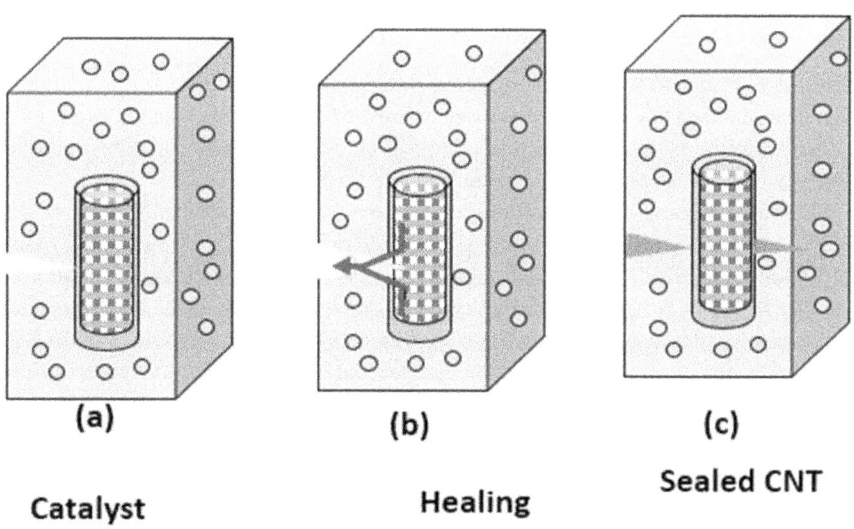

FIGURE 4.9 Scheme describing process using carbon nanotubes in concept of the self-healing [118].

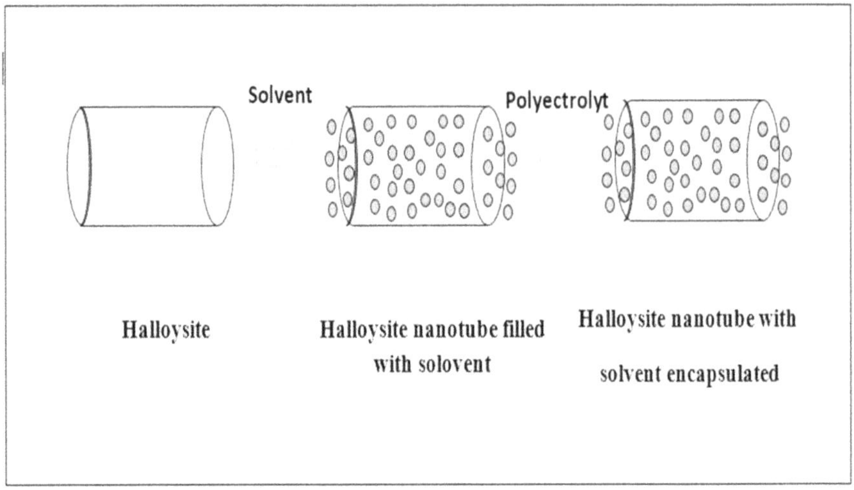

FIGURE 4.10 Scheme describing process the encapsulation of solvent into HNTs [120].

of the part [119]. The work of Melo et al. [120] concerns the solvent encapsulation in light of halloysite and the possibility of preparing self-healing epoxy protective films. In this technique, Dimethyl sulfoxide was inserted into the HNTs by vacuum infiltration, followed by additional encapsulation with polyelectrolytes to plug the ends of the HNTs, as shown in Figure 4.10. This method effectively uses the solvent containing the HNTs in the epoxy to provide self-healing protection.

4.8 INNOVATIVE TECHNIQUES FOR CORROSION INHIBITION BASED ON SMART ORGANIC FILMS

Consumers prefer to use organic coatings based on nanocontainers more than traditional products. They are looking to use multifunctional nanocontainers that have a load capacity, power to ensure the tolerance between the nano-containers and the matrix as well as the optimizing products inside the film [121].

These metal corrosion protectors create a covering., especially capsules that can react to pH variations. To inhibit pitting corrosion in the steel, it was necessary to use pH-sensitive microcapsules containing compounds in the epoxy coating for protection of RCl [121]. A research laboratory [121] demonstrated the corrosive resistance of triethanolamine coated on defective coated steel in simulated concrete pore (SCCP) solutions. Chen et al. [122] have demonstrated the coating power of microcapsules to improve the durability of reinforced concrete. However, other research has shown the corrosion-inhibiting power of layered double hydroxide (LDH)-based smart products. Mei et al. [123] manufactured epoxy film mixed with seawater Mg-Al LDH intercalated with aminobenzoate, and inhibition of the steel by epoxy film was proved in the saturated calcium hydroxide and simulated concrete pore solution. Research has shown the inhibitory effect of Ca-LDH cement film on steel [123]. HNT-based com2 films were prepared by a research laboratory [124] using the method of LBL self-assembly. Data collected showed that their corrosion resistance to steel in a polyaniline–SiO_2 composite (PSC) solution. Yang et al. [125] demonstrated the inhibitory power of nanoreactors able to control corrosion by regulated releasing or adsorbing protons directly at the site of corrosion. A nanoreactor consists of a nano-compartment surrounded by a membrane that reacts with pH, wrapped in graphene/polyamine oxide. To demonstrate the protective power of the film, researchers [126,127] added an inorganic/organic hybrid protector (benzimidazolium (BI)/prideodymium) to the sol–gel silane film to obtain an active corrosion inhibition action.

4.9 CONCLUSION

In this work, we have defined metal corrosion smart coating inhibitors in industry with their different corrosion inhibitors for different metals. We have also defined the different corrosion protectors that can inhibit metal corrosion in industry. Smart coating nanohybrids have good mechanical, thermal, and electrical properties. Their use as corrosion inhibitors for various metals gives good inhibition rates. The use of these inhibitors ensures uniform coating distribution. This ensures good adhesion to the surface and reduces the diffusion of corrosive products. Therefore, the advantage of using these coatings as metal corrosion inhibitors is that they can impart special properties to the coating. However, there are disadvantages in their use that need to be overcome, such as production cost, which is higher than that of other inhibitors. One way of prolonging the life of various structural and industrial metals would be to manufacture new self-healing protective coatings capable of repairing scratches and eliminating corrosion. Intelligent coatings are important because they can react to stimulus and trigger the healing process. These intelligent coatings are therefore becoming very interesting materials for inhibition of metal corrosion, particularly in hard-to-reach areas. For the most part, they are intelligent, designed as an environmentally friendly alternative to toxic chromate films. Self-healing process damages surface metals on a scale of micro to millimeter, making them commercially unattractive for industrial applications. Some of these intelligent coatings are produced using expensive products or are obtained in complicated steps, making them commercially unviable [119].

REFERENCES

1. ISO 8044:2024 – Corrosion of metals and alloys – Vocabulary, standard Edition 6, 2024.
2. Barbara Shaw and Robert Kelly, What is corrosion? *The Electrochemical Society Interface*, 2006, 15 (1), pages: 24–26, Doi: 10.1149/2.F06061IF.
3. Francisco Presuel-Moreno, Marton A. Jakab, Nicolas Tailleart, Michael Goldman and John R. Scully, Corrosion-resistant metallic coatings, *Materials Today*, 2008, 11, pages: 14–23, Doi: 10.1016/S1369-7021(08)70203-7.
4. Tianci Chang, Xun Cao, Ning Li, Shiwei Long, Xiang Gao, Live R. Dedon, Guangyao Sun, Hongjie Luo and Ping Jin, Facile and Low-Temperature Fabrication of Thermochromic Cr2O3/ VO2 Smart Coatings: Enhanced Solar Modulation Ability,

High Luminous Transmittance and UV-Shielding Function, *ACS Applied Materials and Interfaces*, 9 (31), 2017, pages: 26029–26037, Doi: 10.1021/acsami.7b07137.
5. Liana Maria Muresan, Nanocomposite coatings for anti-corrosion properties of metallic substrates, *Materials*, 2023, 16, pages: 1–21, Doi: 10.3390/ma16145092.
6. Ahmad Mehdi, Catherine Reye and Robert Corriu, From molecular chemistry to hybrid nanomaterials. Design and functionalization, *Chemical Society Reviews*, 2011, 40, pages: 563–574, Doi: 10.1039/b920516k.
7. Margarita Darder, Pilar Aranda and Eduardo Ruiz-Hitzky, Bionanocomposites: A new concept of ecological, bioinspired, and functional hybrid materials, *Advances in Materials*, 2007, 19, pages: 1309–1319, Doi: 10.1002/adma.200602328.
8. Eduardo Ruiz-Hitzky, Margarita Darder, Bernd Wicklein, Fidel Antonio Castro-Smirnov and Pilar Aranda, Clay-based biohybrid materials for biomedical and pharmaceutical applications, *Clays and Clay Minerals*, 67, 2019, pages: 44–58, Doi: 10.1007/s42860-019-0005-0.
9. Biao Wen, Yang Li, Congcong Liang, Yanli Chen, Yunfeng Zhao, Qiang Wang, Recent progress on porous carbons for carbon capture, *Langmuir*, 2024, 40, pages: 8327–8351, Doi: 10.1021/acs.langmuir.3c03876.
10. Tanmay Tiwari, Gun Anit Kaur, Pravin Kumar Singh, Sapna Balayan, Anshuman Mishra, Ashutosh Tiwari, Review emerging bio-capture strategies for greenhouse gas reduction: Navigating challenges towards carbon neutrality, *Science of the Total Environment*, 2024, 929, pages: 1–27, Doi: 10.1016/j.scitotenv.2024.172433.
11. Piotr Mendrela, Wojciech Stanek and Tomasz Simla, Thermo-ecological cost – System evaluation of energy-ecological efficiency of hydrogen production from renewable and non-renewable energy resources, *International Journal of Hydrogen Energy*, 2024, 50, pages: 1–14, Doi: 10.1016/j.ijhydene.2023.06.150.
12. Tapas Kuillaa, Sambhu Bhadra, Dahu Yao, Nam Hoon Kim, Saswata Bose and Joong Hee Lee, Recent advances in graphene based polymer composites, *Progress in Polymer Science*, 35, 2010, pages: 1350–13755, Doi: 10.1016/j.progpolymsci.2010.07.005.
13. Wenwen Zhang, Ke Zheng, Jing Ren and Yimin Fan, Strong- ductile and lightweight bionanocomposites constructed by bioinspired hierarchical assembly. *Composites Communications*, 17, 2020, pages: 97–103, Doi: 10.1016/j.coco.2019.11.017.
14. Blessy Joseph, Saravanan Krishnan, V. K. Sagarika, Abhimanyu Tharayi, Nandakumar Kalarikkal and Sabu Thomas, Bionanocomposites as industrial materials, current and future perspectives: A review. *Emergent Materials*, 2020, pages: 1–13, Doi: 10.1007/s42247-020-00133-x.
15. Siti Maznah Kabeb and Nurul Huda Abu Bakar, Polymer blend nanocomposites for polymer electrolyte membrane fuel cell (PEMFC) applications, *Micro and Nano Technologies*, 2023, pages: 479–493, Doi: 10.1016/B978-0-323-99549-8.00002-9.
16. Disha Mishra, Pardeep Singh, Versha Pandey, Anisha Yadav and Puja Khare, The emerging role of nanotechnology in agri-food sector: Recent trends and opportunities, smart nanoarchitectures, technologies, *Challenges, and Applications Micro and Nano Technologies*, 2024, pages: 1–19, Doi: 10.1016/B978-0-323-99682-2.00014-1.
17. Nishu Sudesh Kumar, Smart and innovative nanotechnology applications for water purification, *Hybrid Advances*, 2023, 3, Doi: 10.1016/j.hybadv.2023.100044.
18. Roger Sanchis-Gual, Marc Coronado-Puchau, Talal Mallah and Eugenio Coronado, Review hybrid nanostructures based on gold nanoparticles and functional coordination polymers: Chemistry, physics and applications in biomedicine, catalysis and magnetism, *Coordination Chemistry Reviews*, 2023, 480, pages: 1–23, Doi: 10.1016/j.ccr.2023.215025.

19. Vera Regina Leopoldo Constantino, Mariana Pires Figueiredo, Vagner Roberto Magri, Denise Eulálio, Vanessa Roberta Rodrigues Cunha, Ana Clecia Santos Alcântara and Gustavo Frigi Perotti. Review biomaterials based on organic polymers and layered double hydroxides nanocomposites: Drug delivery and tissue engineering, *Pharmaceutics*, 2023, 15, pages: 1–54, Doi: 10.3390/pharmaceutics1502041.
20. Pedro Gomez-Romero, Anukriti Pokhriyal, Daniel Rueda-García, Leandro N. Bengoa and Rosa M. González-Gi, Hybrid materials: A metareview, *Chemistry of Materials*, 2024, 36, pages: 8–27, Doi: 10.1021/acs.chemmater.3c01878.
21. Mariam Odetallah and Christian Kuss, A review of chemically induced intercalation and deintercalation in battery materials, *Energy Technology*, 2023, 11, pages: 1–23, Doi: 10.1002/ente.202201060.
22. Minwang Laipana, Lichen Xiange, Jingfang Yua, Benjamin R. Martine, Runliang Zhu, Jianxi Zhu, Hongping He, Abraham Clearfield and Luyi Sun, Layered intercalation compounds: Mechanisms, new methodologies, and advanced applications, *Progress in Materials Science*, 2020, 109, pages: 1–37, Doi: 10.1016/j.pmatsci.2019.100631.
23. Jin-Ho Choy, Soon-Jae Kwon and Gyeong-Su Park, High-T_c parsuperconductors in the two dimensional limit: $[(Py\text{-}C_nH_{2n+1})_2HgI_4]\text{-}Bi_2Sr_2Ca_{m-1}Cu_mO_y$ ($m = 1$ and 2), *Science*, 1998, 280, pages: 1589–1592, Doi: 10.1126/science.280.5369.1589.
24. Minoru Osada, Emerging ceramic nanosheets: From tailored synthesis to electronic applications, *Journal of the Ceramic Society of Japan*, 2023, 131, pages: 851–862, Doi: 10.2109/jcersj2.2312.
25. Muhammad Tawalbeh, Hafsah Adeel Khan and Amani Al-Othman, Review: Insights on the applications of metal oxide nanosheets in energy storage systems, *Journal of Energy Storage*, 2023, 60, pages: 1–19, Doi: 10.1016/j.est.2023.106656.
26. Jyoti Bhattacharjee and Subhasis Roy, Exploration of molybdenum oxide compounds-A review, *Catalysis Research*, 2024, 4, pages: 1–28, Doi: 10.21926/cr.2403011.
27. Ajit Pal Singh, Rahul Saxena, Suyash Saxena and Neelesh Kumar Maurya, The future of protection: Unleashing the power of nanotech against corrosion, *Asian Journal of Current Research*, 2024, 9, pages: 23–44, Doi: 10.56557/ajocr/2024/v9i38727.
28. Lewis Christopher Yule, Cameron Liam Bentley, Geoff West, Barbara Anne Shollock and Patrick Richard Unwinn, Scanning electrochemical cell microscopy: A versatile method for highly localised corrosion related measurements on metal surfacesn, *Electrochimica Acta*, 2019, 298, pages: 80–88, Doi: 10.1016/j.electacta.2018.12.054.
29. Aditya Venkatraman, Ryan Michael Katona, Demitri Maestas, Matthew Roop, Philip Noell and David Montes de Oca Zapiain, An active learning framework for the rapid assessment of galvanic corrosion, *Materials Degradation*, 2024, 8, pages: 1–11, Doi: 10.1038/s41529-024-00476-4.
30. Jeramy B. Decker, Kyle M. Rollins and Jared C. Ellsworth, Corrosion effects on the mechanical properties of spun pile materials, *Applied Sciences*, 2023, 13, pages: 1–16, Doi: 10.3390/app13031507.
31. Yingwei Liu and Xinghao Li, Effect of oxygen differential aeration on iron corrosion mechanism, *International Journal of Electrochemical Science*, 2021, 16, pages: 1–14, Doi: 10.20964/2021.08.31.
32. Pratyush Mishra, Denizhan Yavas, Abdullah Alshehri, Pranav Shrotriya, Ashraf Bastawros, Kurt R. Hebert, Model of vacancy diffusion-assisted intergranular corrosion in low-alloy steel, *Acta Materialia*, 2021, 220, pages: 1–12, Doi: 10.1016/j.actamat.2021.117348.
33. Jiaxiu Hu, Lianmin Zhang, Aili Ma, Pingli Mao and Yugui Zheng, Effect of cavitation intensity on the cavitation erosion behavior of 316L stainless steel in 3.5 wt.% NaCl solution, *Metals Journal*, 2022, 12, pages: 1–12, Doi: 10.3390/met12020198.

34. Majid Laleh, Anthony E. Hughes, Anthony E. Hughes, Ian Gibson and Mike Y. Tan, A critical review of the corrosion characteristics of additively manufactured stainless steels, *International Materials Reviews*, 2022, 66, pages: 1–37, Doi: 10.1080/09506608.2020.1855381.
35. Leila Fathyunes and M. A. Mohtadi-Bonab, A review on the corrosion and fatigue failure of gas turbines, *Metals*, 2023, 13, pages: 1–25, Doi: 10.3390/met13040701.
36. Sandeep Kumar Dwivedi and Manish Vishwakarma, Hydrogen fracture in different materials: A review, *International Journal of Hydrogen Energy*, 2018, 43, pages: 21603–21616, Doi: 10.1016/j.ijhydene.2018.09.201.
37. Harshal Mehta, Gurpreet Kaur, Ganga Ram Chaudhary and Uma Batra, Evaluation of biocorrosion and antifouling properties of gold metallosurfactant monolayer on galvanised steel in simulated marine environments inoculated with halophiles, *Corrosion Science journal*, 2021, 179, pages: 1–16, Doi: 10.1016/j.corsci.2020.109102.
38. Sebastián Feliu, Electrochemical impedance spectroscopy for the measurement of the corrosion rate of magnesium alloys: Brief review and challenges, *Metals Journal*, 2020, 10, pages: 775–797, Doi: 10.3390/met10060775.
39. Francesco Fanicchia and Sigrún Nanna Karlsdottir, Research and development on coatings and paints for geothermal environments: A review, *Advanced Materials Technologies*, 2023, 8, pages: 1–54, Doi: 10.1002/admt.202202031.
40. Khalid Bouiti, Nabil Lahrache, Ichraq Bouhouche, Najoua Labjar, Ghita Amine Benabdallah and Souad El Hajjaji, Phytochemicals/plant extracts as corrosion inhibitors for different metals/alloys in phosphoric acid, *Phytochemistry in Corrosion Science Book*, 2024, 1st Edition, pages: 208–229, CRC Press. Doi: 10.1201/9781003394631-1.
41. Khalid Bouiti, Hefdh aldeen Al-sharabi, Meryem Bensemlali, Fatima Bouhlal, Brahim Abidi, Najoua Labjar, Said Laasri and Souad El Hajjaji, Effect of temperature on corrosion inhibition by ethanolic extract of *Eriobotrya japonica* seeds in chloride medium 1M, *The European Physical Journal Applied Physics*, 2022, 97, pages: 1–11, Doi: 10.1051/epjap/2022220070.
42. Khalid Bouiti, Hefdh Aldeen Al-sharabi, Fatima Bouhlal, Brahim Abidi, Najoua Labjar, Meryem Bensemlali and Souad El Hajjaji. Response surface methodology for optimizing corrosion inhibition: Investigating the synergistic effect of *Eriobotrya japonica* extract and potassium iodide, *Euro-Mediterranean Journal for Environmental Integration*, 2024, 9, pages: 469–481, Doi: 10.1007/s41207-023-00457-0.
43. Youssra El Hamdouni, Fatima Bouhlal, Houssam Kouri, Mohammed Chellouli, Mohammed Benmessaoud, Abdelouahed Dahrouch, Najoua Labjar and Souad El Hajjaji, Use of omeprazole as inhibitor for C38 steel corrosion in 1.0 M H3PO4 medium, *Journal of Failure Analysis and Prevention*, 20, pages: 1–9, Doi: 10.1007/s11668-020-00862-5.
44. Kouri Hossam, Fatima Bouhlal, Lina Hermouche, Imane Merimi, Houda Labjar, Abdelkarim Chaouiki, Najoua Labjar, Serghini-Idrissi Malika, Abdelouahed Dahrouch, Mohamed Chellouli, Belkheir Hammouti and Souad El Hajjaji, Understanding corrosion inhibition of C38 steel in HCl Media by Omeprazole: Insights for experimental and computational studies, *Journal of Failure Analysis and Prevention*, 2020, 1, pages: 213–227, Doi: 10.1007/s11668-020-01042-1.
45. Fatima Bouhlal, Najoua Labjar, Farah Abdoun, Aimad Mazkour, Malika Serghini-Idrissi, Mohammed El Mahi, El Mostapha Lotfi, Abdelmajid Skalli and Souad El Hajjaji, Chemical and electrochemical studies of the inhibition performance of hydro-alcoholic extract of used coffee grounds (HECG) for the corrosion of C38

steel in 1M hydrochloric acid, *Egyptian Journal of Petroleum*, 2020, 29, pages: 45–52, Doi: 10.1016/j.ejpe.2019.10.003.
46. Aimad Mazkour, Souad El Hajjaji, Najoua Labjar, El Mostapha Lotfi, Mohammed El Mahi, Corrosion inhibition effect of 5-azidomethyl-8- hydroxyquinoline on AISI 321 stainless steel in phosphoric acid solution, *International Journal of Electrochemical Science*, 2021, 16, pages: 1–25, Doi: 10.20964/2021.03.03.
47. Hefdh Aldeen Al-sharabi, Khalid Bouiti, Fatima Bouhlal, Najoua Labjar, Abdel Wahed Dahrouch, Mohammed El Mahi, El Mostapha Lotfi, Bouchaib El Otmani, Ghita Amine Benabdellah and Souad El Hajjaj, *Citrullus colocynthis* ethanolic extract as an ecological inhibitor of carbon steel C38 corrosion in hydrochloric medium, *Journal of Bio- and Tribo-Corrosion*, 2022, pages: 1–11, Doi: 10.1007/s40735-022-00675-z.
48. Nabil Lahrache, Abdelouahed Dahrouch, Khalid Bouiti, Najoua Labjar and Souad El Hajjaji, Corrosion prevention: Present and past advances and future direction, *Carbon Allotropes*, 2023, pages: 1–11, CRC Press. Doi: 10.1201/9781003323976-3.
49. Mohamed Dalimi, Najoua Labjar, Ghita Amine Benabdallah, Mohamed El Yadri,and Souad El Hajjaji, Characterization methods for nanomaterials, *Innovations in Nanomaterials-Based Corrosion Inhibitors*, IGI Global Scientific Publishing, 2024, pages: 1–33, Doi: 10.4018/979-8-3693-3088-3.ch003.
50. Anjali Peter and Sanjay Kumar Sharma, *Corrosion Protection of Metals and Alloys Using Graphene and Biopolymer Based Nanocomposites Book*, CRC Press, 2021, 1st Edition, pages: 1–11, Doi: 10.1201/9781315171364.
51. Najoua Labjar and Souad El Hajjaji, *Functionalized Nanocomposites as Corrosion Inhibitors*, American Chemical Society Publications, 2022, 1418, pages: 141–154, Doi: 10.1021/bk-2022-1418.ch006.
52. Bin Wang, Yucheng Zhou, Dan Yang, Liujie Xu Xiuqing Li, Shizhong Wei, Effect of Al_2O_3 particles on the corrosion behavior of molybdenum alloys in molten zin, *Corrosion Science*, 2023, 220, pages: 1–15, Doi: 10.1016/j.corsci.2023.111266.
53. N. Wint, S. L. Wijesinghe, Wen Yan, Wai K. Ong, Lin Yin Wu, Gregory Williams and Hugh N. McMurray, The sacrificial protection of steel by zinc-containing sol-gel coatings, *Journal of The Electrochemical Society*, 2019, 166, pages: 434–444, Doi: 10.1149/2.1071913jes.
54. Lara Velasco Davoise, Ana M. Díez-Pascual and Rafael Peña Capilla, Application of graphene-related materials in organic solar cells, *Materials*, 2022, 15, pages: 1–33, Doi: 10.3390/ma15031171.
55. Iman Khosravi, Zahra Shahryari, Seyed Mohammadali Jazaeri Moghadas, Hamidreza Torbati Sarraf and Mahdi Yeganeh, The application of graphene oxide as corrosion barrier, *Corrosion Protection at the Nanoscale Micro and Nano Technologies*, 2020, Chapter 8, Micro and Nano Technologies, pages: 127–140, Doi: 10.1016/B978-0-12-819359-4.00008-8.
56. Gan Cui, Zhenxiao Bi, Ruiyu Zhang, Jianguo Liu, Xin Yu and Zili Li, Review A comprehensive review on graphene-based anti-corrosive coatings, *Chemical Engineering Journal*, 2019, 373, pages: 104–121, Doi: 10.1016/j.cej.2019.05.034.
57. Xuemei Li, Yuyang Long, Limin Ma, Jidong Li, Jun Yin and Wanlin Guo, Coating performance of hexagonal boron nitride and graphene layers, *2D Materials*, 2021, 8, pages: 1–18, Doi: 10.1088/2053-1583/abe777.
58. Mir Mahdi Karimi Azar, Hamed Shooshtari Gugtapeh and Milad Rezaei, Evaluation of corrosion protection performance of electroplated zinc and zinc-graphene oxide nanocomposite coatings in air saturated 3.5 wt. % NaCl solution, *Colloids and Surfaces A*, 2020, 601, pages: 1–11, Doi: 10.1016/j.colsurfa.2020.125051.

59. Yuhui Xie, Mingzhi Chen, Delong Xie, Li Zhong and Xinya Zhang, A fast, low temperature zinc phosphate coating on steel accelerated by graphene oxide, *Corrosion Science*, 2017, 128, pages: 1–8, doi: 10.1016/j.corsci.2017.08.033.
60. Chen Hu, Yujun Yang, Yuqing Lin, Linlin Wang, Rongyang Ma, Yanli Zhang, Xiaoli Feng, Junrong Wu, Lili Chen, Longquan Shao, GO-based antibacterial composites: Application and design strategies, *Advanced Drug Delivery Reviews*; 2021, 178, pages: 1–31, Doi: 10.1016/j.addr.2021.113967.
61. Omid Razavizadeha and Mohammad Ghorbani, Surface modification of carbon steel by ZnO-graphene nano-hybrid thin film, *Surface and Coatings Technology*, 2019, 363, pages: 1–11, Doi: 10.1016/j.surfcoat.2019.01.094.
62. Asma Mariam Elsharif, Ismail Abdulazeez, Monerah Amina Almarzooq and Shamsuddeen Abubakar Haladu, Synthesis, and experimental evaluation of novel 4-(-3-(2-hydroxyethoxy)-3-oxopropenyl)-1,2-phenylene nanohybrid derivatives as potential corrosion inhibitors for mild steel in 1 M HCl, *Journal of Industrial and Engineering Chemistry*, 2022, 116, pages: 474–488, Doi: 10.1016/j.jiec.2022.09.037.
63. Muhammad Nazim Lakhana, Abdul Hananb Yuan Wangc, Hiang Kwee Leed and Hamidreza Arandiyan, Integrated MXenes and metal oxide electrocatalysts for oxygen evolution reaction: Synthesis, mechanisms, and advances, *Chemical Science*, 2024, pages: 1–42, Doi: 10.1039/D4SC04141K.
64. Mingjun Cui, Siming Ren, Jia Chen, Shuan Liu, Guangan Zhang, Haichao Zhao, Liping Wang and Qunji Xue, Anticorrosive performance of waterborne epoxy coatings containing water-dispersible hexagonal boron nitride (h-BN) nanosheets, *Applied Surface Science*, 2017, 397, pages: 77–86, Doi: 10.1016/j.apsusc.2016.11.141.
65. Yuxin Liu, Baomin Fan, Bin Xu and Biao Yang, Ambient-stable polyethyleneimine functionalized $Ti_3C_2T_x$ nanohybrid corrosion inhibitor for copper in alkaline electrolyte, *Materials Letters*, 2023, 337, pages: 1–4, Doi: 10.1016/j.matlet.2023.133979.
66. Liangsen Liu, Mingyue Zhao, Xiaoyuan Pei, Shengkai Liu, Shigang Luo, Minjie Yan, Ruiqi Shao, Ying Sun, Wen Xu and Zhiwei Xu, Improving corrosion resistance of epoxy coating by optimizing the stress distribution and dispersion of SiO_2 filler, *Progress in Organic Coatings*, 2023, 179, pages: 1–11, Doi: 10.1016/j.porgcoat.2023.107522.
67. Iling Aema Wonnie Ma, Gerard Ong and Ammar Shafaamri, Evaluation of natural oil polyol hydrophobic acrylic-based coating incorporated with SiO_2 nanoparticles for enhanced corrosion protection, *Pigment & Resin Technology*, 2023, pages: 1–10, Doi: 10.1108/PRT-07-2022-0092.
68. Yunfeng Liu, Lili Chen, Yongfan Tang, Xingde Zhang and Zhengsong Qiu, Synthesis and characterization of nano SiO_2@octadecylbisimidazoline quaternary ammonium salt used as acidizing corrosion inhibitor, *Reviews on Advanced Materials Science*, 2022, 61, pages: 186–194, Doi: 10.1515/rams-2022-0006.
69. Taoli Xiao, Chenchen Du and Yongle Liu, Electrochemical evaluation on corrosion behavior of SAF 2507 duplex stainless steels in blended concrete with metakaolin and ultrafine slag admixtures, *International Journal of Electrochemical Science*, 2021, 16, pages: 1–11, Doi: 10.20964/2021.06.15.
70. Randis Randis, Djarot Budi Darmadi, Femiana Gapsari, Achmad As'ad Sonief, Ekemini David Akpan and Eno Emmanuel Ebenso, The potential of nanocomposite-based coatings for corrosion protection of metals: A review, *Journal of Molecular Liquids*, 2023, 390, pages: 1–17, Doi: 10.1016/j.molliq.2023.123067.
71. Muntathir AlBeladi, Mohammed Hassan Geesi, Yassine Riadi, Mustapha Alahiane, Talal Abdullah Aljohani, Avni Berisha, Arianit Reka, Abdellah Kaiba and Oussama Ouerghi, Synergistic inhibitive effect of a hybrid zinc oxide-benzalkonium chloride

composite on the corrosion of carbon steel in a sulfuric acidic solution, *Green Processing and Synthesis*, 2023, 12, pages: 1–19, Doi: 10.1515/gps-2023-0139.

72. Agnieszka Slosarczyk, Izabela Klapiszewska, Dagmara Skowronska, Marcin Janczarek, Teofil Jesionowski and Łukasz Klapiszewski, Review a comprehensive review of building materials modified with metal and metal oxide nanoparticles against microbial multiplication and growth, *Chemical Engineering Journal*, 2023, 466, pages: 1–20, Doi: 10.1016/j.cej.2023.143276.
73. Amir Mostafaei and Farzad Nasirpouri, Epoxy/polyaniline–ZnO nanorods hybrid nanocomposite coatings: Synthesis, characterization and corrosion protection performance of conducting paints, *Progress in Organic Coatings*, 2014, 77, pages: 146–159, Doi: 10.1016/j.porgcoat.2013.08.015.
74. Chunming Yang, Yuxuan Lu, Le Zhang, Zhijie Kong, Tianyi Yang, Li Tao, Yuqin Zou and Shuangyin Wang, Defect engineering on CeO_2-based catalysts for heterogeneous catalytic applications, *Small Structure*, 2021, pages: 1–25, Doi: 10.1002/sstr.202100058.
75. Bingting Bao, Yakun Sun, Xinxu Li, Lingyun Li and Yan Yu, Tuning the UV absorbing ability of CeO_2 nanoparticles with F^- doping, *Flat Chemistry*, 2023, 39, pages: 1–9, Doi: 10.1016/j.flatc.2023.100494.
76. Shaidatul Najihah Matussin, Mohammad Hilni Harunsani and Mohammad Mansoob Khan, CeO_2 and CeO_2-based nanomaterials for photocatalytic, antioxidant and antimicrobial activities, *Journal of Rare Earths*, 2023, 41, pages: 167–181, Doi: 10.1016/j.jre.2022.09.003.
77. Mingkai Chen, Yanjun Ma, Lei Chen, Dingjun Zhang, Huidi Zhou and Jianmin Chen, Lubrication, wear resistance and corrosion protection on aluminum alloy toward multifunctional coatings with nano-CeO_2 inhibition effect, *Progress in Organic Coatings*, 2024, 186, pages: 1–13, Doi: 10.1016/j.porgcoat.2023.108066.
78. Hao Li, Qin-Hao Zhang, Xian-Ze Meng, Hao-Jie Yan, Hui-Song Hu, Lian-Kui Wu and Fa-He Cao, Active/passive protection and anti-UV waterborne epoxy coatings based on low defect functionalized PDA-GO-CeO_2 material for excellent corrosion control, *Chemical Engineering Journal*, 2024, 479, pages: 1–17, Doi: 10.1016/j.cej.2023.147859.
79. Ubong Eduok, Omar Faye, Ahmed Tiamiyu and Jerzy Szpunar, Fabricating protective epoxy-silica/CeO_2 films for steel: Correlating physical barrier properties with material content, *Materials and Design*, 2017, 124, pages: 58–68, Doi: 10.1016/j.matdes.2017.03.062.
80. Bahram Ramezanzadeha, Ghasem Bahlakehb and Mohammad Ramezanzadeh, Polyaniline-cerium oxide (PAni-CeO_2) coated graphene oxide for enhancement of epoxy coating corrosion protection performance on mild steel, *Corrosion Science*, 2018, 137, pages: 111–126, Doi:10.1016/j.corsci.2018.03.038.
81. Wenjie Xu, Cong Yang, Wenming Su, Nongjun Zhong and Xinnian Xia, Effective corrosion protection by PDA-BN@CeO_2 nanocomposite epoxy coatings, *Colloids and Surfaces A: Physicochemical and Engineering Aspects*, 2023, 657, pages: 1–10, Doi: 10.1016/j.colsurfa.2022.130448.
82. Mir Saman Safavi and Mohamadreza Etminanfar, A review on the prevalent fabrication methods, microstructural, mechanical properties, and corrosion resistance of nanostructured hydroxyapatite containing bilayer and multilayer coatings used in biomedical applications, *Journal of Ultrafine Grained and Nanostructured Materials*, 2019, 52, pages: 1–17, Doi: 10.22059/JUFGNSM.2019.01.01.
83. Sepideh Pourhashem, Farhad Saba, Jizhou Duana, Alimorad Rashidie, Fang Guan, Elham Garmroudi Nezhad and Baorong Hou, Polymer/Inorganic nanocomposite

coatings with superior corrosion protection performance: A review, *Journal of Industrial and Engineering Chemistry*, 2020, 88, pages: 29–57, Doi: 10.1016/j.jiec.2020.04.029.

84. Shuai Zhang, Yanling Zhang and Shaowen Wu, Effects of ZnO, FeO and Fe_2O_3 on the spinel formation, microstructure and physicochemical properties of augite-based glass ceramics, *International Journal of Minerals, Metallurgy and Materials*, 2023, 30, pages: 1207–1217, Doi: 10.1007/s12613-022-2489-1.
85. Anxiang Li and Aiping Zhu, Preparation of Fe_3O_4/PANI nanocomposite and its metal anticorrosive activity, *Progress in Organic Coatings*, 2021, 161, pages: 1–9, Doi: 10.1016/j.porgcoat.2021.106477.
86. Yu Liu, Qiangliang Yu, Meirong Cai, Bo Yu and Feng Zho, Effects of nanoadditives on the anticorrosion performance of nanocomposite coatings: A review, *Nano Materials*, 2024, 7, pages: 12249–12272, Doi: 10.1021/acsanm.4c01347.
87. Fatma Sezgi Eraslan and Ridvan Gecu, Chemical composition optimization of Al_2O_3-TiO_2 composite coatings for enhanced wear and corrosion resistance, *Surface and Coatings Technology*, 2023, 474, pages: 1–12, Doi: 10.1016/j.surfcoat.2023.130053.
88. Mohamed Ezzat Mahmoud, Mohammed Abdelaziz Khalifa, Rehab Mahmoud El-Sharkawy and Mahues Rashad Youssef, Effects of Al_2O_3 and BaO nano-additives on mechanical characteristics of high-density polyethylene, *Materials Chemistry and Physics*, 2021, 262, pages: 1–11, Doi: 10.1016/j.matchemphys.2021.124251.
89. Xin Liang, Cheng Hua, Mingrui Zhang, Yu Zheng, Shijie Song, Meng Cai, Yu Huang, Can He and Xiaoqiang Fan, Preparation of anticorrosive epoxy nanocomposite coating modified by polyethyleneimine nano-alumina, *Coatings*, 2023, 13, pages: 1–11, Doi: 10.3390/coatings13030561.
90. Bo Wang, Shicheng Wei, Wei Huang, Yujiang Wang, Yi Liang, Junqi Li, Ruidong Shi, Lei Guo and Junfeng Xue, Preparation, anticorrosion mechanism, and application of nano-Ti polymer-containing coating, *European Polymer Journal*, 2023, 195, pages: 1–17, Doi: 10.1016/j.eurpolymj.2023.112200.
91. Yueping Ye, Haibin Chen, Jingshen Wu and Lin Ye, High impact strength epoxy nanocomposites with natural nanotubes, *Polymer*, 2007, 48, pages: 6426–6433, Doi: 10.1016/j.polymer.2007.08.035.
92. Abdel Salam Hamdy, Irmgard Doench and Hans Möhwald, The effect of alkaline etching time on the anticorrosion performance of vanadia film formed on high strength AA2024 in chloride media, *Journal of Materials Science*, 2012, 47, pages: 3784–3792, Doi: 10.1007/s10853-011-6232-y.
93. Abdel Salam Hamdy, Irmgard Doench and Hans Möhwald, Chrome-free self-healing protective coating for AA2024 in chloride containing solution, *Proceedings of Corrosion Nace*, 2012, pages: 11–15.
94. Mohammad BinSabt, Ahmed Abdel Nazeer, Metwally Adkour and Fakhreia Al-Sagheer, Hydrothermally modified PVA/ZnS-NCQD nanocoating for stainless steel corrosion protection in saline water, *RSC Advances*, 2016, 6 (9), pages: 6888–6895.
95. Miaomiao Cui, Demian Ifeanyichukwu Njoku, Bowen Li, Lihui Yang, Zuankai Wang, Baorong Hou and Ying Li, Corrosion protection of Aluminium Alloy 2024 through an epoxy coating embedded with smart microcapsules: The responses of smart microcapsules to corrosive entities, *Corrosion Communications*, 2021, 1, pages: 1–9, Doi: 10.1016/j.corcom.2021.06.001.
96. Artemis Kontiza and Ioannis Andreas Kartsonakis, Smart composite materials with self-healing properties: A review on design and applications review, *Polymers*, 2024, 16, pages: 1–17, Doi: 10.3390/polym16152115.

97. Tatyana Nesterova, Kim Dam-Johansen and Søren Kiil, Synthesis of durable microcapsules for self-healing anticorrosive coatings: A comparison of selected methods, *Progress in Organic Coatings*, 2011, 70, pages: 342–352, Doi: 10.1016/j.porgcoat.2010.09.032.
98. Ioannis Andreas Kartsonakis and George Kordas, Synthesis and characterization of cerium molybdate nanocontainers and their inhibitor complexes, *Journal of the American Ceramic Society,* 2010, 93 (1), pages: 65–73, Doi: 10.1111/j.1551-2916.2009.03310.x.
99. Kai Xu, Xuelian Xiao, Linjing Wang, Ming Lou, Fangming Wang, Changheng Li, Hui Ren, Xue Wang and Keke Chang, Data-driven materials research and development for functional coatings review, *Advanced Science*, 2024, 1, pages: 1–34, Doi: 10.1002/advs.202405262.
100. Dmitry Grigoriev, Elena Shchukina, Aiym Tleuova, Saule Aidarova and Dmitry Shchukin, Core/shell emulsion micro- and nanocontainersfor self-protecting water based coating, *Surface & Coatings Technology*, 2016, 2, pages: 1–35, Doi: 10.1016/j.surfcoat.2016.01.002.
101. Jehan Kothari and Jude Oluwaseun Iroh, Self-healing poly(urea formaldehyde) microcapsules: Synthesis and characterization, *Polymers*, 2023, 15, pages: 1–26, Doi: 10.3390/polym15071668.
102. Saman Zehra, Mohammad Mobin, Ruby Aslam, Mosarrat Parveen and Afroz Aslam, Nanocontainer-loaded smart functional anticorrosion coatings, *Smart Anticorrosive Materials*, 2023, 23, pages: 481–497, Doi: 10.1016/B978-0-323-95158-6.00003-5.
103. Barbara Di Credico, Marinella Levi and Stefano Turri, An efficient method for the output of new self-repairing materials through a reactive isocyanate encapsulation, *European Polymer Journal*, 2013, 49, pages: 2467–2476, Doi: 10.1016/j.eurpolymj.2013.02.006.
104. Jiaxin Wena, Jinglei Leia, Jinlong Chena, Jianjun Goua, Ying Lib and Lingjie Li, An intelligent coating based on pH-sensitive hybrid hydrogel for corrosion protection of mild steel, *Chemical Engineering Journal*, 2020, 392, pages: 123742, Doi: 10.1016/j.cej.2019.123742.
105. Zhengwei Mao, Lie Ma, Changyou Gao and Jiacong Shen, Preformed microcapsules for loading and sustained release of ciprofloxacin hydrochloride, *Journal of Controlled Release*, 2005, 104, pages: 193–202, Doi: 10.1016/j.jconrel.2005.02.005.
106. Amanda Renee Jones, Alicia Cintora, Scott Richard White and Nancy Ruth Sottos, Autonomic healing of carbon fiber/epoxy interfaces, *Applied Materials Interfaces*, 2014, 6, pages: 6033–6039, Doi: 10.1021/am500536t.
107. Yago González-Garcíaa, Jose Manuel Carlos Mola, Thomas Musselleb, Inge De Graeveb, Guido Van Asschec, Gert Scheltjensc, Bart Van Melec and Hugo Terryn, A combined mechanical, microscopic and local electrochemical evaluation of self-healing properties of shape-memory polyurethane coatings, *Electrochimica Acta*, 2011, 56, pages: 9619–9626, Doi: 10.1016/j.electacta.2011.03.081.
108. Alexander Lutz, Otto van den Berg, Jonas Van Damme, Karen Verheyen, Erwin Bauters, Iris De Graeve, Filip Evert Du Prez and Herman Terryn, A shape-recovery polymer coating for the corrosion protection of metallic surfaces, *Applied Materials and Interfaces*, 2015, 7, pages: 175–183, Doi: 10.1021/am505621x.
109. Erika Diane Rodriguez, Xiaofan Luo and Patrick Thomas Mather, Linear/network poly(ε-caprolactone) blends exhibiting shape memory assisted self-healing (SMASH), *Applied Materials and Interfaces*, 2011, 3, pages: 152–161, Doi: 10.1021/am101012c.

110. Guadalupe Rivero, Le-Thu Thi Nguyen, Xander Koen David Hillewaere and Filip Evert Du Prez, One-pot thermo-remendable shape memory polyurethanes, *Macromolecules*, 2014, 47 (6), pages: 2010–2018, Doi: 10.1021/ma402471c.
111. Stefaan D'hollandera, Guido Van Asscheb, Bart Van Meleb and Filip Du Prez, Novel synthetic strategy toward shape memory polyurethanes with a well-defined switching temperature, *Polymer*, 2009, 50, pages: 4447–4454, Doi: 10.1016/j.polymer.2009.07.021.
112. Jean-Baptiste Jorcina, Gill Scheltjensa, Yves Van Ingelgema, Els Tourwéa, Guy Van Asscheb, Iris De Graevea, Bruno Van Meleb, Herman Terryna and Annick Hubin, Investigation of the self-healing properties of shape memory polyurethane coatings with the 'odd random phase multisine' electrochemical impedance spectroscopy, *Electrochimica Acta*, 2010, 55, pages: 6195–6203, Doi: 10.1016/j.electacta.2010.01.027.
113. Chi Yua, Shiunchin C. Wanga, Marek Sosnowskib and Zafar Iqbal, Plasma-enhanced chemical vapor deposition of polyperinaphthalene thin films, *Synthetic Metals*, 2008, 158, pages: 425–429, Doi: 10.1016/j.synthmet.2008.03.004.
114. Giuseppe Lanzara, Yong Yoon, Hongyu Liu, Shuang Peng and Woo-In Lee, Carbon nanotube reservoirs for self-healing materials, *Nanotechnology*, 2009, 20 (33), pages: 1–13, Doi: 10.1088/0957-4484/20/33/335704.
115. Jing Li, Zhenglin Tao, Jincan Cui, Shuling Shen and Hanxun Qiu, Facile Fabrication of Dual Functional Graphene Oxide Microcapsules Carrying Corrosion Inhibitor and ncapsulating Self-Healing Agent, Polymers, 2022, 14, pages 1-9, Doi: 10.3390/polym14194067.
116. Elshad Abdullayev and Yuri Lvov, Halloysite Clay Nanotubes for Controlled Release of Protective Agents, *Journal of Nanoscience and Nanotechnology*, 2011, 11, pages: 10007–10026, Doi: 10.1166/jnn.2011.5724.
117. Dmitri Fix, Daria Valeryevna Andreeva, Yuri Mikhailovich Lvov, Dmitry Gennadievich Shchukin and Helmuth Möhwald, Application of inhibitor-loaded halloysite nanotubes in active anti-corrosive coatings, *Advanced Functional Materials*, 2009, 19 (11), pages: 1720–1727, Doi: 10.1002/adfm.200800946.
118. Marya Rajia, Mounir El Achabyb, Rachid Bouhfida and Abou el kacem Qaiss, Self-healing based on composites and nanocomposites materials: From synthesis to application and modeling, *Polymer Nanocomposite-Based Smart Materials*, Elsevier Ltd, 2020, pages: 41–60, Doi: 10.1016/B978-0-08-103013-4.00003-0.
119. Ahmed Abdel Nazeer and Metwally Madkour, Potential use of smart coatings for corrosion protection of metals and alloys: A review, *Journal of Molecular Liquids*, 2018, 253, pages: 11–22, Doi: 10.1016/j.molliq.2018.01.027.
120. José Daniel Diniz Melo, Ana Paula Cysne Barbosa, Maria Carolina Brito Costa and Gudson Nicolau de Melo, Encapsulation of solvent into halloysite nanotubes to promote self-healing ability in polymers, *Advanced Composite Materials*, 2014, 23, pages: 507–519, Doi: 10.1080/09243046.2014.915116.
121. Changkyu Kim, Ahmad Ivan Karayan, Jose Milla, Marwa Hassan and Homero Castaneda, Smart coating embedded with pH-responsive nanocapsules containing a corrosion inhibiting agent, *ACS Applied Materials & Interfaces*, 2020, 12, pages: 6451–6459, Doi: 10.1021/acsami.9b20238.
122. Yixi Chen, Chris Xia, Zachary Shepard, Nicholas Smith, Nicholas Rice, Amy Megan Peterson and Aaron Sakulich, Self-healing coatings for steel-reinforced concrete, *ACS Sustainable Chemistry & Engineering*, 2017, 5, pages: 3955–3962, Doi: 10.1021/acssuschemeng.6b03142.

123. Yanan Wei, Huihui Yuan, Haibing Zheng, Chong Wang, Linwen Yu and Weihua Li, Ca-LDH-modified cementitious coating to enhance corrosion resistance of steel bars, *Journal of Building Engineering*, 2022, 51, pages: 1–12, Doi: 10.1016/j.jobe.2022.104301.
124. Lin Zhang, Wenya Du, Amit Nautiyal, Zhen Liu and Xinyu Zhang, Recent progress on nanostructured conducting polymers and composites: Synthesis, application and future aspects, *Science China Materials*, 2018, 61 (3), pages: 303–352, Doi: 10.1007/s40843-017-9206-4.
125. Huiping Yang, Yinning Lin, Qian Mo, Zhaoquan Li, Fan Yang and Xinchun Li, Monitoring Enzymatic Reaction Kinetics and Activity Assays in Confined Nanospace, *Analytical Chemistry*, 2024, 96, pages: 12404–12413, Doi: 10.1021/acs.analchem.4c01901.
126. Yao Huang, Panjun Wang, Weimin Tan, Wenkui Hao, Lingwei Ma, Jinke Wang, Tong Liu, Fan Zhang, Chenhao Ren, Wei Liu and Dawei Zhang, Photothermal and pH dual-responsive self-healing coating for smart corrosion protection, *Journal of Materials Science & Technology*, 2022, 107, pages: 34–42, Doi: 10.1016/j.jmst.2021.08.044.
127. Reza Samiee, Bahram Ramezanzadeh, Mohammad Mahdavian, Eiman Alibakhshi and Ghasem Bahlakeh, Designing a non-hazardous nano-carrier based on graphene oxide@ polyaniline-praseodymium (III) for fabrication of the active/passive anti-corrosion coating, 2020, *Journal of Hazardous Materials*, 398, pages: 1–17, Doi: 10.1016/j.jhazmat.2020.123136.

5 Understanding Anticorrosion Coating Failure and Degradation Pathways

Bharathi Kannan Raja, M. Arthanareeswari, and Arockia Selvi J

5.1 INTRODUCTION

Coatings act as a functional barrier of metals in diverse environments as metallic equipment is involved in widespread areas including marine infrastructures, industrial applications, transportation, construction, etc. [1]. Hence, they are subjected to harsh environmental conditions like prolonged immersion in aqueous water, laid underground, might undergo radiation exposure in industrial areas like UV, moisture exposure, high-temperature corrosive liquids, chemical exposure, acid rains, microorganisms in case of underground metallic equipment or in marine applications [2]. In marine environments, there are several factors like pH change, presence of dissolved oxygen, and bacterial and other microorganisms which trigger corrosion reactions and lead to system failures and expensive maintenance [3]. To get higher efficiency without substantial failures, these issues have to be addressed which also helps develop innovative metallic components and infrastructure. One strategy entails modifying the metal surface through the application of protective coatings designed to both inhibit fouling and prevent corrosion [4]. Coatings primarily serve as barrier protection for materials, among other roles, and are typically subjected to atmospheric conditions, varying temperature ranges, and electromagnetic radiation. These environmental factors can lead to the degradation of the coating matrices used in industrial facilities and products, initially resulting in microscopic cracks that may not be immediately visible[5]. Over time, these microcracks can evolve into visible fractures, ultimately leading to coating failure. Such failures can result in significant damage costs and promote severe corrosion reactions on the underlying metallic substrate. This situation can lead to costly damage and significant corrosion of the underlying metallic substrate [6]. Coating failures can incur substantial costs associated with the repair, replacement, and maintenance of materials and structures. Historical data spanning 70 years indicates that corrosion-related expenses account for up to (approx.) 6% of a country's GDP. Steel, being the most widely utilized material across various industrial sectors, faces significant challenges, as it is prone to rapid erosion. Notably, around half of the global steel production is allocated for replacing

 DOI: 10.1201/9781003571162-5

corroded steel. Globally, one ton of steel rusts every 90 seconds, consuming energy equivalent to what an average household uses over 3 months [7, 8]. Furthermore, corrosion and coating failures can compromise the safety and integrity of critical infrastructure, such as hot-rolled bridges, buildings, pipelines, and cold-formed steel connections, leading to potential accidents, structural failures, and loss of life [8]. The failures of protective coatings adversely affect the load-bearing capacity and overall performance of structures, rendering them incapable of meeting essential durability standards. Moreover, corrosion hinders energy efficiency and initiatives to mitigate carbon emissions, therefore presenting significant barriers to the sustainable development of infrastructure. Consequently, an effective coating should be both cost-efficient and possess a range of essential characteristics, including minimal moisture penetration, impact resistance, durability, chemical stability, strong adhesion to the substrate, favorable surface morphology, environmental safety, flexibility, and ease of application.

This chapter reviews the latest developments in the understanding of the mechanisms behind coating failures and the pathways of degradation. The first section outlines the introduction, importance, and historical data of the corrosion failure. In the second section, there will be a detailed description of the coating failure mechanisms and their types with the factors which cause and influences failures in coatings and degradation pathways. The third section deals with case studies of a few important coating failures which were reported globally. The fourth section explores the role of metal–organic frameworks (MOFs) in corrosion integrity. The fifth section, concluded by the future research directions and the emerging trends, highlights the importance of predictive molding and emerging trends in understanding coating degradation mechanisms.

5.2 COATING FAILURES AND DEGRADATION MODELING

Anticorrosion coatings are a widely adopted strategy for mitigating metal corrosion and extending the service life of metal materials [9–12]. These coatings provide protection against corrosive environments through three primary mechanisms: (1) they act as a barrier, preventing the diffusion of corrosive agents to the metal-coating interface [12–14]; (2) inhibitors and pigment particles within the coatings may hydrolyze, passivate, or chemically engage with corrosive ions to safeguard the substrate from corrosion [15, 16]; and (3) the coatings exhibit strong adhesion and cohesion to the metal surface, minimizing the risk of delamination due to the accumulation of corrosion byproducts [17, 18]. In the upcoming subsections, several types of coating failures are discussed in detail.

5.2.1 Forms of Coating Failure Mechanisms

Despite the effectiveness of protective coatings, the performance of coatings can degrade over time, which may ultimately result in coating failure. The coating failures in corrosive environmental conditions primarily occurs due to the dispersion of corrosive agents, including ions, oxygen, and moisture. This diffusion can result in corrosion at the interface between the coating and the metal substrate, potentially compromising the bond between coating and the metal surface. This section offers an in-depth examination of various forms of metallic corrosion. It includes an overview of general corrosion and delves into specific types, such as localized corrosion (including pitting and crevice corrosion), stress corrosion cracking, and intergranular corrosion along with other practical environmental conditions through which the metals may undergo.

Recent research identifies several distinct physical stages of coating failures and degradation. For example, **pitting** corrosion failure is a notable type of localized corrosion that begins in a small spot on a material and progressively develops, creating deeper pits on the surface that can be hard to notice [19]. During this process, pits or holes may form in hemispherical or cup shapes as described in Figure 5.1a. The covered area, which is exposed to impurities or water, acts as the anode, while the exposed area functions as the cathode [20]. This corrosion type is driven by an electrochemical reaction. Materials like steel, aluminum, and iron were particularly susceptible for pitting type of corrosion, making it a serious concern. Even though stainless steel, which contains chromium and nickel, is generally resistant to corrosion, it can still suffer from pitting due to localized damage to its protective oxide layer [21]. **Crevice corrosion** occurs in confined spaces, typically in stagnant and low-oxygen conditions, such as gaps between metal surfaces or under deposits like sand, dirt, or mud. In these crevices, the oxygen level is lower than in the surrounding area, causing the crevice to become the anode and the outer area to act as the cathode (Figure 5.1b). This type of corrosion presents substantial challenges, especially in marine environments. Stainless steels are particularly affected, and the presence of chloride compounds further complicates the management of crevice corrosion [22, 23]. **Intergranular corrosion** – intergranular attack occurs when the edges of a metal surface are more susceptible to corrosion than its center. In this type of corrosion, a narrow zone at the grain boundaries of a metal alloy experiences significant corrosion, while the grains themselves show little to no damage as represented in Figure 5.1d [24]. This corrosion can mask the overall resistance of the material under different conditions and is typically evaluated using an IGA Test [25]. Additionally, impurities like carbon, nitrogen, oxygen, manganese, and sulfur do not affect this type of corrosion, but the presence of silicon and phosphorus does influence it [26]. **Galvanic corrosion** occurs on specific areas of a metal surface due to the presence of an anodic section (the more reactive metal) and a cathodic section (the less reactive metal) (see Figure 5.1c); this type of corrosion occurs when two different metals are in contact and bathed in an electrolyte. In this situation, the less reactive metal (the cathode) is protected, while the more reactive metal (the anode) tends to corrode [27].

Blistering – presence of blisters under the coating surface due to the presence of gases or moisture, stress, contamination, or cathodic disbandment, which may occur during their service life. When these issues arise, the adhesive strength between the coating and the substrate diminishes, leading to peeling failure. Consequently, the coatings lose their effectiveness in protecting the underlying substrate [28, 29]. **Adhesion loss** – detachment of coating from the substrate due to weak adhesion or presence of any contaminants at the interface (Figure 5.2b) [29,–30]. **Cracking** – the

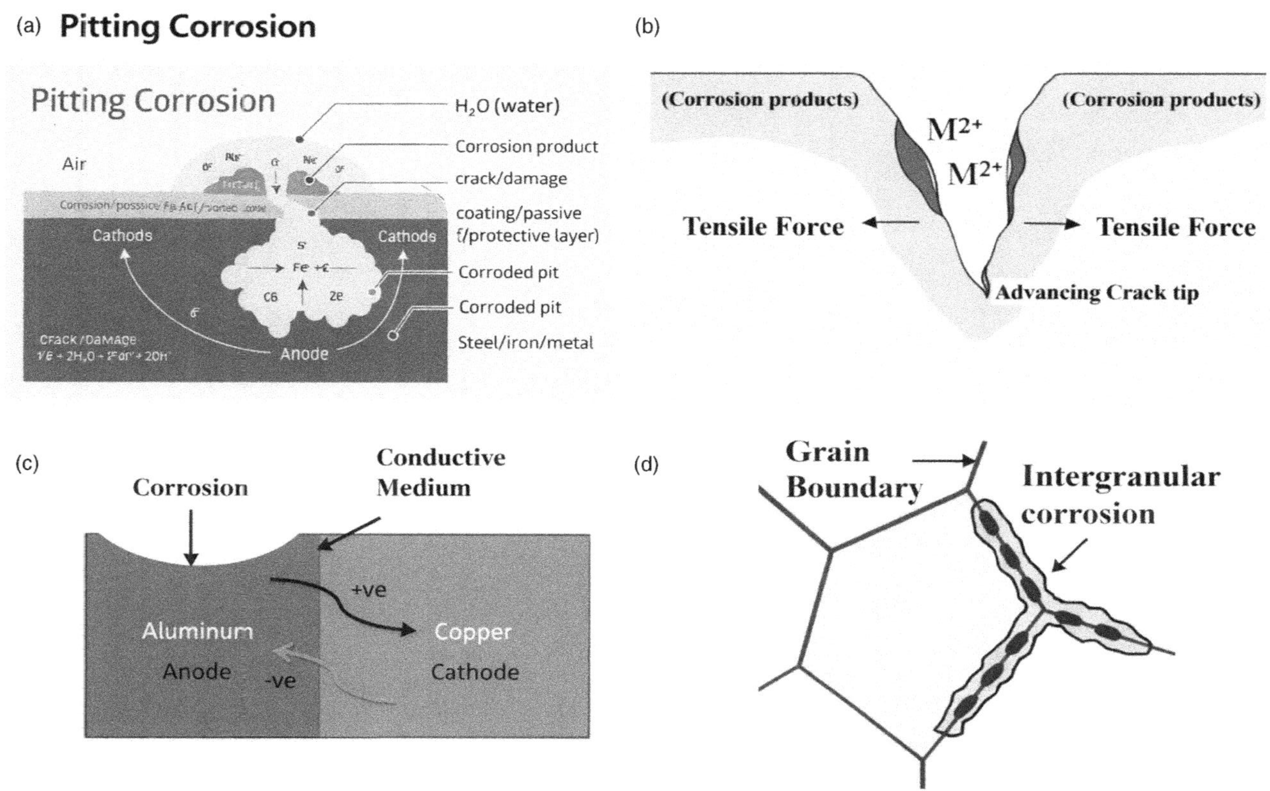

FIGURE 5.1 Schematic illustration of mechanism behind the (a) pitting corrosion, (b) cervices corrosion, (c) galvanic and (d) intergranular coating failures.

FIGURE 5.2 Coating failures due to (a) Temperature (*Source:* Newsletter on Heat exchangers © 2024 Tempco srl), (b) Adhesion loss (*Source:* Fundamentals of Corrosion and Corrosion Control, ©2025 Australian Mining), (c) Salinity (*Source:* Anti-Corrosion Coating Failures, © 2021 SealXpert), (d) Fatigue (*Source:* Types of anti-rust anti corrosion coatings © 2025 Performance Painting), (e) High strength bolt friction (*Source:* Paint & Coating Failures, © 2025 Matergenics Coatings), and (f) Mechanical stress (*Source:* Anti-Corrosion Coating Failures, © 2021 SealXpert).

presence of fissures in the coating due to harsh environmental conditions or even aging of coating material as described in Figure 5.1b. The deterioration of coatings frequently exposes the metal substrate to corrosion, leading to reduced service life and potential safety risks. The reliable performance of coatings is crucial to ensure their durability and effectiveness [13, 31]. In addition to these, **fatigue failures** might arise when the substrate is subjected to high mechanical stress, which might lead to fatigue failures. High-strength bolt friction corrosions (see Figure 5.2d–f) in civil and maritime constructions are caused due to these coating failures. A few examples are shown in Figure 5.2 [32].

Moreover, environmental factors such as temperature (Figure 5.2a), pH, salinity in case of marine application (Figure 5.2c), moisture, UV radiation, etc. (see Figure 5.3) greatly influence the anticorrosive coating. Recently, Imran Mir Chohan observed the effects of salinity, pH, and temperature of the steel in an aqueous medium; from his studies, it is obvious that when salinity increases from 33 to 66 ppt, corrosion rate increases by 28%, This behavior can be explained by the decrease in oxygen solubility that occurs as water salinity rises, and higher salinity increases the number of ions in water, boosting its electrical conductivity. This allows electric currents to flow more easily, speeding up the electrochemical reactions that lead to increase in the rate of corrosion [33, 34]. With the decrease in pH from 8.5 to 7, corrosion rate increases by 42%, This observation aligns with the basic principles of corrosion chemistry, which indicates that an increase in pH typically leads to a lower corrosion rate in

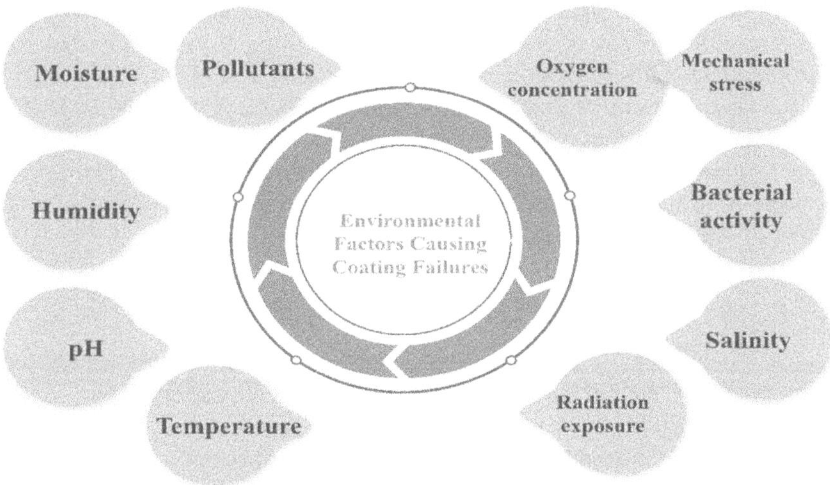

FIGURE 5.3 Schematic representation of environmental factors which causes coating failures.

metals. This reduction in corrosion is attributed to the development of a passive film on the metal's surface when exposed to higher pH conditions [35, 36]. The rise in temperature caused the corrosion rate to increase by 76% [37]. This finding highlights a clear correlation between higher temperatures and increased corrosion rates. This relationship is consistent with the well-established notion that elevated temperatures accelerate corrosion processes. The increased kinetic energy of molecules in warmer water speeds up the chemical reactions that drive corrosion [38]. These inferences were also observed in other works reported [39, 40] and have proven the fact that how the environmental conditions influence the anticorrosive coatings. Coating failure in MOF-based composites can arise from several key factors, including UV degradation, which compromises mechanical integrity and protective properties, especially under outdoor exposure. Increased surface roughness from environmental conditions can indicate damage and promote adhesion loss or moisture ingress. Chemical bonding issues, influenced by the loading percentage of MOFs, can lead to delamination, while mechanical stress concentrations may initiate cracks due to differing mechanical properties between MOFs and the polymer matrix. Thermal instability, moisture absorption causing swelling, harsh environmental conditions, and inadequate dispersion of MOFs can further exacerbate coating degradation [41]. Understanding these factors is essential to enhance the durability and effectiveness of MOF-based coatings in various applications [41].

5.3 CASE STUDIES OF COATING FAILURES THAT OCCURRED GLOBALLY

Coating failures and corrosion present a considerable economic burden worldwide, with annual direct and indirect expenses surpassing 2.5 trillion USD – about 3.4%

of the world's GDP. Despite its widespread implications, many sectors encounter significant difficulties in accurately measuring the distinct costs associated with corrosion, highlighting a lack of empirical data in this area [42, 43]. Throughout history, many corrosion-related incidents have gone unrecognized due to liability concerns or because the evidence was lost in the resulting catastrophic events; however, some have garnered widespread media attention. There have been severe corrosion incidents that resulted in fatalities and led to significant economic losses [44], which are listed in Table 5.1.

5.4 ROLE OF MOFS IN PREVENTING AND MITIGATING COATING FAILURES

MOFs represent a significant new class of porous materials characterized by a "microcapsule" structure filled with cavities. In the domain of anticorrosion organic coatings, MOFs integrate heteroatoms such as nitrogen, sulfur, and oxygen, which can participate in adsorption reactions. Furthermore, the existence of unsaturated metal sites enables these frameworks to adsorb metal ions from surfaces, forming protective films that improve the corrosion resistance of the coatings [45, 46]. MOFs have garnered attention for their substantial surface area, elevated porosity, and excellent compatibility, positioning them as a viable choice for advanced coatings in many applications. MOFs are known for its robust structure and significant adsorption capabilities, serves

TABLE 5.1
Coating Failures and Corrosion Accidents Which Happened Worldwide

Incident	Location	Year	Reason	Loss
Bhopal Accident	India	1984	Lacks in coatings and corrosion effecting pipelines, valves, and other safety apparatus at Union Carbide India Ltd.	3000 causalities and 500,000 were injured.
Sinking of the ship *Erika*	Brittany, France	1999	Coating deterioration and corrosion on the primary deck coaming, as well as the portside and starboard inert gas system risers.	19,800 tons of heavy fuel oil spills and huge economic loss.
Prudhoe Bay Oil Spill	Prudhoe Bay, Alaska	2006	Malfunction of the pipeline corrosion monitoring and leak detection system.	267,000 gallons of crude oil spillage.
Rupture of a Natural Gas Transmission Pipeline	Pennsylvania, USA	2016	Corrosion of gas transmission pipelines and two of the circumferential welds.	Several homes were destroyed.
Ohio State Fair accident	Ohio, USA	2017	A pendulum-type thrill ride failed during operation due to rust.	One causality and several serious injuries.

as an excellent filler in coatings. MOFs contain metal ions coordinated to organic ligands, creating a porous structure that can dynamically respond to environmental changes. This dynamic nature allows for the reformation of bonds when the coating is damaged. MOFs can act as carriers for corrosion inhibitors or healing agents [47, 48]. When the coating is compromised, these agents can be released from the MOF structure to prevent further corrosion and initiate healing. Certain MOFs have the ability to absorb and release moisture [49]. This property can be harnessed to facilitate self-healing processes by maintaining an appropriate environment for healing agents [50]. Zibin chen reported a work on self-healing coating system designed for the protection of aluminum alloys focusing on corrosion prevention with MOF-based nanofiller. This nanofiller consists of ZnMg MOF nanoparticles serving as the structural framework, salicylaldoxime corrosion inhibitors encapsulated within the MOF, and a polydopamine layer coating the MOF surface. The work clearly explains the self-healing property. When the aluminum alloy substrate experiences damage, Al^{3+} ions are released, which can interact with the nanofiller, leading to its breakdown and proclamation of corrosion inhibitors to mitigate corrosion promptly. Following this, Zn^{2+} and Mg^{2+} ions from the nanofiller co-precipitate with Al^{3+} to create a dense layered double hydroxide (LDH) film of AlZnMg, enhancing the protective capabilities at the exposed surface of the alloy [51]. Another study introduces a novel coating derived from ZIF-67 (zeolitic imidazolate frameworks) where MOFs enhanced with nanosheets of g-C_3N_4, containing built-in cerium (Ce^{3+}) and benzotriazole (BTA) corrosion inhibitors. The attention was on creating nano pigments of Ce-BTA/ZI/GN and epoxy coatings are incorporated into them to improve corrosion resistance for mild steel (MS) in marine environments. During the application of the epoxy, microcracks and pores can develop, allowing corrosive agents to penetrate and instigate redox reactions in the metal substrate. This results in free corrosion products (FeOOH), worsening the decay. The adding g-C_3N_4 improves the coating's barrier properties, limiting the pathways for corrosion. The ZIF-67 component enhances the surface area and dispersion stability, although the epoxy may react with its 2-methylimidazole ligand increasing crosslink density and reducing the diffusion of aggressive ions. Pores and microcracks can lead to localized micro-galvanic cells, where oxidation occurs in anodic regions, lowering local pH from hydrolysis of metal ion. Concurrently, reduction of oxygen generates hydroxide ions, causing local alkalinity. Upon release, Ce^{3+} and BTA form protective layers such as BTA-Fe-BTA complexes, cerium hydroxide, and cerium oxide that obstruct the intrusion of corrosion ions, thereby enhancing overall corrosion resistance [47]. The self-healing mechanisms is shown in Figure 5.4a and b. The ionic nature of some MOFs allows them to facilitate ion exchange reactions that can contribute to the healing process. This can help in repairing microcracks or defects in the coating, thereby restoring barrier properties. Consequently, in recent years, the application of MOFs has become a prominent area of research in anticorrosion coatings. ZIFs are a type of MOFs that proved to be thermodynamically stable, pH-responsive, highly porous, and can be synthesized easily [52, 53]. In the study of Ramenzanzadeh, anticorrosive properties of the ZIF-8 dissociated extract with 3.5% NaCl as corrosive medium on steel substrates and found that it exhibited excellent corrosion resistance, ZIF-8 dissociated compounds recombines largely with corrosion products, resulting in the development

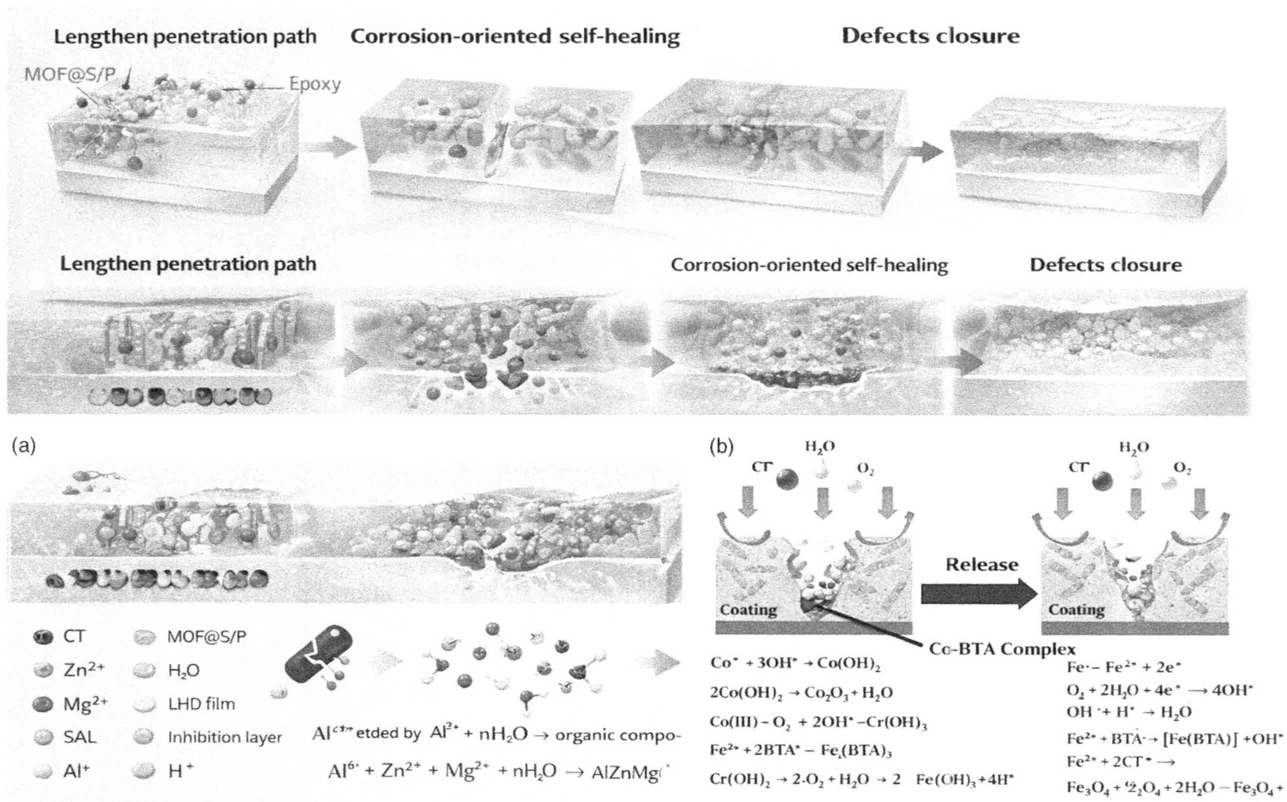

FIGURE 5.4 Self-healing mechanism of (a) ZnMg MOF nanofillers (reprinted with permission from Ref 52 © 2024 Elsevier B.V.). (b) ZIF-67 (reprinted with permission from Ref 48 © 2018 Elsevier B.V.).

of a dense protective layer that extends the erosion trajectory of the corrosive agent [54]. Guo incorporated BTA and ZIF-7, which led to the production of pH-responsive composite. The nanocomposite particles were shown to rapidly dissolve and release BTA within 10 minutes under acidic conditions. BTA subsequently adsorbs onto the metal substrate, forming a protective film that enables the composite coating to attain 99.4% inhibitory efficiency in acidic conditions [55]. In conclusion, the integration of MOFs in anticorrosive coatings significantly enhances protective performance, as evidenced by the above studies.

For instance, Liu Tong [56] presents a self-repairing and corrosion-detecting coating that integrates $CaCO_3$ microcontainers filled with tung oil (TO) and 1,10-phenanthrolin-5-amine (APhen). ZIF-8 acts as nano-stoppers at the openings of the mesoporous $CaCO_3$ microcontainers (see Figure 5.5). TO is a polymerizable botanical oil, while APhen functions as both a corrosion inhibitor and a corrosion indicator, exhibiting a red coloration upon interaction with Fe^{2+}. The microcontainers, labeled ZIF-8@Ca-TO-APhen, can modulate the release of TO and APhen in reaction to pH fluctuations. The microcontainers were subsequently amalgamated with epoxy resin and employed as a coating on carbon steel. The self-healing properties and corrosion-detection capacities of the coating were evaluated using electrochemical analysis when the damaged coated steel was exposed to an aqueous chloride solution, alongside visual inspections and corrosion monitoring during salt spray testing. As another example, Liu [57] regulated the development of UiO-67 nanomaterials (a variant of Zr-MOF) along the c-axis to produce two-dimensional layered nanosheets, referred to as 2D-UiO-67. We successfully generated a high-synergy functionalized composite material, GO@2D-UiO-67, utilizing these nanosheets as a precursor for the first time. Figure 5.5ca depicts the schematic representation of the preventive mechanism, whereas Figure 5.5dc illustrates the synthesis technique. Subsequently, we developed an innovative epoxy organic coating that offers enduring corrosion protection for aluminum alloy surfaces utilizing this composite material. The experimental results suggested that the superior corrosion protection is likely due to the improved barrier effect of the two-dimensional nanosheets, heightened crosslinking density, and less corrosion activity following the breakdown of the MOF organic ligands.

5.5 EMERGING TRENDS AND FUTURE PERSPECTIVES

Current research is focused on enhancing the lifespan of MOFs in anticorrosion applications through (I) tailoring the framework to improve stability and resistance to degradation [58]; (II) combining MOFs with other materials to enhance protective properties and durability [59, 60]; and (III) developing novel coating formulations that maximize the efficacy and lifespan of MOFs [61]. Future research might focus on self-healing coatings that can autonomously repair minor damages, thereby extending the lifespan of protective layers. While traditional coatings may fail, integrating nanomaterials like MOFs can improve mechanical strength, thermal stability, and corrosion resistance. As industries move toward more sustainable practices, the development of eco-friendly coatings that minimize environmental impact while maintaining high performance will be paramount. MOFs are being increasingly

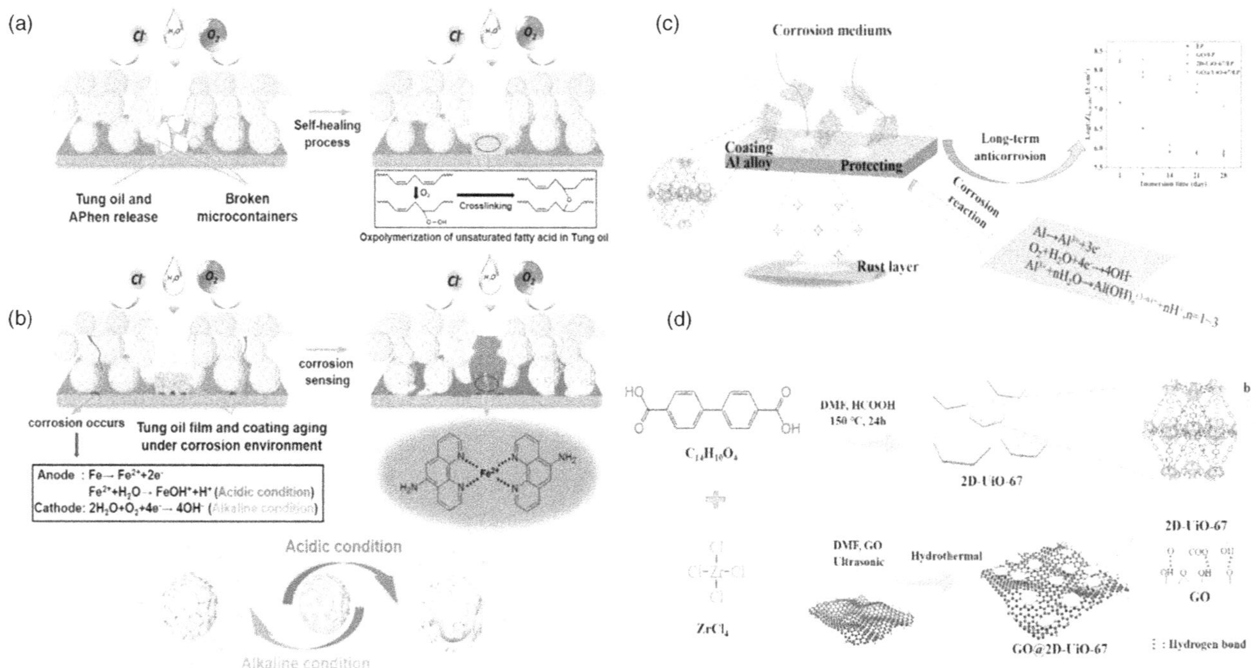

FIGURE 5.5 Schematic illustration of (a) self-healing and corrosion prevention mechanism of (b) ZIF-8@Ca-TO-APhen microcontainers (reprinted with permission from Ref. 57 © 2022 Elsevier B.V.) and the (c) synthesis strategy and (d) prevention mechanism of 2D-UiO-67 nanosheets and GO@2D-UiO-67 composite materials. (Reprinted with permission from Ref. 58 © 2024 Elsevier B.V.)

explored as advanced materials for coatings due to their unique structural properties and behavior. Their implementation in coatings can provide innovative solutions to combat corrosion.

5.6 CONCLUSION

This chapter provides a thorough exploration of the complexities surrounding coating failures and degradation pathways, with a particular emphasis on the role of MOFs in anticorrosive coatings. By detailing the types of coating failures, the underlying mechanisms, and the multifactorial influences on degradation, it highlights the significance of specific MOF properties for enhancing coating durability. The examination of corrosion mechanisms and the factors affecting the lifespan of MOF-based coatings, coupled with insights into various analytical techniques, lays a solid foundation for understanding and addressing these challenges. Furthermore, the discussion of mitigation strategies and illustrative case studies emphasizes the urgent need for continued research in this area. The proposed future research directions advocate for predictive modeling and innovative approaches to deepen our understanding of coating degradation mechanisms, ultimately aiming to enhance the resilience and performance of coatings in diverse industrial applications.

REFERENCES

[1] S.L. Lawal, S.A. Afolalu, T.C. Jen, E.T. Akinlabi, Corrosion control and its application in marine environment: A review, *Solid State Phenomena*, 355 (2024) 61–73.

[2] A. Borgaonkar, G.J.S. McNamara, Environmental impact assessment of anticorrosion coating life cycle processes for marine applications, *Sustainability*, 16 (2024) 5627.

[3] E.A. Aydinsoy, Z.Z. Aghamaliyev, D.B. Aghamaliyeva, V.B. Abbasov, A systematic review of corrosion inhibitors in marine environments: Insights from the Last 5 Years, *Processes of Petrochemistry and Oil Refining*, 25 (2024) 793.

[4] T. Rasitha, N.G. Krishna, B. Anandkumar, S. Vanithakumari, J. Philip, A comprehensive review on anticorrosive/antifouling superhydrophobic coatings: Fabrication, assessment, applications, challenges and future perspectives, *Advances in Colloid and Interface Science*, 324 (2024) 103090.

[5] M. Yan, Y.X. Lan, J.-M. Yeh, Application of graphene and its derivatives in anticorrosion coatings: Research advances and future perspectives, *Journal of the Taiwan Institute of Chemical Engineers*, 154 (2024) 105130.

[6] H. Shao, Z. Chen, J. Chang, X. Yin, Y. Chen, Y. Liu, K. Zhang, W.J.C.P. Yang, Gum Arabic microgel–based biomimetic waterborne anticorrosive coatings with reinforced water and abrasive resistances, *Carbohydrate Polymers*, 342 (2024) 122408.

[7] R. Bingham, D. Nuttall, EUROCORR 2017 in combination with the 20th international corrosion congress and the process safety congress 2017: corrosion control for safer living. Part 2, *Corrosion Engineering, Science and Technology: The International Journal of Corrosion Processes and Corrosion Control* 53 (2018) 163–172.

[8] K. Roy, H.H. Lau, Z. Fang, R. Masood, T.C.H. Ting, J.B. Lim, V.C.C. Lee, Effects of corrosion on the strength of self-drilling screw connections in cold-formed

steel structures-experiments and finite element modeling, *Structures*, 36 (2022) 1080–1096.

[9] M. Xu, C.C. Lam, D. Wong, E. Asselin, Evaluation of the cathodic disbondment resistance of pipeline coatings–A review, *Progress in Organic Coatings*, 146 (2020) 105728.

[10] D.-H. Xia, C.-M. Deng, D. Macdonald, S. Jamali, D. Mills, J.-L. Luo, M.G. Strebl, M. Amiri, W. Jin, S. Song, Electrochemical measurements used for assessment of corrosion and protection of metallic materials in the field: A critical review, *Journal of Materials Science & Technology*, 112 (2022) 151–183.

[11] R.P. Nicolai, R. Dekker, J.M. Van Noortwijk, S. A comparison of models for measurable deterioration: An application to coatings on steel structures, *Reliability Engineering & System Safety*, 92 (2007) 1635–1650.

[12] F. Meng, L. Liu, E. Liu, H. Zheng, R. Liu, Y. Cui, F. Wang, Synergistic effects of fluid flow and hydrostatic pressure on the degradation of epoxy coating in the simulated deep-sea environment, *Progress in Organic Coatings*, 159 (2021) 106449.

[13] M. Zhang, H. Xu, A.L.P. Zeze, X. Liu, M. Tao, Coating performance, durability and anti-corrosion mechanism of organic modified geopolymer composite for marine concrete protection, *Cement and Concrete Composites*, 129 (2022) 104495.

[14] D. Thomas, E. Philip, R. Sindhu, S.B. Ulaeto, A. Pugazhendhi, M.K. Awasthi, Developments in smart organic coatings for anticorrosion applications: A review, *Biomass Conversion and Biorefinery*, 12 (2022) 4683–4699.

[15] N. Hoang, T.A. Khoa, L.T. Nhung, P.M. Phuong, T.D. Binh, T.T.X. Hang, N.V. Chi, T.-D. Nguyen, ZnAl alloy as an effective pigment in silicate coatings for the corrosion protection of steel, *Coatings*, 12 (2022) 1046.

[16] C. Qi, K. Dam-Johansen, C.E. Weinell, H. Bi, H. Wu, Enhanced anticorrosion performance of zinc rich epoxy coatings modified with stainless steel flakes, *Progress in Organic Coatings*, 163 (2022) 106616.

[17] X. Cao, F. Huang, C. Huang, J. Liu, Y.F.J.C.S. Cheng, Preparation of graphene nanoplate added zinc-rich epoxy coatings for enhanced sacrificial anode-based corrosion protection, *Corrosion Science*, 159 (2019) 108120.

[18] T. An, Y. He, Y. Feng, Experimental study on compression behavior of fiber-reinforced resin-based composite stiffened panels in hygrothermal environment, *Strength of Materials*, 52 (2020) 655–661.

[19] Z. Che, R.I. Revilla, C. Li, J. Liu, W. Liu, B. Wouters, K. Marcoen, X. Cheng, C. Liu, X. Li, Role of Te-RE alloying on the passive film and pitting corrosion behavior of 316L stainless steel, *Corrosion Science*, 240 (2024) 112457.

[20] P.-C. Lu, P.-F. Zhang, H. Feng, H.-B. Li, S.-C. Zhang, H.-C. Zhu, Z.-H. Jiang, Revealing the impact of inclusion types on pitting corrosion behavior of high-nitrogen stainless bearing steels with different Ce contents: Induction effect of oxides/sulfides and inhibition effect of nitrides, *Corrosion Science*, 236 (2024) 112247.

[21] S. Sen-Britain, S. Cho, S. Kang, Z. Qi, S. Khairallah, D. Rosas, V. Som, T.T. Li, S. Roger Qiu, Y. Morris Wang, Critical role of slags in pitting corrosion of additively manufactured stainless steel in simulated seawater, *Nature Communications*, 15 (2024) 867.

[22] Y.-J. Kim, C.B. Bahn, S.H. Baek, W. Choi, G.D. Song, Crevice chemistry and corrosion in high temperature water: A review, *Nuclear Engineering and Technology*, 56 (2024) 3112–3122.

[23] U. Kivisakk, Crevice corrosion testing of tubes, in U. Kivisakk, B. Espelid, and D. Feron (eds.) *Methodology of Crevice Corrosion Testing for Stainless Steels in Natural and Treated Seawaters*, CRC Press, 2024, pp. 21–29.

[24] J. Qin, W. Zhao, Z. Yan, R. Wang, Z. Yu, Z. Shi, C. Liu, B. Wang, Intergranular corrosion and stress corrosion cracking properties evaluation and mechanism study of multi-microalloyed 2519 Al alloy, *Corrosion Science*, 236 (2024) 112288.

[25] J.M. Kolotyrkin, Pitting corrosion of metals, *Corrosion*, 16 (1967) 12–19.

[26] W. Zhang, G. Frankel, Transitions between pitting and intergranular corrosion in AA2024, *Electrochimica Acta*, 48 (2003) 1193–1210.

[27] A.I. Ikeuba, C.N. Njoku, O.O. Ekerenam, D.I. Njoku, I.I. Udoh, E.F. Daniel, P.C. Uzoma, I.-I.N. Etim, B.O. Okonkwo, A review of the electrochemical and galvanic corrosion behavior of important intermetallic compounds in the context of aluminum alloys, *RSC Advances*, 14 (2024) 31921–31953.

[28] B.-S. Liu, Y.-H. Wei, W.-Y. Chen, L.-F. Hou, C.-L. Guo, Blistering failure analysis of organic coatings on AZ91D Mg-alloy components, *Engineering Failure Analysis*, 42 (2014) 231–239.

[29] D.Y. Perera, On adhesion and stress in organic coatings, *Progress in Organic Coatings*, 28 (1996) 21–23.

[30] W.J. Soer, W. Ming, C.E. Koning, R.A.T.M. van Benthem, J.M.C. Mol, H. Terryn, Barrier and adhesion properties of anti-corrosion coatings based on surfactant-free latexes from anhydride-containing polymers, *Progress in Organic Coatings*, 65 (2009) 94–103.

[31] G. Cao, Y. Yun, H. Xu, G. Yuan, J. Hu, G. Shao, A mechanism assessment for the anti-corrosion of zirconia coating under the condition of subcritical water corrosion, *Corrosion Science*, 152 (2019) 54–59.

[32] L. Lu, P. Zhu, S. Ding, Z. Ma, R. Li, S. Nie, W. Wang, G.M. Raftery, Impact of anti-corrosion coatings and maintenance on high-strength bolt friction connections in C4 marine environment, *Structures*, 68 (2024), 107098.

[33] X. Bai, B. He, P. Han, R. Xie, F. Sun, Z. Chen, Y. Wang, X. Liu, Corrosion behavior and mechanism of X80 steel in silty soil under the combined effect of salt and temperature, *RSC Advances*, 12 (2022) 129–147.

[34] M. Ascencio, M. Pekguleryuz, S. Omanovic, Corrosion behaviour of polypyrrole-coated WE43 Mg alloy in a modified simulated body fluid solution, *Corrosion Science*, 133 (2018) 261–275.

[35] A.K. Chaudhari, V.J. Singh, A review of fundamental aspects, characterization and applications of electrodeposited nanocrystalline iron group metals, Ni-Fe alloy and oxide ceramics reinforced nanocomposite coatings, *Journal of Alloys and Compounds*, 751 (2018) 194–214.

[36] Z. Wang, L. Zhang, Z. Zhang, M. Lu, Combined effect of pH and H2S on the structure of passive film formed on type 316L stainless steel, *Applied Surface Science*, 458 (2018) 686–699.

[37] I.M. Chohan, A. Ahmad, N. Sallih, N. Bheel, W.M. Salilew, A.H. Almaliki, Effect of seawater salinity, pH, and temperature on external corrosion behavior and microhardness of offshore oil and gas pipeline: RSM modelling and optimization, *Scientific Reports*, 14 (2024) 16543.

[38] H.M. Irshad, I.U. Toor, H.M. Badr, M.A.J.S. Samad, Evaluating the Flow Accelerated corrosion and erosion–Corrosion behavior of a pipeline grade carbon steel (AISI 1030) for sustainable operations, *Sustainability*, 14 (2022) 4819.

[39] A.S. Darmawan, W.A. Siswanto, P.I. Purboputro, A.D. Anggono, M. Masyrukan, A. Hamid, Effect of increasing salinity to corrosion resistance of 5052 aluminum alloy in artificial seawater, *Materials Science Forum*, 961 (2019), 107–111.

[40] D.H. Abdeen, M.A. Atieh, B. Merzougui, Corrosion behaviour of 316L stainless steel in CNTs–water nanofluid: Effect of temperature, *Materials*, 14 (2020) 119.

[41] A.R. Ghohrodi, M. Ramezanzadeh, B. Ramezanzadeh, Investigating the thermo-mechanical and UV-shielding properties of a nano-porous Zr (IV)-type metal-organic framework (MOF) incorporated epoxy composite coating, *Progress in Organic Coatings*, 164 (2022) 106693.

[42] B. Hou, X. Li, X. Ma, C. Du, D. Zhang, M. Zheng, W. Xu, D. Lu, F. Ma, The cost of corrosion in China, *npj Materials*, 1 (2017) 4.

[43] R.E. Ricker, Cost of corrosion, *Science*, 252 (1991) 1232–1232.

[44] M.J. Mazumder, Global impact of corrosion: Occurrence, cost and mitigation, *Global Journal of Engineering Sciences*, 5 (2020) 1–5.

[45] G. Cai, P. Yan, L. Zhang, H.-C. Zhou, H.-L. Jiang, Metal–organic framework-based hierarchically porous materials: Synthesis and applications, *Chemical Reviews*, 121 (2021) 12278–12326.

[46] M. Zhao, Y. Ban, Z. Chang, Y. Zhou, K. Yang, Y. Wang, N. Cao, W. Yang, Pyrazine-interior-embodied MOF-74 for selective CO_2 adsorption, *AICHE Journal*, 68 (2022) e17528.

[47] A. Hasanzadeh, A.R. SaadatAbadi, ZIF-67 metal–organic framework decorated g-C3N4 nanosheets with inbuilt organic–inorganic corrosion inhibitor for self-healing epoxy anti-corrosion coating, *Surface and Coatings Technology*, 492 (2024) 131218.

[48] X. Li, B. Gong, Y. Jin, Z. Chen, W. Yang, Intelligent water-borne coatings with self-reporting, self-healing, and robust barrier performance for the active corrosion protection of Q235 steels, *Progress in Organic Coatings*, 189 (2024) 108366.

[49] Y. Song, X. Wang, J. Tang, G.L. Li, Dynamic polymer/metal–organic framework hybrid microcapsules for self-healing anticorrosion coatings, *ACS Applied Materials & Interfaces*, 16 (2024) 68478–68486.

[50] R. Mohammadkhani, K. Sharifi, M. Fedel, B. Ramezanzadeh, Fabricating epoxy composite coating having self-healing/barrier anti-corrosion functions utilizing ion-exchange/pH-sensitive phosphate-doped ZIF8 MOF decorated Zn-Al-LDH nano-layers, *Surface and Coatings Technology*, 477 (2024) 130284.

[51] Z. Chen, J. Wang, L. Lu, T. Chowwanonthapunya, W. Chen, Z. Li, L. Ma, D. Zhang, Corrosion-oriented self-healing coating based on ZnMg MOF for efficient corrosion protection of aluminum alloy via in-situ LDH film formation, *Chemical Engineering Journal*, 500 (2024) 156898.

[52] X. Zhang, X. Tang, C. Zhao, Z. Yuan, D. Zhang, H. Zhao, N. Yang, K. Guo, Y. He, Y. He, A pH-responsive MOF for site-specific delivery of fungicide to control citrus disease of *Botrytis cinerea*, *Chemical Engineering Journal*, 431 (2022) 133351.

[53] R. Majidi, I. Danaee, L. Vrsalović, D. Zarei, Development of a smart anticorrosion epoxy coating containing a pH-sensitive GO/MOF nanocarrier loaded with 2-mercaptobenzothiazole corrosion inhibitor, *Materials Chemistry and Physics*, 308 (2023) 128291.

[54] M. Ramezanzadeh, B. Ramezanzadeh, M. Mahdavian, G. Bahlakeh, Development of metal-organic framework (MOF) decorated graphene oxide nanoplatforms for anti-corrosion epoxy coatings, *Carbon*, 161 (2020) 231–251.

[55] Y. Guo, J. Wang, D. Zhang, T. Qi, G.L. Li, pH-responsive self-healing anticorrosion coatings based on benzotriazole-containing zeolitic imidazole framework, *Colloids and Surfaces A: Physicochemical and Engineering Aspects*, 561 (2019) 1–8.

[56] T. Liu, D. Zhang, R. Zhang, J. Wang, L. Ma, P. Keil, A. Mol, X. Li, Self-healing and corrosion-sensing coatings based on pH-sensitive MOF-capped microcontainers for intelligent corrosion control, *Chemical Engineering Journal*, 454 (2023) 140335.

[57] B. Liu, L. Xiong, X. Fan, Anti-corrosion design of layered zirconium-based metal–organic frameworks (Zr-MOF) enhanced epoxy nanocomposite coatings, *Applied Surface Science*, 677 (2024) 161051.

[58] H. Pezeshk-Fallah, H. Yari, M. Mahdavian, B. Ramezanzadeh, S.A. Haddadi, T.H. Mekonnen, Yolk–shell structured carbon hollow spheres (CHS)/metal-organic framework nanoparticles-holding epoxy composite inheriting smart anti-corrosion feature, *Applied Materials Today*, 37 (2024) 102091.

[59] C. Yang, Y. Zhu, T. Wang, X. Wang, Y.J. Wang, Research progress of metal organic framework materials in anti-corrosion coating, *Journal of Polymer Engineering*, 44 (2024) 1–12.

[60] V.C. Anadebe, V.I. Chukwuike, K.C. Nayak, E.E. Ebenso, R.C. Barik, Combined electrochemical, atomic scale-DFT and MD simulation of Nickel based metal organic framework (Ni-MOF) as corrosion inhibitor for X65 pipeline steel in CO2-saturated brine, *Materials Chemistry and Physics*, 312 (2024) 128606.

[61] I.K. Nwokolo, H. Shi, A.I. Ikeuba, F. Liu, S. Ahmed, W. Zhang, Epoxy coating containing CoMOF@ MBT metal-organic framework for active protection of aluminum alloy, *Surface and Coatings Technology*, 477 (2024) 130349.

6 Introduction to Metal-Organic Frameworks in Corrosion Mitigation

Alok Kumar Yadav and Sourav Maity

6.1 INTRODUCTION

Since last five decades many materials, including porous ranging from zeolites, coordination polymers, and metal–organic framework (MOF), are garnering attention due to their impeccable industrial applications. The most contrasting and interesting characteristics that make them impeccable are their porosity that allow a variety of foreign molecules easily into their cavities. MOF is colloquially also known as porous coordination polymers (PCPs), a special type of porous crystalline material having a significant application in analytical and bioanalytical area [1]. MOF is not known only as PCPs but also by many other names like microporous coordination polymers (MCPs) and porous coordination networks (PCNs) [2]. These are basically a new type of nonporous materials having ultralow density, high surface area, and regular pore structure. In comparison to traditional porous materials like active carbon and zeolites, MOFs possess more flexibility for creating a new structure and surface chemistry and make them trustworthy for application in gas sorption/separation, catalysis, and sensors [3]. These are manufactured by formation of chemical bonds between organic ligands as linkers and metal ions as nodes, which ultimately leads to generation of crystalline structure having high porosity and large surface area, meant for wide applications in various fields of science and technology [4] (Figure 6.1).

The application for these types of materials was raised in early 1990 due to their reticular design and synthesis [5]. The credit for developing the first MOF is given to Omar M. Yaghi and his team member in 1995, and in 1998 they designed the first MOF $Zn_4O(BDC)_3{\cdot}(DMF)_8(C_6H_5Cl)$ named as MOF-5 which composed of tetrahedral l $[Zn_4O]^{6+}$ with 1,4-benzenedicarboxylate (BDC) ligand to form a compact three-dimensional (3D) cubic unit structure. Many other MOFs like isoreticular MOFs (IRMOF-1 to IRMOF-16) were synthesized having a morphological form of MOF-5 topology [6]. Many other zeolites were also discovered like zeolitic imidazolate framework (ZIF), zeolite-like MOF (ZMOF), metal azolate frameworks (MAFs), metal peptide framework (MPF), and metal biomolecule framework (bioMOF) based on their pore size, particle dimension, surface area, and many other factors [7]. Due to the diverse choice of metal and linkers, it is quite simple and easy to design and synthesize a large variety of MOF material. MOF-5 $[Zn_4O(BDC)_3]$ was synthesized

DOI: 10.1201/9781003571162-6

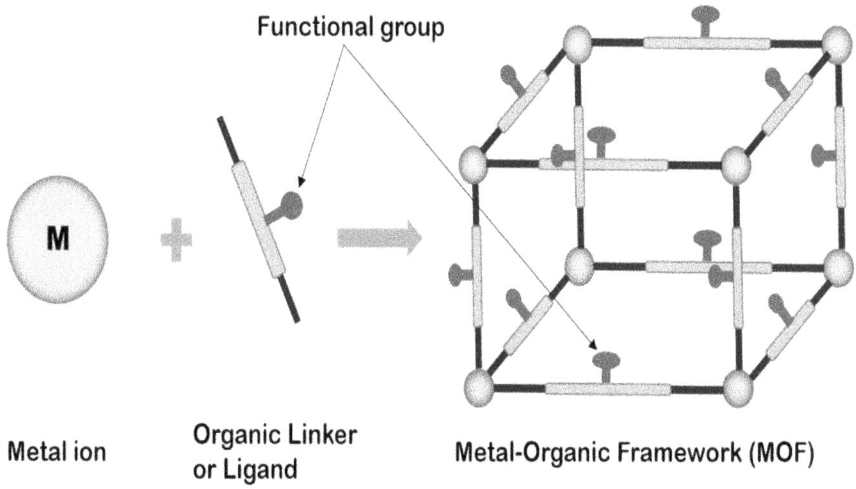

FIGURE 6.1 Schematic illustration of the formation of MOFs.

from the combination of Zn^{2+} ions and 1,4-dicarboxybenzene (H_2BDC). After analyzing the structure through single-crystal X-ray diffraction, it was reported to be the first MOF having richness in stable framework and rich porosity [8].

6.2 BASIC STRUCTURES AND CHARACTERISTICS OF METAL ORGANIC FRAMEWORKS (MOFS)

MOFs consist of both organic as well as inorganic units. The organic unit (linkers/bridging ligands) primarily consist of carboxylates or anions like mainly phosphonate, sulfonate, and heterocyclic compounds [1] (Figure 6.2). A variety of organic molecules are used as linkers in the synthesis of MOFs like 2,4-aminoterepthalic acid, 4-pyridinecarboxylic acid, 1,2-bis-(4-pyridyl)ethane, 1,3,5-benzene tricarboxylic acid, and many more. The inorganic units are the metal ions or cluster termed as secondary building units (SBUs).

The geometry of MOFs is determined by the coordination number, coordination geometry of the metal ions, and the nature of its functional group. A variety of SBU's geometry with different points of extension such as octahedron (six points), trigonal prism (six points), square paddle-wheel (four points), and triangle (three points) is observed in MOF. MOF with large space may result in the formation of interpenetrating structures and hence become very much important to avoid interpenetration after carefully choosing the organic linkers. MOFs are highly porous and crystalline materials. They are characterized by exceptionally high surface areas, tunable structures, and modular designs that allow for the fine-tuning of their properties for specific applications [9] (Figure 6.3). The large porosity and high thermal and chemical stability allow their application in the adsorption as well as separation of gases [4, 10], catalysis [11, 12], microelectronics [13], drug delivery [1, 14], bioimaging

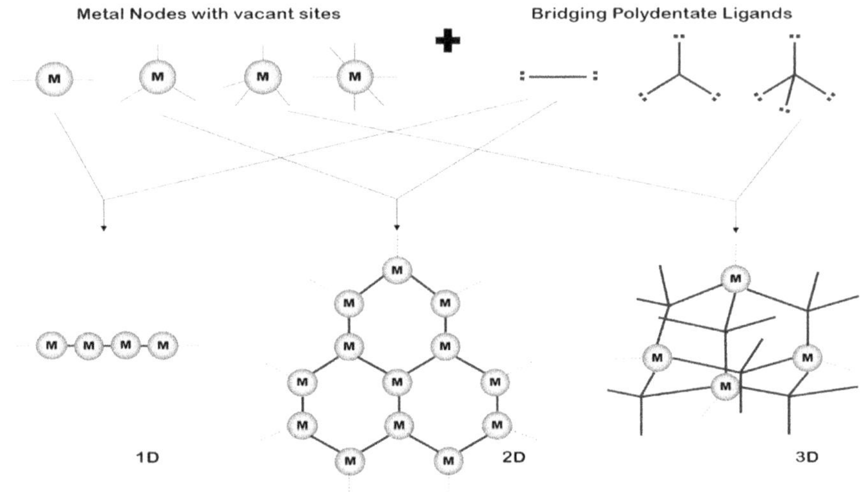

FIGURE 6.2 Formation of MOFs from different metal nodes and bridging.

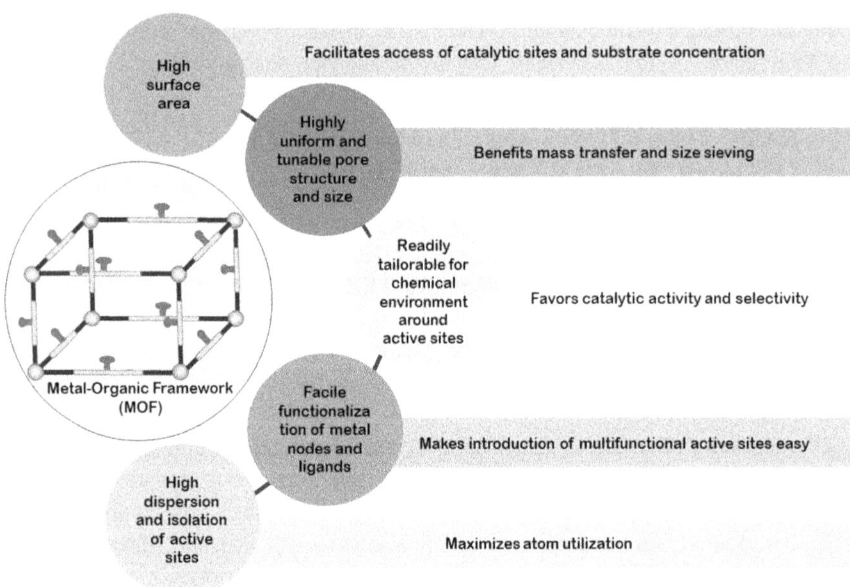

FIGURE 6.3 Some unique physical and chemical characteristics of MOFs.

[15, 16], conductivity [17, 18], energy storage [19, 20], magnetism [21, 22], and many more applications. Many MOFs like UiO-66 (University of Oslo), HKUST (Hong Kong University of Science and Technology), MIL (Materials Institute Lavoisier), SNU (Seoul National University), JUC (Jilin University China), ITQMOF

(Instituto de Tecnología Química MOF), POST (Pohang University of Science and Technology), and CUK (Cambridge University KRICT) were named after the universities as change in the nomenclature system. The present chapter outlines the types of MOFs, different methods for synthesis of MOFs, factors influencing their synthesis, and their potential applications in various fields.

6.3 CLASSIFICATION OF METAL ORGANIC FRAMEWORK (MOF)

Since the last two decades, various approaches were applied for the synthesis of MOFs with different crystal structure, high surface area, particle size, and pore size distribution. When combined with specific materials like metal oxides, quantum dots, polymers, polyoxometalates, enzymes, and others, MOFs can be prepared with desirable functional diversity and desirability. The linkers and bridging ligands can be divided into following types:

6.3.1 Isoreticular MOF

It is a series of MOFs first discovered by Yaghi et al. [23]. These are in the series of self-assembled MOFs having different pore size and similar topologies which are synthesized by selecting different ligand and connected with $Zn_4O_6^+$ tetrahedral [16]. The major drawback of this type of MOF is that when exposed to moist air, it starts decomposing gradually, and stability is another major drawback from the application point of view [24].

6.3.2 Zeolitic Imidazolate Frameworks

ZIFs consist of imidazolate linkers and metal ions. These MOFs have a unique porous structure as well as better thermal and chemical stability, which make it suitable for varied applications [25]. The application of ZIF is limited due to poor hydro stability of its metal–ligand coordination bonds in specific humid and aqueous condition [26].

6.3.3 Porous Coordination Networks

PCPs/MOFs synthesized by combining metal ions with organic bridging ligands has intriguing porous materials due to high regularity in structure and diversity, can be easily modified, has good porosity and flexibility in structure [23, 27–29]. Some of the PCNs are PCN-333, PCN-224, and PCN-57 [30]. Some of the PCNs are applied as DNA sensors like PCN-222 [31].

6.3.4 Porous Coordination Polymers

These types of polymers are usually made by cross-linking/clustering of organic ligands that leads to the formation of 3D structure [32]. These materials are characterized by their high porosity, tunable structure, and very high diverse chemical functionality, which ultimately leads to becoming applicable in various fields of analytical and bioanalytics [33]. The major key features of these types of polymers are high porosity, pore size of tunable nature, and structural versatility, which can range from 1D to 3D [34].

The detailed comparative analysis of these polymers is presented in Table 6.1.

6.4 METHODS FOR SYNTHESIS OF MOF'S

As we know, MOFs usually consist of two components: metal ions and organic ligands. The synthesis of MOFs is usually conducted by mixing a significant proportion of metal ions with organic linkers for the creation of porous and crystalline material. Methods developed for synthesis of MOFs had attracted immense attraction due to their varied and geometrically complex structure with immense application in many fields related to porous and crystalline material. The methods included for the synthesis of these geometrically complex MOFs are mainly microwave-assisted methods, sonochemical methods, electrochemical methods, mechanochemical methods, the hydrothermal method, solvothermal method, etc. (Figure 6.4).

6.4.1 Microwave-Assisted Method

This method is applied mainly for the synthesis of MOFs under aqueous conditions. Using microwaves, small metal and oxide particles are produced [35]. This approach is mainly applied in organic chemistry for the synthesis of nanosized metal oxides [36]. For design of nanosized metal crystals, the temperature of the solution is raised by microwave for around 24 hours. In this method, a Teflon vessel is used where a substrate mixture is blended with a solvent and sealed for around 24 hours under the microwave condition for appropriate time and temperature. Here, microwave plays a very crucial role in the conversion of electromagnetic energy into thermal energy and heating the liquid mixture [37]. Under the influence of oscillating and electromagnetic field, the polar molecules will try to align themselves and change their orientation. Further, the application of appropriate frequency leads to collision of the molecules and ultimately raises the temperature of the system and system kinetic energy. Microwave-assisted MOF synthesis is mainly utilized for the rapid crystallization and formation of nanoscale products [38].

Various organic–inorganic hybrid materials were formed by the present methods, with chromium trimesate being one among them synthesized under the aqueous conditions [39]. This method used for the synthesis of MOFs under aqueous conditions was around 20 times faster than the conventional method as employed earlier in the synthesis of such MOFs. The same method was employed to synthesize other MOFs namely chromium terephthalate under the influence of microwave condition. The outcome of the MOFs was reported to be giant pores, very large surface area, and having very small nanoscale dimension [40]. The application of MOFs comes out as having high adsorption of benzene, which can be employed further for the removal of sorptive contaminants. The crystal size of the MOFs decreases significantly with increases in the concentration of water and pH. Decrease in the size with dilution can be explained with the smaller rate of the crystal formation as compared to the nucleation rate.

TABLE 6.1
A Comprehensive Classification of MOFs [16, 26, 33]

Feature	IRMOFs	ZIFs	PCNs	PCPs
Definition	A type of MOF that has similar or identical topologies but having different metallic centers or organic linkers	MOF having characteristics of zeolite-like structure and imidazolate-based ligands	Metal centers linked by organic ligands to form highly porous structure	Class of materials consisting of metallic centers with coordination of organic ligands, which leads to the formation of highly ordered network with marginal porosity
Framework type	3D	3D	3D or 2D	1D, 2D, or 3D
Porosity	Having high surface area and tunable porosity	Having high surface area and tunable porosity	Ultra-high surface area and tunable porosity	Highly porous and varying pore size based on metal–ligand chemistry.
Metal center	Namely, d block elements are taken like Zn, Cu, or Ni.	Zn and Co are most commonly used in these MOFs	Any metals from d block, lanthanides, or actinoids	Metal from d block (e.g., Zn, Cu, Ni) and transition metal or rare earth metal.
Ligands	Organic or similar ligands like terephthalic acid (BDC) are used.	Imidazolate ligands	Organic ligands like carboxylates, phosphonates are used	Organic ligands like carboxylates, phosphonates, bipyridines are used.
Stability	Moderately stable in nature	Highly stable in both air and water	Stable	Variation in stability based on metal–ligand bindings
Applications	Storage of gas, catalysis, etc.	Gas storage, catalysis, etc.	Sensing application and many more	Gas storage, sensing, etc.
Pore size range	Microporous to mesoporous	Microporous	Can be reversed to microporous to mesoporous or vice versa	Microporous to mesoporous

(*continued*)

TABLE 6.1 (Continued)
A Comprehensive Classification of MOFs [16, 26, 33]

Feature	IRMOFs	ZIFs	PCNs	PCPs
Example	IRMOF-1 (made of terephthalic acid and Zn)	ZIF-8 (Zn imidazolate framework)	PCN-250 (using lanthanide metals and organic linkers)	MIL-53 (based on Al and organic carboxylate ligands)
Thermal stability	Moderate	High thermal stability	Good stability	Ranging from moderate to high
Chemical stability	Moderate to low	Excellent stability in aqueous and acidic/basic condition	Good chemical stability	Stable

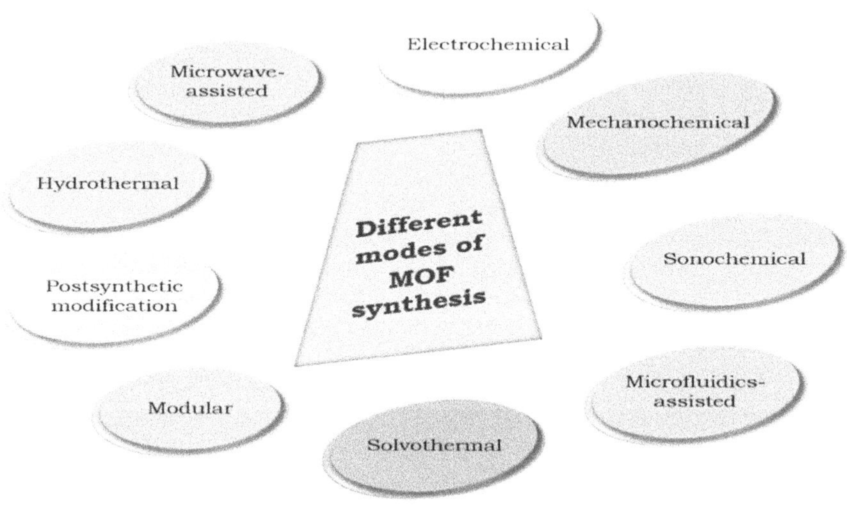

FIGURE 6.4 Different methods for synthesis of MOFs.

6.4.2 Sonochemical Method

Under this method of synthesis, MOFs are usually conducted at room temperature. On applying ultra-radiation, it decreases the time for crystallization. Here in this method, a cyclic mechanical vibrator is used for the synthesis of MOFs having a frequency range of 20 kHz to 10 MHz [41]. Without using the external cooling system, a mixture of metal salt and organic linker is being added in a pyrex reactor with attached sonicator bar with variable power output [35]. The formation of crystal occurs at temperature of 4000–5000 K and an ambient pressure of 1000 bar. The sonicator was used for the first time for the synthesis of MOFs in 2008. Applying the sonication method, IRMOF having cubical topology of the acetylenedicarboxylic acid was synthesized.

6.4.3 Electrochemical Method

This method is employed for the construction of a variety of MOFs. BASF synthesized first HKUST-1 in 2005 using electrochemical method [42]. In this method, synthesis is carried out in the absence of any metal salt and any anions like nitrate/byproducts are avoided in it. Here, in this method, metal ions are added in the electrochemical gradient method rather than solution of metal salts [38]. Applying a suitable voltage to dissolve and particles responsible for MOF creation are present outside of the electrode surface. The metal ions dissolved at ambient voltage respond instantly with the organic linkers present in the solution and lead to the synthesis of MOFs.

6.4.4 Solvothermal Method

This method is commonly used for the preparation of MOFs due to its ease, convenience of usage, crystallinity, and high usage and having enough yield [43]. In this method, organic ligands and metal salt precursors are dissolved in a certain ratio in particular solvent and later on loaded in the glass tubes or stainless-steel-coated Teflon autoclaved with proper sealing in oven for heating. The reaction for synthesis of this kind of MOFs is usually carried out at around 100–200°C that should remain above the boiling point of the solvent. Such high temperature condition will provide a conducive environment for the generation of sufficient pressure to dissolve the reactants and hence promote the nucleation rate which ultimately leads to the MOF's crystal formation of desirability within 2–4 days. In solvothermal method, most commonly used solvents for the synthesis of MOFs are mainly N,N-dimethylformamide, N,N-diethyl formamide, glycol, acetonitrile, acetone, ethanol, methanol, etc. If other than these organic solvents water is used, then it will become hydrothermal method [44]. The merits of solvothermal method include simple operation and equipment requirement, minimum reaction time, and high rate of crystal formation, thus making it the most suitable method for the synthesis of MOFs. This method is conducted in the sealed container having high temperature and pressure, and having complicated environment and reaction progress is difficult to monitor.

6.4.5 Solid-State Synthesis

It is generally considered as the green method for the synthesis of where reactants in solid state are mixed in particular ratio with following of heating or grinding for the integration of the reaction. Mixing of the zinc oxide (ZnO) with 2-methylimidazole and heating at 100°C for 24 hours formation of highly porous material and stable MOF's MAF-4 (ZIF-8) will take place [45]. Similarly, when the ZnO and 2-methylimidazole are mixed and applied with hard grinding and ball milling, it leads to the formation of nonporous framework like $Zn(Im)_2$ as product [45]. When a small amount of the solvent and NH_4NO_3 are added and subjected to milling, it leads to the formation of porous ZIF products as MOFs. Such addition of solvents enhances the mobility of the molecules and promotes the progress of the reaction. However, this solid-state synthesis method is employed for the synthesis of MOFs with relatively low yield.

Apart from these, other methods are also mentioned for the synthesis of MOFs, such as diffusion method, spray drying method, isothermal method, and chemical vapor deposition, which are widely being practiced for the synthesis of MOFs of desirability [46].

6.4.6 Spray Drying Method

This method involves atomization of the liquid precursor solution or suspension into the fine droplets with use of the hot air or gas. In this method, solvent will get evaporated very easily and leaves behind solid particle as powder form which will get collected. This method is employed heavily in industrial operation for drying the liquids and suspension to form powders. The major advantages of the spray drying method for the MOF synthesis is that it has scalability for industrial operation, capable of synthesizing MOF of particular size, rapid in nature for formation of the MOFs of desirability, and solvent can be recovered very easily [47].

6.4.7 Diffusion Method

It is a technique for MOF's synthesis where the growth of the material occurs through the diffusion of the metal ions and organic ligands into the solvent under controlled conditions. This method is quite simple, straightforward, inexpensive in nature, and capable of producing well-defined crystals of high quality. The diffusion method is mainly divided into two types: solvent diffusion method (liquid–liquid diffusion) and vapor diffusion method. The solvent diffusion method (liquid–liquid diffusion) involves the diffusion of two liquid phases consisting of metal salts and organic linkers together into another solvent capable of diffusing them. In this method, MOFs are usually synthesized at the interphase. Likewise, in the vapor diffusion method, a volatile solvent containing metal ions and organic ligands is used and its synthesis takes place in the closed vessel. The metal and the ligand solution later on are separated apart in the reaction chamber and its diffusion occurs in the vapor phase. In another method, gel diffusion method is also practiced where agarose or silica gel is employed for the slow diffusion of the metal ions and the ligands [44].

6.4.8 Isothermal Method

In this method, the synthesis of MOFs is usually conducted while keeping the temperature constant throughout the reaction. In this method, metal and the organic ligands are dissolved in the solvent and the solution is maintained at fixed temperature for a certain period of time to allow the formation of the MOFs. The major advantage of using this technique for the synthesis of the MOF's is the precise control of the process for the synthesis of the crystal, capable of operating at the lower temperature as compared to the solvothermal method and having high purity of the MOFs [48, 49].

Presently, attention is devoted to the synthesis of MOFs for upgrading the crystallization of MOFs by rapid reactions. Merits and demerits of various synthesis methods for MOFs are presented in Table 6.2.

6.5 APPLICATION OF MOF'S

Three acting properties in the MOFs mainly include crystalline nature, porosity, and presence of strong metal–ligand interactions. A rare combination of such makes it an exceptional and valuable material for the application in various fields of analysis. Their high surface area, low density, and many other factors make it desirable for application in many fields like gas storage, drug delivery, super capacitors, and catalysis and sensing [55] (Figure 6.5).

A summarized representation of the applications of MOFs with their merits and examples is presented in Table 6.3.

MOFs have been utilized for anticorrosive coatings because of their porous structure and strong ability to be applied with corrosion inhibitors (Figure 6.6). MOFs exert their effects primarily via two mechanisms: barrier formation and self-healing [9]. The formation of tightly cross-linked gel-type layers provides a physical barrier to the penetration of corrosive species such as water and chloride ions, thereby shielding the underlying metal interface. Because of their multifunctional properties, such as being functional substrate supports and nanofillers, MOFs are receiving increasing attention for their application in anticorrosive coatings [63].

MOFs can be a functional substrate to support corrosion inhibitors or active agents. MOFs can encapsulate corrosion inhibitors (chromates, phosphates, etc.) and release these agents gradually dependent on their high surface area and tunable porosity. This framework allows for a controlled release of the inhibitor, providing durable protection of metal substrates from corrosive environments and minimizing degradation [9].

When added as nanofillers in coatings, MOFs improve the mechanical properties of the coating. The nanoscale and high surface area of the MOFs enhance the coating's adhesion, hardness, and durability. It can also act as a barrier, decreasing the permeability of water or oxygen as an example of corroding species, thus increasing the protective role of the coating. MOFs also have a synergistic enhancement for the performance of coating, especially when used with other nanoparticles (e.g., graphene) [64].

MOFs are substantially improving anticorrosive coatings by providing a functional support for the less soluble inhibitor release, and exhibiting nanofiller properties resulting in enhanced mechanical and protective attributes of the coating (Figure 6.7).

6.6 FUTURE CHALLENGES

MOFs show enormous potential but face challenges before such structures can see widespread commercial and industrial applications. One of the main challenges to their further use, for instance, in anticorrosive coatings is their stability, because many, in particular, MOFs are sensitive to the environment (moisture, temperature, aggressive chemicals) and may degrade their structure with time. Scalability remains

TABLE 6.2
Brief Summary of Different Methods for MOF Synthesis [50–54]

Method	Description	Merits	Demerits
Solvothermal synthesis	This is one of the most common methods for synthesizing MOFs, where metal salts and organic ligands are dissolved in a solvent, and the mixture is heated under pressure in a sealed vessel (autoclave)	• *High-Quality Crystals:* Provides high-quality, well-defined crystals. • *Wide Variety of MOFs:* Suitable for a wide range of MOFs with different metal ions and ligands. • *High Yield:* Can produce large quantities of MOFs.	• *High Energy Consumption:* Requires high temperatures and pressure, leading to high energy costs. • *Long Reaction Times:* Synthesis times can range from several hours to days. • *Solvent Waste:* Involves the use of solvents, which can lead to environmental concerns
Hydrothermal synthesis	Similar to solvothermal synthesis, but uses water as the solvent. The reaction is carried out in an autoclave under high temperature and pressure.	• *Environmentally Friendly:* Uses water, a green solvent, which reduces environmental concerns compared to organic solvents. • *High Crystallinity:* The high pressure and temperature conditions promote the formation of highly crystalline MOFs. • *Scalable:* Suitable for large-scale production.	• *High Energy Consumption:* Like solvothermal synthesis, it requires high temperatures and pressures. • *Limited to Water-Stable MOFs:* This method is limited to MOFs that are stable in aqueous environments. • *Long Reaction Times:* Synthesis can take a long time, sometimes several days.
Electrochemical synthesis	Similar to solvothermal synthesis, but uses water as the solvent. The reaction is carried out in an autoclave under high temperature and pressure.	• *Low Temperature:* Typically performed at room temperature, reducing energy costs. • *Fast Reaction Time:* The electrochemical method is typically faster than traditional solvothermal synthesis. • *Controlled Structure:* Allows fine control over the morphology and particle size of MOFs.	• *Limited Scalability:* This method may not be as easily scalable for large-scale production. • *Narrow Range of Materials:* Limited to specific metal ions and ligands. • *Requires Specialized Equipment:* Requires precise electrochemical setups and monitoring.

Microwave-assisted synthesis	In this method, microwave radiation is used to rapidly heat the reaction mixture, leading to faster synthesis of MOFs compared to conventional heating.	• *Fast Reaction Time:* Synthesis can be completed in a matter of minutes to hours. • *Uniform Heating:* Microwaves provide uniform heating, leading to better-controlled reaction conditions. • *Energy Efficient:* More energy-efficient than conventional heating methods.	• *Limited to Small-Scale Reactions:* Currently, the method is more suited for small-scale laboratory synthesis. • *Not Suitable for all MOFs:* Some MOFs may not be amenable to microwave synthesis due to their instability under microwave irradiation. • *Potential for Over-Heating:* Requires careful control to prevent over-heating and degradation of the MOFs.
Room-temperature synthesis	This method involves synthesizing MOFs at ambient temperature without the need for heating or high-pressure conditions.	• *Energy Efficient:* No high temperatures or pressures are required. • *Simple and Fast:* The reaction can take place in a matter of hours at room temperature. • *Environmentally Friendly:* No need for toxic or volatile solvents.	• *Lower Crystallinity:* MOFs synthesized at room temperature may exhibit lower crystallinity compared to those synthesized at higher temperatures. • *Limited Materials:* Not all MOFs can be synthesized under room-temperature conditions. • *Challenges in Scalability:* Scaling up the process while maintaining consistency can be difficult.

(*continued*)

TABLE 6.2 (Continued)
Brief Summary of Different Methods for MOF Synthesis [50–54]

Method	Description	Merits	Demerits
Solvent-free synthesis (mechanochemical synthesis)	In mechanochemical synthesis, MOFs are synthesized without the use of solvents, using mechanical energy to induce the reaction between metal salts and organic ligands. This is usually achieved by grinding the solid components together in a ball mill.	• *Environmentally Friendly:* No use of solvents, making it a green synthesis method. • *Energy Efficient:* Lower energy consumption compared to solvothermal or hydrothermal methods. • *Scalable:* Can be scaled up to industrial production.	• *Low Yield:* Can result in lower yields compared to solvent-based methods. • *Limited to Specific MOFs:* Not all MOFs can be synthesized via mechanochemical methods. • *Crystallinity Issues:* The crystallinity of MOFs can be compromised due to the lack of solvents.
Template-assisted synthesis	This involves the use of a template (such as a nanoparticle or a porous material) around which the MOF is grown. The template can later be removed to leave behind the MOF structure.	• *Control Over Morphology:* This method allows fine control over the shape and size of the MOF. • *Highly Crystalline MOFs:* Often results in well-defined, highly crystalline MOFs. • *Versatility:* Can be used for a wide range of MOFs.	• *Complex Process:* The synthesis process is often more complicated and requires additional steps for template removal. • *Template Cost:* The use of templates can increase the cost of synthesis. • *Template Contamination:* If not properly removed, templates can contaminate the final product.

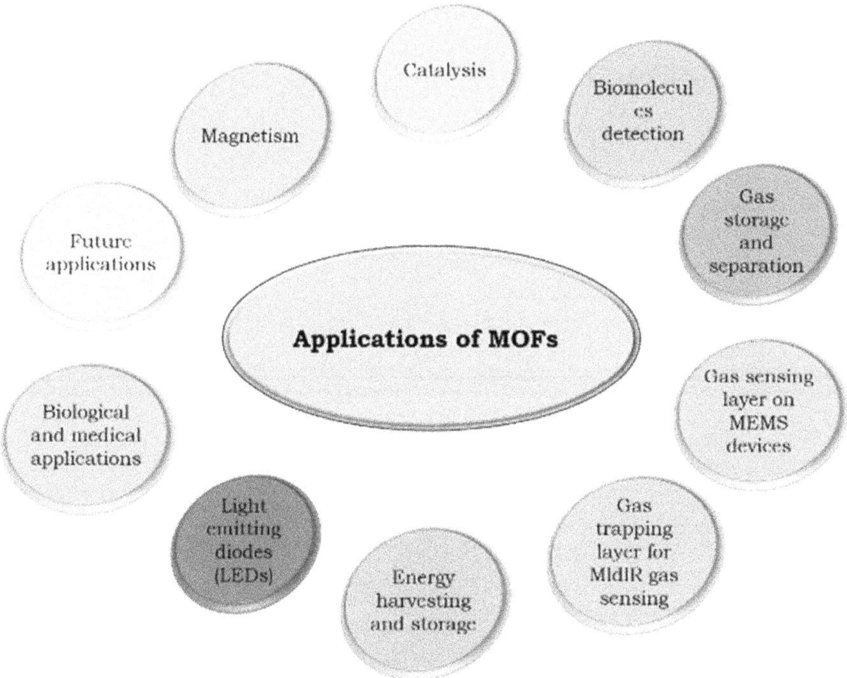

FIGURE 6.5 Different applications of MOFs.

a problem, as the synthesis of MOFs can be complex and expensive, which hinders large-scale production. Moreover, processability emerges as a challenge because incorporating MOFs into coatings without losing crucial properties such as porosity and structural integrity requires sophisticated methodologies. MOFs are very tunable, but turning them into the ideal functionalization for applications like corrosion protection is also a challenge. Concerns regarding their environmental impact and the necessity for standardized safety regulations persist, highlighting the need for further research to ensure their safe and sustainable deployment.

6.7 CONCLUSION

The synthesis of different types of MOFs using different methods plays a vital role in determining the factors influencing their synthesis and their potential applications. Although various methods of the synthesis of MOF are reported but the choice of synthesis methodology depends on the requirement and application perspective of the researcher. Here, among all the methods, only hydro conventional method is seems to be suitable for the synthesis of MOFs due to its various merits over other methods like electrochemical, mechanochemical, and many others. The potential application of MOFs is reported in various fields like next-generation rechargeable batteries and novel super capacitors. These MOFs with well-regulated morphology and flexible structure, high surface area, and huge pore volume make them suitable

TABLE 6.3
Different Applications of MOFs

Sl. No.	Application	Merit	Examples of MOFs	References
1	Gas Storage (Hydrogen)	• High surface area • Reversible adsorption • Low temperature and pressure operation	ZIF-8, MOF-5	[50]
2	Gas Storage (Methane)	• High storage capacity • Efficient at cryogenic temperatures • Tunable pore size	HKUST-1, MOF-74	[56]
3	CO_2 Capture and Sequestration	• High selectivity for CO_2 • High adsorption capacity • Reversible adsorption	UIO-66, Mg-MOF-74	[57]
4	Oxygen Storage and Delivery	• Controlled release • High storage capacity • Biocompatibility for medical use	Fe-MOFs, Co-MOF-74	[58]
5	Ammonia Storage	• High capacity • Selectivity for ammonia • Reversible adsorption	MIL-101	[59]
6	Catalysis	• High surface area • Tunable catalytic sites • Can support both homogeneous and heterogeneous catalysts	MOF-5, ZIF-8, MIL-101	[60]
7	Drug Delivery	• High selectivity for contaminants • Effective in removing heavy metals and organic toxins	MIL-100(Fe), ZIF-8	[61]
8	Supercapacitors	• High surface area • Tunable pore structure • High energy storage capacity	Ni-MOF, Co-MOF-74	[62]

for the designing of energy storage devices. However, some drawbacks associated with MOFs are that they are unstable in air and retard their electrochemical capacity. Some research was also conducted to improve their electrochemical capacity by adding certain types of heteroatom by doping or metal oxide to enhance their capacity to a certain extent. Another drawback associated with the MOFs is that they are water-unstable and therefore water-resistant MOFs need to be developed. MOFs are widely applied in the biomedical field like drug delivery system.

Research in the field of MOFs is expanding in both academic and industry. MOFs are the new type of material that has incredible adaptability and capability to address the concern of the 21st century. In the next few years, nanoscale-based MOFs will be developed that can be applicable in various fields.

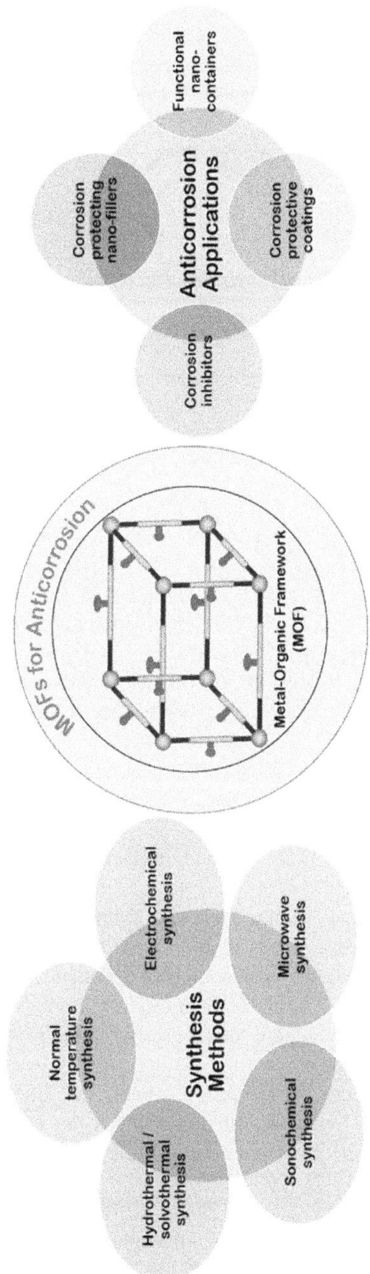

FIGURE 6.6 MOFs applications and types of synthesis required for corrosion protection.

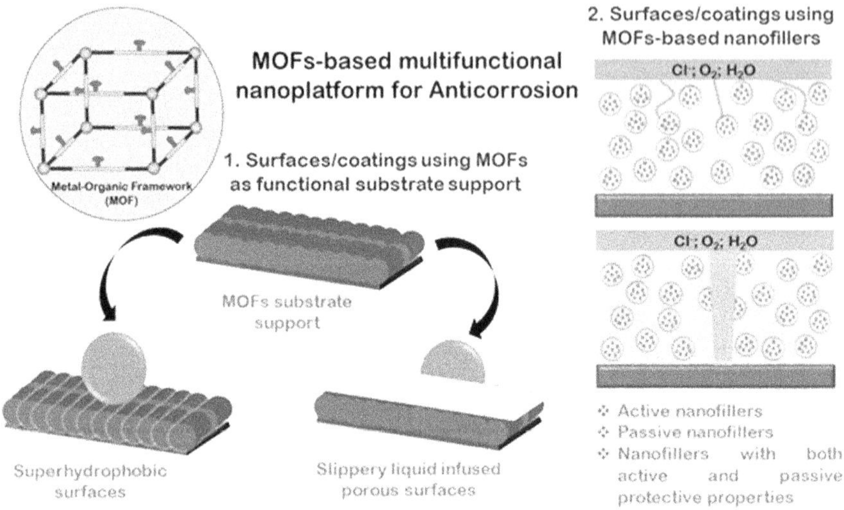

FIGURE 6.7 MOFs-based anticorrosion surfaces/coatings.

REFERENCES

[1] D. Li, A. Yadav, H. Zhou, K. Roy, P. Thanasekaran, C. Lee, Advances and applications of metal–organic frameworks (MOFs) in emerging technologies: a comprehensive review, *Glob. Challenges*. 8 (2024) 2300244.

[2] W. Xia, A. Mahmood, R. Zou, Q. Xu, Metal–organic frameworks and their derived nanostructures for electrochemical energy storage and conversion, *Energy Environ. Sci*. 8 (2015) 1837–1866.

[3] W. Chaikittisilp, K. Ariga, Y. Yamauchi, A new family of carbon materials: synthesis of MOF-derived nanoporous carbons and their promising applications, *J. Mater. Chem. A*. 1 (2013) 14–19.

[4] H.-C. Zhou, J.R. Long, O.M. Yaghi, Introduction to metal–organic frameworks, *Chem. Rev*. 112 (2012) 673–674.

[5] B.F. Hoskins, R. Robson, Infinite polymeric frameworks consisting of three dimensionally linked rod-like segments, *J. Am. Chem. Soc*. 111 (1989) 5962–5964.

[6] Z. Mai, D. Liu, Synthesis and applications of isoreticular metal–organic frameworks IRMOFs-n (n= 1, 3, 6, 8), *Cryst. Growth Des*. 19 (2019) 7439–7462.

[7] J.-R. Li, J. Sculley, H.-C. Zhou, Metal–organic frameworks for separations, *Chem. Rev*. 112 (2012) 869–932.

[8] C. Volkringer, D. Popov, T. Loiseau, G. Férey, M. Burghammer, C. Riekel, M. Haouas, F. Taulelle, Synthesis, single-crystal X-ray microdiffraction, and NMR characterizations of the giant pore metal–organic framework aluminum trimesate MIL-100, *Chem. Mater*. 21 (2009) 5695–5697.

[9] C. Zhou, M. Pan, S. Li, Y. Sun, H. Zhang, X. Luo, Y. Liu, H. Zeng, Metal organic frameworks (MOFs) as multifunctional nanoplatform for anticorrosion surfaces and coatings, *Adv. Colloid Interface Sci*. 305 (2022) 102707.

[10] H. Daglar, H.C. Gulbalkan, G. Avci, G.O. Aksu, O.F. Altundal, C. Altintas, I. Erucar, S. Keskin, Effect of metal–organic framework (MOF) database selection on the

assessment of gas storage and separation potentials of MOFs, *Angew. Chemie Int. Ed.* 60 (2021) 7828–7837.
[11] A. Bavykina, N. Kolobov, I.S. Khan, J.A. Bau, A. Ramirez, J. Gascon, Metal–organic frameworks in heterogeneous catalysis: recent progress, new trends, and future perspectives, *Chem. Rev.* 120 (2020) 8468–8535.
[12] V. Pascanu, G. González Miera, A.K. Inge, B. Martín-Matute, Metal–organic frameworks as catalysts for organic synthesis: a critical perspective, *J. Am. Chem. Soc.* 141 (2019) 7223–7234.
[13] Y. Cui, Y. Yue, G. Qian, B. Chen, Luminescent functional metal–organic frameworks, *Chem. Rev.* 112 (2012) 1126–1162.
[14] H.D. Lawson, S.P. Walton, C. Chan, Metal–organic frameworks for drug delivery: a design perspective, *ACS Appl. Mater. Interfaces.* 13 (2021) 7004–7020.
[15] Y. Liu, T. Jiang, Z. Liu, Metal–organic frameworks for bioimaging: strategies and challenges, *Nanotheranostics.* 6 (2022) 143.
[16] K. Wang, M. Cai, D. Yin, X. Hu, H. Peng, R. Zhu, M. Liu, Y. Xu, C. Qu, J. Ni, IRMOF-8-encapsulated curcumin as a biocompatible, sustained-release nano-preparation, *Appl. Organomet. Chem.* 36 (2022) e6680.
[17] L.S. Xie, G. Skorupskii, M. Dinca, Electrically conductive metal–organic frameworks, *Chem. Rev.* 120 (2020) 8536–8580.
[18] C. Li, L. Zhang, J. Chen, X. Li, J. Sun, J. Zhu, X. Wang, Y. Fu, Recent development and applications of electrical conductive MOFs, *Nanoscale.* 13 (2021) 485–509.
[19] B. He, Q. Zhang, Z. Pan, L. Li, C. Li, Y. Ling, Z. Wang, M. Chen, Z. Wang, Y. Yao, Freestanding metal–organic frameworks and their derivatives: an emerging platform for electrochemical energy storage and conversion, *Chem. Rev.* 122 (2022) 10087–10125.
[20] I. Hussain, S. Iqbal, C. Lamiel, A. Alfantazi, K. Zhang, Recent advances in oriented metal–organic frameworks for supercapacitive energy storage, *J. Mater. Chem. A.* 10 (2022) 4475–4488.
[21] G.M. Espallargas, E. Coronado, Magnetic functionalities in MOFs: from the framework to the pore, *Chem. Soc. Rev.* 47 (2018) 533–557.
[22] A.E. Thorarinsdottir, T.D. Harris, Metal–organic framework magnets, *Chem. Rev.* 120 (2020) 8716–8789.
[23] O.M. Yaghi, M. O'Keeffe, N.W. Ockwig, H.K. Chae, M. Eddaoudi, J. Kim, Reticular synthesis and the design of new materials, *Nature.* 423 (2003) 705–714.
[24] S. Yuan, L. Feng, K. Wang, J. Pang, M. Bosch, C. Lollar, Y. Sun, J. Qin, X. Yang, P. Zhang, Stable metal–organic frameworks: design, synthesis, and applications, *Adv. Mater.* 30 (2018) 1704303.
[25] B. Chen, Z. Yang, Y. Zhu, Y. Xia, Zeolitic imidazolate framework materials: recent progress in synthesis and applications, *J. Mater. Chem. A.* 2 (2014) 16811–16831.
[26] X. Wei, T. Chen, S. Chen, Q. Jia, N.A. Mazlan, A. Lewis, N. Radacsi, Y. Huang, Improving the hydrostability of zeolitic imidazolate framework coatings using a facile silk fibroin protein modification method, *Nano Res.* 17 (2024) 6247–6257.
[27] S. Kitagawa, R. Kitaura, S. Noro, Functional porous coordination polymers, *Angew. Chemie Int. Ed.* 43 (2004) 2334–2375.
[28] D. Bradshaw, J.B. Claridge, E.J. Cussen, T.J. Prior, M.J. Rosseinsky, Design, chirality, and flexibility in nanoporous molecule-based materials, *Acc. Chem. Res.* 38 (2005) 273–282.
[29] G. Férey, Hybrid porous solids: past, present, future, *Chem. Soc. Rev.* 37 (2008) 191–214.

[30] P. Tong, J. Liang, X. Jiang, J. Li, Research progress on metal–organic framework composites in chemical sensors, *Crit. Rev. Anal. Chem.* 50 (2020) 376–392.
[31] P. Ling, J. Lei, H. Ju, Porphyrinic metal–organic framework as electrochemical probe for DNA sensing via triple-helix molecular switch, *Biosens. Bioelectron.* 71 (2015) 373–379.
[32] N. Yanai, K. Kitayama, Y. Hijikata, H. Sato, R. Matsuda, Y. Kubota, M. Takata, M. Mizuno, T. Uemura, S. Kitagawa, Gas detection by structural variations of fluorescent guest molecules in a flexible porous coordination polymer, *Nat. Mater.* 10 (2011) 787–793.
[33] J.-W. Ye, X.-Y. Li, H.-L. Zhou, J.-P. Zhang, Optimizing luminescence sensitivity and moisture stability of porous coordination frameworks by varying ligand side groups, *Sci. China Chem.* 62 (2019) 341–346.
[34] H.J. Buser, D. Schwarzenbach, W. Petter, A. Ludi, The crystal structure of Prussian blue: Fe4 [Fe (CN) 6] 3. xH2O, *Inorg. Chem.* 16 (1977) 2704–2710.
[35] Y.-R. Lee, J. Kim, W.-S. Ahn, Synthesis of metal–organic frameworks: a mini review, *Korean J. Chem. Eng.* 30 (2013) 1667–1680.
[36] A. Rehman, S. Farrukh, A. Hussain, E. Pervaiz, Synthesis and effect of metal–organic frame works on CO_2 adsorption capacity at various pressures: a contemplating review, *Energy Environ.* 31 (2020) 367–388.
[37] M. Safaei, M.M. Foroughi, N. Ebrahimpoor, S. Jahani, A. Omidi, M. Khatami, A review on metal–organic frameworks: synthesis and applications, *TrAC Trends Anal. Chem.* 118 (2019) 401–425.
[38] N. Stock, S. Biswas, Synthesis of metal–organic frameworks (MOFs): routes to various MOF topologies, morphologies, and composites, *Chem. Rev.* 112 (2012) 933–969.
[39] S.-H. Jhung, J.-H. Lee, J.-S. Chang, Microwave synthesis of a nanoporous hybrid material, chromium trimesate, *Bull. Korean Chem. Soc.* 26 (2005) 880–881.
[40] K.M.L. Taylor-Pashow, J. Della Rocca, Z. Xie, S. Tran, W. Lin, Postsynthetic modifications of iron-carboxylate nanoscale metal–organic frameworks for imaging and drug delivery, *J. Am. Chem. Soc.* 131 (2009) 14261–14263.
[41] J.H. Bang, K.S. Suslick, Applications of ultrasound to the synthesis of nanostructured materials, *Adv. Mater.* 22 (2010) 1039–1059.
[42] L. Zhang, F. Liand, L. Luo, Preparation methods of metal organic frameworks and their capture of CO_2, *IOP Conference Series: Earth and Environmental Science*, IOP Publishing (2018) 42104.
[43] C.-W. Huang, V.-H. Nguyen, S.-R. Zhou, S.-Y. Hsu, J.-X. Tan, K.C.-W. Wu, Metal–organic frameworks: preparation and applications in highly efficient heterogeneous photocatalysis, *Sustain. Energy Fuels.* 4 (2020) 504–521.
[44] J. Bedia, V. Muelas-Ramos, M. Peñas-Garzón, A. Gómez-Avilés, J.J. Rodríguez, C. Belver, A review on the synthesis and characterization of metal organic frameworks for photocatalytic water purification, *Catalysts.* 9 (2019) 52.
[45] U. Mueller, M. Schubert, F. Teich, H. Puetter, K. Schierle-Arndt, J. Pastre, Metal–organic frameworks—prospective industrial applications, *J. Mater. Chem.* 16 (2006) 626–636.
[46] M. Rubio-Martinez, C. Avci-Camur, A.W. Thornton, I. Imaz, D. Maspoch, M.R. Hill, New synthetic routes towards MOF production at scale, *Chem. Soc. Rev.* 46 (2017) 3453–3480.
[47] A. Carné-Sánchez, I. Imaz, M. Cano-Sarabia, D. Maspoch, A spray-drying strategy for synthesis of nanoscale metal–organic frameworks and their assembly into hollow superstructures, *Nat. Chem.* 5 (2013) 203–211.

[48] X. Wang, L. Wu, Z. Wang, Y. Feng, Y. Liu, H. Dai, Z. Wang, J. Deng, Photothermal synergistic catalytic oxidation of ethyl acetate over MOFs-derived mesoporous N-TiO2 supported Pd catalysts, *Appl. Catal. B Environ.* 322 (2023) 122075.

[49] A. Singh, A.K. Singh, J. Liu, A. Kumar, Syntheses, design strategies, and photocatalytic charge dynamics of metal–organic frameworks (MOFs): a catalyzed photo-degradation approach towards organic dyes, *Catal. Sci. Technol.* 11 (2021) 3946–3989.

[50] H. Li, M. Eddaoudi, M. O'Keeffe, O.M. Yaghi, Design and synthesis of an exceptionally stable and highly porous metal–organic framework, *Nature.* 402 (1999) 276–279.

[51] J. Hungerford, K.S. Walton, Room-Temperature synthesis of metal–organic framework isomers in the tetragonal and kagome crystal structure, *Inorg. Chem.* 58 (2019) 7690–7697.

[52] D. Chen, J. Zhao, P. Zhang, S. Dai, Mechanochemical synthesis of metal–organic frameworks, *Polyhedron.* 162 (2019) 59–64.

[53] M.V. Varsha, G. Nageswaran, Direct electrochemical synthesis of metal organic frameworks, *J. Electrochem. Soc.* 167 (2020) 155527.

[54] G. Mani, Graphene-metal organic framework composite based electrochemical sensors for toxic chemicals, *Graphene-Based Electrochem. Sensors Toxic Chem.* 82 (2020) 276–308.

[55] C. Pettinari, F. Marchetti, N. Mosca, G. Tosi, A. Drozdov, Application of metal–organic frameworks, *Polym. Int.* 66 (2017) 731–744.

[56] W. Zhou, Methane storage in porous metal–organic frameworks: current records and future perspectives, *Chem. Rec.* 10 (2010) 200–204.

[57] H.R. Abid, H. Tian, H.-M. Ang, M.O. Tade, C.E. Buckley, S. Wang, Nanosize Zr-metal organic framework (UiO-66) for hydrogen and carbon dioxide storage, *Chem. Eng. J.* 187 (2012) 415–420.

[58] J.B. DeCoste, M.H. Weston, P.E. Fuller, T.M. Tovar, G.W. Peterson, M.D. LeVan, O.K. Farha, Metal–organic frameworks for oxygen storage, *Angew. Chemie Int. Ed.* 53 (2014) 14092–14095.

[59] G. An, X. Xia, S. Wu, Z. Liu, L. Wang, S. Li, Metal–organic frameworks for ammonia-based thermal energy storage, *Small.* 17 (2021) 2102689.

[60] J. Klinowski, F.A.A. Paz, P. Silva, J. Rocha, Microwave-assisted synthesis of metal–organic frameworks, *Dalt. Trans.* 40 (2011) 321–330.

[61] A. Bhattacharjee, S. Gumma, M.K. Purkait, Fe_3O_4 promoted metal organic framework MIL-100 (Fe) for the controlled release of doxorubicin hydrochloride, *Microporous Mesoporous Mater.* 259 (2018) 203–210.

[62] C. Wang, X. Li, W. Yang, Y. Xu, H. Pang, Solvent regulation strategy of Co-MOF-74 microflower for supercapacitors, *Chinese Chem. Lett.* 32 (2021) 2909–2913.

[63] P. Choudhary, S.K. Ola, I. Chopra, V. Dhayal, D.S. Shekhawat, Metal–organic framework (MOF)/graphene–oxide (GO) nanocomposites materials: a potential formulation for anti-corrosive coatings-a review, *Mater. Today Proc.* 79 (2023) 172–178.

[64] M. Ramezanzadeh, B. Ramezanzadeh, M. Mahdavian, G. Bahlakeh, Development of metal–organic framework (MOF) decorated graphene oxide nanoplatforms for anti-corrosion epoxy coatings, *Carbon N. Y.* 161 (2020) 231–251.

7 Corrosion Inhibition Mechanisms of Metal-Organic Framework Coatings

Samariddin Rakhimov, Utkir Madatov, Khasan Berdimuradov, Yusufboy Rajabov, Jaykhun Mamatov, Elyor Berdimurodov, Anvar Khamidov and Fotima Sobirova

7.1 INTRODUCTION

Corrosion is a pervasive challenge affecting a wide range of metals and alloys, leading to structural degradation, reduced service life, and significant economic costs. This phenomenon is driven by chemical or electrochemical reactions between a material and its environment, resulting in the dissolution or oxidation of the metallic substrate [1–3]. Traditional methods to combat corrosion involve coatings such as paints, metallic overlays, and organic inhibitors; however, these solutions often face limitations in harsh conditions and may fail to provide long-term protection. As modern industrial applications increasingly demand higher durability and reliability, there is a growing impetus to develop advanced materials capable of mitigating corrosion with greater efficiency and specificity [4, 5].

One of the promising strategies to address this challenge is the utilization of metal–organic frameworks (MOFs) as protective coatings. MOFs are porous crystalline materials composed of metal nodes coordinated to organic ligands, creating robust, customizable frameworks. The unique characteristics of MOFs—namely their high surface area, tunable pore sizes, and versatile chemical functionalities—make them well-suited for use as corrosion-inhibiting coatings [6–9]. By tailoring the choice of metal centers and organic linkers, the resulting MOF can be designed to interact selectively with corrosive species, thus mitigating their penetration to the underlying substrate. Furthermore, MOFs can encapsulate or immobilize corrosion inhibitors within their porous network, creating multifunctional, "smart" coatings capable of responding adaptively to environmental triggers [10, 11].

In addition to their intrinsic structural properties, MOFs can be engineered to exhibit synergistic inhibition behaviors. For example, integrating corrosion inhibitors into the pores of an MOF allows the controlled release of active species at critical moments, thereby extending the service life of the coating and enhancing its

 DOI: 10.1201/9781003571162-7

protective action. Moreover, the highly ordered porous architecture of MOFs can serve as a barrier by reducing the permeation of water, oxygen, and other corrosive agents to the metal surface [12, 13]. This dual capability serving as both a physical barrier and a platform for inhibitor loading illustrates the versatile nature of MOFs as next-generation anticorrosion coatings.

Understanding the mechanisms by which MOF coatings inhibit corrosion is crucial for optimizing their performance and guiding the rational design of new frameworks [14, 15]. A comprehensive grasp of these inhibition processes can foster breakthroughs in material design, enabling coatings that are not only more effective but also tailored to specific service environments. For instance, identifying the roles of pore structure, metal–ligand chemistry, and interfacial bonding with the substrate can inform strategies to minimize defect formation and maximize adhesion [16, 17]. This deeper mechanistic insight ultimately drives the development of advanced coating systems with enhanced durability, self-healing capabilities, and environmentally friendly profiles.

Related to these considerations, the exploration of MOF coatings for corrosion inhibition stands at the forefront of materials science research. By clarifying the underlying inhibition mechanisms—ranging from their barrier effect to their ability to incorporate and release active inhibitors—it becomes possible to exploit the full potential of MOFs in preventing corrosion under an array of operating conditions [18, 19]. This chapter provides an in-depth examination of these mechanisms, laying the groundwork for future innovations that will shape the next generation of protective coatings.

7.2 FUNDAMENTALS OF INHIBITION MECHANISMS

The remarkable effectiveness of MOF coatings in corrosion prevention is rooted in their multifaceted inhibition mechanisms. These mechanisms are a direct consequence of the unique structural and chemical properties of MOFs, which enable them to act as both passive and active barriers against corrosion [20, 21]. Among these, the barrier effect is a primary mechanism that significantly enhances the corrosion resistance of coated substrates (Figure 7.1).

7.2.1 Barrier Effect

The barrier effect is one of the most fundamental mechanisms by which MOF coatings inhibit corrosion. This effect relies on the physical blocking of corrosive species, such as water, oxygen, chloride ions, and other aggressive chemicals from reaching the underlying metallic substrate [22, 23]. The highly ordered and customizable porous structure of MOFs allows for the creation of a dense and impermeable coating that reduces the diffusion of corrosive agents through the coating layer.

Physical Blocking of Corrosive Species: The intrinsic porosity of MOFs, characterized by their high surface area and tunable pore sizes, plays a pivotal role in this blocking effect. By carefully selecting the metal nodes and

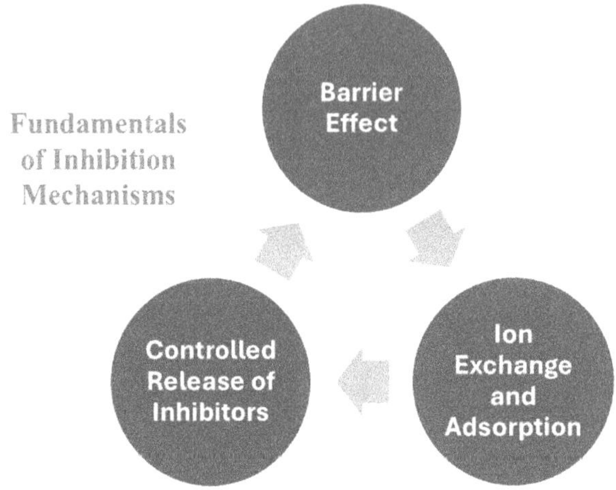

FIGURE 7.1 Fundamentals of inhibition mechanisms.

organic linkers, MOFs can be designed with pore sizes small enough to prevent the passage of larger corrosive molecules while still allowing the coating to maintain structural stability [24, 25]. For instance, hydrophobic ligands can be incorporated to repel water molecules, further enhancing the coating's resistance to environmental exposure. Additionally, the uniformity and compactness of the MOF layer are critical in minimizing defects or voids, which can serve as pathways for corrosive agents [26, 27]. These coatings act as a dense shield that delays or entirely prevents the penetration of water and ions, thus protecting the substrate from direct chemical or electrochemical attack.

Influence of MOF Porosity and Layer Thickness: While the porosity of an MOF enhances its capacity to act as a barrier, excessive porosity or irregular pore distribution can sometimes reduce its effectiveness. Balancing the porosity to ensure effective blocking without compromising the mechanical integrity of the coating is therefore a key consideration in the design of MOF coatings. The thickness of the MOF layer is another critical factor influencing its barrier performance. A thicker layer generally provides a longer diffusion path for corrosive agents, thereby delaying their arrival at the substrate surface [28–30]. However, excessively thick coatings may lead to increased brittleness or detachment, which could compromise the coating's protective function. Optimizing the layer thickness to achieve a balance between durability and protective capability is essential for maximizing the efficiency of MOF coatings. In conclusion, the barrier effect provided by MOF coatings represents a powerful and passive mechanism for corrosion inhibition. By physically blocking the ingress of corrosive species and leveraging their tunable porosity and layer thickness, MOFs offer a robust solution for protecting metals in a variety of environments [31–33]. Subsequent sections will explore additional mechanisms, such as ion exchange, adsorption, and the controlled release of inhibitors, which complement the barrier effect to provide comprehensive corrosion protection.

Ramezanzadeh et al. [34] explored the synthesis and application of ZIF-8 on graphene oxide (GO@ZIF-8) to enhance epoxy coatings' anticorrosion properties. GO@ZIF-8 demonstrated improved barrier properties by resisting water, oxygen, and ion diffusion, achieving a 79% increase in specific surface area. The coatings provided active corrosion protection with 79% inhibition efficiency, achieved through the controlled release of zinc ions and imidazole molecules (Figure 7.2). Additionally, the composite showed improved adhesion to steel substrates and resistance to delamination under wet and cathodic conditions. Its smart coating behavior allowed self-healing through inhibitor release in corrosive environments. Computational insights from molecular dynamics and DFT simulations validated the material's interactions and effectiveness, highlighting its potential for scalable, durable, and intelligent anticorrosion coatings in industrial applications.

7.2.2 Ion Exchange and Adsorption

In addition to serving as physical barriers, MOF coatings exhibit a dynamic capability to mitigate corrosion through ion exchange and adsorption mechanisms. These processes involve the interaction of corrosive ions with the MOF's internal structure, where they are either trapped or replaced by less aggressive species [35, 36]. The capacity for ion exchange and adsorption is a result of the highly tunable chemistry and porosity of MOFs, which enables them to act as active corrosion inhibitors.

Trapping or Exchanging Corrosive Ions within MOF: MOFs possess highly porous architectures with well-defined channels and cavities, providing ample space for the accommodation of ions. This structural feature allows MOFs to trap corrosive ions, such as chloride (Cl^-) and sulfate (SO_4^{2-}), which are common culprits in initiating and accelerating corrosion processes. Once trapped within the framework, these ions are immobilized and prevented from reaching the underlying metal substrate. In addition to trapping ions, MOFs can perform ion exchange reactions, replacing aggressive ions with less reactive species. For instance, chloride ions present in a corrosive environment may be exchanged for hydroxide (OH^-) or phosphate (PO_4^{3-}) ions preloaded within the MOF structure. This exchange neutralizes the corrosive effects of the trapped ions and provides an additional layer of protection. The ion exchange capability depends on the type of metal nodes and the coordination chemistry of the organic linkers used in the MOF structure [37–39].

Role of Functional Groups in Capturing Aggressive Species: The functional groups present in MOF linkers play a crucial role in enhancing their adsorption and ion exchange properties. These functional groups—such as carboxyl (–COOH), hydroxyl (–OH), amine ($-NH_2$), and sulfonate ($-SO_3H$)—serve as active sites for binding corrosive ions through various interactions, including electrostatic attraction, hydrogen bonding, and coordination bonding [40, 41]. For example,

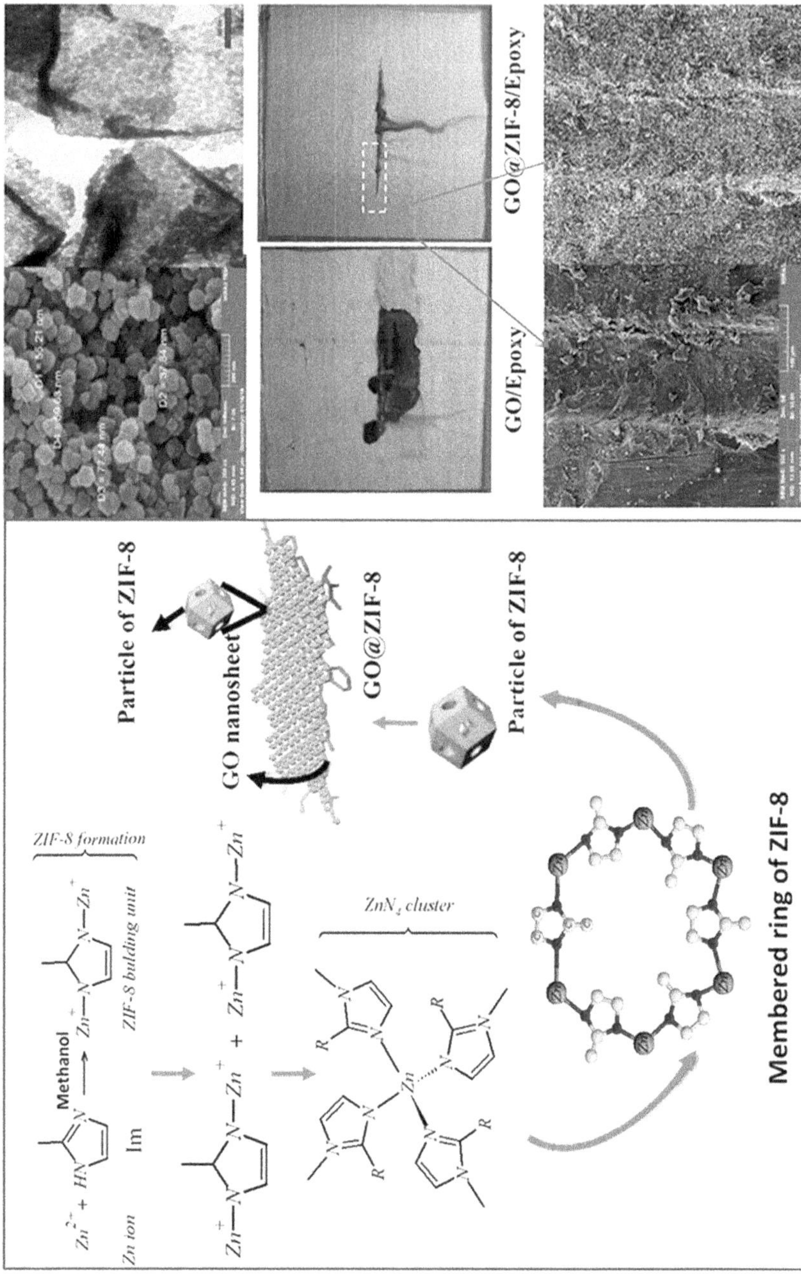

FIGURE 7.2 MOF decorated graphene oxide nanoplatforms for anticorrosion epoxy coatings [34].

an MOF containing amine-functionalized linkers can effectively capture chloride ions through electrostatic interactions, preventing their diffusion to the substrate surface. Moreover, the incorporation of hydrophilic or hydrophobic functional groups into the MOF can be tailored to target specific corrosive species [42, 43]. Hydrophilic functional groups can enhance the adsorption of polar species, while hydrophobic groups can repel water, further reducing the exposure of the metal to corrosive agents. This versatility enables MOFs to adapt to diverse corrosive environments and provide customized protection.

Synergistic Protection: The ability of MOFs to adsorb and exchange ions complements their physical barrier effect, creating a synergistic mechanism that enhances corrosion resistance. By actively interacting with corrosive ions, MOFs not only prevent direct contact with the metal substrate but also neutralize aggressive species in the surrounding environment. This dynamic protective capability distinguishes MOFs from traditional coatings, which typically rely solely on passive blocking [44–46]. Therefore, the ion exchange and adsorption mechanisms of MOF coatings provide a powerful and adaptive approach to corrosion inhibition. Through the selective trapping or replacement of corrosive ions and the strategic use of functional groups, MOFs offer tailored solutions for mitigating corrosion in complex and challenging environments. Subsequent sections will explore how these mechanisms integrate with the controlled release of inhibitors to create multifunctional protective systems.

Figure 7.3 illustrates the dual protection mechanism of UiO-Schiff Base. Zn-based epoxy coatings. Passive protection arises from the dense nano-reservoirs, which block water, oxygen, and chloride ion penetration, enhancing barrier properties. Active protection is pH-sensitive, releasing zinc ions (Zn^{2+}) and organic inhibitors (SP and Tannic Acid). At anodic sites (acidic conditions), inhibitors form protective complexes with iron, while at cathodic sites (alkaline conditions), Zn^{2+} reacts with hydroxyl ions to form stable deposits. This targeted release ensures localized corrosion inhibition and self-healing in damaged areas, offering robust, long-term protection in saline environments [19].

7.2.3 Controlled Release of Inhibitors

One of the most innovative features of MOF coatings in corrosion protection is their ability to incorporate and regulate the release of corrosion-inhibiting agents. This controlled release mechanism transforms MOFs from passive protective layers into smart, responsive systems capable of delivering inhibitors precisely when and where they are needed [19]. Such functionality not only enhances the durability of the coatings but also minimizes the environmental impact of using excessive inhibitors.

Incorporation and Regulated Release of Corrosion-Inhibiting Agents: The porous structure of MOFs provides an ideal platform for encapsulating corrosion inhibitors within their frameworks. These inhibitors, which can include organic molecules, ions, or small particles, are stored within the cavities of the MOF

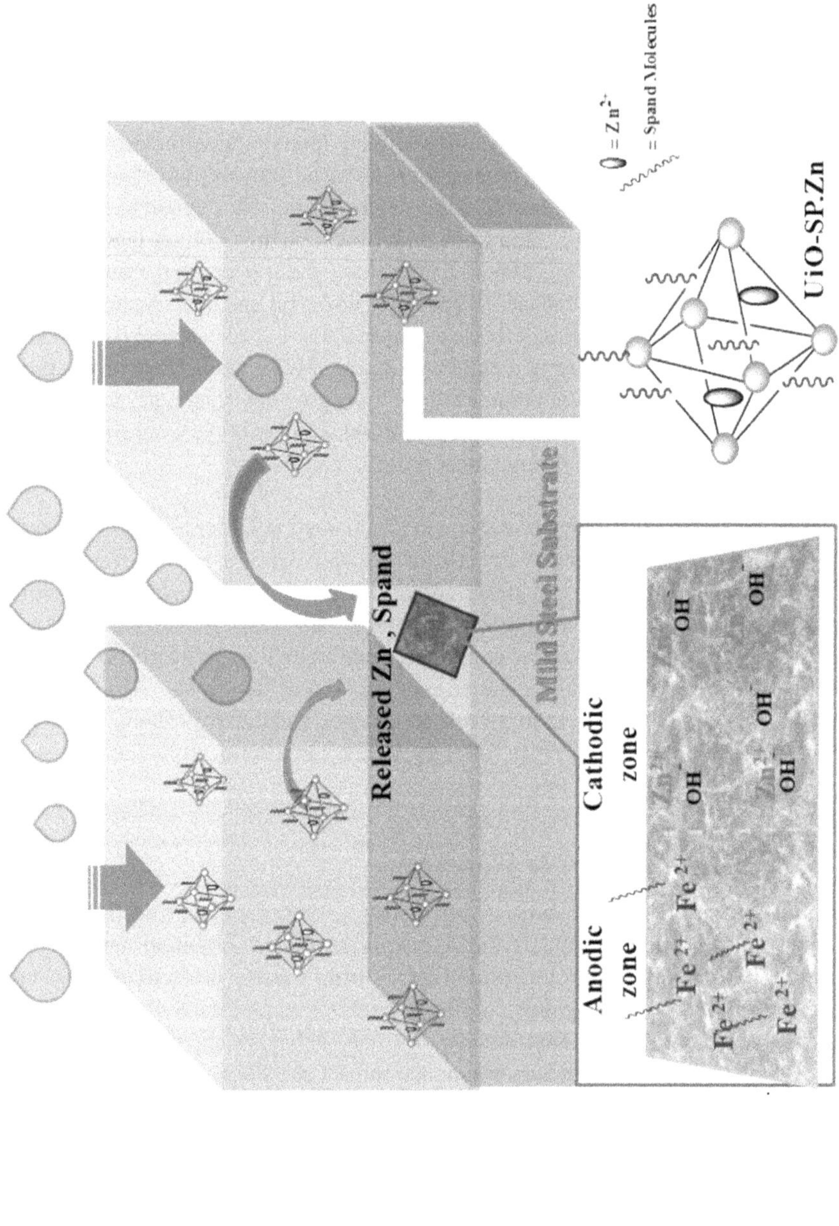

FIGURE 7.3 Active and passive protection exhibition by UiO-66 smart MOF-based coating [19].

during synthesis or post-synthetic modification. Once integrated into the MOF, the inhibitors remain immobilized until specific conditions trigger their release. The controlled release is often governed by environmental stimuli such as pH, humidity, or ionic concentration. For example, in an acidic environment where corrosion is more likely to occur, the MOF coating can release inhibitors that neutralize the acid or form a passivation layer on the metal surface. Similarly, the presence of chloride ions—a common cause of corrosion in marine and industrial settings—can trigger the release of inhibitors that counteract the aggressive effects of these ions [48–50]. The release mechanism typically relies on the degradation or partial dissolution of the MOF in response to environmental changes. Alternatively, it may involve reversible interactions, such as ion exchange, that allow the inhibitors to be gradually released while maintaining the structural integrity of the coating.

Smart Coatings Responsive to Local Environmental Changes: What sets MOF-based coatings apart from conventional systems is their ability to act as smart coatings that respond dynamically to the local environment. This responsiveness ensures that inhibitors are released only when needed, reducing the wastage of active agents and extending the lifetime of the protective layer. For instance, pH-responsive MOFs can release inhibitors in acidic conditions commonly associated with corrosion. Similarly, coatings designed to react to high ionic concentrations can release inhibitors selectively in the presence of chloride or sulfate ions [51–53]. Some MOFs are also sensitive to electrochemical potentials, releasing inhibitors in areas experiencing localized corrosion or pitting.

Smart coatings provide several benefits, including:

1. *Localized Protection:* Targeted inhibitor release in regions experiencing the highest risk of corrosion.
2. *Efficiency:* Minimizing inhibitor consumption by releasing agents only under specific conditions.
3. *Sustainability:* Reducing environmental contamination associated with excess use of corrosion inhibitors.

Synergistic Role in Corrosion Inhibition: The controlled release of inhibitors complements the barrier effect and ion exchange mechanisms of MOF coatings, resulting in a multi-layered protective strategy. By combining these approaches, MOFs offer a comprehensive solution to corrosion mitigation, addressing both the prevention and repair of localized damage. Therefore, the ability to incorporate and release corrosion-inhibiting agents in a controlled manner elevates the functionality of MOF coatings [54, 55]. This smart, responsive approach not only enhances protection but also aligns with sustainability goals, making MOF coatings a highly attractive option for advanced corrosion resistance applications. Subsequent sections will delve into the factors influencing the efficiency of these mechanisms and strategies for optimizing MOF coatings.

7.3 FACTORS INFLUENCING INHIBITION

The performance of MOF coatings as corrosion inhibitors depends on key factors like chemical stability, interfacial adhesion, and coating microstructure. Chemical stability ensures the MOF's resistance to harsh environments, including acidic or alkaline conditions, preventing structural degradation. Robust metal nodes, strong ligands, and post-synthetic modifications enhance stability. Interfacial adhesion ensures the coating remains intact under stress, with strong MOF–substrate bonds achieved through surface treatments and functionalization [56, 57]. Coating microstructure, including uniformity, low defect density, and optimal thickness, prevents pathways for corrosive agents while maintaining durability (Figure 7.4). Addressing these factors improves MOF coatings' reliability and effectiveness for long-term corrosion protection.

7.3.1 Chemical Stability of MOFs

The chemical stability of MOFs is a critical factor influencing their effectiveness as corrosion inhibitors. Their ability to withstand harsh conditions, including variations in pH, temperature, and exposure to aggressive ions, determines their long-term performance. The pH of the environment plays a significant role in the structural integrity of MOFs. Acidic conditions often result in the protonation of organic linkers, weakening the coordination bonds between metal nodes and ligands, potentially leading to structural disintegration. However, MOFs with high-valence metal centers, such as Zr^{4+} or Ti^{4+}, and robust ligands, such as carboxylates, demonstrate better acid resistance. For example, zirconium-based MOFs such as UiO-66 maintain their structural stability in moderately acidic conditions, making them suitable for protective coatings. In alkaline environments, hydrolysis of metal–ligand bonds can occur, especially in MOFs with divalent metal nodes like Zn^{2+} or Cu^{2+}. To counter this, incorporating ligands with strong metal-binding affinity, such as phosphate-functionalized

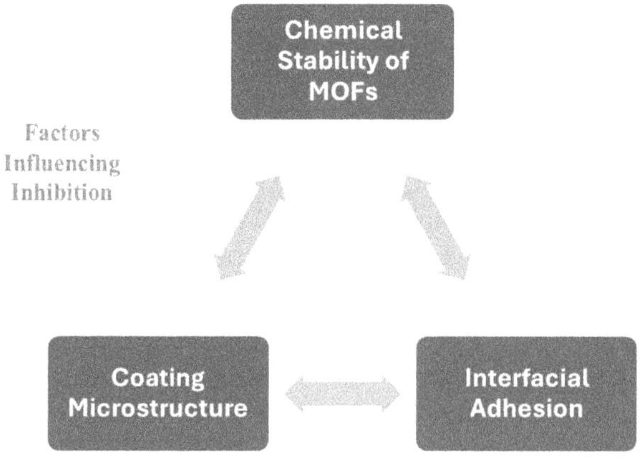

FIGURE 7.4 Factors influencing inhibition.

ligands, enhances stability [58, 59]. The presence of aggressive species, particularly chloride ions, also poses a threat to MOF stability as these ions can destabilize coordination bonds. Strategies such as using hydrophobic ligands to repel water and chloride ions or applying protective polymer or hybrid material layers can mitigate these effects. To enhance stability further, researchers focus on selecting high-valence or transition metal nodes that form strong bonds, employing rigid and hydrophobic ligands, and utilizing post-synthetic modifications like surface functionalization or hybridization. These approaches ensure MOFs retain their structural integrity in corrosive environments, making them effective and reliable corrosion inhibitors.

7.3.2 Interfacial Adhesion

The effectiveness of MOF coatings in corrosion protection relies not only on their intrinsic properties but also on their ability to form strong and durable bonds with the underlying substrate. Interfacial adhesion, the interaction between the MOF layer and the metal surface, plays a pivotal role in ensuring the stability and functionality of the coating. A robust MOF–substrate bond prevents the delamination of the coating, minimizes the ingress of corrosive agents, and ensures that the protective properties of the MOF are maintained over time. The strength of interfacial adhesion is determined by several factors, including the chemical and physical interactions between the MOF and the substrate. Chemically, strong covalent or coordinate bonds between the metal surface and the MOF's metal nodes or functional groups enhance adhesion [1, 17]. For instance, substrates pre-treated with functionalization layers or primers can provide active sites for MOF attachment, improving the bonding at the interface. Physically, the surface roughness of the substrate can also influence adhesion. A roughened surface increases the contact area and mechanical interlocking, further enhancing the stability of the coating. Weak interfacial adhesion can lead to coating detachment, exposing the substrate to corrosive agents and undermining the protective effectiveness of the MOF. Delamination often results from environmental stresses such as thermal expansion, mechanical abrasion, or chemical attack. Therefore, optimizing the MOF–substrate interface is critical for the long-term performance of the coating. Techniques such as substrate pre-treatment (e.g., etching, plasma cleaning) and post-synthetic modifications of MOFs (e.g., adding surface-functional groups) are commonly employed to strengthen this interaction. Therefore, strong interfacial adhesion is essential for the durability and effectiveness of MOF coatings in corrosion protection. By ensuring a robust MOF–substrate bond, the coating remains intact, providing consistent protection against corrosive agents and environmental stresses. Future research aimed at improving adhesion through advanced surface treatments and functionalized MOF designs will further enhance the application potential of these coatings.

7.3.3 Coating Microstructure

The microstructure of MOF coatings is a crucial determinant of their performance as corrosion inhibitors. Key aspects such as uniformity, defect density, and layer thickness directly impact the coating's ability to provide effective protection against corrosive agents. Ensuring an optimal microstructure is essential for maximizing the coating's efficiency and durability in challenging environments. Uniformity in the MOF coating is critical for forming a continuous protective layer that effectively isolates the substrate from its surroundings. Inhomogeneous coatings, characterized by uneven thickness or irregular deposition, create weak points that can serve as entry paths for corrosive species such as water, oxygen, and chloride ions. Achieving uniformity often involves optimizing the deposition process, whether by chemical vapor deposition, electrochemical methods, or spray-coating techniques, to ensure consistent coverage across the entire substrate. Defect density is another key parameter that influences the coating's performance [19, 20]. Defects such as pinholes, cracks, or voids act as vulnerabilities that allow corrosive agents to penetrate through the coating and reach the substrate, accelerating the corrosion process. Reducing defect density requires careful control over synthesis parameters, including reaction conditions, precursor quality, and post-treatment processes. Additionally, hybrid approaches that combine MOFs with other materials, such as polymers or nanoparticles, can help to fill defects and enhance the overall integrity of the coating. Layer thickness also plays a significant role in determining the effectiveness of MOF coatings. A thicker coating typically provides a longer diffusion path for corrosive species, delaying their penetration to the substrate. However, excessively thick layers can lead to brittleness, cracking, or detachment, which compromise the protective properties of the coating. An optimized thickness ensures a balance between mechanical robustness and effective protection [18, 22]. This balance is often achieved through iterative testing and precise control of the deposition process. Therefore, the microstructure of MOF coatings, encompassing uniformity, defect density, and layer thickness, is integral to their ability to inhibit corrosion. By focusing on these parameters during the synthesis and application of MOFs, researchers can develop coatings with enhanced performance and reliability. Future advancements in deposition techniques and material design are expected to further refine the microstructural properties of MOF coatings, paving the way for their broader adoption in industrial corrosion protection applications.

7.4 RECENT PROGRESS AND FUTURE DIRECTIONS

Recent advancements in MOF coatings have significantly enhanced their potential as corrosion inhibitors, with novel compositions, hybrid coatings, and smart functionalities leading the way. These developments, coupled with emerging trends and efforts toward large-scale applications, are shaping the future of MOF-based corrosion protection [60, 61]. Innovations in MOF design have introduced new compositions with enhanced stability, functionality, and adaptability for corrosion inhibition. High-valence metal nodes (e.g., Zr^{4+}, Ti^{4+}) paired with functionalized organic linkers have resulted in MOFs with exceptional resistance to harsh environments. Additionally, hybrid coatings that combine MOFs with polymers, nanoparticles, or other materials have demonstrated improved mechanical strength, defect sealing, and multifunctionality [36, 38]. These hybrid systems address the limitations of pure MOF coatings, such as brittleness or susceptibility to environmental degradation. Smart MOF coatings capable of responding dynamically to environmental changes are an exciting area of progress. These coatings incorporate mechanisms like pH sensitivity, ionic responsiveness, or electrochemical activation to release corrosion inhibitors on demand. Self-healing materials, enabled by MOFs that regenerate protective layers or repair defects autonomously, are also gaining attention [62, 63]. These features enhance the longevity and efficiency of coatings, making them highly suitable for applications in extreme conditions. For MOF coatings to transition from laboratory research to widespread industrial use, challenges related to scalability, cost, and manufacturing must be addressed. Advances in scalable synthesis methods, such as continuous flow or green chemistry approaches, are critical for reducing production costs. Furthermore, ensuring compatibility with existing industrial processes, such as spray or dip-coating, will facilitate broader adoption [36, 43]. The development of cost-effective and environmentally friendly MOFs tailored for specific industries, such as automotive, marine, and infrastructure, also represents a promising direction. Therefore, the recent progress in MOF compositions, hybrid systems, and smart functionalities underscores their growing potential as corrosion inhibitors. Future research focusing on scalability, sustainability, and integration with industrial applications will be pivotal in advancing MOFs from innovative materials to practical solutions for corrosion protection.

7.5 CONCLUSION

MOF coatings represent a transformative approach to corrosion protection, combining passive and active mechanisms for superior performance. Their intrinsic properties, including tunable porosity, high surface area, and chemical versatility, enable MOFs to serve as physical barriers and active inhibitors. Innovations such as pH-responsive inhibitor release and self-healing functionalities underscore their potential as "smart" materials tailored for challenging environments. This chapter highlights the critical role of factors like chemical stability, interfacial adhesion, and microstructure in optimizing the effectiveness of MOF coatings. While significant progress has been made, challenges related to scalability, cost, and integration with industrial processes remain. Addressing these challenges through advancements in material design and synthesis will unlock the full potential of MOFs, paving the way for their widespread adoption as sustainable and efficient corrosion inhibitors in diverse industrial sectors.

REFERENCES

1. Bicalho, H.A., et al., Tuning the photophysical and photochemical properties of rare-earth cluster-based metal–organic frameworks. *ACS Applied Electronic Materials*, 2024. 6(10): pp. 7055–7064.
2. Bolghari, A.J., et al., Achieving a high-performance anti-corrosive polyvinyl ester based on bisphenol a coating by addition of polyaniline-Cu-based metal organic framework composite. *Polymer Composites*, 2024. 46: pp. 387–399.
3. Bose, S., et al., Development of a multiparticulate metal–organic framework/textile fiber swatch. *ACS Applied Materials and Interfaces*, 2024. 17: pp. 17813–17822.

4. Chakraborty, A., and T.K. Maji, Integration of metal–organic frameworks and clay toward functional composite materials. *Dalton Transactions*, 2024. 54: pp. 433.
5. Chen, H., Q. Liu, and S. Cao, Carbon-coated iron selenide derived from double-framework as an advance anode for Na-ion battery. *Journal of Colloid and Interface Science*, 2023. 652: pp. 619–626.
6. Chen, Z., et al., Corrosion-oriented self-healing coating based on ZnMg MOF for efficient corrosion protection of aluminum alloy via in-situ LDH film formation. *Chemical Engineering Journal*, 2024. 500: pp. 156898.
7. Ciria-Ramos, I., et al., Evaluation of triphenylene-based MOF ultrathin films for lithium batteries. *Dalton Transactions*, 2023. 52(21): pp. 7196–7207.
8. Ihamdane, R., et al., Essential oil of *Origanum vulgare* as a green corrosion inhibitor for carbon steel in acidic medium. *Arabian Journal for Science and Engineering*, 2023. 48(6): pp. 7685–7701.
9. Dewangan, Y., et al., N-hydroxypyrazine-2-carboxamide as a new and green corrosion inhibitor for mild steel in acidic medium: Experimental, surface morphological and theoretical approach. *Journal of Adhesion Science and Technology*, 2022. 36(23–24): pp. 2644–2664.
10. Dalakoti, S., et al., Cu-trimesate and mesoporous silica composite as adsorbent showing enhanced CO_2/CH_4 and CO_2/N_2 selectivity for biogas and flue gas separation. *Microporous and Mesoporous Materials*, 2025. 381: pp. 113354.
11. Dehghan-Chenar, S., H.R. Zare, and Z. Mohammadpour, Chitosan-ImH@γ-CD: a pH-sensitive smart bio-coating to enhance the corrosion resistance of magnesium alloys in bio-implants. *RSC Advances*, 2024. 14(45): pp. 33301–33310.
12. Fattah-alhosseini, A., et al., Enhancing corrosion resistance in Mg–based alloys through MOF–incorporated coatings: A comprehensive review. *Applied Surface Science Advances*, 2024. 21: pp. 100607.
13. Fattah-alhosseini, A., S. Zaheri, and M. Karbasi, A review of corrosion resistance in steel alloys with MOF-embedded coatings. *Nano-Structures and Nano-Objects*, 2024. 40: pp. 101384.
14. Berdimurodov, E., et al., New and green corrosion inhibitor based on new imidazole derivate for carbon steel in 1 M HCl medium: Experimental and theoretical analyses. *International Journal of Engineering Research in Africa*, 2022. 58: pp. 11–44.
15. Daoudi, W., et al., New chlorophenyl-imidazole derivative as a novel corrosion inhibitor in the gas and oil industry. *Journal of Physics and Chemistry of Solids*, 2023. 179: pp. 111409.
16. Frank, F., B. Baumgartner, and B. Lendl, Metal–organic frameworks combined with mid-infrared spectroscopy for the trace analysis of phosphates in water. *Sensors and Actuators B: Chemical*, 2024. 399: pp. 134778.
17. Huang, L., et al., Nanocellulose-based polymeric nanozyme as bioinspired spray coating for fruit preservation. *Food Hydrocolloids*, 2023. 135: pp. 108138.
18. Huang, Y., et al., A smart sol-gel coating incorporating pH-responsive BTA-ZIF-8 MOF assembled hexagonal boron nitride for active/passive corrosion protection of 1060 aluminum alloy. *Surface and Coatings Technology*, 2023. 474: pp. 130072.
19. Keramatinia, M., B. Ramezanzadeh, and M. Mahdavian, Unlocking the potential of a zirconium-based MOF for advanced protective coatings: Delve into the UIO-66 smart release functioning for a bio-based Zn2+ complex. *Surface and Coatings Technology*, 2024. 481: pp. 130578.

20. Keshmiri, N., et al., A novel approach towards controlled growth of metal–organic framework ZIF-8 thin film on steel with excellent corrosion protection. *Journal of Industrial and Engineering Chemistry*, 2023. 117: pp. 157–171.
21. Kumari, S., R.N. Ehrman, and J.J. Gassensmith, Expanding past ZIF-8: Biomimetic mineralization using other MOFs. *Matter*, 2023. 6(8): pp. 2570–2573.
22. Lai, Z., et al., Low-cost flocculant coated MIL-88B as highly efficient and stable fenton-like catalyst for phenol removal. *Journal of Inorganic and Organometallic Polymers and Materials*, 2024. 35: pp. 1743–1750.
23. Lee, B.W., et al., Facile synthesis of biomass-derived degradable poly(ketal-ester) elastomers using dual-acidic catalysts. *ACS Applied Polymer Materials*, 2024. 6(6): pp. 3507–3516.
24. Dagdag, O., et al., Synthesis, physicochemical properties, theoretical and electrochemical studies of tetraglycidyl methylenedianiline. *Journal of Molecular Structure*, 2022. 1265: pp. 133508.
25. Ech-chebab, A., et al., Anticorrosion properties of the epoxy polymer TGETBAU for mild steel in a solution of HCl (1.0 M): Experimental and computational approaches. *Journal of Molecular Structure*, 2023. 1284: pp. 135441.
26. Lee, M., et al., Supramolecular metal–organic framework for the high stability of aqueous rechargeable zinc batteries. *ACS Nano*, 2024. 18(33): pp. 22586–22595.
27. Li, W., et al., Constructing dual-ligand Ce-MOF on graphene oxide modified with polydopamine endowing polyurethane coating with long-term smart anti-corrosion and mechanical robustness. *Journal of Colloid and Interface Science*, 2025. 680: pp. 173–190.
28. El-Aouni, N., et al., Experimental and computational insights into destruction protection of E24 carbon metal by new trifunctional sulfur-phosphorus epoxy polymer. *Journal of the Taiwan Institute of Chemical Engineers*, 2024. 155: pp. 105281.
29. Li, Y.M., et al., Optimization design of the multidimensional heterostructure toward lightweight, broadband, highly efficient, and flame-retarding electromagnetic wave-absorbing composites. *ACS Applied Materials and Interfaces*, 2024. 16(38): pp. 51333–51345.
30. Liu, X., et al., Integrated MOF-74 coatings on magnesium for corrosion control, cytocompatibility, and antibacterial properties. *Inorganic Chemistry*, 2023. 62(26): pp. 10269–10278.
31. Long, S.J., et al., In situ construction of iron-rich metal–organic frameworks on wool surface for enhanced flame retardancy and low smoke generation. *Colloids and Surfaces A: Physicochemical and Engineering Aspects*, 2024. 692: pp. 134034.
32. Ma, K., et al., Protection against chemical warfare agents and biological threats using metal–organic frameworks as active layers. *Accounts of Materials Research*, 2023. 4(2): pp. 168–179.
33. Ma, Y., et al., Design synthesis and excellent anticorrosion properties of ZIF-8/$ZnMoO_4$ heterojunction nanomaterials. *ACS Applied Nano Materials*, 2024. 7(18): pp. 21951–21963.
34. Ramezanzadeh, M., et al., Development of metal–organic framework (MOF) decorated graphene oxide nanoplatforms for anti-corrosion epoxy coatings. *Carbon*, 2020. 161: pp. 231–251.
35. Majidi, R., et al., Development of a smart anticorrosion epoxy coating containing a pH-sensitive GO/MOF nanocarrier loaded with 2-mercaptobenzothiazole corrosion inhibitor. *Materials Chemistry and Physics*, 2023. 308: pp. 128291.

36. Mondal, S.K., et al., Design and development of UiO-67-coated PIM-1-based composites and demonstration of their detoxification performance. *Chemical Engineering Journal*, 2024. 493: pp. 152269.
37. Nwokolo, I.K., H. Shi, and F. Liu, MOF-based protective coatings for metal corrosion protection: A critical review of design/synthesis, performance, and mechanism. *Materials Science and Engineering: B*, 2025. 313: pp. 117932.
38. Pineda-Vásquez, T., et al., From e-waste to high-value materials: Sustainable synthesis of metal, metal oxide, and MOF nanoparticles from waste printed circuit boards. *Nanomaterials*, 2024. 14(1): pp. 69.
39. Qian, X., et al., Application of Y-MOF-CNT-derived Y2O3-C@CNT composites in lithium-sulfur battery separators. *Langmuir*, 2024. 40(44): pp. 23529–23537.
40. Arrousse, N., et al., Corrosion protection studies of different alloys in 1 M HCl by benzimidazole derivative: Combined molecular dynamic simulations/DFT. *Journal of Environmental Chemical Engineering*, 2023. 11(3), p. 109642.
41. Rbaa, M., et al., Synthesis of new halogenated compounds based on 8-hydroxyquinoline derivatives for the inhibition of acid corrosion: Theoretical and experimental investigations. *Materials Today Communications*, 2022. 33: pp. 104654.
42. Qiu, S., et al., Epoxy polymers containing cetal-organic fFramework-derived layered double hydroxide and black phosphorus for fire resistance, smoke suppression, and wear resistance. *ACS Applied Nano Materials*, 2024. 7(12): pp. 14727–14736.
43. Ramu, A.G., et al., Advancements in integrating MOFs into micro-arc oxidation coatings on Mg alloys: A perspective on PEO-MOF coatings as innovative corrosion inhibitors. *Journal of Magnesium and Alloys*, 2024. 12: pp. 4363–4394.
44. Ren, J., et al., Amorphous MOF as smart artificial solid/electrolyte interphase for highly-stable Zn-ion batteries. *Chemical Engineering Journal*, 2023. 462: pp. 142270.
45. Rezvani Niya, S., et al., Fabrication of green and smart anti-corrosion PDA@ bio-MOF-11@nano-curcumin coating for mild steel. *ChemistrySelect*, 2024. 9(47): pp. 3888.
46. Romero-Ángel, M., et al., Vapor-assisted conversion of heterobimetallic titanium-organic framework thin films. *Chemistry of Materials*, 2023. 35(24): pp. 10394–10402.
47. Sabzehmeidani, M.M., and M. Kazemzad, Recent advances in surface-mounted metal–organic framework thin film coatings for biomaterials and medical applications: A review. *Biomaterials Research*, 2023. 27(1): pp. 115.
48. Song, Y., et al., Dynamic Polymer/Metal–Organic Framework Hybrid Microcapsules for Self-Healing Anticorrosion Coatings. *ACS Applied Materials and Interfaces*, 2024. 16(49): pp. 68478–68486.
49. Wahba, A.M., and M.M.I. Helal, Synthesis, characterization and application of two metal–organic frameworks (MOF) as a corrosion inhibitor for aluminium in a hydrochloric acid environment. *Case Studies in Chemical and Environmental Engineering*, 2024. 10: pp. 100805.
50. Wang, H., et al., Constructing a rapid ion-transport anode interface protective layer for zinc ion batteries to suppress solvation and improve surface electronic structure. *Chemical Engineering Journal*, 2024. 485: pp. 149544.
51. Wang, P., et al., A Cu-Zn bimetallic organic framework as protective interlayer for dendrite-free Zn deposition in zinc-based flow batteries. *Electrochimica Acta*, 2024. 504: pp. 144949.
52. Wen, X., et al., Preparation of Co-MOF doped rosin-based alkyd resin coating and its corrosion protection property. *Corrosion and Protection*, 2023. 44(11): pp. 7–11.

53. Wu, C., et al., Polymer coating assisted shaping of metal–organic framework particles into pellets with enhanced methane uptake. *Sustainable Energy and Fuels*, 2024. 8(8): pp. 1636–1640.
54. Wu, W., et al., Superhydrophobic anticorrosive fluoroethylene vinyl ether coating nanocomposited with metal–organic framework derived ferrite nanocapsules and MXene. *Advanced Composites and Hybrid Materials*, 2024. 7(2): pp. 60.
55. Yang, K., et al., Green construction of MBI corrosion-resistant interfaces modified NZVI@MOFs-regulated 3D PAN cryogel film to enhance Cr(VI) removal. *Separation and Purification Technology*, 2024. 333: pp. 125902.
56. Yin, C., et al., Hierarchical spheroidal MOF-derived MnO@C as cathode components for high-performance aqueous zinc ion batteries. *Journal of Colloid and Interface Science*, 2023. 642: pp. 513–522.
57. Yue, M., et al., Carbon-coated MOF-derived porous SnPS3 core–shell structure as superior anode for sodium-ion batteries. *Small*, 2024. 20: pp. e2405262.
58. Zhao, Q., W. Zhao, and Y. Wu, Constructing bi-functional Ce-MOF on carbon fiber endowing epoxy coating with excellent anti-corrosion and erosion wear resistance. *Carbon,* 2023. 214: pp. 118374.
59. Zhou, J., et al., Submersible voltammetric sensing probe for rapid and extended remote monitoring of opioids in community water systems. *Microchimica Acta*, 2024. 191(8): pp. 463.
60. Haldhar, R., et al., Corrosion inhibitors: Industrial applications and commercialization, in Eliboev, I., et al. (eds.), *ACS Symposium Series*. 2021. pp. 219–235.
61. Salim, R., et al., Inhibition behavior of new ecological corrosion inhibitors for mild steel, copper and aluminum in acidic environment: Theoretical and experimental investigation. *Journal of Molecular Liquids*, 2024. 393.
62. Eliboev, I., et al., Supramolecular corrosion protection: Eco-friendly synthesis and efficacy of a β-cyclodextrin/o-phenylenediamine complex. *Journal of the Taiwan Institute of Chemical Engineers*, 2023. 147: pp. 104944.
63. Berdimurodov, E., et al., MOFs-based corrosion inhibitors, in Viswanathan S. Saji (ed.), *Supramolecular Chemistry in Corrosion and Biofouling Protection*. 2021. pp. 287–305.

8 Synthesis and Characterization Techniques for Anticorrosive MOF Coatings

Ahmad Ramazani Saadatabadi and Ali Mehri

8.1 INTRODUCTION

Metal–organic frameworks (MOFs) are nanoporous materials with structures composed of metal ions, such as Zn, Cr, Cu, Zr, and Al, or their clusters and organic ligands [1]. During the last two decades, MOFs have attracted much attention due to their unique properties, which include high surface area, tunable porosity, and unique chemical behavior. After being discovered in 1995 by Omar M. Yaghi, MOFs have also undergone intensive development, functionalization, and diverse field applications [2, 3]. One-, two-, and three-dimensional configuration properties transfer easily due to changes either with the size of a pore, capacity of absorption, or chemical behavior through their variation of either metal ion change or ligand that renders them prominent among up-to-date materials in vast, practically unlimited fields as protection against corrosion included (Figure 8.1) [4, 5].

MOFs have tremendous anticorrosion potential due to their excellent thermal and chemical resistance, ability to enhance barrier properties, and self-healing characteristics of coatings. So far, four main strategies have been proposed for using MOFs in this field: use as corrosion inhibitors, nanofiller additives in polymer coatings, fabrication of continuous membranes, and nanocontainers to encapsulate corrosion inhibitors [6–12]. Accordingly, some methods, such as solvothermal, sonochemical, and mechanochemical, are proposed to prepare MOF-based anticorrosion materials. Each method has merits and disadvantages. The mechanochemical approach employs a solvent-free technique, which has some advantages. However, its major disadvantage is that it consumes a very high amount of energy [13].

Some types of MOFs, such as Zeolitic Imidazolate Framework-8 (ZIF-8), University of Oslo-66 (UiO-66), and Materials of Institute Lavoisier-53 (MIL-53), exhibit hydrophobic performance with good water stability and corrosion inhibition properties [14, 15]. Among them, zeolite imidazole frameworks, or ZIFs, have gained much attention due to their special properties, such as ultra-high stability under harsh conditions, anticorrosion properties, and modifiability of polymer matrices [10, 16].

DOI: 10.1201/9781003571162-8

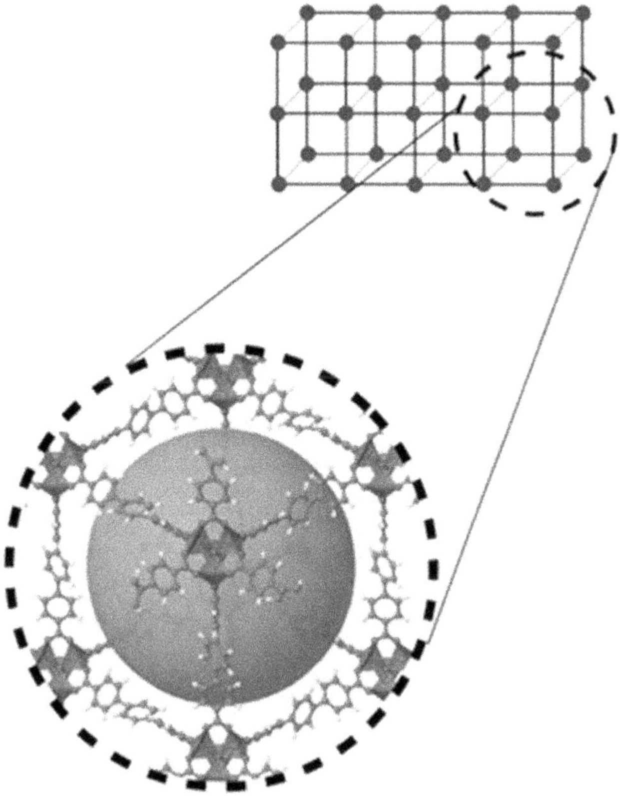

FIGURE 8.1 A general illustration of the MOF network schematically.

It was reported that MIL-88A (Fe) MOF used as anticorrosion nanofillers could effectively improve the active protection ability of epoxy/polyamide coatings [17]. ZIFs might also act as effective inhibitors of metal corrosion due to the presence of heteroaromatic species and abundant π-systems in the structure with inherent imidazole/benzimidazole ligands [10].

This chapter aims to provide a general overview of the synthesis and characterization techniques of MOF coatings, map a path for further research, and explore ways to utilize these materials more widely in coating and anticorrosion industrial applications.

8.2 MOF FABRICATION METHODS FOR ANTICORROSION COATINGS

The synthesis of MOF materials is possible using a variety of methods. Generally, any device capable of producing controlled heat and appropriate mixing conditions can be used for this process. Combining three main components is essential in this process. The key components of MOF synthesis are a metal salt, an organic ligand,

and a suitable solvent. Synthesis is designed to achieve specified structural units and inhibit the deterioration of organic ligands [18, 19].

Several key parameters must be determined to develop new MOFs. In this regard, preserving secondary structural units (SBUs) is recognized as one of the most critical aspects. The construction of new organic ligands and the optimization of synthesis conditions to improve metal–organic bonds have attracted much attention and have become a research focus. Ligand properties, including geometry, chirality, bond angles, volume, and length of ligand, can significantly affect the structure of MOFs [20].

When synthesizing MOFs, we need to prepare and select appropriate models and have a good prediction of how the constituent components will be assembled in the final product. The concept of SBU as a structural component significantly contributes to predicting MOF structure. Mineral SBUs are metal clusters or coordination spheres connected by organic bonds to form MOFs [21] (Figure 8.2).

These materials can be synthesized using two general categories of methods: liquid phase and solid phase [22]. Liquid-phase methods include techniques such as normal (ambient) temperature synthesis [23, 24], hydrothermal or solvothermal synthesis [8], electrochemical (EC) [25], sonochemical [26], microwave [27], mechanochemical [28], and diffusion [29]. These techniques are generally carried out by combining solutions containing metal salts and organic ligands or by adding a solvent to a reaction containing solid salts and ligands. Conversely, solid-phase synthesis methods have received less attention because of the difficulties in producing single crystals and confirming the structure in these methods [22].

Six main methods for synthesizing MOFs in anticorrosion applications have been identified, and their respective advantages and disadvantages have been carefully assessed. Table 8.1 provides a summary of various MOFs, their synthesis methods, and applicable conditions.

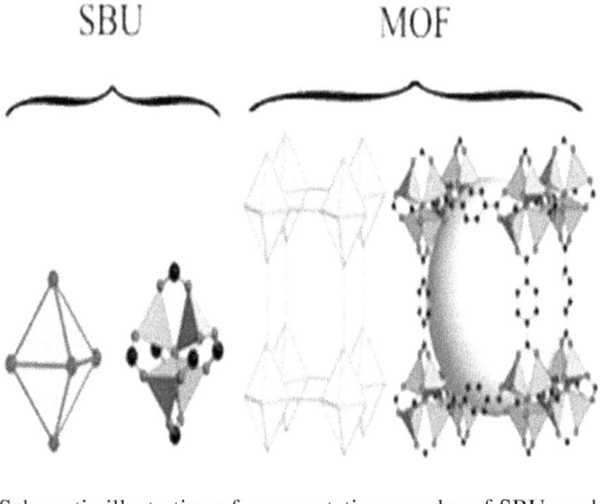

FIGURE 8.2 Schematic illustration of representative samples of SBUs and MOFs. MOF is an abbreviation for the metal–organic framework, and SBU represents a SBU.

TABLE 8.1
Summary of Various MOFs and Their Synthesis Methods [30]

MOF	Condition and Solvent	Synthesis Method	Finding	Ref.
UiO-66(Ce)	25°C, 1 hours (with terephthalic as linker), 24 hours (with modulator), water	Normal temp	The use of acetic acid as a modulator enhances crystallinity while requiring no additional energy input	[23]
ZIF-8	25°C, 2 hours, methanol	Normal temp	Zinc nitrate and 2-methylimidazole are combined in methanol, resulting in high thermal and chemical stability	[31]
CPO-27-Ni MOF	25°C, 4 hours, water	Normal temp	Achieves a high BET surface area through water-based synthesis with stepwise optimization, emphasizing green synthesis	[32]
IRMOF-0	25°C, 24 hours, DMF solvent	Normal temp	Acetylene dicarboxylate, a thermally sensitive linker, exhibits double interpenetration with an extremely small pore size	[33]
MOF-2	25°C, 2 hours, water and DMF	Normal temp	Synthesis is achieved by mixing zinc acetate and terephthalic acid without requiring pH controllers. However, the thermogram indicated weight loss attributed to the two solvents and the decomposition of the linker	[34]
UiO-66	100°C, 1 atm, 24 hours, water	Hydrothermal	Excellent water stability, scalable manufacturing, eco-friendly, and gentle synthesis	[35]
Fe-MOF	High temp and pressure, 36 hours, water	Hydrothermal	High porosity and functionality, with potential applications in gas storage and catalysis	[36]
MOF-74	25–30°C, 1 atm, 6 hours, water	Hydrothermal	Exceptional gas storage capacity, high BET surface area	[37]
MIL-100(FE)	150°C, sealed reactor, 6 hours, DMF	Solvothermal	Synthesized high-quality MOFs of regular particle size, high surface areas, and high catalytic activity	[38]

(continued)

TABLE 8.1 (Continued)
Summary of Various MOFs and Their Synthesis Methods [30]

MOF	Condition and Solvent	Synthesis Method	Finding	Ref.
Zr-MOF	150°C, sealed vessel, 18 hours, DMF	Solvothermal	Tailored functionality, chemical stability	[39]
HKUST-1	110°C, 18 hours; water, ethanol, DMF	Solvothermal	Utilized a dual solvent system to boost porosity and framework stability; TGA, contact angle, and SEM analysis verified enhanced hydrothermal stability through amino acid functionalization	[40]
MOF-801	120°C, high pressure, 24 hours, DMF	Solvothermal	Exceptional selectivity towards light alkanes renders it a material of interest for storing and purifying such molecules and high thermal and chemical stability	[41]
Zr-MOF	120°C, Microwave irradiation, 2.5 hours, DMF	Microwave-assisted	Optimized UiO-67 synthesis under microwave conditions by adjusting modulators to improve the surface area and pore volume; increasing the modulator amount further enhanced these properties	[42]
Ln-MOFs	Microwave irradiation, 5 minutes,	Microwave-assisted	Improved luminescence and moderate thermal stability, with emphasis on rapid crystallization and precise size control of the frameworks	[42]
MIL-100(Fe)	10 minutes running with power 1080 W, water	Sonochemical	Exhibited a high surface area of 1033.8 m^2/g and a pore volume of 0.79 cm^3/g, achieving a reaction yield of over 72% under 1080 W power at a 0.1 M concentration for 10 minutes	[43]
HKUST-1	The sonicator operated for 30 minutes at 60% amplitude.	Sonochemical	Achieved precise particle size control between 1 and 4 μm with reproducible size distributions; the mechanism involves constructive crystal growth and deconstructive sono fragmentation	[44]
$[Zn_4(oba)_3(DMF)_2]$	Heated at 100°C (sealed), Frequency of 40 kHz, DMF	Sonochemical	SEM images of such MOFs display plate-like structures, suggesting effective dye removal and controlled nanoplate formation	[45]

TABLE 8.1 (Continued)
Summary of Various MOFs and Their Synthesis Methods [30]

MOF	Condition and Solvent	Synthesis Method	Finding	Ref.
Cu-BTC	Variable pH, buffer solution, room temperature, graphite as anode, copper wire as cathode	EC	Copper dissolves anodically and also reacts with trimesic acid in a buffered medium to deposit Cu-BTC layers on graphite electrodes that display high mechanical stability, subsequently coating electrodes with MOF layers to provide more efficient EC sensing	[46]
MOF-5	Low current, room temperature, water medium, copper as cathode, zinc as anode	EC	Synthesized using a galvanic displacement reaction in which Zn^{2+} ions displace terephthalic acid to yield a highly porous structure that is of interest in gas storage applications	[47]
ZIF-8	Platinum as anode and zinc sheet as cathode, ethanol medium, controlled voltage	EC	Utilizes reductive electrosynthesis, during which Zn is reduced at the cathode and in situ reacts with anodic imidazole to enable rapid and direct formation on conductive substrates for use in electronics	[47]
ZIF-L	Room temperature	Mechano-chemical	Utilized liquid-assisted resonant acoustic mixing (LA-RAM) to scale up synthesis, enabling scalability without excessive solvent use	[48]
ZIF-8	Room temperature	Mechano-chemical	Entails solvent-free mechanochemical grinding of organic linkers and a zinc source with a strict control of particle size and crystallinity	[49]
UiO-66	RT, liquid assisted	Mechano-chemical	Utilizes minimal solvent to allow grinding, providing fast reaction kinetics, high crystallinity, and improved scalability without excessive solvent use	[48]

Source: Jomy and Prabhu [30]." and "*Source:* Jomy and Prabhu [117].

8.2.1 Conventional Synthesis

The traditional and common method of synthesis of MOFs falls into two categories: the solvothermal synthesis method and the non-solvothermal (normal temperature) method. Solvothermal refers to the use of any solvent and is considered a more general method than hydrothermal, which uses water as the solvent. Non-solvothermal synthesis is carried out at a temperature below the solvent's boiling point in open flasks and at atmospheric pressure. Conversely, solvothermal synthesis is carried out at or above the solvent's boiling point in special closed reactors and at high pressure, which is created by evaporation of the solvent or using a pump [50, 51].

8.2.1.1 Normal Temperature Synthesis

MOF synthesis does not require special equipment and can be conducted at room temperature or with slight heating. In this conventional method, metal elements, organic ligands, and solvents are selected, and pH and temperature are adjusted. The mixture is then stirred continuously for a specified time to produce high-yield MOFs. Precipitation/crystallization is achieved by optimizing the reactant concentration to facilitate nucleation, which can often be achieved by increasing the temperature, evaporating the solvent, and cooling the solution [50].

Modifying the synthetic parameters precisely can regulate the nanoscale particle size of MOFs. Multiple factors influence their formation at normal temperatures, such as the metal source, the nature of the organic ligand, the concentration of the reactant, the pH, and the conditions of the solvent. In general, lower activation energy is needed to prepare MOFs at ambient temperature without harming the nucleation kinetics. The pH plays a vital role in the growth of MOF, whereas the concentration of the reactant affects the nucleation process more significantly. Hence, the latter is crucial in the normal temperature synthesis of MOFs [24, 52].

David et al. produced MOFs synthesized by this method, including MOF-5, ZIF-8, MOF-74, MOF-177, and MOF-199 (Hong Kong University of Science and Technology -1). These MOFs demonstrate that these synthetic methods are effective for MOFs containing copper(II) and zinc(II) [33].

For MOFs with robust chemical properties, such as ZIFs, normal temperature synthesis allows for their facile fabrication under conditions of preformed functional molecules [53].

Recent research has shown that ZIFs synthesized at room temperature can be used for corrosion protection. Specifically, Ramezanzadeh et al. [54] have discussed the application of a new type of ZIF-67 that acts as a corrosion inhibitor in enhancing the anti-corrosion performance of epoxy coatings. They synthesized ZIF-67 nanoparticles at room temperature using cobalt and 2-methylimidazole. These nanoparticles of ZIF-67@APS (aminopropyl triethoxysilane (APS)) loaded with 3-APS had the best performance in enhancing steel coating protection efficiency by up to 81% reduction of its corrosion rate.

Gao et al. [55] prepared ZnG@ZIF-8 composite materials, which were synthesized at normal temperature using zinc gluconate (ZnG), a green corrosion inhibitor, as the sole metal source. Additionally, they investigated the anticorrosion performance of an epoxy coating incorporating ZnG@ZIF-8 as nanofillers and showed that this coating exhibits stable anticorrosion ability for magnesium alloy.

Normal temperature synthesis has advantages and disadvantages, including low energy consumption, safety in operation, ease, and so on. However, this effective method has apparent shortcomings, including being time-consuming and extreme sensitivity to surroundings and solvents or reactants, rendering it unsuitable for preparing all types of MOFs.

8.2.1.2 Hydrothermal or Solvothermal Synthesis

Hydrothermal and solvothermal are the most widely used methods in synthesizing MOFs, especially the solvothermal method, which makes up 70% of MOFs [56].

Hydrothermal methods refer to homogeneous or heterogeneous chemical reactions carried out in water as a polar protic solvent at temperatures exceeding the boiling point and under pressures greater than one atmosphere. Conversely, solvothermal methods usually use organic solvents with high boiling points, and the reaction is carried out in closed vessels at temperatures and pressures above the solvent's boiling point. These methods are often carried out in steel autoclaves and under controlled temperature and pressure conditions [57] (Figure 8.3).

Changing the reaction conditions can tune the nucleation rate and crystal growth of MOFs in hydrothermal and solvothermal methods [58]. This process depends highly on the compositional parameters (such as solvent type, pH, metal sources, organic ligands, and reactant concentrations) and process parameters (such as

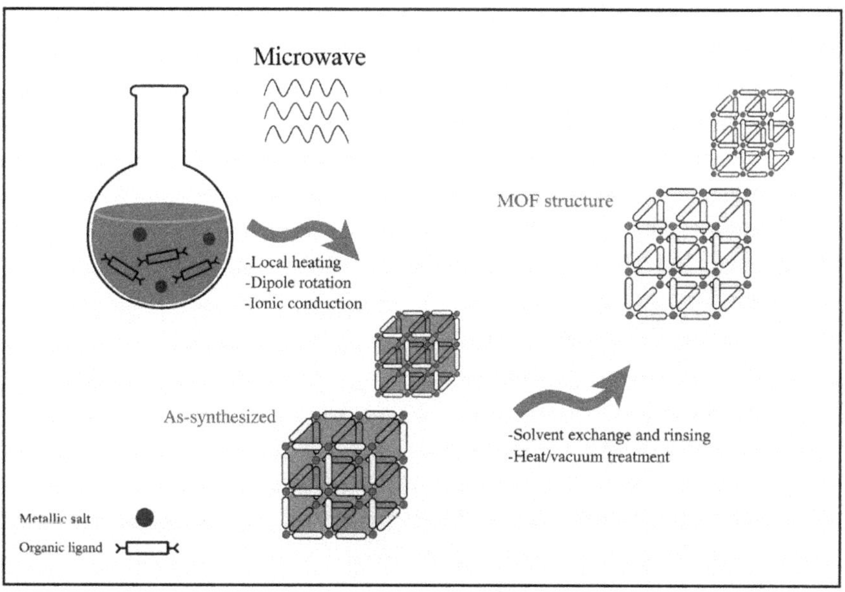

FIGURE 8.3 Solvothermal synthesis of MOF structures.

temperature, reaction time, and pressure). The reactant concentration must exceed the critical nucleation concentration to achieve transparent MOF crystals. Under these conditions, the difference between the critical nuclei can be overcome by reducing the reactant concentration through solvent evaporation or adjusting the reaction temperature [20].

In general, the formation process of MOFs depends on factors such as gradual changes in reactant temperature, slow diffusion of reactants, solution layering, and controlled solution evaporation. Solvents used in solvothermal or hydrothermal methods usually act as hydrogen proton donors and must have strong polarity and high dielectric constants. Therefore, the selection of solvents must be made with careful consideration of their chemical properties [59].

These processes' most commonly used solvents include water, methanol, ethanol, dimethylformamide (DMF), and acetone. These solvents provide favorable conditions for dissolving heavy organic compounds and crystal growth due to their high polar properties and ability to form hydrogen bonds. In some cases, a combination of solvents is used to optimize the solubility of the starting materials. However, using some hazardous solvents, such as DMF, poses limitations due to environmental and safety concerns [60].

Synthesis of MOF nanoparticles (e.g., fillers, containers) via hydrothermal or solvothermal methods can further enhance the anti-corrosion effect. For example, Ramezanzadeh et al. [61] reported the synthesis of NH_2-UiO, UiO-66, and glycidyl methacrylate-modified NH_2-UiO materials via a thermal solution process. The researchers systematically investigated the effect of Zr-MOF nanofillers on the anti-corrosion performance of epoxy coatings. The results showed that organic compounds and Zr ions are corrosion inhibitors and endow the coating with a smart and pH-sensitive property with desirable mechanical properties.

It is worth noting that even when similar reaction mixtures, including metal ions, ligands, and solvents, are used for the synthesis of MOFs, slight changes in process parameters such as reaction time, temperature, and pressure can lead to differences in the coordination numbers between metal ions and ligands during crystal growth. These changes may affect the properties of MOFs, including their structure, particle size, yield, and morphology [62].

MOFs with dense structures can usually be produced by carefully adjusting the reaction temperature. However, longer reaction times may have side effects, such as the destruction of MOFs' crystal structure and the formation of weak pores [62]. Fine-tuning key factors such as process parameters and composition is very important for producing MOF-based anticorrosion materials via solvothermal and hydrothermal synthesis methods.

Hydrothermal or solvothermal synthesis methods have significant advantages, which are mainly reflected in two main attributes: (1) this method allows the production of dispersed products and optimizes the synthesis parameters such as the choice of reaction temperature, precursors, and type of solvent, the structure and composition of MOFs can be readily adjusted; (2) this method can produce new MOFs with diverse functionalities and effectively prevents contamination during the synthesis process [59].

Due to their numerous advantages, hydrothermal and solvothermal methods are practical tools for synthesizing new MOFs with diverse properties and applications. However, this method also has its disadvantages. A primary limitation is MOFs' languid nucleation kinetics, which require an intricate synthesis process, high energy consumption, and an extended reaction time. In addition, environmental problems in the use of solvents are an important area for future research [31, 59].

8.2.2 Microwave-Assisted Synthesis

Microwave technology has been widely considered an innovative and alternative method for synthesizing MOF compounds. This technology is based on electromagnetic radiation with a frequency range of 300 MHz to 300 GHz and a wavelength spanning from 1 mm to 1 m, which has a longer wavelength and penetration than other electromagnetic waves, such as infrared [63]. This method's basic operation mechanism is based on the energy transfer through molecular dipole rotation and ionic conduction. This mechanism increases the material's internal and external temperature due to the molecules' intense vibration and the friction between them, which is particularly efficient for the synthesis of MOFs. Due to their high dipole moments, water and ionic liquids (ILs) are known as the most suitable solvents in this technology [64, 65] (Figure 8.4).

One of the key advantages of microwave technology is the significant reduction in reaction time compared to conventional methods such as hydrothermal and solvothermal. For example, MOFs such as MIL-100 and MOF-5 can be synthesized

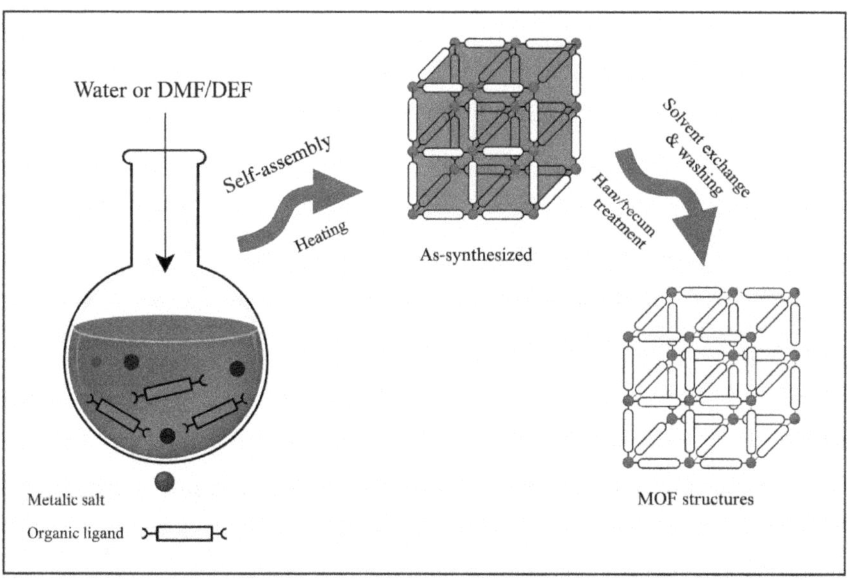

FIGURE 8.4 Microwave-assisted synthesis of MOF structures.

by this method in a shorter time than conventional methods without negatively affecting the structure and physical properties of the materials. This indicates that the microwave method can effectively optimize the time and energy consumption for the production of MOFs. In addition to reducing the time, this technology allows for better control over MOFs' structural and morphological properties. For example, these compounds' particle size distribution, phase control, and porous structure are optimized. In addition, the synthesis of MOFs with a fine porous structure is complicated by conventional methods such as hydrothermal but is easily possible by utilizing microwave radiation [63, 64].

To achieve uniform (110)-oriented MOF coatings for corrosion protection, Wang et al. [66] introduced a novel synthesis strategy to produce uniform ZIF-7 films by directing epitaxial growth on the surface via microwave irradiation (carried out in a single-mode microwave oven at 90°C for 5 minutes). The results showed that single-mode microwave irradiation, controlled by epitaxial deposition, is key to creating ZIF-7 films with desirable microstructures. The obtained films showed exceptional ability to protect aluminum plates against corrosion. This study provides a practical approach to producing high-quality MOF films and a new avenue for developing MOF-based anticorrosion materials using microwave synthesis.

Finally, the advantages and disadvantages of the microwave method can be summarized as follows: (1) A high heating rate accompanies it and can significantly reduce the synthesis time (in some cases, it takes only a few seconds), which leads to increased energy efficiency and a significant reduction in energy consumption. (2) The synthesis process can be exactly controlled because the products cool down quickly after the microwave energy is turned off. (3) This method is considered environmentally friendly due to the significant reduction in environmental pollution. However, the microwave method also has limitations. First, microwave energy equipment is substantially more expensive than conventional thermal technologies, and it requires sophisticated equipment to generate microwave radiation. Second, this method often has poor reproducibility due to its dependence on several factors. The products produced are strongly affected by the microwave frequency and power, as well as the uniformity of the microwave fields, which makes it difficult to create similar conditions in different devices. Third, the materials and size of the reactors in the microwave process have limitations, as microwave efficiency is affected by the reactor material, and its limited penetration depth imposes limitations on absorption into the environment. Similar to other advanced technologies, further research is needed to develop and improve the versatility of this method in producing different types of MOFs [31, 59].

8.2.3 Sonochemical Synthesis

The sonochemical method is one of the efficient, simple, and environmentally friendly approaches for synthesizing MOFs. This method, based on ultrasound waves (in the range of 20 kHz to 10 MHz), has unique features such as short reaction time, high efficiency, and energy saving. Sonochemistry facilitates chemical and physical changes in molecules by creating the phenomenon of cavitation, including the formation, growth, and collapse of bubbles in the liquid phase. In this process, hot spots with high temperature and pressure (about 5000 K and 500 bar) are created in a very short time, which can activate chemical reactions and accelerate crystal nucleation [56] (Figure 8.5). These features have made the synthesis of MOFs such as MOF-177, MOF-5, Fe-MIL-53, and ZIF-8 widely used in this method. It has been reported that adjusting the ultrasonic power levels can affect the arrangement of MOFs and lead to the production of particles with smaller sizes [67, 44].

Among the outstanding advantages of the sonochemical method, the remarkable reduction of crystallization time and the possibility of controlling the particle size distribution by variation of the power of ultrasound waves can be outlined. For example, Zare et al. [68] synthesized a copper-based MOF (Cu-BCT) by this method and used it as a carrier of corrosion inhibitor (2-aminobenzothiazole) to improve the properties of silica sol–gel coating. Results showed that the compactness and corrosion resistance of composite coatings are better than that of the traditional coating. The anticorrosion efficiency of coating in the 3.5 wt% sodium chloride environment increased to about 97% (Figure 8.6).

The main advantages of the sonochemical method can be summarized in three aspects: (1) the simplicity of the process and the high efficiency of the reactions under ultrasound irradiation, (2) the reduction of the temperature and reaction time, which is beneficial for commercial applications and large-scale production, and (3) the high selectivity of the products produced under the influence of the process. This route is not so versatile and permits synthesizing only a certain type of MOF. With regard to the general routes such as hydrothermal, solvothermal, and microwave methods, there are a number of situations when such limitations may raise obstacles for their applications. Actually, all these specific features make sonochemistry one of the

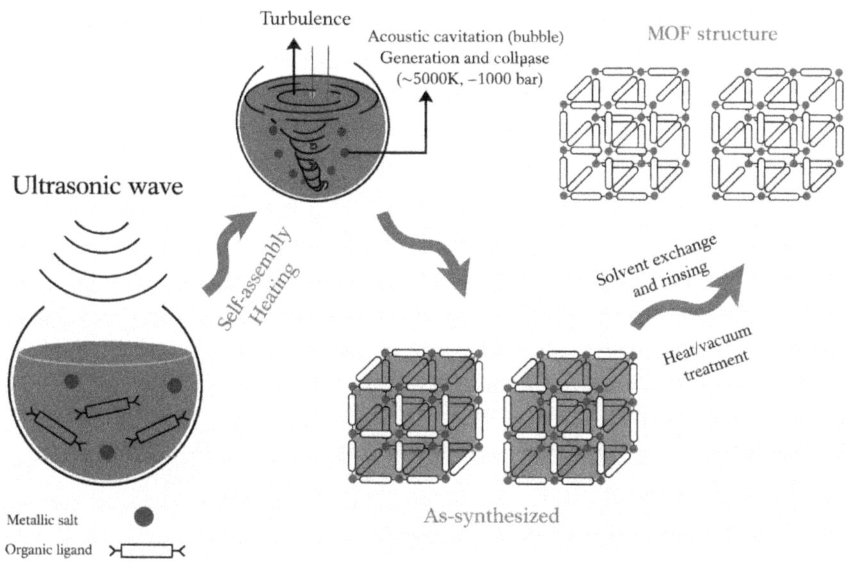

FIGURE 8.5 Sonochemical synthesis of MOF structures.

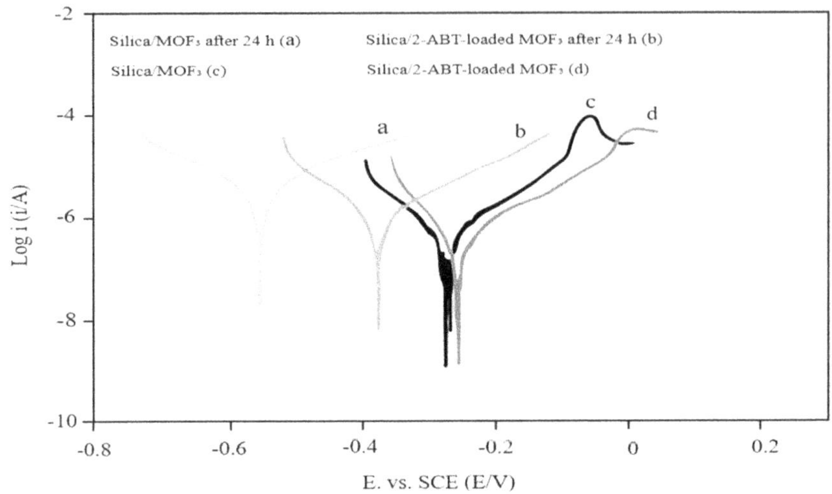

FIGURE 8.6 PDP curves of composite coatings containing silica/MOF and silica/2-ABT-loaded MOF both before and after 24 hours of immersion in a 3.5 wt% NaCl solution [68].

powerful, practical tools for the design and fabrication of advanced materials such as MOFs [59, 63].

8.2.4 EC Synthesis

EC synthesis is an efficient and rapid technique for producing MOF materials that allows for the direct and controlled deposition of these materials onto substrate surfaces. This method has been considered a suitable industrial approach due to its high product purity, precise process control, and the absence of the need for counterions from metal salts. In this method, metal ions are directly produced through the anodic dissolution of the metal substrate, which eliminates interfering ions and improves the process efficiency [69, 70] (Figure 8.7).

The EC method includes three main mechanisms: anodic deposition, cathodic deposition, and electrophoretic deposition, which are used to deposition MOFs. In anodic deposition, the synthesis mixture contains electrolytes and organic ligands, but metal salts are not used because metal ions are directly supplied through anodic dissolution. During this process, metal ions are generated near the anode, and MOF crystals are uniformly nucleated and grown on the metal surface. This method was first reported by Muller et al. for the synthesis of HKUST-1. Anodic deposition has demonstrated the ability to produce MOF coatings based on various metals, including Zn, Cu, Fe, and Al. Due to the optimized deposition conditions, the coatings produced by this method exhibit favorable mechanical characteristics, including strong adhesion and hardness. In the cathodic method, organic ligands and metal ions are used in the mixed solution, and the desired conductive substrate acts as the cathode. Near the cathode surface, hydroxide anions are produced, and the pH increases, leading to protonated ligands forming. These ligands react with metal ions to form an MOF

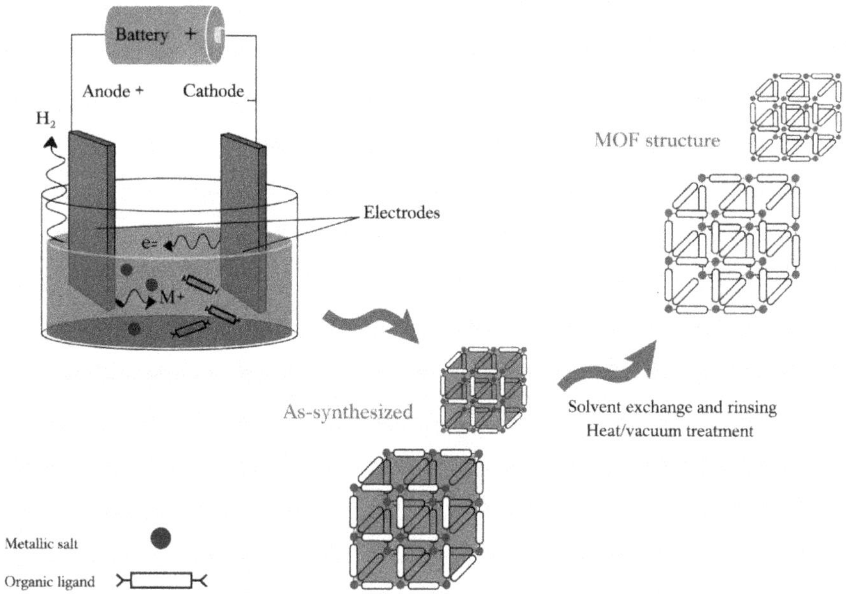

FIGURE 8.7 EC synthesis of MOF structures.

film on the cathode. Numerous reports of cathodic deposition of MOFs based on zinc, copper, and other metals indicate that the supplied potential is essential for determining coating quality. The substrate mainly used determines the coating's reduction potential, cohesion, and adhesion [69, 71, 72].

MOF is known to be the first compound to be synthesized using the EC technique. In this method, the EC bath was filled with a mixture of benzene tricarboxylate and methanol, and copper plates served as the anode and cathode electrodes. By applying an electric current of 1.3 A and a voltage between 12 and 19 V, a turquoise-colored deposit of Cu-MOF was formed within 150 minutes. Other structures produced by this technique include HKUST-1 and Zn-imidazolate [73].

In addition, MOF coatings can be produced by simultaneous cathodic and anodic EC deposition processes. Ameloot et al. [25] introduced a technique for synthesizing UiO-66 films by cathodic and anodic deposition and comprehensively investigated the formation mechanisms of these films. The results showed that anodic deposition, due to the formation of oxide bridges, provides higher adhesion in MOFs. However, using this EC method to produce MOF films for corrosion protection has rarely been considered in existing studies.

Finally, the advantages and disadvantages of EC deposition approaches are summarized. The advantages of these methods include the following: (1) EC techniques such as cathodic and anodic deposition for producing MOFs are simple and cost-effective methods that usually offer high yields. These techniques allow the preparation of multifunctional MOF coatings in a short time and under mild conditions, which helps to avoid film cracking caused by high temperatures.

(2) MOF with elevated-density crystals can be produced using only a tiny amount of reactants, and the film synthesis process is continuous. (3) The spot-on nature of these techniques allows the creation of randomly oriented films with controllable substrate and thickness that remain unmodified [74].

Nevertheless, these methods also have disadvantages that should not be overlooked. First, the deposition of MOFs is limited to specific substrate regions that must be exposed to the solution. Second, the conductivity of the substrate is essential for EC deposition, and the non-uniform distribution of current on the substrate surface makes it difficult to achieve homogeneous MOF layers. Third, in the electrophoretic synthesis method, consisting of two stages, the inherent characteristics of EC deposition, such as mild conditions and shorter time, are less evident, and there is also the possibility of microcracks. Therefore, the brittleness of MOF coatings produced by this method should be given special attention. However, the problem of structural discontinuity of MOF films obtained by electrophoretic deposition can be reduced by fabricating MOF composite coatings. Overall, the flexibility and efficiency of EC deposition make this method a powerful tool for fabricating MOF coatings for diverse applications. These coatings will receive more extensive research in the future to protect metals against corrosion [59, 75].

8.2.5 Mechanochemical Synthesis

Mechanochemical synthesis has been introduced as an innovative and environmentally friendly method for producing MOFs. This method reduces environmental risks by eliminating harmful organic solvents and is more economically viable. Using mechanical force, chemical bonds between metal precursors and organic ligands are stimulated in the solid state, and the products are produced in a short time (10–60 minutes) with good yields [76] (Figure 8.8). In 2006, the first MOF was synthesized using this method by Pichen et al. [77]. This method does not require high temperature or pressure, and its byproducts, such as water and acetic acid, are low hazard, making it a suitable alternative to the solvothermal methods. In addition, it is possible to use metal oxides instead of metal salts, which emphasizes the flexibility of this method.

In this context, another attempt has been made to see the mechanochemical syntheses of $Cu_3(BTC)_2$, better known as HKUST-1, by ball milling: This process consists of milling the reactants $Cu(OAc)_2{\cdot}H_2O$ and H_3BTC without additional solvent and was followed to completion after 15 minutes at 25 Hz-the material produced was comparable in thermal stability and porosity to samples from traditional solvent-based methods. This work indicates that liquid-assisted grinding (LAG) can improve the crystal structure and reduce mesopores originating from cracks formed during initial milling steps. However, continuous milling is required in the mechanochemical process, and such a process is not as self-sustaining as other MOF reactions. It was also reflected in the reaction that a little content added to the solvent did not affect the reaction rate in a big way but produced a product with better crystallinity [78]. Finally, high consumption of energy, which makes the chemical reaction happen quickly, can be a serious disadvantage of this method.

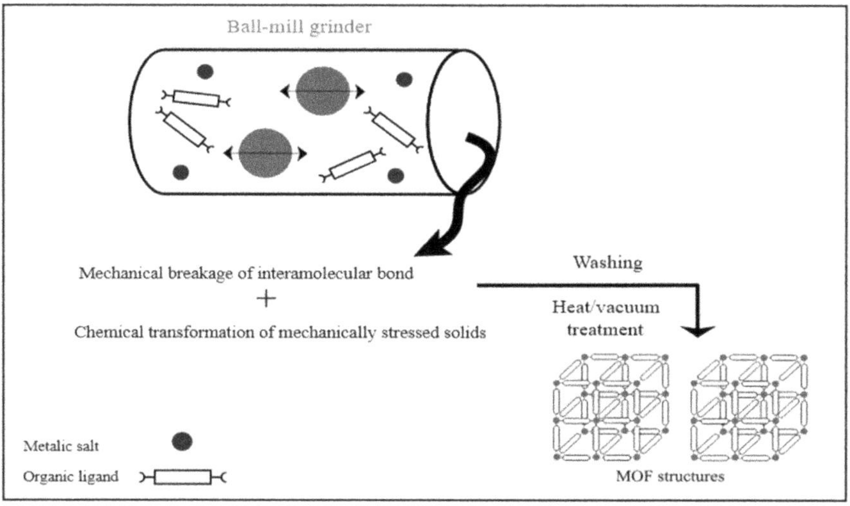

FIGURE 8.8 Mechanochemical synthesis of MOF structures.

8.2.6 Diffusion Method

The diffusion method is used to prevent sediment formation and produce powder samples that result from rapidly mixing materials. Powder samples consisting of microcrystals are unsuitable for X-ray diffraction (XRD) analysis. The diffusion method is divided into two modes: liquid infiltration and gas infiltration. In the liquid-phase diffusion method, an immiscible solvent is used to dissolve metal ions and organic ligands, and MOF crystals are formed due to surface contact reactions of metal ions and organic ligands. In the gas-phase diffusion technique, a solution of volatile organic ligands serves as the solvent. MOFs are formed when sufficient mixing occurs between the organic ligand solution and the metal ion solution [79, 20].

In a different approach, Yanhu Wei et al. [28] prepared NPs in cyclodextrin-based MOFs through simple diffusion methods. Such NPs are produced by using RbOH or CsOH to prepare CD-MOFs and reduce metal salts with hydroxide ions. This results in the controlled release of the nanoparticles during MOF dissolution.

8.3 CHARACTERIZATION TECHNIQUES FOR MOFS ANTICORROSION COATINGS

MOFs have a native function, porosity, and surface character, making them ideal coatings. Constituent materials and guest molecule interactions define their function. Hence, an in-depth analysis of their physicochemical character and such interactions is critical for new mechanisms and process optimizations [80].

The review below lists significant MOF characterization techniques for studying crystalline, chemical, mechanical, thermal, and EC behavior. Due to their complex structures, various techniques must be adopted to study factors such as crystallinity, porosity, and specific surface area. XRD, for example, detects crystallinity and purity of phases, while scanning electron microscopy (SEM) and transmission electron microscopy (TEM) reveal morphology and crystallite dimensions.

The chemical behavior of MOF coatings is also essential. Various techniques—adsorption, thermal durability, and surface and porosity tests—help define such factors. DRIFTs, for example, can both be utilized to determine elemental composition, and proper techniques must then be selected for each use. Combining experimental data with computational methodologies can sharpen MOF characterization and optimize synthesis and design methodologies.

Accurate structure determination is critical in knowing the functional behavior of MOFs for future technological applications. Because MOFs have a general periodical structure, X-ray diffraction (XRD) methodologies, namely single-crystal X-ray diffraction (SC-XRD) and powder X-ray diffraction (PXRD), have become universally applicable for molecular structure determination and crystallinity analysis at an atom level or below. SC-XRD, with its high resolution, can produce a lot of information but at a high demand for big, high-quality single crystals, and MOFs have a tendency towards difficulty in crystallization through complex crystallization processes and have a variety of polymorphs or domains. In instances when proper single crystals cannot be attained, PXRD comes in handy as an alternative [81, 82].

Diffraction information in PXRD is reduced in one out of three dimensions (the 2θ value) of three dimensions in three-dimensional reciprocal space, and therefore, significant peak overlap can arise, with a deleterious effect in terms of determination of position and intensity of a peak [81 83]. PXRD is subsequently used for confirming the purity of phases with a comparison with experimental and calculated (based on SC-XRD information or computational simulation) patterns [84].

TEM and SEM are two powerful tools for MOF characterization. TEM utilizes a high-energy electron beam for 10^6 times enlargement, generating high-magnification and high-resolution micrographs and diffraction information beneficial for confirming crystallinity, morphology, defects, and specific facet presence. It can even generate three-dimensional micrographs through electron tomography, providing information regarding size distribution and inner structures [85].

Conversely, SEM operates at a micro- to nanoscale, utilizing an electron beam to produce high-resolution MOF surface micrographs. Backscattered electrons reveal surface topography and secondary electrons reveal surface morphology. Secondary electrons yield information about contrast at an atomic level. In most instances, SEM is accompanied by energy-dispersive X-ray spectroscopy (EDS), an X-ray specific-emitting detectability technique. Both qualitative and quantitative analyses of elemental composition, such as metal distribution in MOFs, can be conducted [86, 87].

Many MOFs' insulating characterization can cause charging and cloud SEM micrographs. To counteract this, a thin conductive layer (e.g., gold, osmium) is most commonly sprayed onto the sample in an attempt to inhibit charge build-up. In addition, where dimensions permit, optical microscopy can also provide general information about the dimensions and shape of a crystal beforehand for SEM analysis [88]. Together, these tools give researchers a complete picture of MOF character—from structure to elemental composition—enabling them to streamline synthesis and application strategies for improvement.

Thermogravimetric analysis (TGA) is an important tool for characterizing thermal and accessible-pore volumes in MOFs. In TGA, a sample is taken through a programmed temperature increase in a defined atmosphere (e.g., N_2, air, O_2), and its real-time weight loss is measured. Mention of a specific carrier gas is significant, for an MOF's decomposition mechanism can depend enormously in a range of environments.

Normally, TGA plots will exhibit a sequence of several discrete processes below about 150°C; most MOFs will initially have an initial desolvation, with a detectable loss in mass via evaporating solvents. There then arises a level, a plateau, when the framework is free of solvents, and, at a temperature higher, a secondary loss in mass, indicative of degradation of MOF structure, in many cases, is detected [89]. By taking note of these observations, both MOF thermal range of stability and level of extraction of solvents can be determined, and information useful for MOFs' development for use in coatings can be attained.

Diffuse Reflectance Infrared Fourier Transform Spectroscopy (DRIFTS) is a routine tool in MOF investigations for its ease in the analysis of powdered samples with minimum preparation of samples. In DRIFTS, diffuse-reflected IR radiation scattered in all directions is detected; an ellipsoid mirror reflects such scattered radiation, and it is processed to monitor IR-active functional groups. It is a convenient tool for investigating MOFs' behavior under variable temperatures or under selective gas exposures, such as CO and CO_2, and can disclose information about MOFs' stability and adsorption behavior [90, 91].

Practically, MOFs can be analyzed in powdered form or in a dilution in a matrix, most frequently potassium bromide (KBr). Gentle grinding with a mortar and a pestle is a routine form of dilution. Nevertheless, care must be taken in case MOFs can become mechanically stressed, and, in such a case, care must be taken when checking IR stretches between 3500 and 3800 cm^{-1} not to swamp out desired signals with broad –OH bands of water, in the case of a sample, or in a KBr matrix [88].

Aqueous stability is a significant property for assessing such compounds in scenarios where contact with water cannot be avoided. Metal–ligand breakage and the forming of a more stable species are involved in degradation. MOF chemical stability is sensitive to inherent factors such as metal ion charge density, metal cluster connectivity, and ligand character (e.g., basicity, hydrophobicity, and configuration). All of these factors, in combination, will assess whether an MOF will maintain its integrity in an aquatic environment. Hence, a uniform analysis of MOF performance and suitability under acid, base, and neutral environments is significant in an attempt to make proper comparisons between materials and compositions [80].

For routine testing, we recommend suspending a known mass of activated MOF (e.g., 30 mg) in an appropriate volume of water (e.g., 15 mL) and measuring the pH before and after the exposure. Where acid or base stability is of interest, solutions with accurately measured pH should be prepared using HCl or NaOH, taking care to avoid buffers or counterions (e.g., phosphates, fluoride) that may coordinate with metal ions and confound results. After at least 12 hours at room temperature, MOF can then be filtered, and its pH can be reanalyzed for any acid and base shifts in acidity and basicity. Any crystallinity, porosity, and chemical composition shifts detected can reveal degradation pathways, possibly through hydrolysis but increasingly through strong coordination with electrolyte species [88]. By utilizing such routine protocols, researchers can effectively assess MOF aqueous stability and inform the selection and development of frameworks for use in corrosion applications.

Other tests applied also include the standard test, the Salt Spray Test (SST); this tests for resistance by exposing materials or coatings to a fine mist of saline solution and thereby simulates exposure in severe saline environments. Other EC techniques include the following: (1) potentiodynamic polarization (PDP) is one of the valuable techniques used to establish the rate of corrosion by studying the corrosion potential (E_{corr}) and current density (i_{corr}) to passivation and susceptibility to localized corrosion. (2) EC impedance spectroscopy (EIS) is an effective technique for investigating charge-transfer processes and processes at electrodes in MOFs. With a three-electrode arrangement in a potentiostat, EIS watches a small, periodic perturbation in a system and examines its behavior in a linear manner. Nyquist and Bode plots, in general, illustrate generated spectra, and such plots exhibit how circuit elements—resistance (R), capacitance (C), and inductance (L)—behave at a range of frequencies. Software fitting then processes and refines such information, and exciton lifetimes can then be analyzed. A key variable in EIS performance is porosity, which affects charge transport and even enables the estimation of pore dimensions [80, 92]. Overall, EIS is important for the coating integrity and protective performance of MOF coatings during service. These methods give a complete view of corrosion resistance and its mechanisms.

Below, we will briefly discuss the mechanical properties, thermal stability, and anticorrosion properties of MOF coatings using each of these characterization techniques, emphasizing data analysis performed in the literature.

8.3.1 Mechanical Properties

From an engineering perspective, the mechanical properties of MOFs under vacuum or pressure are crucial to their practical and industrial applications. In some cases, the porous network's instability in vacuum conditions may result in partial structural failure or significant phase transformations [93]. Solvent exchange and solvent evacuation are common approaches to activating MOFs and fully inhibiting structural collapse [94].

There is no single helpful technique for assessing the structural stability of MOFs. Depending on the operating environment, distinct standards must be considered. For example, a few catalysis methods require acidic/basic environments or oxidizing/reducing conditions [95].

The mechanical stability of an MOF is of interest to industrial applications. These structures may behave quite differently under different conditions. For example, MOFs containing carboxylate bases and high-valence metal ions demonstrate long-term stability in acidic environments but are prone to degradation in the presence of coordinating anions like Co_3^{2-} and under basic conditions. Conversely, $C_5H_4NO_2$-based structures with low-valence metal ions exhibit strong stability in alkaline solutions but are easily degraded in acidic environments [96].

Lal et al. [97] propose a novel approach for improving the mechanical properties of MOFs by covalently crosslinking MOPs. The authors used Cu_{24} isophthalate MOPs with functionalized alkyl chains bearing terminal alkene groups and underwent olefin metathesis to access flexible all-MOF materials. Key results included increased hardness and elasticity with increased crosslink density. This method overcomes common weaknesses, such as poor adhesion and non-homogeneous dispersion of MOF–polymer composites. As the MOFs maintain structural integrity with functional cavities, this method paves the way for robust and flexible applications of MOFs. In another study, embedding polymer chains within MOF increases its hardness and reduces the potential degradation of MOF [98].

From a different perspective, organic coatings, especially epoxy-based coatings, have poor mechanical properties despite their diverse applications [99]. MOFs have been shown to help overcome this shortcoming. MOFs can transform epoxy coatings' brittle and crack-prone nature, forming a robust structure with high toughness and crack resistance [100]. They can maintain the coating's early properties when the surface is exposed to scratches and electrolyte penetration [61].

The mechanical properties of the two types of added MOFs, NH_2-UiO-66 and modified GMA-UiO-66, to epoxy resin were investigated to ascertain their impacts on mechanical strength. According to the findings, adding both MOFs enhanced the elongation at break, absorbed energy until the break, and thus enhanced tensile strength compared with pristine epoxy. Interestingly, the modified MOF GMA-UiO-66 improved the coating's mechanical performance [101]. Another study confirmed such a trend, showing that the strength and durability of epoxy coatings were also enhanced with Sn-MOF@PANI [102].

8.3.2 Thermal Stability

The thermal stability of an MOF is defined by its ability to resist high-temperature conditions without irreversible physical or chemical transformations. During thermal treatment, the structure can undergo degradation processes such as dehydrogenation of linkers, dehydration of metal-oxo clusters, amorphization, and either graphitization or melting [103]. These degradation phenomena usually develop gradually with the increase in temperature. In yet other cases, significant changes occur only above the decomposition point when the release of guest molecules and/or combustion, along with the failure of ligand-metal links and the subsequent ignition of the organic spacer [104].

Elements from rare earth, alkaline earth, alkali, transition metals, and the main group can serve as metallic nodes in MOF architectures. Within the structure, these nodes may appear as polynuclear metal-oxo clusters, layered arrangements, chain-like networks, or even single metallic centers. Importantly, the inherent characteristics of a metal ion, such as coordination number, interactions with the organic linker, oxidation state, and ionic radius, play a crucial role in determining the thermal stability of the MOF [105, 106]. The intrinsic properties of the organic linker form the second major factor governing the thermal stability of MOF frameworks. Linkers that contain nitrogen and/or oxygen donor atoms can be grouped into two primary categories. These linkers may be further classified as aromatic or aliphatic based on their chemical configuration. A single ligand molecule can incorporate multiple features or all of them at once, typically referred to as a "multifunctional" spacer [104]. Consequently, predicting the thermal stability of MOF-based materials can be quite complex.

Nevertheless, numerous investigated frameworks utilize only one type of organic spacer with a uniform functional group or occasionally two distinct organic ligands for coordination. As drawn from the earlier discussion, an ideal MOF design for high thermal stability often exhibits dense, defect-free packing, potent ligand–metal bonding, metal ions with stable oxidation states, and relatively short aromatic linkers. However, because MOF-based compounds must also meet various other requirements linked to their intended applications, a compromise typically must be struck among stability, functionality, and porosity [105, 107].

In a study on the thermal stability of MOFs, researchers compared how graphene oxide (GO) nanoparticles behave on their own versus when paired with ZIF-67, both with and without an added layer of APS. They heated these materials through three temperature ranges and found that pure GO broke down more quickly than either of the ZIF-67 composites, which is a clear sign that incorporating ZIF-67 boosts GO's heat resistance.

Moreover, the APS-modified ZIF-67/GO outperformed its unmodified counterpart. Specifically, upon heating to 900°C, GO nanoparticles retained only 8% of their initial mass, whereas both ZIF-67/GO and ZIF-67/GO@APS demonstrated higher residual weights—in fact, the APS-functionalized composite surpassed the unmodified one by 20% [108]. In a separate comparison of three MOFs with nickel, copper, and cobalt as metal ions, the cobalt-based MOF displayed superior thermal stability relative to the others [12].

8.3.3 Anticorrosion Properties

As noted, MOFs can be directly dispersed in a solution as corrosion inhibitors or loaded with active corrosion-inhibiting agents to passively enhance the protection of surfaces exposed to such a solution. They can also be incorporated into organic coatings, where they actively heal corrosion-initiating defects, enhancing the coating's protective efficiency. Interestingly, as one study demonstrated, MOFs themselves can form a surface coating when derived from the transformation of a previously applied layered double hydroxide (LDH) layer [61, 51, 110–112]. In the section on fabricating MOFs, the anticorrosion properties of each method were discussed, But further on, specifically, the EC data extracted from the tests are presented.

The coatings of PPY/ZIF-67 are well electrodeposited on 430 stainless steel for the effective protection of materials in acidic chloride environments, ensuring about 99% corrosion inhibition efficiency throughout all 744-hour tests. Such a high level of protection should come true by the synergistic anodic protection of the PPY inner film, active inhibition by ZIF-67 nanoparticles, and the compact structure of coatings hampering corrosive species [113]. Ramazani et al. [114] developed epoxy coatings with ZIF-67-decorated g-C3N4 nanosheets containing BTA and Ce^{3+} inhibitors. These coatings reduced the corrosion rate by 50% and increased the resistance of mild steel in marine conditions by fivefold (Figure 8.9).

Open circuit potential (OCP) measurements demonstrated that applying ZIF-67, ZIF-67/GO, or ZIF-67/GO@APS solutions to steel surfaces yielded more positive E_{OCP} values than the saline-only control. All three solutions enhanced the steel's corrosion resistance relative to the untreated sample; however, PDP tests indicated that the APS-modified variant offered the highest inhibition efficiency [108, 116]. Similar results were observed for ZIF-8 and ZIF-8/GO, where the release of zinc ions and 2-methylimidazole molecules from the MOF effectively inhibited corrosion [8]. The corrosion performance of carbon steel with the Cu-MOF/GO composite was comprehensively evaluated. The coatings exhibited excellent corrosion resistance in 3.5% NaCl under EC testing. Among them, this addition of 3% Cu-MOF/GO gave the highest impedance magnitude of 23,804 Ω cm^2 and polarization resistance of 55,097 Ω cm^2, which could protect the carbon steel even after immersion for 30 days. That would indicate that combining MOFs with GO significantly enhances the anticorrosion performance of coatings, with great potential for industrial corrosion prevention in applications, as demonstrated here [102]. During the curing stage of an organic coating, micro-pores can form, potentially diminishing the coating's corrosion resistance. By incorporating MOFs into the coating, these pores can be filled, reducing the entry of corrosive substances and extending the coating's protective lifespan [54]. Notably, impedance measurements on an epoxy layer containing 2-mercapto benzimidazole-loaded ZIF-8 assembled on a GO nanosheet were four times higher

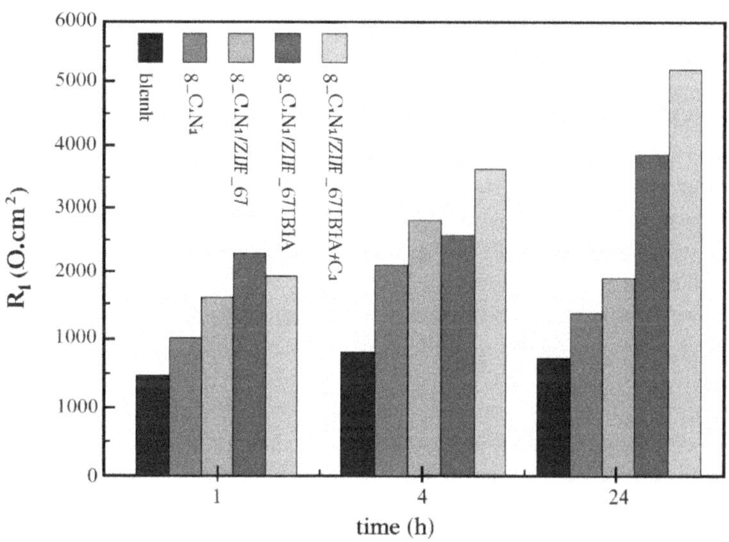

FIGURE 8.9 Samples immersed in blank salt solution and containing extracts of g-C_3N_4, g-C_3N_4/ZIF-67 g-C_3N_4/ZIF-67/BTA, and C_3N_4/ZIF-67/BTA/Ce [115].

TABLE 8.2
EC Information Obtained from Different Studies Related to MOFs [117]

Protection Approach	Substrate	i_{corr} (A/cm^2)	E_{corr} (V)	Impedance (Ω cm^2)	Ref.
GO@ZIF-8	Steel	2.875×10^{-6}	−724.2	—	[8]
EP coating/ZIF-67@APS	Steel	—	—	10.7	[54]
EP cating/ZIF67/GO@APS	Steel	—	—	10.7	[108]
Vinyl phosphonic acid (VP)-2 mM	Steel	5.94×10^{-6}	−0.491	-	[111]
2-Mercaptobenzimidazole-ZIF-8/GO/EP coating	Steel	—	—	3.30×10^{9}	[112]
$\{[Nd_2(2,3\text{-}Hpyza)_3].4(H_2O)\}_n$	Cu	18.35×10^{-6}	0.779	—	[118]
Folic acid-modified HKUST-1 incorporated into a polycaprolactone coating.	Mg	1.10×10^{-10}	−1.33	—	[119]

than those of pure epoxy after 60 days in NaCl solution [112]. Table 8.2 presents to facilitate a direct comparison of EC test results across various studies.

8.4 FUTURE CHALLENGES

MOFs have drawn considerable attention toward corrosion protection in terms of their specific structure and flexible functionality. Despite that, a variety of factors at present restrict widespread application. One of them is MOF preparation, which involves careful manipulation of solvents, pressures, and temperatures and, therefore, is cumbersome in its preparation processes. Even when high-purity MOFs can be synthesized via solvothermal and hydrothermal processes, both processes require a lot of energy and time, high volumes cannot be prepared, and production costs rise. Compared to conventional corrosion inhibitors, MOFs are even more expensive regarding production protocols and raw materials, thus limiting their use in budget-sensitive industries.

Apart from cost and scalability, long-term durability in real-use environments is a key concern. Effective inhibition of corrosion is exhibited in laboratory experiments, but long-term performance and behavior in a range of environments have not yet been well established in detail. Long-term, sustained contact with corrosive species can degrade their integrity and inhibitor-release mechanism. Therefore, long-term, thorough investigations and field testing must confirm MOFs' performance for actual use.

In addition to these challenges, deeper knowledge of MOF formation mechanisms and structural characteristics is critical for developing new anticorrosion materials. Notably, certain organic ligands serve inherently as corrosion inhibitors (e.g., the imidazole ligands in ZIFs), while in other cases, ligand exchange methods allow for effective inhibitor loading. Because different MOFs utilize distinct mechanisms, selecting the most suitable frameworks for specific corrosion scenarios requires a thorough understanding of their individual properties. Theoretical investigations and computational simulations can even shed new light on MOF fabrication processes, routes for corrosion inhibition, and loading strategies for inhibitors. Meanwhile, new tools, including machine learning, can accelerate MOF design, fabrication, and optimization by establishing predictive structure–performance relations.

Lastly, translating laboratory MOF studies into commercial applications will necessitate added efforts toward overcoming key factors such as ease of large-scale synthesis, overall cost-effectiveness, and environmentally friendly processes. Inventing new MOF materials and atom-economical reaction engineering to minimize waste are exciting avenues for overcoming such requirements and unlocking additional industriality. To make actual breakthroughs, one will have to integrate all such requirements—cost-effectiveness, bulkability, long-term durability, and sustainability—into MOF's fundamental design and optimization for corrosion protection. Cross-disciplinary collaboration will become imperative in overcoming current impediments and unlocking MOF anticorrosion technology's full potential.

8.5 CONCLUSION

MOFs, with their unique pore structures, very high accessible surface, variable chemistry, and exceptional stabilities, have emerged as a highly versatile class of materials in corrosion mitigation. The current chapter has discussed several synthesis routes from traditional solvothermal and non-solvothermal methods to certain newer alternatives that include microwave-assisted, sonochemical, EC, and mechanochemical processes, all of which have their respective advantages and challenges. The edits in MOFs with the selection of proper metal ions, organic ligands, and synthesis conditions allow fine-tuning in properties such as pore size, active site density, and interfacial adhesion in coatings.

PXRD, SEM, TGA, DRIFTS, and EC techniques provide extensive characterizations and highlight the contribution of MOFs to improving coatings by sealing micropores, releasing corrosion inhibitors, or forming protective films. All this is consistent evidence that the MOF-based coatings provide enhanced corrosion resistance, mechanical strength, and improved thermal stability over the unmodified ones. The synergy of MOFs with polymers or graphene derivatives helps establish an effective barrier layer that reduces the permeation of corrosive ions, thus enhancing the service life. Further efforts must be directed toward greener synthesis routes, better scalability, and consideration of environmental factors in this regard. Amidst such challenges, MOF-based coatings have great promise for the next generation in a variety of applications pertaining to industrial corrosion protection.

REFERENCES

[1] M. Eddaoudi *et al.*, "Modular chemistry: Secondary building units as a basis for the design of highly porous and robust metal–organic carboxylate frameworks," *Acc Chem Res*, vol. 34, no. 4, pp. 319–330, 2001.

[2] D. J. Tranchemontagne, J. L. Mendoza-Cortés, M. O'keeffe, and O. M. Yaghi, "Secondary building units, nets and bonding in the chemistry of metal–organic frameworks," *Chem Soc Rev*, vol. 38, no. 5, pp. 1257–1283, 2009.

[3] J. R. Long and O. M. Yaghi, "The pervasive chemistry of metal–organic frameworks," *Chem Soc Rev*, vol. 38, no. 5, pp. 1213–1214, 2009.

[4] H. Ghasempour *et al.*, "Metal–organic frameworks based on multicarboxylate linkers," *Coord Chem Rev*, vol. 426, p. 213542, 2021.

[5] S. A. Mazari *et al.*, "An overview of catalytic conversion of CO_2 into fuels and chemicals using metal organic frameworks," *Process Saf and Environ Protect*, vol. 149, pp. 67–92, 2021.

[6] H. Chen, F. Wang, H. Fan, R. Hong, and W. Li, "Construction of MOF-based superhydrophobic composite coating with excellent abrasion resistance and durability for self-cleaning, corrosion resistance, anti-icing, and loading-increasing research," *Chem Eng J*, vol. 408, p. 127343, 2021.

[7] K. Cao, Z. Yu, and D. Yin, "Preparation of Ce-MOF@ TEOS to enhance the anti-corrosion properties of epoxy coatings," *Prog Org Coat*, vol. 135, pp. 613–621, 2019.

[8] M. Ramezanzadeh, B. Ramezanzadeh, M. Mahdavian, and G. Bahlakeh, "Development of metal-organic framework (MOF) decorated graphene oxide nanoplatforms for anti-corrosion epoxy coatings," *Carbon*, vol. 161, pp. 231–251, 2020.

[9] K. Cao, Z. Yu, D. Yin, L. Chen, Y. Jiang, and L. Zhu, "Fabrication of BTA-MOF-TEOS-GO nanocomposite to endow coating systems with active inhibition and durable anticorrosion performances," *Prog Org Coat*, vol. 143, p. 105629, 2020.

[10] J. Cao, C. Guo, X. Guo, and Z. Chen, "Inhibition behavior of synthesized ZIF-8 derivative for copper in sodium chloride solution," *J Mol Liq*, vol. 311, p. 113277, 2020.

[11] A. Mesbah, S. Jacques, E. Rocca, M. François, and J. Steinmetz, "*Compact Metal–Organic Frameworks for Anti-Corrosion Applications: New Binary Linear Saturated Carboxylates of Zinc*," 2011, Wiley Online Library.

[12] S. Kumaraguru, R. Pavulraj, and S. Mohan, "Influence of cobalt, nickel and copper-based metal-organic frameworks on the corrosion protection of mild steel," *Trans IMF*, vol. 95, no. 3, pp. 131–136, 2017.

[13] J. Ding *et al.*, "The synthesis of MOF derived carbon and its application in water treatment," *Nano Res*, vol. 15, no. 8, pp. 6793–6818, 2022.

[14] X. Liu *et al.*, "Probe into metal-organic framework membranes fabricated via versatile polydopamine-assisted approach onto metal surfaces as anticorrosion coatings," *Corros Sci*, vol. 177, p. 108949, 2020.

[15] M. Zhang, L. Ma, L. Wang, Y. Sun, and Y. Liu, "Insights into the use of metal–organic framework as high-performance anticorrosion coatings," *ACS Appl Mater Interfaces*, vol. 10, no. 3, pp. 2259–2263, 2018.

[16] S. Zafari, M. Niknam Shahrak, and M. Ghahramaninezhad, "New MOF-based corrosion inhibitor for carbon steel in acidic media," *Met Mater Int*, vol. 26, pp. 25–38, 2020.

[17] N. Alipanah, H. Yari, M. Mahdavian, B. Ramezanzadeh, and G. Bahlakeh, "MIL-88A (Fe) filler with duplicate corrosion inhibitive/barrier effect for epoxy coatings: Electrochemical, molecular simulation, and cathodic delamination studies," *J Ind Eng Chem*, vol. 97, pp. 200–215, 2021.

[18] A. Dhakshinamoorthy, Z. Li, and H. Garcia, "Catalysis and photocatalysis by metal organic frameworks," *Chem Soc Rev*, vol. 47, no. 22, pp. 8134–8172, 2018.

[19] P. S. Sharanyakanth and M. Radhakrishnan, "Synthesis of metal-organic frameworks (MOFs) and its application in food packaging: A critical review," *Trends Food Sci Technol*, vol. 104, pp. 102–116, 2020.
[20] R. Luebke *et al.*, "Versatile rare earth hexanuclear clusters for the design and synthesis of highly-connected ftw-MOFs," *Chem Sci*, vol. 6, no. 7, pp. 4095–4102, 2015.
[21] A. E. Platero-Prats, V. A. de la Peña-O'Shea, D. M. Proserpio, N. Snejko, E. Gutierrez-Puebla, and A. Monge, "Insight into the SBU condensation in mg coordination and supramolecular frameworks: A combined experimental and theoretical study," *J Am Chem Soc*, vol. 134, no. 10, pp. 4762–4771, 2012.
[22] N. Al Amery, H. R. Abid, S. Al-Saadi, S. Wang, and S. Liu, "Facile directions for synthesis, modification and activation of MOFs," *Mater Today Chem*, vol. 17, p. 100343, 2020.
[23] J. M. Yassin, A. M. Taddesse, and M. Sánchez-Sánchez, "Room temperature synthesis of high-quality Ce (IV)-based MOFs in water," *Micropor Mesopor Mat*, vol. 324, p. 111303, 2021.
[24] S. Dai, F. Nouar, S. Zhang, A. Tissot, and C. Serre, "One-step room-temperature synthesis of metal (IV) carboxylate metal—Organic frameworks," *Angew Chem*, vol. 133, no. 8, pp. 4328–4334, 2021.
[25] I. Stassen *et al.*, "Electrochemical film deposition of the zirconium metal–organic framework UiO-66 and application in a miniaturized sorbent trap," *Chem Mater*, vol. 27, no. 5, pp. 1801–1807, 2015.
[26] Y.-R. Lee, M.-S. Jang, H.-Y. Cho, H.-J. Kwon, S. Kim, and W.-S. Ahn, "ZIF-8: A comparison of synthesis methods," *Chem Eng J*, vol. 271, pp. 276–280, 2015.
[27] X. Kang *et al.*, "Microwave-assisted hydrothermal synthesis of MOFs-derived bimetallic CuCo-N/C electrocatalyst for efficient oxygen reduction reaction," *J Alloys Compd*, vol. 795, pp. 462–470, 2019.
[28] T. Friščić *et al.*, "Real-time and in situ monitoring of mechanochemical milling reactions," *Nat Chem*, vol. 5, no. 1, pp. 66–73, 2013.
[29] Y. Wei, S. Han, D. A. Walker, P. E. Fuller, and B. A. Grzybowski, "Nanoparticle core/shell architectures within MOF crystals synthesized by reaction diffusion," *Angew Chem-Int Ed*, vol. 51, no. 30, p. 7435, 2012.
[30] J. Jomy and D. Prabhu, "Metal organic frameworks as a corrosion inhibitor-highlighting their structural properties along with synthetic methods and its mechanism-A review," *Curr Res Green and Sustain Chem*, p. 100438, 2024.
[31] Y.-R. Lee, J. Kim, and W.-S. Ahn, "Synthesis of metal-organic frameworks: A mini review," *Korean J Chem Eng*, vol. 30, pp. 1667–1680, 2013.
[32] L. Garzon-Tovar, A. Carné-Sánchez, C. Carbonell, I. Imaz, and D. Maspoch, "Optimised room temperature, water-based synthesis of CPO-27-M metal–organic frameworks with high space-time yields," *J Mater Chem A Mater*, vol. 3, no. 41, pp. 20819–20826, 2015.
[33] D. J. Tranchemontagne, J. R. Hunt, and O. M. Yaghi, "Room temperature synthesis of metal-organic frameworks: MOF-5, MOF-74, MOF-177, MOF-199, and IRMOF-0," *Tetrahedron*, vol. 64, no. 36, pp. 8553–8557, 2008.
[34] N. Getachew, Y. Chebude, I. Diaz, and M. Sanchez-Sanchez, "Room temperature synthesis of metal organic framework MOF-2," *J Porous Mater*, vol. 21, pp. 769–773, 2014.
[35] Z. Hu, Y. Peng, Z. Kang, Y. Qian, and D. Zhao, "A modulated hydrothermal (MHT) approach for the facile synthesis of UiO-66-type MOFs," *Inorg Chem*, vol. 54, no. 10, pp. 4862–4868, 2015.

[36] Q. Xia *et al.*, "State-of-the-art advances and challenges of iron-based metal organic frameworks from attractive features, synthesis to multifunctional applications," *Small*, vol. 15, no. 2, p. 1803088, 2019.

[37] A. R. Millward and O. M. Yaghi, "Metal–organic frameworks with exceptionally high capacity for storage of carbon dioxide at room temperature," *J Am Chem Soc*, vol. 127, no. 51, pp. 17998–17999, 2005.

[38] A. B. D. Nandiyanto, "Nano metal-organic framework particles (ie MIL-100 (Fe), HKUST-1 (Cu), Cu-TPA, and MOF-5 (Zn)) using a solvothermal process," *IJoST*, vol. 4, no. 2, pp. 220–228, 2019.

[39] S. B. Peh *et al.*, "Cluster nuclearity control and modulated hydrothermal synthesis of functionalized Zr 12 metal–organic frameworks," *Dalton Trans*, vol. 48, no. 21, pp. 7069–7073, 2019.

[40] R. Rani, A. Deep, B. Mizaikoff, and S. Singh, "Enhanced hydrothermal stability of Cu MOF by post synthetic modification with amino acids," *Vacuum*, vol. 164, pp. 449–457, 2019.

[41] H. Liu *et al.*, "Investigation on a Zr-based metal–organic framework (MOF-801) for the high-performance separation of light alkanes," *Chem Commun*, vol. 57, no. 96, pp. 13008–13011, 2021.

[42] P. P. Bag, X.-S. Wang, and R. Cao, "Microwave-assisted large scale synthesis of lanthanide metal–organic frameworks (Ln-MOFs), having a preferred conformation and photoluminescence properties," *Dalton Trans*, vol. 44, no. 26, pp. 11954–11962, 2015.

[43] B. Le Thanh, T. H. P. Nguyễn, D. D. LA, and T. P. Nguyen, "Study on some factors affecting the process of synthetic MIL-100 (Fe) by sonochemical synthesis method," *J Mil Sci Technol*, vol. 90, pp. 79–86, 2023.

[44] M. R. Armstrong, S. Senthilnathan, C. J. Balzer, B. Shan, L. Chen, and B. Mu, "Particle size studies to reveal crystallization mechanisms of the metal organic framework HKUST-1 during sonochemical synthesis," *Ultrason Sonochem*, vol. 34, pp. 365–370, 2017.

[45] N. Abdollahi, M. Y. Masoomi, A. Morsali, P. C. Junk, and J. Wang, "Sonochemical synthesis and structural characterization of a new Zn (II) nanoplate metal–organic framework with removal efficiency of Sudan red and Congo red," *Ultrason Sonochem*, vol. 45, pp. 50–56, 2018.

[46] P. Arul, N. S. K. Gowthaman, S. A. John, and M. Tominaga, "Tunable electrochemical synthesis of 3D nucleated microparticles like Cu-BTC MOF-carbon nanotubes composite: Enzyme free ultrasensitive determination of glucose in a complex biological fluid," *Electrochim Acta*, vol. 354, p. 136673, 2020.

[47] M. V. Varsha and G. Nageswaran, "Direct electrochemical synthesis of metal organic frameworks," *J Electrochem Soc*, vol. 167, no. 15, p. 155527, 2020.

[48] H. M. Titi, J.-L. Do, A. J. Howarth, K. Nagapudi, and T. Friščić, "Simple, scalable mechanosynthesis of metal–organic frameworks using liquid-assisted resonant acoustic mixing (LA-RAM)," *Chem Sci*, vol. 11, no. 29, pp. 7578–7584, 2020.

[49] S. Tanaka, "Mechanochemical synthesis of MOFs," in H. W. (Ben) Liu and T. Begley, (eds.), Metal-organic frameworks for biomedical applications, Elsevier. pp. 197–222, 2020.

[50] V. V. Butova, M. A. Soldatov, A. A. Guda, K. A. Lomachenko, and C. Lamberti, "Metal-organic frameworks: Structure, properties, methods of synthesis and characterization," *Russ Chem Rev*, vol. 85, no. 3, p. 280, 2016.

[51] S. Kumar, S. Jain, M. Nehra, N. Dilbaghi, G. Marrazza, and K.-H. Kim, "Green synthesis of metal–organic frameworks: A state-of-the-art review of potential environmental and medical applications," *Coord Chem Rev*, vol. 420, p. 213407, 2020.

[52] P. Silva, S. M. F. Vilela, J. P. C. Tomé, and F. A. A. Paz, "Multifunctional metal–organic frameworks: From academia to industrial applications," *Chem Soc Rev*, vol. 44, no. 19, pp. 6774–6803, 2015.
[53] B. Ren *et al.*, "Rational design of metallic anti-corrosion coatings based on zinc gluconate@ ZIF-8," *Chem Eng J*, vol. 384, p. 123389, 2020.
[54] S. M. Lashgari, H. Yari, M. Mahdavian, B. Ramezanzadeh, G. Bahlakeh, and M. Ramezanzadeh, "Application of nanoporous cobalt-based ZIF-67 metal-organic framework (MOF) for construction of an epoxy-composite coating with superior anti-corrosion properties," *Corros Sci*, vol. 178, p. 109099, 2021.
[55] A. Fattah-alhosseini, Z. Sangarimotlagh, M. Karbasi, and B. Dikici, "Enhancing corrosion resistance in Mg–based alloys through MOF–incorporated coatings: A comprehensive review," *Appl Surf Sci Adv*, vol. 21, p. 100607, 2024.
[56] C. Dey, T. Kundu, B. P. Biswal, A. Mallick, and R. Banerjee, "Crystalline metal-organic frameworks (MOFs): Synthesis, structure and function," *Acta Crystallogr B Struct Sci Cryst Eng Mater*, vol. 70, no. 1, pp. 3–10, 2014.
[57] S. Bhattacharjee, J.-S. Choi, S.-T. Yang, S. B. Choi, J. Kim, and W.-S. Ahn, "Solvothermal synthesis of Fe-MOF-74 and its catalytic properties in phenol hydroxylation," *J Nanosci Nanotechnol*, vol. 10, no. 1, pp. 135–141, 2010.
[58] B. Li *et al.*, "A strategy toward constructing a bifunctionalized MOF catalyst: Post-synthetic modification of MOFs on organic ligands and coordinatively unsaturated metal sites," *Chem Comm*, vol. 48, no. 49, pp. 6151–6153, 2012.
[59] L. Jiang, Y. Dong, Y. Yuan, X. Zhou, Y. Liu, and X. Meng, "Recent advances of metal–organic frameworks in corrosion protection: From synthesis to applications," *Chem Eng J*, vol. 430, p. 132823, 2022.
[60] V. R. R. Marthala *et al.*, "Solvothermal synthesis and characterization of large-crystal all-silica, aluminum-, and boron-containing ferrierite zeolites," *Chem Mater*, vol. 23, no. 10, pp. 2521–2528, 2011.
[61] M. Ramezanzadeh, B. Ramezanzadeh, G. Bahlakeh, A. Tati, and M. Mahdavian, "Development of an active/barrier bi-functional anti-corrosion system based on the epoxy nanocomposite loaded with highly-coordinated functionalized zirconium-based nanoporous metal-organic framework (Zr-MOF)," *Chem Eng J*, vol. 408, p. 127361, 2021.
[62] S. A. Younis, N. Bhardwaj, S. K. Bhardwaj, K.-H. Kim, and A. Deep, "Rare earth metal–organic frameworks (RE-MOFs): Synthesis, properties, and biomedical applications," *Coord Chem Rev*, vol. 429, p. 213620, 2021.
[63] N. A. Khan and S. H. Jhung, "Synthesis of metal-organic frameworks (MOFs) with microwave or ultrasound: Rapid reaction, phase-selectivity, and size reduction," *Coord Chem Rev*, vol. 285, pp. 11–23, 2015.
[64] Y. Yoo, Z. Lai, and H.-K. Jeong, "Fabrication of MOF-5 membranes using microwave-induced rapid seeding and solvothermal secondary growth," *Micropor Mesopor Mater*, vol. 123, no. 1–3, pp. 100–106, 2009.
[65] J.-Y. Choi, J. Kim, S.-H. Jhung, H.-K. Kim, J.-S. Chang, and H. K. Chae, "Microwave synthesis of a porous metal-organic framework, zinc terephthalate MOF-5," *Bull Korean Chem Soc*, vol. 27, no. 10, pp. 1523–1524, 2006.
[66] S. Chen *et al.*, "Facile fabrication of a highly (110)-oriented ZIF-7 film with rod-shaped seeds," *Chem Commun*, vol. 57, no. 17, pp. 2128–2131, 2021.
[67] O. Abuzalat, D. Wong, M. Elsayed, S. Park, and S. Kim, "Sonochemical fabrication of Cu (II) and Zn (II) metal-organic framework films on metal substrates," *Ultrason Sonochem*, vol. 45, pp. 180–188, 2018.

[68] Z. Mohammadpour and H. R. Zare, "The role of embedded 2-ABT@Cu-BTC MOF on the anti-corrosion performance of electro-assisted deposited silica sol-gel composite film," *Mater Chem Phys*, vol. 267, p. 124590, 2021.

[69] R. Ameloot, L. Stappers, J. Fransaer, L. Alaerts, B. F. Sels, and D. E. De Vos, "Patterned growth of metal-organic framework coatings by electrochemical synthesis," *Chem Mater*, vol. 21, no. 13, pp. 2580–2582, 2009.

[70] T. R. C. Van Assche, G. Desmet, R. Ameloot, D. E. De Vos, H. Terryn, and J. F. M. Denayer, "Electrochemical synthesis of thin HKUST-1 layers on copper mesh," *Micropor Mesopor Mater*, vol. 158, pp. 209–213, 2012.

[71] K.-Y. Cheng *et al.*, "Electrochemical synthesis, characterization of Ir–Zn containing coordination polymer, and application in oxygen and glucose sensing," *Dalton Trans*, vol. 43, no. 17, pp. 6536–6547, 2014.

[72] A. Martinez Joaristi, J. Juan-Alcañiz, P. Serra-Crespo, F. Kapteijn, and J. Gascon, "Electrochemical synthesis of some archetypical Zn^{2+}, Cu^{2+}, and Al^{3+} metal organic frameworks," *Cryst Growth Des*, vol. 12, no. 7, pp. 3489–3498, 2012.

[73] N. Campagnol *et al.*, "High pressure, high temperature electrochemical synthesis of metal–organic frameworks: Films of MIL-100 (Fe) and HKUST-1 in different morphologies," *J Mater Chem A Mater*, vol. 1, no. 19, pp. 5827–5830, 2013.

[74] P. Falcaro, R. Ricco, C. M. Doherty, K. Liang, A. J. Hill, and M. J. Styles, "MOF positioning technology and device fabrication," *Chem Soc Rev*, vol. 43, no. 16, pp. 5513–5560, 2014.

[75] X. Zhang, K. Wan, P. Subramanian, M. Xu, J. Luo, and J. Fransaer, "Electrochemical deposition of metal–organic framework films and their applications," *J Mater Chem A Mater*, vol. 8, no. 16, pp. 7569–7587, 2020.

[76] D. Yan *et al.*, "Mechanochemical synthesis of a fluorenone-based metal organic framework with polarized fluorescence: An experimental and computational study," *J Mater Chem C Mater*, vol. 1, no. 5, pp. 997–1004, 2013.

[77] A. Pichon, A. Lazuen-Garay, and S. L. James, "Solvent-free synthesis of a microporous metal–organic framework," *Cryst Eng Comm*, vol. 8, no. 3, pp. 211–214, 2006.

[78] H. Yang, S. Orefuwa, and A. Goudy, "Study of mechanochemical synthesis in the formation of the metal–organic framework $Cu_3(BTC)_2$ for hydrogen storage," *Micropor Mesopor Mater*, vol. 143, no. 1, pp. 37–45, 2011.

[79] J. Yao, D. Dong, D. Li, L. He, G. Xu, and H. Wang, "Contra-diffusion synthesis of ZIF-8 films on a polymer substrate," *Chem Commun*, vol. 47, no. 9, pp. 2559–2561, 2011.

[80] A. F. Payam, S. Khalil, and S. Chakrabarti, "Synthesis and characterization of MOF-derived structures: Recent advances and future perspectives," *Small*, p. 2310348, 2024.

[81] S. Subudhi, S. P. Tripathy, and K. Parida, "Highlights of the characterization techniques on inorganic, organic (COF) and hybrid (MOF) photocatalytic semiconductors," *Catal Sci Technol*, vol. 11, no. 2, pp. 392–415, 2021.

[82] J. Martí-Rujas, "Structural elucidation of microcrystalline MOFs from powder X-ray diffraction," *Dalton Trans*, vol. 49, no. 40, pp. 13897–13916, 2020.

[83] P. Hirschle *et al.*, "Exploration of MOF nanoparticle sizes using various physical characterization methods–is what you measure what you get?," *Cryst Eng Comm*, vol. 18, no. 23, pp. 4359–4368, 2016.

[84] C. E. Wilmer *et al.*, "Large-scale screening of hypothetical metal–organic frameworks," *Nat Chem*, vol. 4, no. 2, pp. 83–89, 2012.

[85] J. Zhang, N. Cheng, and B. Ge, "Characterization of metal-organic frameworks by transmission electron microscopy," *Adv Phys X*, vol. 7, no. 1, p. 2046157, 2022.

[86] T. Wang *et al.*, "Bottom-up formation of carbon-based structures with multilevel hierarchy from MOF–guest polyhedra," *J Am Chem Soc*, vol. 140, no. 19, pp. 6130–6136, 2018.

[87] F. Marpaung *et al.*, "Metal–organic framework (MOF)-derived nanoporous carbon materials," *Chem Asian J*, vol. 14, no. 9, pp. 1331–1343, 2019.

[88] A. J. Howarth, A. W. Peters, N. A. Vermeulen, T. C. Wang, J. T. Hupp, and O. K. Farha, "Best practices for the synthesis, activation, and characterization of metal–organic frameworks," *Chem Mater*, vol. 29, no. 1, pp. 26–39, 2017.

[89] I. A. Lázaro, "A comprehensive thermogravimetric analysis multifaceted method for the exact determination of the composition of multifunctional metal-organic framework materials," *Eur J Inorg Chem*, vol. 2020, no. 45, pp. 4284–4294, 2020.

[90] L. Liu *et al.*, "Ultrasmall FeNi 3 N particles with an exposed active (110) surface anchored on nitrogen-doped graphene for multifunctional electrocatalysts," *J Mater Chem A Mater*, vol. 7, no. 3, pp. 1083–1091, 2019.

[91] Y. Li, Y. Lin, J. Guo, Z. Xu, B. Wang, and T. Zhu, "Carbon consumption and regeneration of oxygen-containing functional groups on activated carbon for flue gas purification," *Environ Sci Pollut Res*, vol. 29, no. 18, pp. 26599–26612, 2022.

[92] M. Ghaderi, A. R. SaadatAbadi, M. Mahdavian, and S. A. Haddadi, "pH-sensitive polydopamine–La (III) complex decorated on carbon nanofiber toward on-demand release functioning of epoxy anti-corrosion coating," *Langmuir*, vol. 38, no. 38, pp. 11707–11723, 2022.

[93] H. Wu, T. Yildirim, and W. Zhou, "Exceptional mechanical stability of highly porous zirconium metal–organic framework UiO-66 and its important implications," *J Phys Chem Lett*, vol. 4, no. 6, pp. 925–930, 2013.

[94] L. D. O'Neill, H. Zhang, and D. Bradshaw, "Macro-/microporous MOF composite beads," *J Mater Chem*, vol. 20, no. 27, pp. 5720–5726, 2010.

[95] B. Van de Voorde *et al.*, "Improving the mechanical stability of zirconium-based metal–organic frameworks by incorporation of acidic modulators," *J Mater Chem A Mater*, vol. 3, no. 4, pp. 1737–1742, 2015.

[96] I. M. Hönicke *et al.*, "Balancing mechanical stability and ultrahigh porosity in crystalline framework materials," *Angew Chem Int Ed*, vol. 57, no. 42, pp. 13780–13783, 2018.

[97] G. Lal *et al.*, "Mechanical properties of a metal–organic framework formed by covalent cross-linking of metal–organic polyhedra," *J Am Chem Soc*, vol. 141, no. 2, pp. 1045–1053, 2018.

[98] T. Iizuka, K. Honjo, and T. Uemura, "Enhanced mechanical properties of a metal–organic framework by polymer insertion," *Chem Commun*, vol. 55, no. 5, pp. 691–694, 2019.

[99] A. Y. Adesina, I. H. Zainelabdeen, M. A. Dalhat, A. S. Mohammed, A. A. Sorour, and F. A. Al-Badour, "Influence of micronized waste tire rubber on the mechanical and tribological properties of epoxy composite coatings," *Tribol Int*, vol. 146, p. 106244, 2020.

[100] M. Motamedi, M. Ramezanzadeh, B. Ramezanzadeh, and M. Mahdavian, "One-pot synthesis and construction of a high performance metal-organic structured nano pigment based on nanoceria decorated cerium (III)-imidazole network (NC/CIN) for effective epoxy composite coating anti-corrosion and thermo-mechanical properties improvement," *Chem Eng J*, vol. 382, p. 122820, 2020.

[101] M. Ramezanzadeh, A. Tati, G. Bahlakeh, and B. Ramezanzadeh, "Construction of an epoxy composite coating with exceptional thermo-mechanical properties using

Zr-based NH_2-UiO-66 metal-organic framework (MOF): Experimental and DFT-D theoretical explorations," *Chem Eng J*, vol. 408, p. 127366, 2021.
[102] S. Ma *et al.*, "Metal-organic framework@ polyaniline nanoarchitecture for improved fire safety and mechanical performance of epoxy resin," *Mater Chem Phys*, vol. 247, p. 122875, 2020.
[103] W. Kleist, M. Maciejewski, and A. Baiker, "MOF-5 based mixed-linker metal–organic frameworks: Synthesis, thermal stability and catalytic application," *Thermochim Acta*, vol. 499, no. 1–2, pp. 71–78, 2010.
[104] H. G. T. Nguyen *et al.*, "Vanadium-node-functionalized UiO-66: A thermally stable MOF-supported catalyst for the gas-phase oxidative dehydrogenation of cyclohexene," *ACS Catal*, vol. 4, no. 8, pp. 2496–2500, 2014.
[105] G. Calleja, J. A. Botas, M. Sánchez-Sánchez, and M. G. Orcajo, "Hydrogen adsorption over Zeolite-like MOF materials modified by ion exchange," *Int J Hydrogen Energy*, vol. 35, no. 18, pp. 9916–9923, 2010.
[106] Y. Zeng, Z. Fu, H. Chen, C. Liu, S. Liao, and J. Dai, "Photo-and thermally induced coloration of a crystalline MOF accompanying electron transfer and long-lived charge separation in a stable host–guest system," *Chem Commun*, vol. 48, no. 65, pp. 8114–8116, 2012.
[107] E. V. Perez, K. J. Balkus Jr, J. P. Ferraris, and I. H. Musselman, "Mixed-matrix membranes containing MOF-5 for gas separations," *J Memb Sci*, vol. 328, no. 1–2, pp. 165–173, 2009.
[108] S. M. Lashgari, H. Yari, M. Mahdavian, B. Ramezanzadeh, G. Bahlakeh, and M. Ramezanzadeh, "Synthesis of graphene oxide nanosheets decorated by nanoporous zeolite-imidazole (ZIF-67) based metal-organic framework with controlled-release corrosion inhibitor performance: Experimental and detailed DFT-D theoretical explorations," *J Hazard Mater*, vol. 404, p. 124068, 2021.
[109] B. Mirzayi, H. Basharnavaz, A. Babapoor, H. Kamali, A. Khodayari, and S. Sohrabnezhad, "Effects of aluminum terephthalate metal-organic framework and its nanocomposites on the corrosion of AM60B magnesium alloy in ethylene glycol solution containing chloride ions," *Mater Chem Phys*, vol. 272, p. 125056, 2021.
[110] H. Tian *et al.*, "Controlled delivery of multi-substituted triazole by metal-organic framework for efficient inhibition of mild steel corrosion in neutral chloride solution," *Corros Sci*, vol. 131, pp. 1–16, 2018.
[111] B. Maranescu, N. Plesu, and A. Visa, "Phosphonic acid vs phosphonate metal organic framework influence on mild steel corrosion protection," *Appl Surf Sci*, vol. 497, p. 143734, 2019.
[112] H. Li, Y. Qiang, W. Zhao, and S. Zhang, "2-Mercaptobenzimidazole-inbuilt metal-organic-frameworks modified graphene oxide towards intelligent and excellent anti-corrosion coating," *Corros Sci*, vol. 191, p. 109715, 2021.
[113] L. Jiang *et al.*, "Electrodeposited ZIF-67 reinforced polypyrrole coating on ferritic stainless steel substrate with durable anticorrosion performance," *Mater Today Chem*, vol. 38, p. 102084, 2024.
[114] A. Hasanzadeh and A. R. SaadatAbadi, "ZIF-67 metal-organic framework decorated g-C_3N_4 nanosheets with inbuilt organic-inorganic corrosion inhibitor for self-healing epoxy anti-corrosion coating," *Surf Coat Technol*, vol. 492, p. 131218, 2024.
[115] A. Hasanzadeh, *Fabrication and Evaluation of Anti-corrosion Properties of Epoxy/ Modified Graphitic Carbon Nitride (g-C_3N_4) Nanocomposite on mild Steel*, Sharif University of Technology, 2023.
[116] S. M. Lashgari, H. Yari, M. Mahdavian, B. Ramezanzadeh, G. Bahlakeh, and M. Ramezanzadeh, "Unique 2-methylimidazole based inorganic building brick

nano-particles (NPs) functionalized with 3-aminopropyltriethoxysilane with excellent controlled corrosion inhibitors delivery performance; Experimental coupled with molecular/DFT-D simulations," *J Taiwan Inst Chem Eng*, vol. 117, pp. 209–222, 2020.

[117] M. H. Shahini, H. E. Mohammadloo, M. Ramezanzadeh, and B. Ramezanzadeh, "Recent innovations in synthesis/characterization of advanced nano-porous metal-organic frameworks (MOFs); current/future trends with a focus on the smart anti-corrosion features," *Mater Chem Phys*, vol. 276, p. 125420, 2022.

[118] W. Wei *et al.*, "Corrosion inhibition performance of nitrogen-containing metal organic framework compounds on copper flakes in dilute HCl medium," *Prot Met Phys Chem Surf*, vol. 56, pp. 638–650, 2020.

[119] Q. Zheng *et al.*, "Metal–organic frameworks incorporated polycaprolactone film for enhanced corrosion resistance and biocompatibility of Mg alloy," *ACS Sustain Chem Eng*, vol. 7, no. 21, pp. 18114–18124, 2019.

9 Performance Evaluation of Anticorrosive Metal-Organic Framework Coatings

Humira Assad, Praveen Kumar Sharma, and Ashish Kumar

9.1 INTRODUCTION

Corrosion is one of the most pervasive and costly problems affecting metal materials across a range of industries, leading to significant economic losses and safety concerns [1, 2]. It is a naturally occurring process in which metals react with environmental elements like oxygen, moisture, and salts, resulting in deterioration of material properties [3–5]. The global cost of corrosion has been estimated to exceed 3% of the world's gross domestic product (GDP), which includes maintenance, repairs, replacements, and productivity losses across industries like construction, marine, automotive, aerospace, and infrastructure. Corrosion not only impacts the structural integrity of metal components but also affects operational efficiency and safety, making it a significant issue for both economic and environmental reasons [6–8]. Historically, corrosion has been mitigated using various protective coatings, such as polymers and paints, with epoxy and polyurethane being among the most commonly used. These coatings are designed to form a physical barrier between the metal and the corrosive environment, preventing direct contact with moisture, salts, and other corrosive agents [9, 10]. However, despite their widespread use, these coatings come with certain limitations. Over time, traditional coatings can degrade due to factors like UV radiation, temperature fluctuations, and mechanical wear. As these coatings deteriorate, they may crack or peel, allowing corrosion to initiate at localized spots. Furthermore, conventional coatings often lack the ability to respond to damage or environmental stress in a proactive way, and their durability in extreme environments, such as marine or industrial settings, is limited as illustrated in Figure 9.1. These challenges have led to the need for more advanced and efficient alternatives to traditional coatings.

In recent years, metal–organic frameworks (MOFs) have garnered significant attention as a potential solution to these challenges [11]. MOFs are a class of materials consisting of metal ions or clusters coordinated to organic ligands, forming a highly ordered, porous structure [12]. The unique properties of MOFs, such as their high surface area, tunable chemistry, and ability to host various chemical species, make

DOI: 10.1201/9781003571162-9

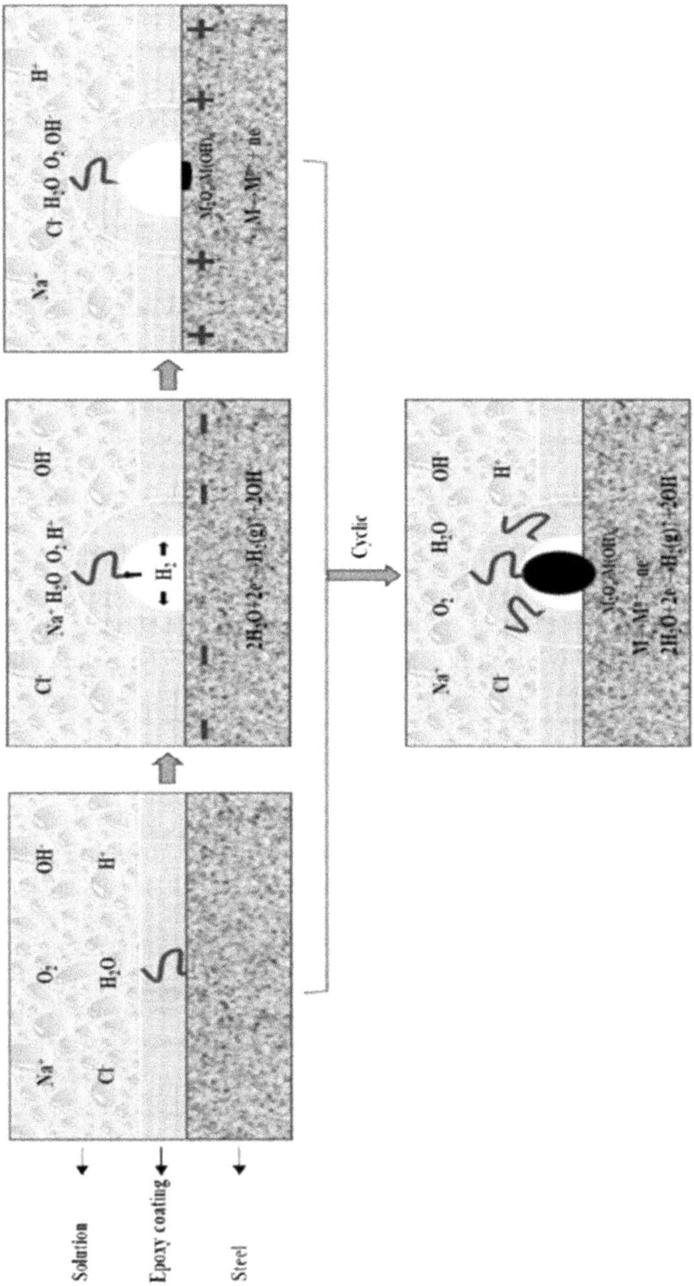

FIGURE 9.1 Mechanism of corrosion failure of the epoxy coating samples [15].

them particularly attractive for applications in areas like gas storage, catalysis, drug delivery, and, more recently, anticorrosion coatings [13, 14].

Unlike conventional coatings, MOFs can be designed to not only act as a physical barrier but also deliver corrosion inhibitors or other protective agents in a controlled manner. Their porosity allows for a large surface area that can be leveraged to trap reactive species, while their structural flexibility enables the integration of corrosion-fighting elements that can be gradually released to enhance protection over time [16]. Despite the promising potential of MOF-based coatings, their practical application is not without challenges. One of the most significant issues is the stability of the MOFs, particularly their sensitivity to moisture. Many MOFs are hygroscopic, meaning they readily absorb water, which can disrupt the delicate metal–organic interactions that form the structure of the framework. This can lead to the degradation of the coating's integrity and a reduction in its anticorrosive performance. Furthermore, while MOFs offer excellent protection against corrosion, the cost and scalability of producing high-quality MOF coatings for industrial use remain obstacles to widespread adoption. The synthesis of MOFs often involves complex processes that may not be easily adapted for large-scale manufacturing, and the cost of producing MOF coatings may currently exceed that of traditional coatings. Therefore, it is critical to evaluate the performance of MOF coatings thoroughly, to assess their long-term effectiveness, durability, and cost-effectiveness in real-world applications.

Performance evaluation is essential for determining the viability of MOF-based coatings as a practical solution to corrosion. A comprehensive evaluation encompasses several key aspects, including corrosion resistance, adhesion strength, durability, and the ability to self-heal in response to damage. Corrosion resistance is the most fundamental property of any anticorrosive coating, as it determines how well the coating can prevent the metal substrate from reacting with corrosive agents. To evaluate corrosion resistance, various electrochemical techniques are employed, such as open circuit potential (OCP) and electrochemical impedance spectroscopy (EIS). These methods provide valuable insights into the protective ability of the coating by measuring its resistance to corrosion under different environmental conditions. Additionally, salt spray tests (SSTs) and neutral salt fog testing are commonly used to simulate real-world exposure to corrosive environments, allowing for an accelerated assessment of the coating's performance. Adhesion strength is another crucial factor for assessing the performance of MOF coatings. A coating may be highly resistant to corrosion, but if it cannot adhere properly to the substrate, it is unlikely to remain intact over time, especially under mechanical stress. The adhesion strength of MOF coatings is often evaluated through pull-off adhesion tests, which measure the force required to detach the coating from the metal surface, and cross-hatch adhesion tests, which examine the coating's integrity under shear stress. These tests help simulate the real-world conditions under which the coating will be used and provide

In this regard, this chapter will focus on the methods used to assess the performance of MOF coatings, including corrosion resistance, adhesion strength, and durability, as well as their self-healing properties. Through an examination of case studies and current research, we will highlight the potential of MOF coatings as a viable solution for corrosion protection, while also discussing the challenges that need to be addressed for their successful implementation in industrial applications.

9.2 IMPORTANCE OF PERFORMANCE EVALUATION OF ANTICORROSIVE MOF COATINGS

The performance evaluation of anticorrosive MOF coatings is crucial for their successful application in protecting metals from corrosion across a range of industries. Corrosion is a pervasive and costly issue that can lead to significant material degradation, safety hazards, and substantial economic losses [17, 18]. While traditional corrosion protection methods, such as paints and polymer coatings, have been employed for years, they are not always durable enough to withstand the harsh environmental conditions that many industrial applications face. This limitation has spurred the search for more advanced, long-lasting solutions, with MOF coatings emerging as a promising candidate due to their unique properties and capabilities. The protection mechanism of one of the MOF-based anticorrosion coatings is shown in Figure 9.2.

However, to fully understand their potential and ensure they meet industry standards, a detailed and rigorous performance evaluation is essential. Below, we discuss the key reasons why performance evaluation is critical for anticorrosive MOF coatings.

- *Ensuring Effectiveness in Corrosion Protection*
  The primary purpose of any anticorrosive coating is to protect the underlying metal substrate from environmental factors, such as moisture, oxygen, salts, acids, and other corrosive agents, which can lead to degradation and failure. Corrosion is not only a problem of aesthetic appearance but can also compromise the structural integrity and safety of materials, leading to catastrophic failures in critical industries such as aerospace, automotive, marine,

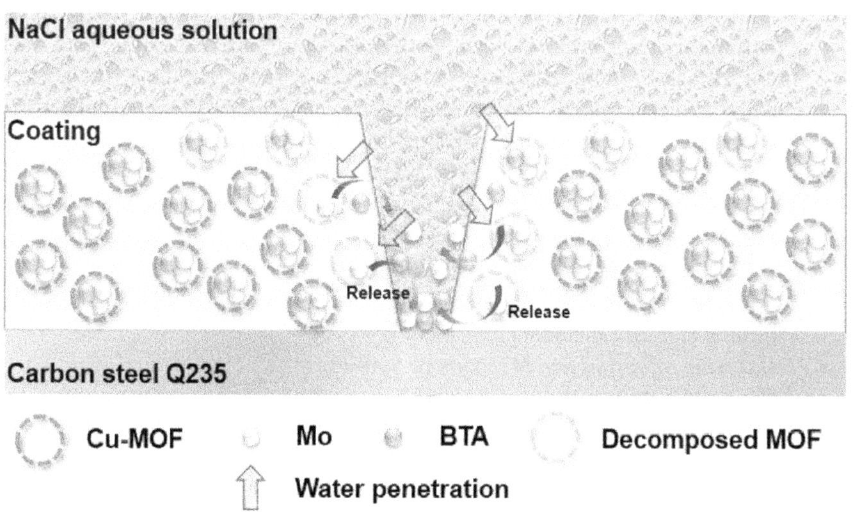

FIGURE 9.2 Schematic representation of the protection mechanism of the BTA–Mo @MOF–199/acrylic coating [19].

and infrastructure. MOF coatings are particularly promising because of their high surface area, porosity, and tunable chemical composition, which can be designed to offer superior protection against corrosion [20]. Performance evaluation techniques, such as OCP, and EIS allow researchers and industry professionals to assess how well these coatings act as a barrier to corrosive agents. By testing corrosion resistance under various conditions, it is possible to determine how effectively MOF coatings can prevent corrosion, providing a clear picture of their reliability and longevity.

- *Optimizing Coating Composition and Structure*
  MOFs are highly tunable materials, meaning their properties can be customized based on the metal ions, organic ligands, and their arrangement within the structure. This flexibility allows for the development of coatings with enhanced corrosion resistance and additional functionalities, such as self-healing properties, corrosion inhibitor release, or UV resistance [21]. However, the performance of a MOF coating depends heavily on the specific combination of these components and the resulting structure. Performance evaluation helps to identify which formulations work best under various environmental conditions and for different substrates. By assessing the coating's response to corrosion in real-world scenarios, researchers can optimize the composition, structure, and processing conditions of MOF coatings to enhance their effectiveness. Without this evaluation, it would be challenging to determine the most suitable MOF formulations for specific applications or environments.
- *Understanding Durability and Longevity*
  While the corrosion resistance of MOF coatings is crucial, their ability to maintain performance over time is just as important. In many industrial applications, coatings are exposed to harsh environments for extended periods, which can include high temperatures, UV radiation, fluctuating humidity, and mechanical wear. Over time, coatings can degrade, reducing their ability to protect the substrate effectively. Accelerated aging tests, such as SST and thermal cycling, allow for the assessment of the coating's durability and performance over extended periods [22]. These tests simulate real-world conditions and help determine how well the MOF coating withstands prolonged exposure to corrosive agents, extreme temperatures, and other environmental stressors. The results of these tests provide valuable insights into the coating's longevity, informing industries about the expected service life of the coating and whether it requires maintenance or reapplication in the future. Without performance evaluation, industries could face unexpected failures or costly maintenance schedules.
- *Assessing Adhesion and Mechanical Stability*
  The bond between the coating and the substrate is critical for ensuring the long-term performance of anticorrosive coatings. Even the most corrosion-resistant coating will fail if it cannot adhere securely to the metal surface. Poor adhesion can lead to coating delamination, cracking, and peeling, which exposes the metal to corrosive elements and accelerates degradation. MOF coatings, with their complex structure and properties, may present unique

challenges in terms of adhesion to substrates. Adhesion strength evaluation techniques, such as the pull-off adhesion test and cross-hatch test, are therefore essential to assess how well the MOF coating bonds with the metal surface under mechanical stress, temperature fluctuations, and environmental exposure [23]. These tests provide valuable data on the mechanical stability of the coating and its ability to withstand abrasion, impact, or stress-induced deformation. Strong adhesion is a key indicator of a successful anticorrosive coating, and performance evaluation ensures that MOF coatings meet the necessary adhesion standards for long-term effectiveness.

- *Enabling Innovation and New Applications*
  The performance evaluation of MOF coatings is not only crucial for assessing their effectiveness in traditional applications but also crucial for exploring new, innovative uses. The versatility of MOFs enables the development of coatings with novel properties, such as self-healing capabilities, where the coating can repair itself upon damage, thus extending its lifespan and maintaining corrosion protection [21]. Evaluating the self-healing properties of MOF coatings requires specialized techniques to assess how well the coating regenerates its protective functions after damage, such as through corrosion inhibitor release or other restorative mechanisms. Furthermore, MOFs have the potential to incorporate additional functionalities, such as resistance to fire, UV radiation, or chemical degradation. Evaluating these new functionalities is essential for understanding how MOF coatings can be applied to emerging industries or extreme environments. Performance evaluation allows researchers and manufacturers to explore the full potential of MOF coatings and identify innovative solutions for corrosion protection that go beyond conventional coatings.
- *Meeting Regulatory and Industry Standards*
  In many industries, anticorrosive coatings must meet specific regulatory standards for safety, environmental protection, and performance [24]. For example, in the automotive and aerospace industries, coatings must adhere to stringent guidelines regarding corrosion resistance, mechanical properties, and durability to ensure the safety and reliability of vehicles and aircraft. The performance evaluation of MOF coatings helps to ensure that these coatings meet or exceed industry standards and regulatory requirements. By testing their performance across various metrics, such as corrosion resistance, adhesion strength, and durability, MOF coatings can be certified for use in critical applications. This evaluation provides both manufacturers and regulatory bodies with the necessary data to approve the coating for commercial use and ensures that products meet the required safety and performance criteria.
- *Overcoming Challenges and Limitations*
  Despite the promising potential of MOF coatings, there are still several challenges to overcome before they can be widely adopted in industrial applications. These include issues related to the scalability of production, the cost of raw materials, and the stability of MOFs under harsh environmental conditions [25]. Performance evaluation is crucial in identifying

the limitations of MOF coatings, allowing researchers to pinpoint areas for improvement and address these challenges. By evaluating the coatings under different environmental conditions and stress tests, it becomes possible to identify weaknesses in the coating's structure or performance. This knowledge can then be used to refine the coating design, improve its stability, and enhance its cost-effectiveness, paving the way for the broader implementation of MOF coatings in corrosion protection.

Hence, the performance evaluation of anticorrosive MOF coatings is a critical step in advancing their application in real-world settings. By assessing factors such as corrosion resistance, adhesion strength, durability, and self-healing properties, researchers can determine the effectiveness and longevity of MOF coatings, optimize their composition and structure, and ensure their suitability for various industrial applications. Performance evaluation also enables the identification of challenges and limitations, guiding further research and development to enhance the practical use of these coatings. Ultimately, thorough and systematic evaluation ensures that MOF coatings provide the necessary protection and performance required for industries that rely on long-term corrosion prevention.

9.3 CORROSION RESISTANCE EVALUATION OF ANTICORROSIVE MOF COATINGS

Organic coatings with anticorrosive MOFs have become viable options in the pursuit of long-lasting and efficient corrosion protection [26]. This section explores the useful performance elements of MOF coatings, providing information on their adhesion, durability, and corrosion resistance. Following the MOF-based material's treatment, it is easy to measure the degree and duration of corrosion inhibition, which offers important hints for the upkeep and refurbishment of numerous metal artifacts, light metal structures, and fragile historical works. Research on MOF-based protective coatings is far more extensive than that of other metals, particularly for iron-based compounds. Humid marine or industrial environments can typically create a protective rust coating on steel surfaces during the well-known atmospheric corrosion of iron, where the rust with goethite and magnetite is the ultimate thermodynamic equilibrium. Regretfully, the rust layers are unable to stop deep-sea corrosion and more gas diffusion. A solid basis for preventing metal corrosion can be established by synthesizing MOFs for preparation and choosing the best MOF-based coating [27]. Following direct contact with the metal substrate, the coating fills in the pits and valleys on the substrate's surface and becomes cross-linked with it. This section explores the useful performance elements of MOF coatings, providing information on their adhesion, durability, and corrosion resistance.

9.3.1 Key Evaluations of Corrosion Resistance

An essential component of assessing the effectiveness of anticorrosive coatings, such as MOF coatings, is corrosion resistance testing. Testing a coating's ability to stop the underlying metal from deteriorating owing to exposure to corrosive conditions, such as moisture, salt, acids, and gases, is the main goal of corrosion resistance testing. This testing aids in assessing a coating's durability and dependability under actual circumstances [28]. As outlined below, a number of electrochemical techniques are frequently employed for corrosion resistance testing.

9.3.1.1 Electrochemical Methods

Corrosion monitoring analyzes the many forms of corrosion in every environment and determines the key factors affecting the rate and spread of corrosion. Thus, in short-term trials, electrochemical testing techniques provide useful information about material corrosion. A number of electrochemical testing techniques have been used to evaluate the corrosion protection capabilities of MOF-based coatings, including OCP, potentiodynamic polarization, EIS, local EIS (LEIS), and scanning vibrating electrode technique (SVET) [29, 30]. In this chapter, only the OCP and EIS tests were used. An equivalent circuit of a physical model that depicts the corrosion process of metal surfaces displays the EIS parameters (Rs, Rc, Rct, etc.). EIS can be used as a reliable, quick methodology for the data obtained since, generally speaking, an ideal protective film in anticorrosion testing should increase Rc (coating resistance) and Rct (charge transfer) while decreasing C_{dl} (double-layer capacitance). A decrease in coating barrier performance is often indicated by an increase in the C_{dl} value and a decrease in the Rc and Rct values [31].

- *Open Circuit Potential*
  OCP represents the thermodynamic stability of the test material's surface and is a useful technique for evaluating the anticorrosion efficacy of MOF-based coatings. OCP changes are caused by variations in the sample surface's oxidation tendency [32]. As immersion duration increases, OCP usually starts out quite positive and progressively moves towards the negative. Researchers can ascertain the coating's thermodynamic stability under particular immersion conditions by measuring the OCP. Superior anticorrosion qualities are indicated by a more positive and stable OCP, which shows how well the coating shields the underlying material from oxidation. The protective efficacy of MOF-based coatings can be improved by a positive shift in OCP, which can indicate the activation of the coating's special self-healing properties. For copper metal protection, Dehghan-Chenar et al. investigated the creation of a curcumin@Cu-BTC-MOF smart anticorrosion coating. They loaded a curcumin inhibitor into Cu-BTC-MOF to create an anticorrosion coating known as Cur@Cu(II)-benzene-1,3,5-tricarboxylate metal–organic framework (Cur@Cu-BTC-MOF) [33]. Comparing the uncoated Cu and Cu-coated Cu-BTC-MOF in Figure 9.3, the E_{OCP} shifted to more negative values in due to the existence of the Cur@Cu-BTC-MOF coating and attained a constant value considerably faster. Based on the shifting trend of the OCP, this indicates that the Cur@Cu-BTC MOF coating is more resilient than the Cu-BTC-MOF coating. According to Kumaraguru et al., who investigated the impact of MOFs based on cobalt, nickel, and copper on mild steel corrosion

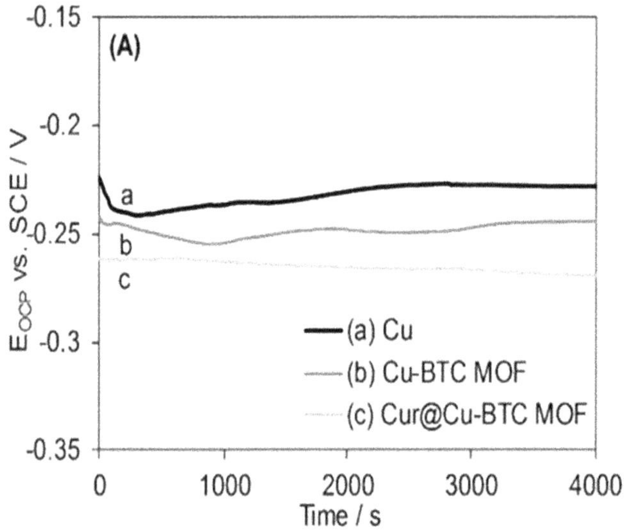

FIGURE 9.3 Variation of open-circuit potential (EOCP) vs. time for uncoated Cu, Cu-BTC MOF, and Cur@Cu-BTC MOF coatings on copper in 3.5% NaCl solution at 25°C [33].

protection, after 50 minutes, the OCP values of mild steel coated with MOF coatings continued to shift in a positive direction, whereas those of bare mild steel shifted in a negative direction. The findings demonstrate the thermodynamic stability of the MOF coatings and their ability to successfully shield the mild steel surface [34].

- *Electrochemical Impedance Spectroscopy*
  EIS is a quick, incredibly effective, and adaptable electrochemical technique used to assess a material's ability to withstand corrosion and a coating's ability to provide protection. It analyzes the corrosion cell's distinct impedance properties over a wide frequency range using a low-amplitude alternating current (AC) signal [35]. In this regard, it assesses the longevity of protective coatings produced by corrosion inhibitors and offers a thorough examination of the corrosive processes involved in the formation and deterioration of such coatings. Additionally, it helps predict the corrosion-prevention processes that these inhibitors provide. One of the main advantages of applying the EIS technique to corrosion investigation over other approaches is that the potential wave's amplitude is usually kept to a minimum, usually about ±10 mV, which ensures that the system being studied is not adversely affected. A potentiostat/galvanostat combined with an advanced impedance analyzer is frequently used in EIS procedures. Three electrodes make up the electrochemical cell configuration: the working electrode, which represents the base metal of interest; a counter electrode, such as platinum or stainless steel wires; and reference electrodes, such as Ag/AgCl [36]. The phase angle (δ) and impedance (|Z|) between the applied potential and the resultant current are examined in connection to the frequency used during the EIS evaluation [37, 38].

Hsia et al. conducted a comprehensive study combining experimental and analytical methods to explore the corrosion mechanism on epoxy-based anticorrosive coatings, incorporating zirconium-based metal–organic frameworks (Zr-MOFs) that are two-dimensional (2D) and water-stable. The electrochemical behavior of carbon steel substrates coated with CRS, EP, BA-ZrBTB/EP, and ZrBTB/EP was examined in a 3.5 wt. % chloride solution using EIS [22]. The measurements were carried out at 25 ± 0.5 °C for two different durations: 30 minutes and 24 hours, in 3.5 wt. % NaCl aqueous solution. The results demonstrated that the hydrophobic BA-ZrBTB/EP composite coatings exhibited significantly improved electrochemical resistance compared to the EP coating. Specifically, the resistance of the BA-ZrBTB/EP coating reached 197.32 kΩ·cm² after 24 hours of immersion in the NaCl solution, compared to just 26.22 kΩ·cm² for the EP coating as shown in Figure 9.4. Furthermore, long-term immersion (24 hours) is generally preferred for evaluating the anticorrosion properties of these materials, as it provides a more stable and accurate assessment. After 24 hours of immersion, the corrosion current density (icorr) values were measured. The observed reduction in icorr values for all samples during extended immersion can be attributed to the gradual formation of a protective passive layer on the material's surface over time. This phenomenon plays a key role in enhancing the material's anticorrosion performance [22, 38].

In a recent study, Ren et al. used the EIS test method to assess the corrosion resistance of a pH-responsive intelligent container system created using the 8-hydroxyquinoline (8-HQ) inhibitor and the MOF@PCN-222 [39]. After 100 days of immersion, the impedance value of 5.0 × 1010 Ω·cm² indicates that 8-HQ@PCN-222 coatings can offer long-term corrosion protection for the Al substrate. The EIS research also revealed that the Nyquist plots in the coatings exhibit capacitive arcs more than in the pure EP coating. To clarify the coatings' performance, the researchers examined a number of EIS parameters taken from the equivalent electrical circuits (EEC). In particular, the evaluation of Rct shed light on the electrochemical properties of the metal/coating interface and clarified the difficulties associated with electron transmission at the Al metal substrates. The porosity of the coating was examined using the measure Rc.

Ramezanzadeh et al. developed a novel corrosion-protective coating that exhibits excellent barrier properties and superior active anticorrosion characteristics. This coating was synthesized through a one-pot method, incorporating zeolitic imidazolate framework-8 (ZIF-8) on graphene oxide (GO) sheets, aimed at protecting mild steel. The electrochemical behavior of mild steel substrates immersed in a 3.5 wt. % chloride solution was investigated using EIS, both with and without the presence of ZIF-8 and GO@ZIF-8 composites. The results showed that the GO@ZIF-8/epoxy coating demonstrated high low-frequency impedance (|Z| at 10 mHz) values exceeding 10^{10} Ω cm² at all immersion times, indicating exceptional barrier and corrosion resistance. Notably, the EIS analysis revealed a 70% improvement in corrosion resistance for the GO@ZIF-8-loaded epoxy composites with an artificial defect, after 72 hours of immersion, compared to the neat epoxy sample. This indicates the promising smart inhibition activity of the GO@ZIF-8/epoxy coating, highlighting its potential for enhanced corrosion protection [25].

To establish self-healing barrier properties for epoxy coating on Al alloy via controlled release, Xiao et al. employed EIS to track the anticorrosion performance of

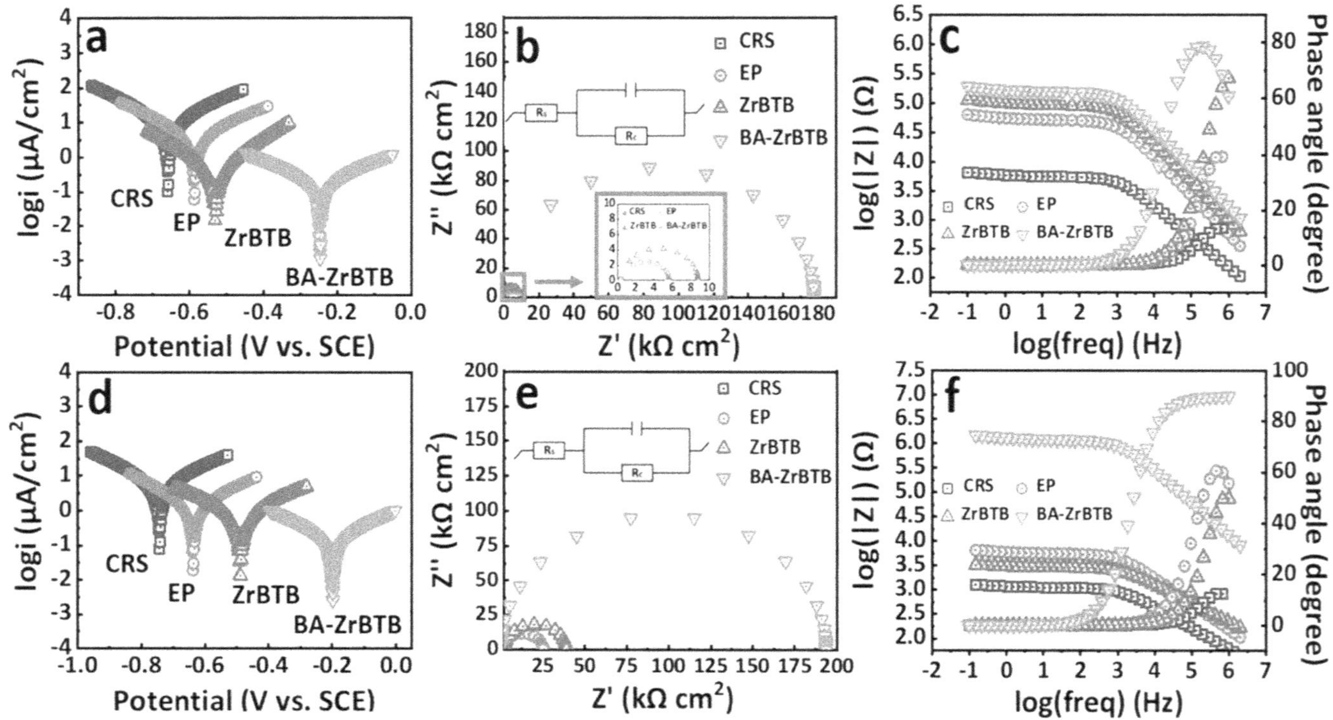

FIGURE 9.4 Electrochemical plots of CRS, EP, BA-ZrBTB/EP, and ZrBTB/EP-coated CRS electrodes in 3.5 wt. % NaCl solution at 25 ± 0.5 °C for (a–c) 30 minutes and (d–f) 24 hours [22].

a pH-responsive cerium-imidazole adorned ZIF-8 [28]. The EIS analysis indicates that the Ce-IM doped ZIF-8 NPs may successfully improve the anticorrosion performance of EP coatings because the |Z| 0.01 Hz values of epoxy coatings containing 0.5% and 3% Ce-IM/ZIF-8 nanoparticles (NPs) are still above 5 × 107 $\Omega \cdot cm^2$ even after 6 cycles, which is higher than that of EP coatings (4.8 × 106 $\Omega \cdot cm^2$). To assess the coatings' barrier qualities, the authors additionally employed the EIS parameters Rc and Rct, a total resistance (Rt = Rc+ Rct), and a total capacitance (Qt = Qc ||Qdl). According to the findings, pure EP coatings exhibited the quickest Rt value decline and Qt value increase. Ce-IM/ZIF-8 NP-containing EP coatings have higher Rt values than EP coatings alone. The EIS technique was used in a number of other studies that have been published in the literature to assess the corrosion protection capacity of MOF-based coatings. The findings indicate that adding MOFs to organic coatings greatly improved the coatings' capacity to inhibit corrosion [40, 41].

Majidi et al. investigated two types of synthesized MOFs composed of copper (Cu) and nickel (Ni) ions coordinated with 2-methylimidazole, which were used as novel nanofillers to enhance the anticorrosion behavior of epoxy (EP) coatings [42]. To assess the anticorrosion properties of these nanocomposite coatings, EIS measurements were conducted using an Autolab potentiostat/galvanostat (PGSTAT302N), with the OCP set across a frequency range of 10^5 Hz to 10^{-2} Hz and a 10 mV peak-to-zero potential. Artificial scratches, measuring 1.5 cm in length, were created with a sharp cutting tool to simulate coating defects before immersion in a 3.5 wt.% NaCl solution. EIS measurements were performed over a 63-day immersion period. The Nyquist plots obtained during this immersion period revealed that the neat EP coating exhibited the smallest semicircle diameter, indicating the poorest anticorrosion performance. Conversely, the Ni-MOF/EP composite showed the largest Nyquist semicircle diameter, indicating the best anticorrosion properties. As the immersion time increased, the diameter of the semicircle decreased for all coatings, signifying increased penetration of the corrosive media. Additionally, the charge transfer resistance (Rct) was calculated from the diameter of the Nyquist plots, and an inverse relationship was observed between Rct and the corrosion process: a higher Rct indicates a slower corrosion process. The Rct values for the Cu-MOF/EP and Ni-MOF/EP coatings were significantly higher than that of the neat EP coating with artificial scratches. These results demonstrate that incorporating MOFs into EP coatings enhances their anticorrosion properties, as evidenced by the larger Rct values and semicircle diameters observed for the Cu-MOF/EP and Ni-MOF/EP composites compared to the neat EP coating [42].

9.3.1.2 Adhesion Strength

Adhesion strength is a critical property of anticorrosion coatings, as it directly influences the effectiveness and durability of the protective layer. It refers to the force required to separate the coating from the substrate, and is typically assessed through methods such as pull-off tests, scratch tests, or tensile tests. A strong adhesion ensures that the coating remains intact under mechanical stresses, environmental conditions, and corrosive exposures, preventing delamination or peeling, which can compromise the coating's protective function. In the case of anticorrosion coatings, particularly those applied to metals like steel, the adhesion strength is affected by factors such as surface preparation, the chemical composition of the coating, and the presence of additives like MOFs. Coatings that incorporate anticorrosive MOFs exhibit superior bonding properties compared to traditional coatings. Extensive adhesion tests consistently show strong adhesion to a variety of surfaces, outperforming coatings without MOFs. Maintaining ongoing protection against corrosive conditions, reducing the chance of premature failure, and guaranteeing the protective layer's lifespan and efficacy all depend on this strong adherence [43, 44]. Therefore, evaluating the adhesion strength of coatings, especially those with added MOFs plays a key role in confirming the reliability and durability of the coating, ensuring it delivers long-term protection for the substrate. To create an active/barrier bi-functional anticorrosion system, Ramezanzadeh et al. looked into creating an epoxy nanocomposite that was loaded with highly coordinated functionalized zirconium-based nanoporous metal–organic framework (Zr-MOF). The study emphasizes how MOF particles improve epoxy polyamide coatings' adhesion and corrosion resistance when put through pull-off and cathodic disbondment testing (CDT) [23]. Figure 9.5 displays the digital image of the adhesion results.

Measurements of the delamination radius for coatings loaded with MOFs and those without particles shown significant improvements after CDT. While coatings containing UIO, UIN, and G-UIN MOF particles demonstrated reduced delamination radii of 9.42 mm, 10 mm, and 8.02 mm, respectively, indicating greater adhesion, unfilled epoxy coatings displayed a delamination radius of approximately 14.86 mm. Notably, the adhesion of the G-UIN/EP sample was improved by 46%, while the UIO/EP and UIN/EP coatings showed improvements of 36.6% and 32.7%, respectively.

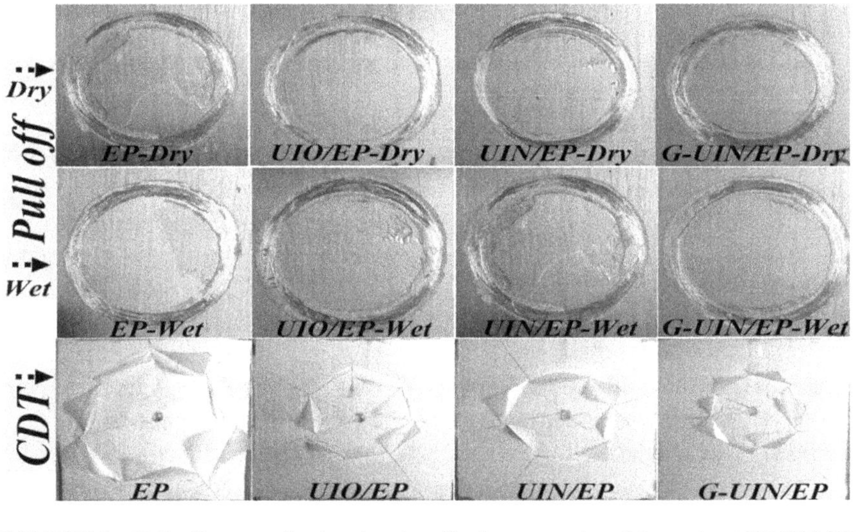

FIGURE 9.5 Pull-off test results showing the adhesion properties of dry and wet EP, UIO/EP, UIN/EP, and G-UIN/EP coatings before and after 14 days of exposure to the SST accelerated chamber. Reprinted with permission from [23].

This improved performance is explained by the release of Zr ions from partially dissolving MOFs, which reduce disbondment by neutralizing pH and blocking cathodic activity on the metal surface. The regulated release of Zr ions and good particle dispersion in the G-UIN/EP sample are responsible for its improved performance, highlighting the promise of MOF-loaded coatings for extended corrosion resistance in hostile alkaline environments [23]. Following SST, the pull-off test results for EP, UIO/EP, UIN/EP, and G-UIN/EP coatings show notable increases in adhesion strength as a result of MOF incorporation, especially in wet conditions. According to estimations of adhesive strength loss (ASL%), unfilled EP coatings had the most adhesion deterioration, with an ASL% of 48%. Poor adhesive resilience in corrosive situations is indicated by this significant loss in the EP sample, which emphasizes the delaminated area under moist conditions. Adhesion characteristics significantly increased for coatings loaded with MOF. ASL values were lower in the UIO/EP and UIN/EP samples by 35% and 22.4%, respectively. Nonetheless, a slight cohesive failure was noted in the UIN/EP sample, especially when it was wet, suggesting some vulnerability to extended exposure. With the lowest ASL% of 21.1% and no discernible cohesive or adhesive failure, the G-UIN/EP sample performed the best. This outcome highlights the efficient dispersion of G-UIN particles within the EP matrix as well as the strong interfacial bonding at the metal/coating contact. G-UIN/EP coatings' improved adhesion stability suggests that they have the potential to provide strong, long-term protection in corrosive situations [23].

Similarly, Lashgari et al. investigated the use of ZIF-67, a nanoporous MOF based on cobalt, to create an epoxy-composite coating with exceptional anticorrosion qualities [45]. Using a pull-off test method, the effect of ZIF-67 and ZIF-67@APS NPs on the epoxy-composite film's adherence to the iron substrate was investigated. The addition of NPs considerably increases adhesion strength in both dry and wet situations, as shown by the pull-off test on EPC, EPC/ZIF-67, and EPC/ZIF-67@APS coatings. To evaluate the effects of the coatings following 7 days of exposure to a salt spray chamber, the adhesion loss percentages were computed. The adhesion capabilities of the EPC coating significantly deteriorated, as seen by the results, which showed an adhesion loss of 58%. The EPC/ZIF-67 and EPC/ZIF-67@APS coatings, on the other hand, showed significantly reduced adhesion losses of 24% and 22%, respectively [45].

The chemical interactions between the epoxy matrix and the 2-methylimidazole ligand, which encourage chemical cross-linking and lessen the migration of corrosive substances into the coating, are responsible for the enhanced adhesion in the coatings modified by NPs. ZIF-67's functionalization with APS improved its compatibility with epoxy and improved its adherence to the metal substrate. Furthermore, ZIF-67 and ZIF-67@APS decreased the disbondment diameter, confirming the enhanced adhesion characteristics, according to CDT. This is partially because of the reaction between cobalt and hydroxyl ions at the cathodic site, which lowers pH and forms a protective coating on the surface. Overall, these results demonstrate how well ZIF-67 and ZIF-67@APS work in epoxy coatings to increase adhesion and corrosion resistance [45].

Furthermore, the impact of metal–organic framework (ZnMOF-BTA) particles on the adhesive qualities of polyurethane coatings for active protection of AA2024–T3′

was also shown by Nwokolo et al. The researchers assessed the adhesive strength of polyurethane coatings incorporating MOF particles using the cross-cut and impact adhesion tests. The findings demonstrated that the MOF particles significantly improved the coatings' adherence to the aluminum substrate. The ZnMOF-BTA coating's resistance to separation from the substrate is evaluated by the cross-cut adhesion test. The findings show that the coating performed well in this test, maintaining the integrity of interfaces, seal regions, and fluid flow while showing very slight separation of tiny flakes at the borders and intersections of the incisions. A fair degree of adhesion is indicated by ASTM class 3B adhesion behavior, which is in line with this performance. According to the impact test, the ZnMOF-BTA-filled coatings showed just a minor depression at the impact site and no signs of cracking in terms of impact resistance. This implies that the coatings exhibit resilience under stress and are not brittle [44].

9.3.1.3 Salt Spray Test

The SST is a widely used accelerated corrosion test to evaluate the effectiveness of anticorrosion coatings. It involves exposing coated samples to a controlled, humid environment with a fine mist of saltwater (typically a 5% sodium chloride solution) at a constant temperature, usually around 35 °C. The test simulates the corrosive effects of seawater or harsh atmospheric conditions on the coated surface. During the SST, coatings are subjected to continuous exposure to the salt-laden mist, which accelerates the onset of corrosion. The test duration can vary, typically lasting from 24 hours to several weeks, depending on the specific testing requirements and the type of coating being evaluated. The primary goal is to assess the coating's ability to resist corrosion, adhesion failure, and degradation under simulated harsh conditions. In the case of MOF-based anticorrosion coatings, the SST provides valuable insights into the enhanced protective properties of coatings incorporating MOFs. MOFs, with their unique porous structure and high surface area, can significantly improve the corrosion resistance of coatings by acting as effective barriers, inhibiting the diffusion of corrosive agents like water and chloride ions to the substrate. Coatings containing MOFs, such as zirconium-based or other MOFs, have shown improved resistance to corrosion under SST conditions compared to traditional coatings. This is due to the ability of MOFs to form protective layers or release corrosion-inhibiting ions over time, which contributes to the longevity and durability of the coating. The SST helps demonstrate the superior performance of MOF-enhanced coatings, highlighting their potential for long-term protection in aggressive, salt-rich environments. Hsia et al. performed SST on various nanocomposite coatings, where fresh and coated plates were exposed to a saline fog chamber [22]. The results, shown in Figure 9.6, were monitored over a period of up to 480 hours (20 days) of exposure.

The samples tested included carbon steel (CS), EP, ZrBTB/EP, and BA-ZrBTB/EP. A 5×5 cm^2 scribed region was applied to each coated sample, except for the CS, which was left uncoated, as it would corrode completely within a few hours. After 96 hours of exposure, rust formation was observed on both the CS and EP samples, indicating their limited protective capability. After 288 hours of exposure, visible corrosion traces appeared near the scribed region of the ZrBTB/EP coating, extending diagonally from the top left to the bottom right. Conversely, the BA-ZrBTB/EP coating

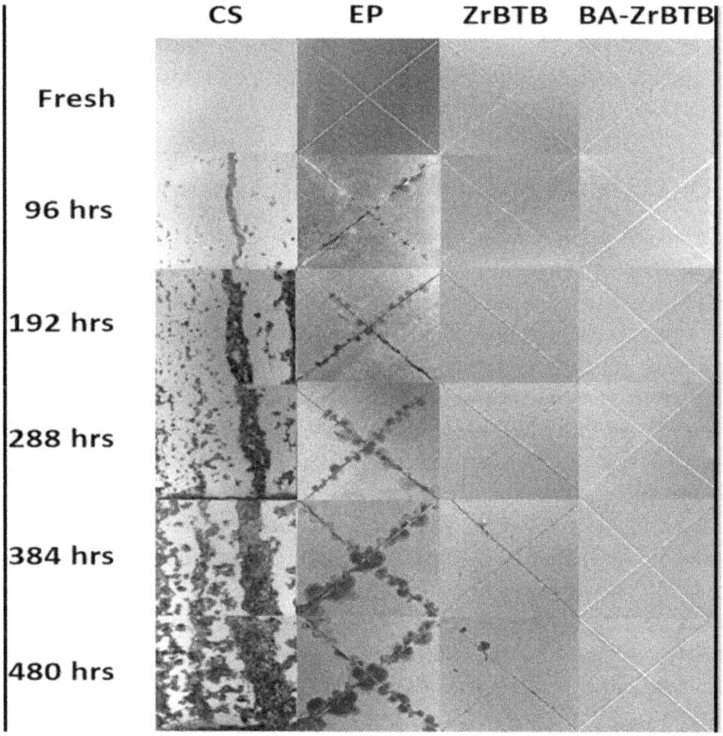

FIGURE 9.6 Images of CS, EP, ZrBTB/EP, and BA-ZrBTB/EP samples following 480 hours of exposure to salt spray corrosion testing [22].

remained unaffected, highlighting its superior resistance to corrosion. However, after 480 hours of salt spray exposure, the scribed area of the BA-ZrBTB/EP coating showed early signs of corrosion penetration, and rust began to form. This suggests that while BA-ZrBTB/EP provides exceptional corrosion resistance over extended periods, it is not completely impervious to corrosion under extreme conditions [22].

Besides, Ramezanzadeh et al. evaluated the performance of UIO, UIN, and G-UIN particles in epoxy coatings by conducting SST on scratched coated samples. The scratched EP, UIO/EP, UIN/EP, and G-UIN/EP coatings were evaluated before and, after 10 days, and after 20 days of exposure to the SST chamber. The EP sample exhibited significant delamination after 10 days of SST exposure, which was attributed to the penetration of electrolyte and Fe-oxide products into the EP/steel interface, weakening the coating's adhesion. The UIO/EP and UIN/EP coatings provided acceptable corrosion protection up to 10 days of exposure, with minimal delamination observed. However, the corrosion protection of the UIO/EP coating began to deteriorate after 20 days of exposure. The UIN/EP sample showed signs of blistering in the non-scratched region after 10 days, indicating a weakness in its water-barrier properties or adhesion, which aligns with previous findings from corrosion studies. In contrast, the G-UIN/EP coating demonstrated excellent corrosion resistance, maintaining integrity even after 20 days of SST exposure. No significant delamination was observed around the scratched area, and no blistering appeared in the non-scratched region, highlighting the superior corrosion resistance of the G-UIN/EP coating [23].

9.4 CONCLUSION AND FUTURE OUTLOOK

The performance evaluation of anticorrosive MOF coatings has revealed their significant potential in enhancing the corrosion resistance of metal substrates. As per the reported literature, the incorporation of MOFs, such as Cu-BTC, ZrBTB etc., into epoxy-based coatings and others significantly improves their protective properties, offering superior corrosion resistance compared to conventional coatings. Through various tests, including OCP, EIS, SST, adhesion strength test etc., it has been demonstrated that MOF-based coatings provide enhanced durability, adhesion, and protection against corrosion under aggressive environmental conditions. These coatings also show additional advantages, such as self-healing properties, sustained release of corrosion inhibitors, and improved mechanical strength, making them highly valuable for industrial applications, particularly in sectors that require resistance to harsh corrosive environments.

However, several challenges remain in the performance evaluation of MOF-based anticorrosive coatings. One key challenge is the variability in the results of corrosion resistance tests. The performance of MOF coatings can differ subject to factors like the type of MOF used, coating thickness, and the technique of application. Standardizing these variables is crucial to obtaining consistent and reliable performance data. Another challenge lies in the long-term evaluation of the coatings under real-world conditions. While MOF-based coatings may perform well in accelerated tests such as SST or EIS, their long-term stability under fluctuating environmental conditions, such as UV exposure, temperature extremes, and humidity, still requires further study. Additionally, the adhesion strength of MOF coatings over extended periods of use is an area that requires more detailed evaluation. While initial results may show good adhesion, the effect of protracted revealing to corrosive environments may lead to deterioration of adhesion over time, which needs to be accurately assessed. Furthermore, assessing the mechanical properties of MOF coatings in real-world applications presents another challenge. Although MOF coatings demonstrate promising corrosion resistance, they must also maintain their mechanical integrity under stress, such as during exposure to mechanical impact, abrasion, or other forms of wear and tear. This aspect of performance evaluation remains underexplored in current research. Another challenge is the environmental impact of certain MOF materials, particularly regarding their long-term stability and potential leaching of metal components under aggressive conditions. Accurate testing and evaluation of these factors are crucial for understanding the sustainability and safety of MOF-based coatings. Therefore, while MOF-based anticorrosive coatings show great promise, future research must focus on addressing these challenges in performance evaluation. Standardization of testing methods, long-term stability analysis, and better understanding of mechanical properties and environmental impact will be essential for advancing MOF coatings from laboratory-scale success to practical, real-world applications.

REFERENCES

[1] C. Verma, A. Thakur, R. Ganjoo, S. Sharma, H. Assad, A. Kumar, M. Quraishi, A. Alfantazi, Coordination bonding and corrosion inhibition potential of nitrogen-rich heterocycles: Azoles and triazines as specific examples, *Coordination Chemistry Reviews*, 488 (2023) 215177.

[2] H. Assad, A. Kumar, Understanding functional group effect on corrosion inhibition efficiency of selected organic compounds, *Journal of Molecular Liquids*, 344 (2021) 117755.

[3] H. Assad, I. Fatma, A. Thakur, A. Kumar, 14 Functionalized thin film coatings for reinforced concrete engineering, in Ashish Kumar, Chandrabhan Verma and Abhinay Thakur (eds.) *Corrosion Mitigation Coatings: Functionalized Thin Film Fundamentals and Applications*, 2023, p. 347. De Gruyter.

[4] H. Assad, S. Kumar, A. Kumar, *Nitrogen, Sulfur, and Nitrogen-based Heterocycles Corrosion Inhibitors, Handbook of Heterocyclic Corrosion Inhibitors*, CRC Press 2024, pp. 198–211.

[5] H. Assad, A. Thakur, A. Bharmal, S. Sharma, R. Ganjoo, S. Kaya, Corrosion inhibitors: Fundamental concepts and selection metrics, *Corrosion Mitigation*, (2022) 19–50.

[6] H. Assad, A. Thakur, A. Kumar, *Chitosan and Its Derivatives as Corrosion Inhibitors, Biopolymers in Sustainable Corrosion Inhibition*, CRC Press, 2024, pp. 88–101.

[7] H. Assad, R. Ganjoo, P.K. Sharma, A. Kumar, Corrosion inhibitors: Introduction, classification and selection criteria, in Jeenat Aslam, Chandrabhan Verma, Ruby Aslam (eds.), *Grafted Biopolymers as Corrosion Inhibitors: Safety, Sustainability, and Efficiency*, 2023, pp. 145–169. Wiley.

[8] H. Assad, A. Kumar, Carbon nanotubes (SWCNTs/MWCNTs) and functionalized carbon nanotubes in corrosion prevention, in Berdimurodov Elyor Tukhlievich, Chandrabhan Verma (eds.), *Corrosion Prevention Nanoscience: Nanoengineering Materials and Technologies*, 2023, pp. 85–98. De Gruyter.

[9] H. Assad, I.A. Lone, A. Sihmar, A. Kumar, A. Kumar, An overview of contemporary developments and the application of graphene-based materials in anticorrosive coatings, *Environmental Science and Pollution Research*, 32 (2023) 1–20.

[10] H. Assad, I. Fatma, A. Kumar, An overview of the application of graphene-based materials in anticorrosive coatings, *Materials Letters*, 330 (2023) 133287.

[11] I.K. Nwokolo, H. Shi, A.I. Ikeuba, F. Liu, S. Ahmed, W. Zhang, Epoxy coating containing CoMOF@ MBT metal-organic framework for active protection of aluminum alloy, *Surface and Coatings Technology*, 477 (2024) 130349.

[12] S. Kitagawa, Metal–organic frameworks (MOFs), *Chemical Society Reviews*, 43 (2014) 5415–5418.

[13] Y. Lei, Z. Qiu, N. Tan, H. Du, D. Li, J. Liu, T. Liu, W. Zhang, X. Chang, Polyaniline/CeO2 nanocomposites as corrosion inhibitors for improving the corrosive performance of epoxy coating on carbon steel in 3.5% NaCl solution, *Progress in Organic Coatings*, 139 (2020) 105430.

[14] H. Assad, A. Assad, A. Kumar, Recent developments in 3D bio-printing and its biomedical applications, *Pharmaceutics*, 15 (2023) 255.

[15] Y. Xu, G. Song, D. Zheng, C. Liu, E. Han, Polarization-accelerated seawater splash simulation for rapid evaluation of protection performance of an epoxy coating on carbon steel, *Materials*, 17 (2024) 3623.

[16] M. Motamedi, M. Ramezanzadeh, B. Ramezanzadeh, M. Mahdavian, One-pot synthesis and construction of a high performance metal-organic structured nano pigment based on nanoceria decorated cerium (III)-imidazole network (NC/CIN) for effective epoxy composite coating anti-corrosion and thermo-mechanical properties improvement, *Chemical Engineering Journal*, 382 (2020) 122820.

[17] H. Assad, S. Kumar, S.K. Saha, N. Kang, I. Fatma, H. Dahiya, P.K. Sharma, A. Thakur, S. Sharma, R. Ganjoo, Evaluating the adsorption and corrosion inhibition capabilities of Pyridinium-P-Toluene Sulphonate on MS in 1 M HCl medium: An experimental and theoretical study, *Inorganic Chemistry Communications*, 153 (2023) 110817.

[18] R. Ganjoo, C. Verma, A. Thakur, A. AlFantazi, H. Assad, S. Sharma, S. Dubey, A. Kumar, Mannich bases: Chemical structure, chemistry, coordination bonding and application in aqueous phase corrosion protection, *Journal of Industrial and Engineering Chemistry*, 131 (2024) 136–166.

[19] Z. Hao, S. Chen, Z. Chen, Z. Lin, W. Li, Water-triggered self-healing composite coating: Fabrication and anti-corrosion application, *Polymers*, 14 (2022) 1847.

[20] Y. Zhang, H. Guo, J. Gao, W. Wei, M. Liu, C. Zheng, P. Deng, Self-lubricated anti-icing MOF coating with long-term durability, *Progress in Organic Coatings*, 151 (2021) 106089.

[21] T. Liu, D. Zhang, R. Zhang, J. Wang, L. Ma, P. Keil, A. Mol, X. Li, Self-healing and corrosion-sensing coatings based on pH-sensitive MOF-capped microcontainers for intelligent corrosion control, *Chemical Engineering Journal*, 454 (2023) 140335.

[22] H.-H. Hsia, Y.-L. Chen, Y.-T. Tai, H.-K. Tian, C.-W. Kung, W.-R. Liu, Two-dimensional metal–organic frameworks/epoxy composite coatings with superior O2/H2O resistance for anticorrosion applications, *ACS Applied Materials & Interfaces*, 16 (2024) 41421–41434.

[23] M. Ramezanzadeh, B. Ramezanzadeh, G. Bahlakeh, A. Tati, M. Mahdavian, Development of an active/barrier bi-functional anti-corrosion system based on the epoxy nanocomposite loaded with highly-coordinated functionalized zirconium-based nanoporous metal-organic framework (Zr-MOF), *Chemical Engineering Journal*, 408 (2021) 127361.

[24] H. Leckie, Corrosion standards and control in the iron and steel industry, in FH Cocks (ed.) *Manual of Industrial Corrosion Standards and Control,* ASTM International, 1973.

[25] M. Ramezanzadeh, B. Ramezanzadeh, M. Mahdavian, G. Bahlakeh, Development of metal-organic framework (MOF) decorated graphene oxide nanoplatforms for anti-corrosion epoxy coatings, *Carbon*, 161 (2020) 231–251.

[26] Y. Chen, L. Wu, A. Atrens, F. Pan, A review of metal-organic framework protective coatings for light metals, *Surface Engineering*, 38 (2022) 807–829.

[27] S. Jiang, W. Li, J. Liu, J. Jiang, Z. Zhang, W. Shang, N. Peng, Y. Wen, ZnO@ ZIF-8 core-shell structure nanorods superhydrophobic coating on magnesium alloy with corrosion resistance and self-cleaning, *Journal of Magnesium and Alloys*, 11 (2023) 3287–3301.

[28] S. Xiao, X. Cao, Z. Dong, X. Ma, X. Zhang, G. Cai, A pH-responsive cerium-imidazole decorated ZIF-8 to achieve self-healing barrier property for epoxy coating on Al alloy by controlled release, *Progress in Organic Coatings*, 163 (2022) 106640.

[29] J. Li, N. Gao, W. Zhang, L. Xu, K.I. Nwokolo, H. Wu, F. Liu, E.-H. Han, A sandwich-like structure hybrid coating of cold galvanizing coating/polydopamine on hot-dip galvanized steel with enhanced adhesion and corrosion resistance, *Science China Technological Sciences*, 66 (2023) 2381–2395.

[30] N. Gao, J. Li, W. Zhang, L. Ma, I.K. Nwokolo, F. Liu, E.-H. Han, Double-layer peelable coating with eminent mechanical properties and anti-permeability, *Progress in Organic Coatings*, 160 (2021) 106517.

[31] Y.H. Kim, Y.S. Kwon, M.Y. Shon, M.J. Moon, Corrosion protection performance of PVDF/PMMA-blended coatings by electrochemical impedance method, *Journal of Electrochemical Science and Technology*, 9 (2018) 1–8.

[32] H. Assad, S.K. Saha, N. Kang, S. Kumar, H. Dahiya, P. Banerjee, A. Thakur, S. Sharma, R. Ganjoo, A. Kumar, Assessment of the inhibitory efficacy of a thiazole derivative as an efficient corrosion inhibitor for augmenting the resistance of MS in acidic environments, *Langmuir*, 40 (2024) 16367–16383.

[33] S. Dehghan-Chenar, H.R. Zare, Z. Mohammadpour, Fabrication of a Curcumin@ Cu-BTC MOF smart anti-corrosion coating for copper, *ChemistrySelect*, 8 (2023) e202303392.

[34] S. Kumaraguru, R. Pavulraj, S. Mohan, Influence of cobalt, nickel and copper-based metal-organic frameworks on the corrosion protection of mild steel, *Transactions of the IMF*, 95 (2017) 131–136.

[35] H. Assad, S.K. Saha, N. Kang, S. Kumar, P.K. Sharma, H. Dahiya, A. Thakur, S. Sharma, R. Ganjoo, A. Kumar, Electrochemical and computational insights into the utilization of 2, 2-dithio bisbenzothiazole as a sustainable corrosion inhibitor for mild steel in low pH medium, *Environmental Research*, 242 (2024) 117640.

[36] H. Assad, S. Kumar, S.K. Saha, N. Kang, H. Dahiya, A. Thakur, S. Sharma, R. Ganjoo, A. Kumar, A research combined experimental and computational approaches of succinylsulfathiazole hydrate as potent corrosion inhibitor for mild steel in acidic medium, *Journal of Molecular Liquids*, 388 (2023) 122739.

[37] A.I. Ikeuba, F.O. Faithpraise, K.I. Nwokolo, F.E. Umo, O.C. Echem, A.T. Ibrahim, H.O. Edet, B.I. Ita, P.C. Okafor, F.C. Asogwa, A combined electrochemical and DFT investigation of ornidazole as a benign anti-corrosion agent for carbon steel materials in acidizing environments, *Results in Materials*, 21 (2024) 100542.

[38] W. Zhang, F. Liu, Y. Li, T. Chen, I.K. Nwokolo, S. Ahmed, E.-H. Han, Modified graphene micropillar array superhydrophobic coating with strong anti-icing properties and corrosion resistance, *Coatings*, 14 (2024) 247.

[39] S. Ren, F. Meng, X. Li, Y. Cui, R. Liu, Y. Liu, X. Hu, L. Liu, F. Wang, A self-healing epoxy composite coating based on pH-responsive PCN-222 smart containers for long-term anticorrosion of aluminum alloy, *Corrosion Science*, 221 (2023) 111318.

[40] J. Li, H. Shi, F. Liu, E.-H. Han, Self-healing epoxy coating based on tung oil-containing microcapsules for corrosion protection, *Progress in Organic Coatings*, 156 (2021) 106236.

[41] I.K. Nwokolo, H. Shi, A.I. Ikeuba, N. Gao, J. Li, S. Ahmed, F. Liu, Synthesis, characterization and investigation of anticorrosion properties of an innovative metal–organic framework, ZnMOF-BTA, on carbon steel in HCl solution, *Coatings*, 12 (2022) 1288.

[42] R. Majidi, A. Farhadi, I. Danaee, N.B. Panah, D. Zarei, S. Nikmanesh, Investigation of synthesized planar Cu-MOF and spherical Ni-MOF nanofillers for improving the anti-corrosion performance of epoxy coatings, *Progress in Organic Coatings*, 183 (2023) 107803.

[43] H. Chen, F. Wang, H. Fan, R. Hong, W. Li, Construction of MOF-based superhydrophobic composite coating with excellent abrasion resistance and durability for self-cleaning, corrosion resistance, anti-icing, and loading-increasing research, *Chemical Engineering Journal*, 408 (2021) 127343.

[44] I.K. Nwokolo, H. Shi, P.C. Uzoma, S. Ahmed, J. Li, F. Liu, Protective coatings containing ZnMOF-BTA metal–organic framework for active protection of AA2024-T3, *Acta Metallurgica Sinica (English Letters)*, 36 (2023) 1893–1908.

[45] S.M. Lashgari, H. Yari, M. Mahdavian, B. Ramezanzadeh, G. Bahlakeh, M. Ramezanzadeh, Application of nanoporous cobalt-based ZIF-67 metal-organic framework (MOF) for construction of an epoxy-composite coating with superior anti-corrosion properties, *Corrosion Science*, 178 (2021) 109099.

10 Surface Modifications and Functionalization of Anticorrosive MOF Coatings

Humira Assad, Praveen Kumar Sharma, and Ashish Kumar

10.1 INTRODUCTION

Humans employ a variety of metals and alloys in their daily lives, yet most of these materials eventually corrode in the workplace [1, 2]. Bridges, pipelines, ships, public buildings, home equipment, etc. can sustain irreparable damage from corrosion, just like natural disasters such as earthquakes and tsunamis. This not only endangers public safety but also results in significant financial losses and degradation of the environment [3, 4]. A major worry in sectors like construction, automotive, aerospace, and marine applications is corrosion, which is the slow deterioration of materials, mostly metals, because of chemical reactions with their surroundings [1, 5–7]. While it is impossible to completely prevent corrosion, corrosion protection techniques can slow its pace [8]. Coatings like paints, polymers, and metallic coatings that act as an obstruction between the substrate and the harsh environment are frequently used in traditional corrosion protection techniques as shown in Figure 10.1 [9–11]. Even though these methods can offer sufficient protection, it is still quite difficult to achieve an environmentally friendly property with long-term protective capabilities. Applications of green nanomaterials with inhibitory and/or barrier properties to improve metals' and alloys' anticorrosion performance have gained increasing interest in recent years [12–14].

To overcome several obstacles faced by traditional corrosion protection techniques, metal–organic frameworks (MOFs), a family of materials distinguished by their distinct porosity architectures made up of metal nodes coordinated to organic linkers, have shown great promise. Numerous tunable features, such as large surface area, tunable porosity, and the capacity to include different functional groups, are made possible by the freedom in designing both the metal and the organic component. Because of these characteristics, MOFs are very appealing for a range of uses, such as catalysis and gas storage. Additionally, several MOF types (such as ZIF-8, MIL-53, and UiO-66) have drawn interest in the field of corrosion protection materials because of their hydrophobic performance, strong water stability, and corrosion inhibitory qualities [16, 17]. They have demonstrated promise in a variety of sectors

DOI: 10.1201/9781003571162-10

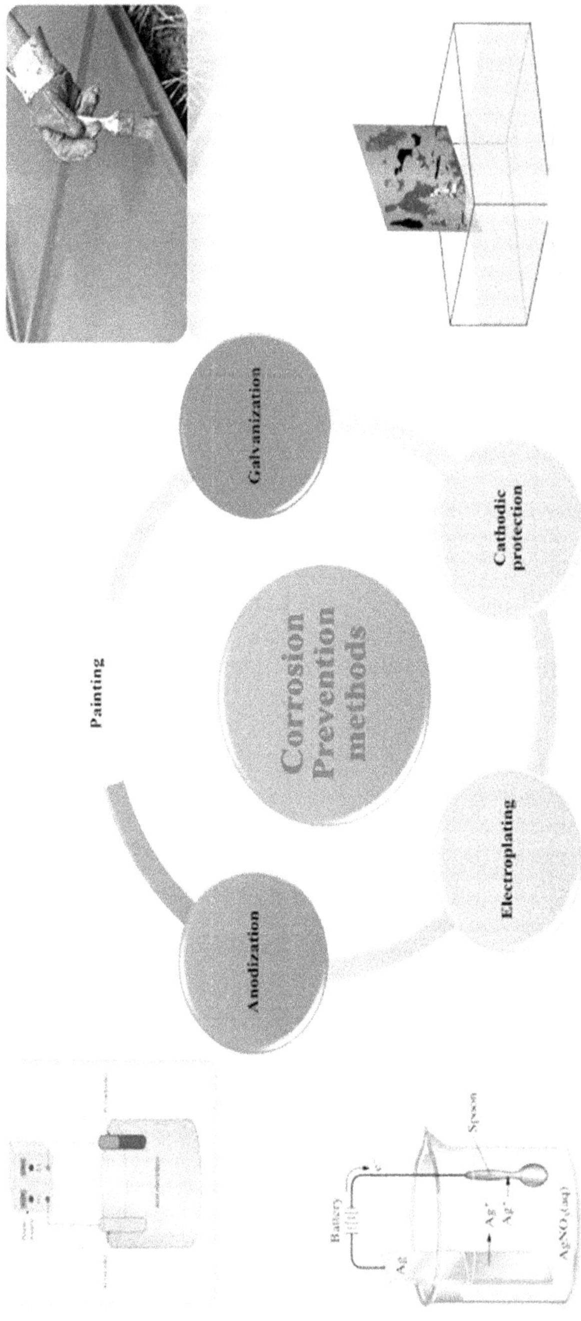

FIGURE 10.1 Several types of corrosion protection methods [15].

where resistance to corrosion is crucial. For example, MOF-based coatings have shown great promise in avoiding corrosion of ship hulls and offshore platforms in the maritime industry, where materials are continuously exposed to saltwater. Similar to this, MOF coatings offer a novel way to reduce material deterioration and extend the life of infrastructure in the oil and gas industry, where equipment and pipelines are vulnerable to corrosion because of exposure to harsh chemicals, moisture, and temperature fluctuations. Materials that are lightweight, strong, and resistant to corrosion are crucial in the automotive and aircraft sectors. Critical components' resistance to corrosion can be preserved or enhanced while weight reduction is possible using MOF coatings. The long-term durability of buildings, bridges, and other infrastructure in construction, especially in environments exposed to high humidity, chemicals, or maritime conditions, depends on corrosion protection, especially in reinforced concrete and steel structures. The application of MOF coatings as environmentally benign substitutes for traditional corrosion protection techniques is further encouraged by the increased interest in sustainable technologies. As many MOFs are made from readily available, non-toxic ingredients, they are a desirable option for the creation of environmentally friendly coatings and corrosion inhibitors.

The use of MOF-based corrosion protection materials does, however, confront several challenges, including poor adherence, unstable protection performance, and weak durability, despite the fact that numerous studies have demonstrated that MOFs have good qualities in anticorrosion domains [17, 18]. Functionalization and surface modification are crucial methods for improving MOF coatings' defensive qualities. Pure MOFs may not always show the required degree of durability or corrosion resistance in hostile environments, while having inherent properties like porosity and chemical tunability. The corrosion protection capabilities of MOFs can be greatly enhanced by altering their surface or functionalizing their structure. To improve the MOF coating's durability, adherence to substrates, and resistance to corrosive chemicals, surface modification entails changing its chemical or physical characteristics. This can be accomplished in a number of ways, including by adding exogenous agents or by nanostructuring, which improves the MOF's resilience to environmental stresses like UV light, moisture, and salt. Conversely, functionalization is the process of adding particular functional groups or guest molecules to the MOF structure, which might give it hydrophobicity, self-healing capabilities, or corrosion-inhibiting qualities. These alterations improve the long-term corrosion prevention capabilities of MOF coatings. All of the strategies offer a viable way to increase the longevity of metallic structures and lower the financial and environmental expenses related to corrosion damage. By modifying MOFs to satisfy the unique requirements of diverse settings, they can be effectively used to counteract the negative impacts of corrosion in a range of industrial settings. The chapter provides a brief overview of the role of MOF-based in corrosion-protective coatings. Moreover, practical modification and functionalization techniques of MOFs to address the current issues of low durability, nonuniformity, and unstable protective performance of MOFs in commercial anticorrosion applications are also briefly explained.

10.2 ROLE OF MOF IN CORROSION PROTECTION

In terms of their long-term potential, MOFs can play a significant role in the development of protective coatings as they provide a thorough understanding of dimensional designs, material alterations, and application perspectives. MOFs have shown great promise in anticorrosive applications because they are extremely versatile materials with significant structural benefits over traditional molecular containers. These structures are composed of metal ions or clusters joined by organic ligands, producing crystalline three-dimensional structures with tunable pore sizes, large specific surface areas, and a variety of morphologies, as shown in Figure 10.2.

Because of their unique architecture, MOFs can act as extremely efficient containers that can hold and release specific compounds, such as corrosion inhibitors. Through the selection of different metal ions and organic ligands, MOFs' modular qualities allow for exact adjustment of properties including pore diameters, channel topologies, and surface chemistry, making them appropriate for a wide range of applications. Their use as solid matrices in anticorrosive coatings, where they not only protect organic materials but also have the potential to be further improved through organic polymer networks, has garnered a lot of interest due to their versatility. A key component of MOFs' effectiveness as containers is their noticeable crystallinity. Because of their strong metal–ligand coordination interactions, which add to their exceptional stability and rigidity, MOFs exhibit highly ordered structures in contrast to conventional molecular containers. In addition to improving stability, this crystallinity increases surface area and porosity, which gives guest molecules more places to interact. Many MOFs have surface areas greater than 1000 m^2/g, which allows them to hold a wide range of molecules and to host only those with particular sizes and shapes, therefore outperforming traditional containers in terms of encapsulation. Furthermore, MOFs may be applied under demanding circumstances, including

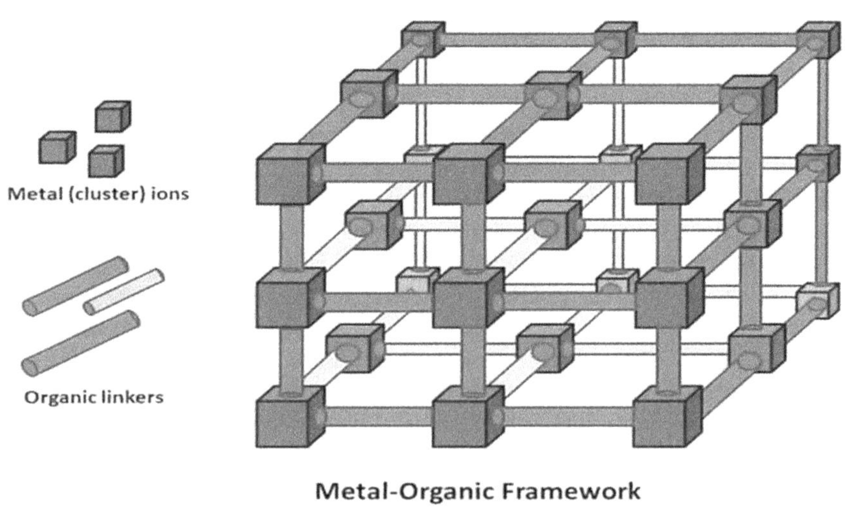

FIGURE 10.2 Three-dimensional structure of metal–organic framework [19].

exposure to different solvents, temperatures, and pressures, because their stability enhanced by strong metal–ligand interactions. This is essential for long-lasting protective applications [20, 21].

Another noteworthy feature of MOFs is their ability to be functionalized, in which a variety of organic ligands are added to provide tailored chemical functionalities that maximize inhibitor encapsulation and controlled release. In situations like corrosion resistance, where MOFs can serve as reservoirs for slow-release inhibitors, providing long-lasting protection, this level of personalization is beneficial. Because of their inherent functional versatility, MOFs can be designed to meet specific needs, such as drug delivery, gas storage, or, as demonstrated here, improving the protective qualities of anticorrosive coatings through controlled inhibitor release. This greatly expands the range of applications for MOFs beyond those of conventional molecular containers [22, 23]. Moreover, metal antioxidants are highly prized in the natural sciences and materials fields. Motivated by this, certain organic ligands have increased antioxidative activity. The secret to attaining superior performance is the development of antioxidative porosity with particular dimensions and retention, where the variety of antioxidative behaviors offers convenient access. These characteristics are perfectly met by MOFs, which are unique materials that can display or partially exhibit beneficial behaviors. Effective resistance against hostile environments is made possible by their diverse porosities, which create perfect barriers to stop the bulk transmission of corrosive species [24, 25]. According to earlier reports, there are four primary methods for protecting MOFs against corrosion, and the chapter in this area shows an increasing tendency over the last few years, ideally in relation to anticorrosion coatings [26–28]. The first application for MOFs is as a corrosion inhibitor. The widespread presence of heteroaromatic and π-system species in MOF structure, along with their supramolecular inherence, is usually believed to be the cause of several MOF species' corrosion prevention properties. For instance, zeolite imidazole frameworks (ZIFs) with built-in imidazole/benzimidazole ligands may be effective at preventing corrosion in metals or alloys [29, 30]. Distributing MOF crystals in inorganic or organic protective coating materials, where they act as nanofillers for the anticorrosion coatings, is the second method of creating MOF-based protective coatings [31]. The active protection ability of epoxy/polyamide coatings can be effectively improved by using MIL-88A (Fe) MOF material as an anticorrosive nanofiller. Additionally, MIL-53 (Al) MOF nanoparticles have been found to successfully increase the corrosion prevention effectiveness of the hybrid sol–gel coating for Al 2024 alloys [32, 33]. The third MOF use involves the fabrication of continuous and intergrown MOF membranes over metal surfaces via in-situ nucleation and/or self-adsorption, which act as direct anticorrosion coatings [34, 35]. The final alternative method for protecting MOFs from corrosion is the creation of composite coating materials that employ MOFs as nanocontainers to encapsulate inhibitors, taking advantage of their huge capacity and controllable release features [27]. For example, the graphene oxide/MOFs hybrid materials are shown to be suitable nanocontainers for polyvinyl butyral (PVB) resin coatings that safeguard pure copper and archaized bronze against corrosion [36]. Additionally, by using MOFs as a host to load the corrosion inhibitors, long-lasting composite coatings with self-healing capabilities can be produced [27].

Nevertheless, research into the methods by which MOFs prevent corrosion is currently ongoing. The type of MOF (connected to the organic ligands and metal ions utilized), the corrosive environment, the composite materials used, and the substrate to be protected are some of the variables that affect how well MOFs protect against corrosion [2]. The majority of MOFs are sensitive to pH and can respond rapidly to their acidic surroundings. Aggressive species induce the organic ligands and metal ions to split when they enter the MOF structure, generating compounds that prevent corrosion. According to a prior study, these materials have the ability to self-reassemble during long-term operation, reforming the MOF network [26]. Furthermore, some MOFs have been shown to limit metal surface corrosion, as discussed above, by forming a hydrophobic layer on the metal surface and concurrently interfering with the metal's cathodic and anodic corrosion processes [35]. It is noteworthy that MOFs, particularly those based on imidazole/benzimidazole ligands or rare earth metals, have the ability to suppress corrosion on their own [2]. The need for organic coatings with longer service lives is growing because of expanding applications and greater awareness of social and environmental sustainability. Organic coatings with anticorrosive pigments are widely acknowledged by industry and academia as anticorrosion emerges as a key performance criterion. MOFs, which are porous crystals made of metal nodes and organic ligands, are incredibly versatile and have a great deal of promise to deliver superior performance in this important area.

10.3 MOFS AS CORROSION PROTECTIVE COATINGS

Long-term corrosion protection may be maintained by using corrosion protective coatings with superior physical barrier qualities that could successfully withstand water and corrosive electrolytes. One of the proposed mechanisms underlying the protective barrier of the PI/FMOF nanocomposite coatings is shown in Figure 10.3.

As the above-mentioned corrosion inhibition capabilities of MOFs were the primary focus of their protective systems, however, in reality, their inherent advantages have not yet been fully utilized. The application of MOF films to metal surfaces, where MOF materials act as direct anticorrosion coatings, is an additional alternate method for protecting metals from corrosive agents. In addition to encapsulating inhibitors as previously mentioned, functional anticorrosion coatings based on MOFs can be further developed to prevent the hostile species from penetrating the corrosive environment by taking use of MOFs' unique organic groups and customizable porosity. For instance, the HKUST-1 coating coated with cetyltrimethylammonium bromide (CTAB) inhibitors has good corrosion resistance, according to the previously cited study by Cao et al. [38]. According to research on implanted metal, the compact and homogeneous topography of the Mg–MOF-74/MgF_2 composite coating provides excellent corrosion protection for magnesium alloy implants in the simulated bodily fluid [34]. As a hydrophobic surface may efficiently repel water and corrosive solutions, anticorrosion coatings with hydrophobic qualities typically have a high barrier effect [39, 40]. As previously noted, a variety of MOF types, such as ZIFs, MIL-53, and UiO-66, have hydrophobic qualities and good water stability, making them

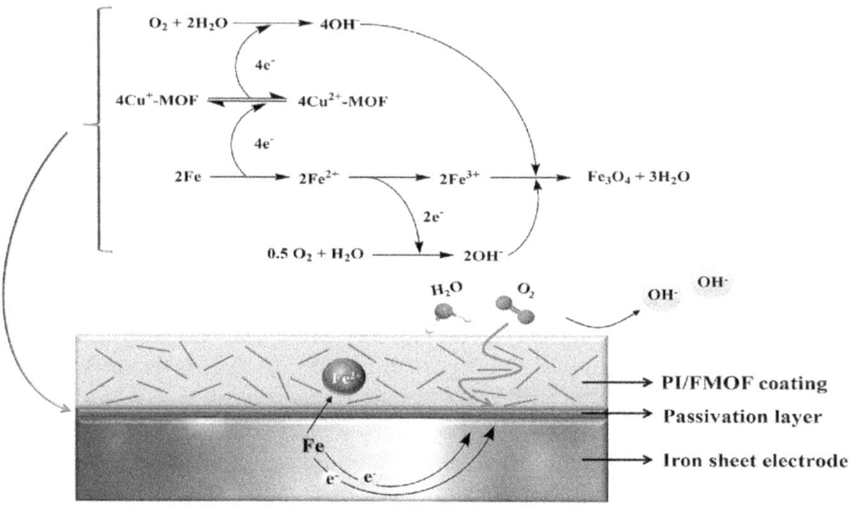

FIGURE 10.3 Illustration of corrosion protection mechanism of the PI/FMOF nanocomposite coatings. Reprinted with permission from [37].

highly promising for use in the corrosion prevention sector [41–43]. Qi et al. described a flexible method for corrosion prevention using MOF coatings (ZIF-9, ZIF-90, and UiO-66) inspired by polydopamine (PDA) [17]. The PDA layer at the substrate/MOF interface facilitates the heterogeneous nucleation and growth of MOF coatings. The lightweight coating framework, well-intergrown MOF crystals, hydrophobicity, and efficient isolation of corrosive species all contribute to the coating's anticorrosion performance, while the substrate, MOF crystal size, and coating thickness all affect the specific corrosion protection efficiency. The scientists discovered that the most effective corrosion protection for Al alloy is provided by the synthesized UiO-66 layer with PDA assistance. Both the morphology and the adhesion of the UiO-66 film could be maintained at the highest level after 7 days of immersion in a 3.5 weight percent NaCl solution with a critical load of 14 N. Consequently, this UiO-66 film can function as a strong and suitable coating for corrosion prevention. This work highlights the qualified and prospective candidates of hydrophobic MOF coatings for anticorrosion purposes and sheds important insight on the production of MOF-based protective coatings. By employing a "twin zinc source" technique, Zhang et al. creatively created a ZIF-90 coating on the surface of Zn–Al alloy, which significantly improves the alloy's resistance to corrosion [43]. The ZIF-90 crystals that make up the coating have a compact and perfect shape, and because of their micron size, peaks and valleys emerge in depth. The ZIF-90 coating's morphology shows the capability to trap a lot of air, which makes it clearly hydrophobic with a water contact angle of about 112.4° as well as water-repellent. Because there are many different sources of metal ions for MOFs, Liu et al. described a novel method for creating a compact ZIF-8 coating by using a ZnAl double hydroxide ($ZnAlCO_3$ LDH) buffer layer as the metal source and heterogeneous nucleation center of ZIF-8 [16]. Contrary to ZIF-8 coatings generated in-situ, the ZnAlCO3 LDH in this study greatly promotes ZIF-8 coating nucleation and growth, and the coating shows excellent adherence on Al substrate. This coating may successfully reduce the corrosion current density of the Al substrate by four orders of magnitude by utilizing the inherent hydrophobicity and water stability of ZIF-8. This indicates a possible candidate for MOF coatings for corrosion protection. Additionally, using a similar technique, they were able to successfully develop a ZIF-8-ZnAl-NO_3 LDH composite coating system on aluminum plate, and it was discovered that the coating had excellent mechanical stability and anticorrosion effectiveness [44]. Additionally, Li et al. created a superhydrophobic LDH/ZIF-8 double-layer composite coating and investigated the coating's anticorrosion capabilities for magnesium alloy [45]. The outer layered ZIF-8 film was created using the dip-coating technique on the electrodeposited underlying LDH layer surface while a low-energy material-fabricated ZIF-8 solution was present. According to their findings, this composite coating has a high level of electrolyte blocking efficacy and long-term corrosion prevention while also demonstrating an effective self-cleaning capability. For the actual use of metallic anticorrosion, the aforementioned works offer a novel method for creating superior MOF-based anticorrosion coatings with strong hydrophobicity.

10.4 CHALLENGES AND NEED FOR MODIFICATION

The need for better protective solutions in challenging conditions has prompted the creation of MOFs for anticorrosive coatings, which is a state-of-the-art development in materials research. MOFs have distinct benefits over conventional coatings because of their large surface area, adjustable porosity, and adaptable chemical functionalities. However, there are many obstacles in the way of realizing their full potential, including preserving compatibility with different coating matrices, attaining scalable production, and guaranteeing chemical and thermal stability. MOFs are becoming essential parts of next-generation anticorrosive coatings because of advancements in synthesis methods (see Figure 10.4), functionalization, and intelligent coating designs. This integration represents a major advancement in increasing the durability and efficacy of protective coatings by promising improved performance, self-healing properties, and multifunctionality.

Stability and Durability: One of the biggest challenges in anticorrosive coatings is making sure MOFs are stable and durable in corrosive environments. Creating MOFs with improved chemical stability and using encapsulation techniques to guard against environmental deterioration are examples of innovations.

Uniform Dispersion: Although it presents a problem, uniform dispersion of MOFs in coating matrices is essential for good performance. To improve MOF dispersion in coating formulations, innovations include sophisticated mixing techniques like sonication or mechanical stirring.

Scale-Up Challenges: There are financial and logistical obstacles to overcome when producing coatings with MOFs on a larger scale for industrial use. To enable large-scale production, innovations concentrate on improving coating formulations and developing scalable MOF synthesis techniques.

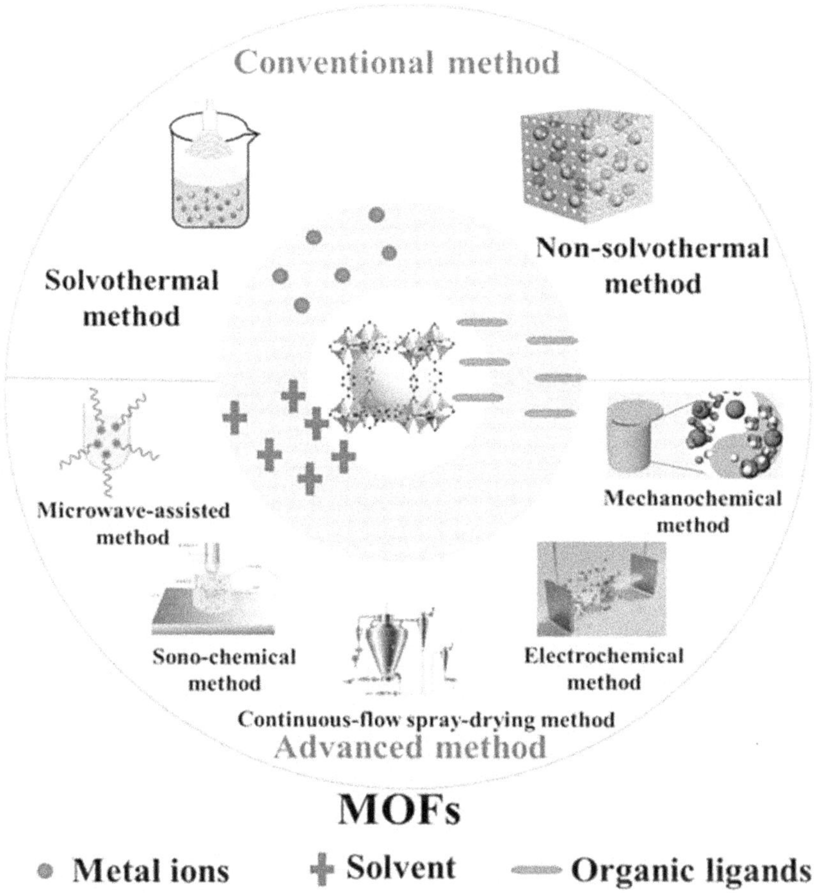

FIGURE 10.4 Representation of different conventional and advanced approaches for fabrication of metal–organic frameworks [46].

Cost Considerations: The price of MOFs may prevent them from being widely used in coatings. To meet economic considerations, innovations investigate alternative sources of MOFs and cost-effective synthesis techniques.

Adaptability to Current Coating Procedures: It is essential to make sure that coatings with integrated MOFs can be easily implemented into current procedures. Creating formulations that work with common coating methods and are simple to incorporate into industrial processes is an example of innovation.

Customizing MOF Properties: It can be difficult to customize MOF properties to meet certain anticorrosive needs. Advances in MOF synthesis methods enable the accurate design of MOFs with characteristics suited to various corrosive conditions. These issues are being addressed by ongoing research and innovation, which could lead to the development of efficient, long-lasting, and sustainable coatings for a range of industrial applications.

10.5 MODIFICATION AND FUNCTIONALIZATION OF MOF-BASED COATINGS

MOF-based thin films have many obstacles that prevent them from being used practically in corrosion protection, despite their promising qualities. There are still problems including poor adherence, uneven protective effectiveness, and limited durability. The following are some important functionalization and modification techniques (see Figure 10.5) that have helped to overcome the obstacles and produce anticorrosive MOFs with improved qualities.

10.5.1 Chemical Modification

The term "chemical modification" refers to any surface treatment that modifies the surface characteristics without the use of mechanical tools. This means that a certain thickness of the surface, which can range from nanometers to micrometers, must undergo chemical changes. Because of this, this type of surface alteration is often described in studies as being made possible by functional layers, one of which

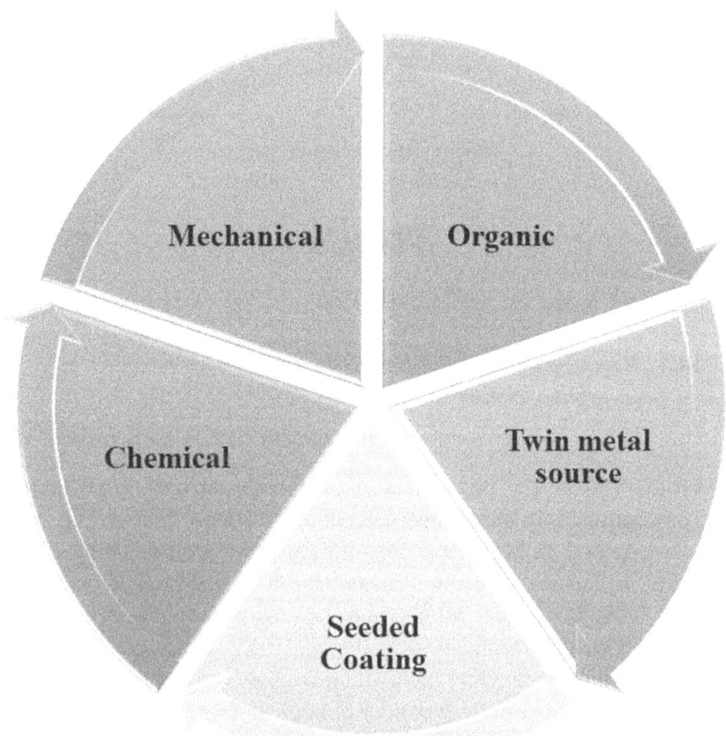

FIGURE 10.5 Several surface modifications and functionalization for effective MOF coating development.

may be the metal oxide of the substrate. However, there are several techniques for applying this metal oxide layer on the surface [47]. Furthermore, this oxide layer is not always the metal oxide of the substrate itself; it can be externally put onto a substrate in a completely different fashion. The resulting coating is known as a conversion coating if the substrate's surface experiences chemical changes and the oxide layer is the substrate. Nevertheless, the coating cannot be categorized as a conversion coating if an external oxide layer is applied [48]. When a metal surface reacts with a chemical solution, the substrate's surface layers oxidize, creating conversion coatings. A porous and sticky oxide layer is produced because of this procedure. Depending on the particular method, different electric currents are required for this operation. Because of their unique qualities and uses, these coatings, which include anodizing, chromating, phosphating, and blackening, can also be based on rare earth elements. Many academics have investigated and assessed the many types of conversion coatings, describing their applications and traits in a number of thorough literature review [49, 50]. On the other hand, MOF crystals that adhere strongly to surfaces and form a film are suitable for a wide range of applications. These coatings' porosity, in conjunction with the metal oxides, offers nucleation sites for the growth of the MOF layer on the outer layer, improving corrosion resistance. Apart from conversion coatings, a variety of external functional layers, such as metal oxides, organic molecules, and LDHs, can develop on a surface and serve as sacrificial templates [51, 52].

Additionally, the chemical etching procedure is one of the techniques included in chemical modification and significantly roughens the surface. Etching is a labor-intensive process that encourages the formation of MOF-based layers by converting a smooth, mirror-like surface into a rough, opaque one. A technique for cultivating MOF thin films (MOF-TFs) on uneven surfaces is described in a report by Cui et al [53]. To do this, they used hydrofluoric acid to roughen stainless steel wires before immersing them in a heated solution that contained the MOF building blocks. After 8 hours, a full layer of HKUST-1 with a thickness of 40 μm has formed because of the attachment and growth of MOF particles encouraged by this rough surface [54].

10.5.2 Mechanical Modification

In the quickly developing field of coatings, MOF-derived thin films have shown promise qualities for a variety of applications; yet, the density of flaws considerably affects the final thin film's quality. Therefore, it is important to consider how this aspect affects the final quality of the manufactured layer [55]. Although the word "defects" is frequently used negatively, in some applications, having defects is advantageous because it directly affects the layer's characteristics. It is often known that having too many flaws without sufficient understanding might have negative consequences. Improving our comprehension of the connection between defects and material properties can greatly increase a researcher's understanding in a variety of applications. According to research, the lowest density of flaws might result in high-quality layers [56]. It is possible to introduce faults in these films in a controlled way, for example, by using "defective linkers" or applying heat treatment [57, 58].

A surface's surface energy rises with roughness. Surface activation may be facilitated by this increased surface energy as well as the presence of surface defects. Chemical or mechanical procedures can produce this type of roughness. The goal is to produce peaks and valleys from rough surfaces; these peaks are sharp, have more energy than their surroundings, and serve as anodic sites [59]. These anodic sites dissolve easily, releasing metal cations required for the production of the MOF layer as the metal node is formed by dissolving the MOF layer's surface. With regard to creating other types of MOFs, this method is limited. This results from the requirement that the metal cation that is obtained from the substrate's dissolution match the MOF's metal node. In this situation, one tactic for surface activation is surface roughness [17].

10.5.3 Organic Modification

There are several methods for improving heterogeneous nucleation-induced MOF nanoparticle production on substrate surfaces. Covalent linkers are a useful technique for creating a chemical bond between the MOF layer and substrate surface. This approach's functionalization method involves π–π interactions, hydrolysis, and hydrogen interactions between the covalent ligand's terminal groups and those on the surface. A new reactive agent is created as a result of these interactions, and it can then interact with metal ions or organic linkers to help form the MOF layer. In practical applications, 3-Aminopropyltriethoxysilane (APTES) is commonly used because it creates a covalent bond between the base material and the MOF layer. The ethoxy groups of APTES are hydrolyzed in this process to produce SiOH groups, which subsequently form hydrogen bonds with hydroxyl groups on the surface. After that, the amino groups engage in an imine condensation process with chemical ligands or metal ions. These processes are somewhat intricate, though. To develop MOF-based anticorrosion coatings, researchers are looking for a simple, effective, and adaptable way to produce MOF fine films on metal surfaces. There is a lot of interest in this field of study [60, 61]. Furthermore, the remarkable properties of polymers, such as their high mechanical resilience, resistance to wear, and resistance to corrosion, which have garnered a lot of attention are what propel their increased use across a range of industrial sectors. Because several of the functional groups on the functionalized surface are close to one another, the metal and polymeric encompassing react. Through in-situ heterogeneous nucleation, this reaction makes it easier to create compact and well-integrated MOF thin films. In a groundbreaking accomplishment, a group of researchers used a PDA coating platform to successfully apply integrated and sticky coatings of ZIF-8, MOF-5, MIL-100(Fe), and HKUST-1 onto chemically resistant membrane surfaces. Despite early difficulties, these coatings have a lot of potential. Zhou et al.'s paper presents a simple and effective synthesis method for producing porous composite films with a hierarchical structure [62]. These films blend microporous materials with the unique characteristics of MOFs. Interestingly, the researchers found that by adding common linkers like pyridine or benzoic acid, PDA-treated "inert" surfaces can be given further functional characteristics. During bulk synthesis, MOFs are incorporated into the mother solution to accomplish this functionalization. PDA's unique capacity to serve as a nucleation center for the coordinated generation of selected MOF compounds on various polymer membranes is the main discovery. Surprisingly, commercial "inert" polymer membranes respond especially well to this method. The described approach creates viable opportunities for increasing and improving the uses of porous polymer membranes in a variety of industries by utilizing the unique characteristics of MOFs. According to the researcher's findings, a group of scientists have effectively created an adaptable approach using multifunctional, mussel-inspired PDA to generate anti-corrosion coatings with MOF coats on metal surfaces. The formation of a stable and adhesive PDA layer on the surface of the metal substrate is the result of dopamine self-polymerizing, and this layer is rich in catechol functional groups that form a metal-chelating bond with the PDA layer, allowing for the in-situ and heterogeneous formation of a MOF-based thin layer [17, 63]. Furthermore, by depositing them on a primary substrate, self-assembled monolayers (SAMs) are a very effective way to produce a surface end with certain termination groups. SAMs produce organic interfaces with pyridine moieties and termination groups like –OH, –COOH, and $-NH_2$. The synthesis of MOF fine films can be initiated by using these chains to bind metal or metal-oxo nodes and organic ligands. See the recent review by Zhuang et al. for further details on this aspect of SAM-triggered MOF nucleation. In the best cases, this can produce aligned MOF fine films with a high degree of structural stability [64]. It has been repeatedly demonstrated that MOF-based thin film growth direction can be controlled by utilizing different functional groups or adjusting their density [65].

10.5.4 Modification Via Twin Metal Source

It is usual practice to create a dual metal source to supply the required MOF metal nodes. Using this technique, MOF metal is produced from a source other than the metal cations present in the bulk solution. The metal ions on the surface of the metal supports may provide heterogeneous nucleation sites that promote continuous film growth. They also aid in the coordination process with the synthesis solution's organic ligands. The formation of an oxide layer on the outside or the disintegration of the metal substrate using electrochemical techniques as an additional source of metal nodes are two ways to accomplish this. Depending on the researcher's inventiveness and method of thinking, the scope of this methodology can be expanded. Gue et al. effectively created a $Cu_3(BTC)_2$ membrane for separation applications directly on copper metal by utilizing the twin metal source approach [66]. Compared to traditional in-situ synthesis techniques, our method produced a denser film with better mechanical strength and compatibility with the support. By adding a Zn–Al LDH buffer layer to the metal surface, Zhang et al. also used a unique approach to establish a dual metal source for the ZIF-8 metal node. This produced a well-developed layer with desired anticorrosion capabilities [16, 43].

10.5.5 Modification Via Seeded Coating

The strategy of seeding desired MOF grains that are anticipated to eventually form a thin film is one of the efficient ways to produce stable, intergrown, and compact MOF-based layers on substrate surfaces. The non-homogeneous nucleation of the MOF layer is facilitated by the production of these grains. This technique promotes the formation of the MOF layer by dispersing MOF nanoparticles across the surface prior to the final stage of growth [67]. This can be accomplished in a number of ways. For example, the substrate can be placed separately in a solution containing metal nodes to create a layer in situ by submerging it in a solution containing precursors. Some metal nodes are deposited on the surface because of this procedure. MOF nanoparticles are then dispersed on the surface after the substrate is transferred to a solution containing an organic ligand. In a two-step process known as seeding, MOF "seeds" are first made and then applied to the interface's base [68]. The seed-adorned interface is then submerged in the MOF nucleation solution to create the MOF film on the substrate's surface. This method's main advantage is that it may be utilized to pattern seed decoration and subsequent film development by utilizing the substrate surface's micro-scale physical features, such as holes and grooves. The possible separation of seeds within the developing arrangement and the substrate surface's low adherence are two disadvantages of seeding techniques [69].

10.6 CONCLUSION AND FUTURE OUTLOOK

The surface modification and functionalization of MOF coatings represent a highly promising strategy for enhancing corrosion protection. By strategically tailoring the surface properties of MOFs, such as improving hydrophobicity, chemical stability, and incorporating corrosion inhibitors, the overall performance of these coatings can be significantly boosted. The inherent versatility of MOFs, in terms of both structure and functionality, allows them to be adapted to various industrial environments, offering a highly effective means of mitigating corrosion across different applications. This chapter explores various surface modification and functionalization techniques aimed at enhancing the stability, uniformity, and overall performance of MOF coatings. It provides an in-depth discussion on strategies to improve the corrosion resistance of MOFs, focusing on their potential as effective corrosion protective coatings. The chapter also highlights the challenges that MOFs face in corrosion protection, addressing issues such as limited stability and poor adhesion to substrates. To overcome these obstacles, the need for targeted modifications is emphasized. A range of modification approaches are discussed, offering tailored solutions to meet the specific requirements of different industrial environments. These strategies collectively present a powerful means to combat corrosion and enhance the applicability of MOF coatings in real-world applications.

While the advancements in MOF coatings present significant opportunities for corrosion protection, there remain challenges that must be addressed to fully realize their potential. Future research should focus on improving the long-term stability, durability, and scalability of these coatings, along with developing cost-effective methods for their large-scale production and implementation. Optimizing these materials for practical use in diverse and challenging environments will be crucial to advancing the field. As our understanding of MOF materials continues to grow, their potential to serve as advanced, sustainable, and efficient solutions for corrosion protection remains a highly promising area for further development.

REFERENCES

[1] H. Assad, S.K. Saha, N. Kang, S. Kumar, P.K. Sharma, H. Dahiya, A. Thakur, S. Sharma, R. Ganjoo, A. Kumar, Electrochemical and computational insights into the utilization of 2, 2-dithio bisbenzothiazole as a sustainable corrosion inhibitor for mild steel in low pH medium, *Environmental Research*, 242 (2024) 117640.

[2] L. Jiang, Y. Dong, Y. Yuan, X. Zhou, Y. Liu, X. Meng, Recent advances of metal–organic frameworks in corrosion protection: From synthesis to applications, *Chemical Engineering Journal*, 430 (2022) 132823.

[3] A. Farhadian, A. Rahimi, N. Safaei, A. Shaabani, M. Abdouss, A. Alavi, A theoretical and experimental study of castor oil-based inhibitor for corrosion inhibition of mild steel in acidic medium at elevated temperatures, *Corrosion Science*, 175 (2020) 108871.

[4] U. Eduok, E. Ohaeri, J. Szpunar, Conversion of imidazole to N-(3-aminopropyl) imidazole toward enhanced corrosion protection of steel in combination with carboxymethyl chitosan grafted poly (2-methyl-1-vinylimidazole), *Industrial & Engineering Chemistry Research*, 58 (2019) 7179–7192.

[5] H. Assad, I. Fatma, A. Kumar, An overview of the application of graphene-based materials in anticorrosive coatings, *Materials Letters*, 330 (2023) 133287.

[6] H. Assad, I.A. Lone, A. Sihmar, A. Kumar, A. Kumar, An overview of contemporary developments and the application of graphene-based materials in anticorrosive coatings, *Environmental Science and Pollution Research*, 32 (2023) 1–20.

[7] H. Assad, S.K. Saha, N. Kang, S. Kumar, H. Dahiya, P. Banerjee, A. Thakur, S. Sharma, R. Ganjoo, A. Kumar, Assessment of the inhibitory efficacy of a thiazole derivative as an efficient corrosion inhibitor for augmenting the resistance of MS in acidic environments, *Langmuir*, 40 (2024) 16367–16383.

[8] H. Assad, A. Kumar, Understanding functional group effect on corrosion inhibition efficiency of selected organic compounds, *Journal of Molecular Liquids*, 344 (2021) 117755.

[9] H. Assad, S. Kumar, A. Kumar, Nitrogen, Sulfur, and Nitrogen-based heterocycles corrosion inhibitors, in Verma, C. (ed.) *Handbook of Heterocyclic Corrosion Inhibitors*, CRC Press, 2024, pp. 198–211.

[10] H. Assad, S. Kumar, S.K. Saha, N. Kang, H. Dahiya, A. Thakur, S. Sharma, R. Ganjoo, A. Kumar, A research combined experimental and computational approaches of succinylsulfathiazole hydrate as potent corrosion inhibitor for mild steel in acidic medium, *Journal of Molecular Liquids*, 388 (2023) 122739.

[11] H. Assad, S. Kumar, S.K. Saha, N. Kang, I. Fatma, H. Dahiya, P.K. Sharma, A. Thakur, S. Sharma, R. Ganjoo, Evaluating the adsorption and corrosion inhibition capabilities of Pyridinium-P-Toluene Sulphonate on MS in 1 M HCl medium: An experimental and theoretical study, *Inorganic Chemistry Communications*, 153 (2023) 110817.

[12] H. Assad, A. Kumar, Carbon nanotubes (SWCNTs/MWCNTs) and functionalized carbon nanotubes in corrosion prevention, in Verma, C., Tukhlievich, B.V. (eds.) *Corrosion Prevention Nanoscience: Nanoengineering Materials and Technologies*, 2023, pp. 85–98.

[13] Z. Sanaei, G. Bahlakeh, B. Ramezanzadeh, Active corrosion protection of mild steel by an epoxy ester coating reinforced with hybrid organic/inorganic green inhibitive pigment, *Journal of Alloys and Compounds*, 728 (2017) 1289–1304.

[14] S.M. Lashgari, H. Yari, M. Mahdavian, B. Ramezanzadeh, G. Bahlakeh, M. Ramezanzadeh, Application of nanoporous cobalt-based ZIF-67 metal-organic framework (MOF) for construction of an epoxy-composite coating with superior anti-corrosion properties, *Corrosion Science*, 178 (2021) 109099.

[15] K. Bijapur, V. Molahalli, A. Shetty, A. Toghan, P. De Padova, G. Hegde, Recent trends and progress in corrosion inhibitors and electrochemical evaluation, *Applied Sciences,* 13 (2023) 10107.

[16] M. Zhang, L. Ma, L. Wang, Y. Sun, Y. Liu, Insights into the use of metal–organic framework as high-performance anticorrosion coatings, *ACS Applied Materials & Interfaces*, 10 (2018) 2259–2263.

[17] X. Liu, T. Yue, K. Qi, B.Y. Xia, Z. Chen, Y. Qiu, X. Guo, Probe into metal-organic framework membranes fabricated via versatile polydopamine-assisted approach onto metal surfaces as anticorrosion coatings, *Corrosion Science*, 177 (2020) 108949.

[18] H. Chen, F. Wang, H. Fan, R. Hong, W. Li, Construction of MOF-based superhydrophobic composite coating with excellent abrasion resistance and durability for self-cleaning, corrosion resistance, anti-icing, and loading-increasing research, *Chemical Engineering Journal*, 408 (2021) 127343.

[19] P.A. Sampaio, E.C. Pereira, P.G. Sá, J.M.T. Alencar Filho, L.R. Ferraz, R.H. Nishimura, A.S. Ferreira, P.J. Rolim Neto, E.S. Araújo, L.A. Rolim, A review on metal–organic frameworks as technological excipients: Synthesis, characterization, *Toxicity, and Application in Drug Delivery Systems, Compounds*, 5 (2025) 1.

[20] F. Seidi, M. Jouyandeh, M. Taghizadeh, A. Taghizadeh, H. Vahabi, S. Habibzadeh, K. Formela, M.R. Saeb, Metal-organic framework (MOF)/epoxy coatings: A review, *Materials*, 13 (2020) 2881.

[21] Z. Zhang, Q. Liu, G. Qi, Facile In situ growth of ZIF-8 nanosheets with enhanced anti-corrosion performance for carbon steel in seawater, *Coatings*, 12 (2022) 318.

[22] S. Kumar, S. Vijayan, K. Goyal, M. Kathuria, S. Gulati, Functionalization strategies of metal–organic frameworks (MOFs): Diverse ways to versatile MOFs, in *Metal-organic Frameworks (MOFs) as Catalysts*, Springer, 2022, pp. 99–123.

[23] L. Jiao, J.Y.R. Seow, W.S. Skinner, Z.U. Wang, H.-L. Jiang, Metal–organic frameworks: Structures and functional applications, *Materials Today*, 27 (2019) 43–68.

[24] H.M.A. El-Lateef, T. El-Dabea, M.M. Khalaf, A.M. Abu-Dief, Recent overview of potent antioxidant activity of coordination compounds, *Antioxidants*, 12 (2023) 213.

[25] W. Park, S. Park, K.-Y. Park, E.-Y. Choi, Long-term antioxidant metal–organic frameworks, *ACS Omega*, 9 (2024) 21484–21493.

[26] M. Ramezanzadeh, B. Ramezanzadeh, M. Mahdavian, G. Bahlakeh, Development of metal-organic framework (MOF) decorated graphene oxide nanoplatforms for anti-corrosion epoxy coatings, *Carbon*, 161 (2020) 231–251.

[27] K. Cao, Z. Yu, D. Yin, L. Chen, Y. Jiang, L. Zhu, Fabrication of BTA-MOF-TEOS-GO nanocomposite to endow coating systems with active inhibition and durable anticorrosion performances, *Progress in Organic Coatings*, 143 (2020) 105629.

[28] S. Kumaraguru, R. Pavulraj, S. Mohan, Influence of cobalt, nickel and copper-based metal-organic frameworks on the corrosion protection of mild steel, *Transactions of the IMF*, 95 (2017) 131–136.

[29] J. Cao, C. Guo, X. Guo, Z. Chen, Inhibition behavior of synthesized ZIF-8 derivative for copper in sodium chloride solution, *Journal of Molecular Liquids*, 311 (2020) 113277.

[30] S. Zafari, M. Niknam Shahrak, M. Ghahramaninezhad, New MOF-based corrosion inhibitor for carbon steel in acidic media, *Metals and Materials International*, 26 (2020) 25–38.

[31] M. Ramezanzadeh, B. Ramezanzadeh, G. Bahlakeh, A. Tati, M. Mahdavian, Development of an active/barrier bi-functional anti-corrosion system based on the epoxy nanocomposite loaded with highly-coordinated functionalized zirconium-based nanoporous metal-organic framework (Zr-MOF), *Chemical Engineering Journal*, 408 (2021) 127361.

[32] N. Alipanah, H. Yari, M. Mahdavian, B. Ramezanzadeh, G. Bahlakeh, MIL-88A (Fe) filler with duplicate corrosion inhibitive/barrier effect for epoxy coatings: Electrochemical, molecular simulation, and cathodic delamination studies, *Journal of Industrial and Engineering Chemistry*, 97 (2021) 200–215.

[33] Y.J. Tarzanagh, D. Seifzadeh, Z. Rajabalizadeh, A. Habibi-Yangjeh, A. Khodayari, S. Sohrabnezhad, Sol-gel/MOF nanocomposite for effective protection of 2024 aluminum alloy against corrosion, *Surface and Coatings Technology*, 380 (2019) 125038.

[34] W. Liu, Z. Yan, X. Ma, T. Geng, H. Wu, Z. Li, Mg-MOF-74/MgF2 composite coating for improving the properties of magnesium alloy implants: Hydrophilicity and corrosion resistance, *Materials*, 11 (2018) 396.

[35] M. Chen, H. Cen, C. Guo, X. Guo, Z. Chen, Preparation of Cu-MOFs and its corrosion inhibition effect for carbon steel in hydrochloric acid solution, *Journal of Molecular Liquids*, 318 (2020) 114328.

[36] Y. Zhao, F. Jiang, Y.-Q. Chen, J.-M. Hu, Coatings embedded with GO/MOFs nanocontainers having both active and passive protecting properties, *Corrosion Science*, 168 (2020) 108563.

[37] X.-Y. Lin, Y.-H. Yu, M.-M. Tse, S.-H. Cheng, Y.-H. Liu, K.-L. Lu, Semiconductive (Cu–S) n metal–organic framework incorporated polyimide nanocomposite coatings forming an oxide barrier for durable anticorrosion effects, *Progress in Organic Coatings*, 180 (2023) 107562.

[38] W. Li, B. Ren, Y. Chen, X. Wang, R. Cao, Excellent efficacy of MOF films for bronze artwork conservation: The key role of HKUST-1 film nanocontainers in selectively positioning and protecting inhibitors, *ACS Applied Materials & Interfaces*, 10 (2018) 37529–37534.

[39] H. Chen, H. Fan, N. Su, R. Hong, X. Lu, Highly hydrophobic polyaniline nanoparticles for anti-corrosion epoxy coatings, *Chemical Engineering Journal*, 420 (2021) 130540.

[40] M.M. Kooshksara, S. Mohammadi, Investigation of the in-situ solvothermal reduction of multi-layered Graphene oxide in epoxy coating by acetonitrile on improving the hydrophobicity and corrosion resistance, *Progress in Organic Coatings*, 159 (2021) 106432.

[41] F. Ahmadijokani, R. Mohammadkhani, S. Ahmadipouya, A. Shokrgozar, M. Rezakazemi, H. Molavi, T.M. Aminabhavi, M. Arjmand, Superior chemical stability of UiO-66 metal-organic frameworks (MOFs) for selective dye adsorption, *Chemical Engineering Journal*, 399 (2020) 125346.

[42] A.A. Meshram, S.M. Sontakke, Synthesis of highly stable nanoscale MIL-53 MOF and its application for the treatment of complex mixed dye solutions and real-time dye industry effluent, *Separation and Purification Technology*, 274 (2021) 119073.

[43] X. Zhang, Y. Zhang, Y. Wang, D. Gao, H. Zhao, Preparation and corrosion resistance of hydrophobic zeolitic imidazolate framework (ZIF-90) film@ Zn-Al alloy in NaCl solution, *Progress in Organic Coatings*, 115 (2018) 94–99.

[44] M. Zhang, Y. Liu, Enhancing the anti-corrosion performance of ZIF-8-based coatings via microstructural optimization, *New Journal of Chemistry*, 44 (2020) 2941–2946.
[45] X. Yin, P. Mu, Q. Wang, J. Li, Superhydrophobic ZIF-8-based dual-layer coating for enhanced corrosion protection of Mg alloy, *ACS Applied Materials & Interfaces*, 12 (2020) 35453–35463.
[46] S. Tang, Y. Wang, P. He, Y. Wang, G. Wei, Recent advances in metal–organic framework (MOF)-based composites for organic effluent remediation, *Materials*, 17 (2024) 2660.
[47] X. Guo, D. Liu, T. Han, H. Huang, Q. Yang, C. Zhong, Preparation of thin film nanocomposite membranes with surface modified MOF for high flux organic solvent nanofiltration, *AIChE Journal*, 63 (2017) 1303–1312.
[48] L. Chen, N. Chen, Optimization of source and detector configurations based on Cramer–Rao lower bound analysis, *Journal of Biomedical Optics*, 16 (2011) 035001–035010.
[49] M. Dehghan, M. Attar, Enhancement of adhesion properties, corrosion resistance and cathodic disbonding of mild steel-epoxy coating systems by vanadium conversion coating, Progress in Color, *Colorants and Coatings*, 13 (2020) 223–238.
[50] M. Shahini, H. Eivaz Mohammadloo, B. Ramezanzadeh, Recent advances in steel surface treatment via novel/green conversion coatings for anti-corrosion applications: A review study, *Journal of Coatings Technology and Research*, 19 (2022) 159–199.
[51] Y. Zhang, C.-H. Chang, Metal–organic framework thin films: Fabrication, modification, and patterning, *Processes*, 8 (2020) 377.
[52] Z.-G. Gu, C. Zhan, J. Zhang, X. Bu, Chiral chemistry of metal–camphorate frameworks, *Chemical Society Reviews*, 45 (2016) 3122–3144.
[53] M. Shahini, H.E. Mohammadloo, M. Ramezanzadeh, B. Ramezanzadeh, Recent innovations in synthesis/characterization of advanced nano-porous metal-organic frameworks (MOFs); current/future trends with a focus on the smart anti-corrosion features, *Materials Chemistry and Physics*, 276 (2022) 125420.
[54] X.-Y. Cui, Z.-Y. Gu, D.-Q. Jiang, Y. Li, H.-F. Wang, X.-P. Yan, In situ hydrothermal growth of metal– organic framework 199 films on stainless steel fibers for solid-phase microextraction of gaseous benzene homologues, *Analytical Chemistry*, 81 (2009) 9771–9777.
[55] C. Zhou, M. Pan, S. Li, Y. Sun, H. Zhang, X. Luo, Y. Liu, H. Zeng, Metal organic frameworks (MOFs) as multifunctional nanoplatform for anticorrosion surfaces and coatings, *Advances in Colloid and Interface Science*, 305 (2022) 102707.
[56] Z.-G. Gu, L. Heinke, C. Wöll, T. Neumann, W. Wenzel, Q. Li, K. Fink, O.D. Gordan, D.R. Zahn, Experimental and theoretical investigations of the electronic band structure of metal-organic frameworks of HKUST-1 type, *Applied Physics Letters*, 107 (2015) 183301.
[57] M. Falahaty marvast, M. Gholambargani, M. Ramezanzadeh, B. Ramezanzadeh, Enhancing self-healing and adhesion bonding properties: Investigating the promising potential of Ce-ZIF8 MOF hybrid thin films as pH-responsive nanocoatings, *Langmuir*, 39 (2023) 9579–9594.
[58] Z. Zhang, H. Yoshikawa, K. Awaga, Monitoring the solid-state electrochemistry of Cu (2, 7-AQDC)(AQDC= anthraquinone dicarboxylate) in a lithium battery: Coexistence of metal and ligand redox activities in a metal–organic framework, *Journal of the American Chemical Society*, 136 (2014) 16112–16115.
[59] W.-J. Liu, Y.-H. Chang, S.-T. Hsu, C.-L. Fern, Y.-T. Chen, S.-Y. Tsao, S.-H. Lin, Exploring the correlation between surface roughness, surface energy, nano-indentation, electrical properties, and magnetic characteristics of annealed Co40Fe40Dy20 thin

films deposited on Si (100) substrates, *Journal of Electronic Materials*, 53 (2024) 4498–4511.

[60] Z.Y. Yeo, S.-P. Chai, P.W. Zhu, A.R. Mohamed, An overview: Synthesis of thin films/ membranes of metal organic frameworks and its gas separation performances, *RSC Advances*, 4 (2014) 54322–54334.

[61] S. Qiu, M. Xue, G. Zhu, Metal–organic framework membranes: From synthesis to separation application, *Chemical Society Reviews*, 43 (2014) 6116–6140.

[62] M. Zhou, J. Li, M. Zhang, H. Wang, Y. Lan, Y.-n. Wu, F. Li, G. Li, A polydopamine layer as the nucleation center of MOF deposition on "inert" polymer surfaces to fabricate hierarchically structured porous films, *Chemical Communications*, 51 (2015) 2706–2709.

[63] N. Wang, Y. Liu, Z. Qiao, L. Diestel, J. Zhou, A. Huang, J. Caro, Polydopamine-based synthesis of a zeolite imidazolate framework ZIF-100 membrane with high H 2/CO 2 selectivity, *Journal of Materials Chemistry A*, 3 (2015) 4722–4728.

[64] J.-L. Zhuang, A. Terfort, C. Wöll, Formation of oriented and patterned films of metal–organic frameworks by liquid phase epitaxy: A review, *Coordination Chemistry Reviews*, 307 (2016) 391–424.

[65] E. Biemmi, C. Scherb, T. Bein, Oriented growth of the metal organic framework Cu3 (BTC) 2 (H2O) 3⊙ x H2O tunable with functionalized self-assembled monolayers, *Journal of the American Chemical Society*, 129 (2007) 8054–8055.

[66] H. Guo, G. Zhu, I.J. Hewitt, S. Qiu, "Twin copper source" growth of metal– organic framework membrane: Cu3 (BTC) 2 with high permeability and selectivity for recycling H2, *Journal of the American Chemical Society*, 131 (2009) 1646–1647.

[67] Y. Zhao, N. Kornienko, Z. Liu, C. Zhu, S. Asahina, T.-R. Kuo, W. Bao, C. Xie, A. Hexemer, O. Terasaki, Mesoscopic constructs of ordered and oriented metal–organic frameworks on plasmonic silver nanocrystals, *Journal of the American Chemical Society*, 137 (2015) 2199–2202.

[68] P. Falcaro, K. Okada, T. Hara, K. Ikigaki, Y. Tokudome, A.W. Thornton, A.J. Hill, T. Williams, C. Doonan, M. Takahashi, Centimetre-scale micropore alignment in oriented polycrystalline metal–organic framework films via heteroepitaxial growth, *Nature Materials*, 16 (2017) 342–348.

[69] P.-C. Guo, T.-Y. Chen, X.-M. Ren, Z. Chu, W. Jin, A guest-dependent thermochromic feature in a metal–organic framework and its thin film on different supports, *Journal of Materials Chemistry A*, 2 (2014) 13698–13704.

11 Industrial Applications of Anticorrosive Metal-Organic Frameworks-Based Coatings

Gamze Tüzün, Burak Tüzün, Dilara Ülger Özbek, and Elyor Berdimurodov

11.1 INTRODUCTION TO METAL-ORGANIC FRAMEWORKS (MOFS): STRUCTURE AND PROPERTIES

Metal–organic frameworks (MOFs) are one of the advanced materials that have attracted great attention in the last few decades and have opened the door to innovative applications in various fields [1]. MOFs are porous structures formed by the combination of metal ions or metal clusters and organic ligands. These materials exhibit unique properties due to their flexibility in chemical structures and high surface areas. The history of MOFs, their chemical and physical properties, classification and design criteria for industrial applications are discussed in detail below. When the studies are examined in detail, it is seen that polymerizable surfactants make the field of corrosion control attractive and allow their widespread use in the protection of metal surfaces, pipes, ship hulls, and industrial equipment [2–4]. There are many previous studies on corrosion prevention; these studies are generally conducted at the molecular level [5–8]. It has been seen that researchers have started to work with nanomaterials to design more effective and more active molecules [9–12].

11.1.1 History and Development of MOFs

The history of MOFs is based on coordination chemistry. The first coordination compound was identified as the iron(III) hexacyanoferate complex at the end of the 18th century [13]. However, the modern definition of MOFs emerged in the 1990s [14]. In 1995, Yaghi and his colleagues synthesized the first MOFs in the modern sense and characterized the crystal structure of these materials. This structure was named MOF-5 and showed how coordination chemistry can be used for the design of porous materials.

Since the 2000s, studies in the field of MOFs have gained momentum. New ligand designs and synthesis methods have made it possible to further optimize both the chemical and physical properties of these materials [15]. During this period, iconic

 DOI: 10.1201/9781003571162-11

structures such as MOF-74, UiO-66, and ZIF-8 have emerged [16]. These stages in the historical development of MOFs have opened the door to new applications in a wide variety of fields such as materials science, catalysis, gas storage, and separation.

11.1.2 Chemical and Physical Properties of MOFs

One of the most striking features of MOFs is their large surface areas. In some MOF structures, surface areas can reach extreme values such as 7000 m^2/g. This high surface area makes MOFs ideal materials for gas storage and separation processes.

Another important feature of MOFs is the flexibility in their porous structures. Pore sizes and shapes can be easily adjusted by the bonding geometry between metal ions and ligands. This customizability allows for the targeted design of MOFs.

Chemical stability is another important feature that increases the application potential of MOFs. For example, Zr(IV)-based MOFs are known for their resistance to acidic and basic conditions [17]. In addition, the thermal stability of MOFs is a factor that expands their application areas; some structures can maintain their stability up to 400°C.

MOFs can contain chemically reactive centers [18]. This allows them to be used as catalysts. Metal centers and ligands directly contribute to the reaction mechanisms, making MOFs suitable for both homogeneous and heterogeneous catalysis applications.

11.1.3 Classification of MOFs: Typical Examples

MOFs are usually classified according to the variety of metal centers and ligands. Here are some of the common classifications:

Zirconium-Based MOFs (Example: UiO-66): They have high chemical and thermal stability. These materials are frequently used for gas storage and catalysis applications [19].

Zinc Imidazolate Frameworks (ZIFs): Structures such as ZIF-8 are known for their hydrophobic properties and chemical resistance. They are widely used in membrane technologies and gas separation applications [20].

Copper-Based MOFs (Example: HKUST-1): They stand out with their high porosity and gas storage capacity. However, they may have limitations in terms of chemical stability [21].

Aluminum-Based MOFs: They are used in adsorption and energy storage applications due to their lightweight structure and environmentally friendly properties [22].

These classifications show the chemical diversity of MOFs and their customizability for different applications.

These classifications show the chemical diversity of MOFs and their customizability for different applications. In addition, the classification of MOFs is presented in Figure 11.1.

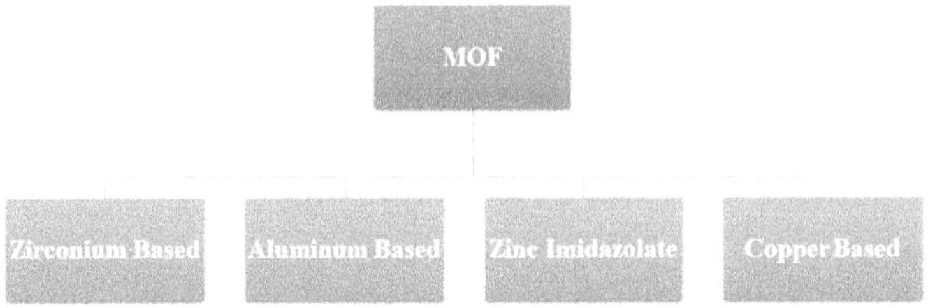

FIGURE 11.1 Classification of MOFs.

11.1.4 MOF Design Criteria for Industrial Applications

The success of MOFs in industrial applications depends on the right design criteria. These are shown in Figure 11.2. Here are some of these criteria:

1. *Stability:* Materials used in industrial processes must be chemically, thermally, and mechanically durable. Therefore, the right metal–ligand combination should be selected depending on the application.
2. *Cost and Scalability:* The materials and methods used in MOF synthesis must be cost-effective. It is also important that they can be easily applied in large-scale production processes.
3. *Porosity and Surface Area:* Depending on the intended application, MOFs may need to have high porosity and surface area. For example, high surface area is a critical factor in gas storage applications.
4. *Functionalization:* The surfaces of MOFs can be chemically modified for a specific application. These modifications can increase the selectivity and activity of MOFs.

These design criteria enable MOFs to be used in a variety of industrial applications, including energy storage, gas separation, catalysis, drug delivery, and sensor technologies in Figure 11.3 [23–26].

11.2 MECHANISMS OF CORROSION AND THE ROLE OF MOFS IN CORROSION INHIBITION

11.2.1 Thermodynamics and Kinetics of Corrosion

Corrosion is the deterioration of metals as a result of chemical or electrochemical reactions with environmental factors [27]. Thermodynamically, this process is due to the tendency of the metal to pass to a lower energy level, that is, the oxidized form. Thermodynamically, corrosion is defined by the change in the free energy of the metal. Reactions with a negative free energy change occur spontaneously. Therefore, the corrosion tendency of metals can be evaluated by means of standard electrode potentials. For example, while active metals (Fe, Zn) have a high tendency to oxidize, noble metals such as gold have a high resistance to corrosion.

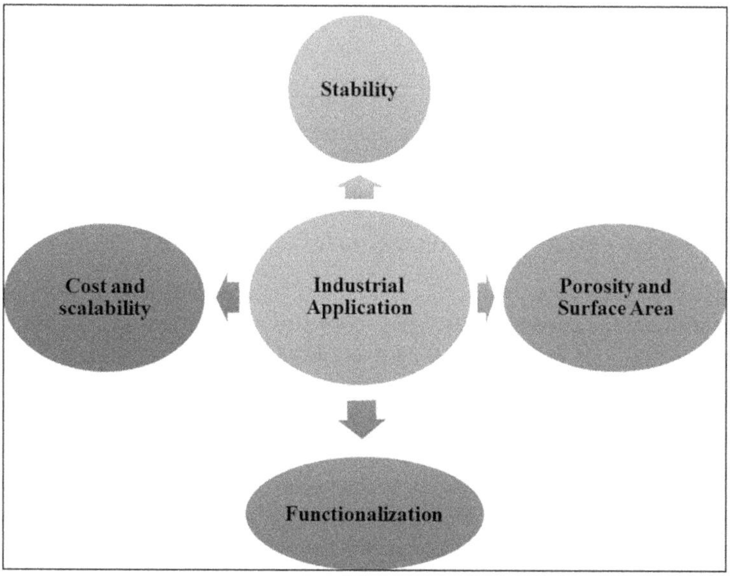

FIGURE 11.2 Industrial field applications of MOFs.

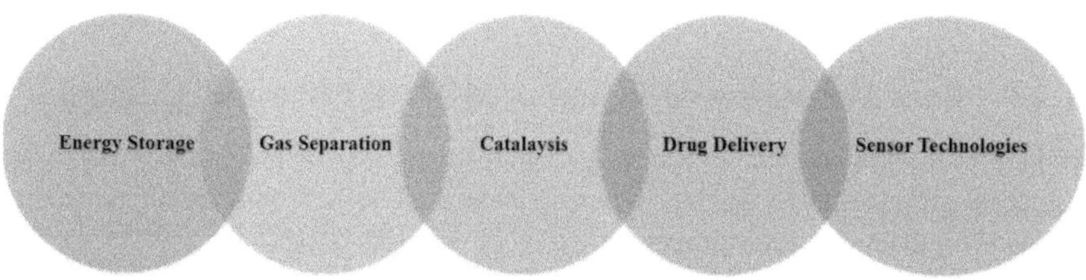

FIGURE 11.3 Areas in which MOFs are used.

Kinetically, corrosion is defined by the anode and cathode reactions that occur on the metal surface [28–31]. In anodic reactions, metal atoms lose electrons and become ionic:

$$M \rightarrow M^{n+} + ne^-$$

Cathodic reactions can occur in various ways, such as the reduction of hydrogen ions or the reduction of oxygen in the presence of water, depending on the properties of the medium:

$$2H^+ + 2e^- \rightarrow H_2$$

$$O_2 + 4H^- + 2e^- \rightarrow 2H_2O$$

The rates of these processes are governed by the properties of the protective surface layer, the characteristics of the solution environment, and the electrochemical polarization conditions.

11.2.2 Solvent Effects and Electrochemical Reactions

The solvent medium directly affects the rate and mechanism of corrosion. Aqueous environments are the environments where corrosion is most frequently observed due to the conductivity of ions. Factors such as pH of the solution, ionic strength, and oxygen content are critical in determining the corrosion mechanism [32].

In acidic environments, high concentrations of hydrogen ions on the metal surface can accelerate corrosion reactions resulting in the release of hydrogen gas. In neutral or basic environments, reduction of oxygen occurs as the main reaction [33]. Some aggressive ions, such as chloride ions in the solution, can disrupt the protective passive layer and lead to localized corrosion types such as pitting.

Electrochemical reactions occur in the form of two half-cells on the metal surface. An electrical connection is established between the anodic and cathodic reactions, creating a galvanic cell mechanism [34]. This process causes potential differences to form in different regions of the metal, which can lead to more complex processes such as galvanic corrosion.

11.2.3 Basic Mechanisms of Action of MOFs in Preventing Corrosion

MOFs are of great interest as a new and innovative class of materials in corrosion protection applications [35]. The basic mechanisms of MOFs in preventing corrosion are based on their structural properties and chemical functionalities.

1. *Creating a Physical Barrier:* MOF coatings prevent direct contact of environmental agents (such as water, oxygen, and ions) with the metal surface by creating a physical barrier on the metal surface. These coatings can limit the passage of water and gases because of their porous structures.
2. *Release of Corrosion Inhibitors:* Some MOFs can store corrosion inhibitors in their structural pores and release them in a controlled manner, allowing the renewal of a protective layer on the metal surface. For example, the release of inhibitors such as benzotriazole from the MOF pores can prevent pitting corrosion on the metal surface.
3. *Modification of Electrochemical Reactivity:* MOFs can slow down anodic and cathodic reactions by promoting the formation of a passive layer on the metal surface. In addition, metal centers in MOF structures can act as catalytic centers that control the rate of electrochemical reactions.
4. *Retention of Aggressive Ions:* MOF structures can selectively retain aggressive ions such as chloride ions in solution. They prevent the interaction of these ions with the metal surface, slowing down corrosive processes.

11.2.4 Interaction of MOF-Based Materials with Reactive Species

The effectiveness of MOFs in corrosion protection mechanisms depends on their interactions with reactive species [36]. Species such as water molecules, oxygen, and chloride ions can affect the chemical stability and performance of MOFs.

1. *Interaction with Water Molecules:* Some MOFs with hydrophilic properties can undergo structural deformation by interacting with water molecules. However, this problem can be overcome with hydrophobic modifications and the resistance of MOFs to water can be increased.
2. *Oxygen and Redox Reactions:* MOFs can control electrochemical processes by entering into redox reactions with oxygen molecules. This mechanism is particularly effective in slowing down cathodic reactions.
3. *Interaction with Chloride Ions:* Chloride ions can be trapped in the pores of MOFs or disrupt ligand–metal bonds. The detrimental effects of these interactions can be minimized by stable ligand design.

11.3 METHODS USED IN MOF SYNTHESIS

MOFs are coordination polymers formed by the combination of metal ions or metal clusters and organic ligands. The synthesis methods of these materials play a critical role in determining their chemical properties and application areas [37, 38]. MOF synthesis methods are categorized in Figure 11.4.

11.3.1 The Main Methods Used for MOF Synthesis

1. *Solvothermal Synthesis:* This is the most widely used method in MOF synthesis. The metal salt and ligand in the solvent are reacted under high temperature and pressure in a closed system. This method allows MOFs with high crystallinity and porosity to be obtained. The solvent selection, temperature, and time determine the properties of the product [39].
2. *Microwave-Assisted Synthesis:* MOF synthesis can be done quickly and energy-efficiently using microwave radiation. This method significantly shortens the reaction time and provides a more homogeneous structure of the product. The microwave method is preferred especially in industrial applications to increase scalability [40].
3. *Ultrasonic-Assisted Synthesis:* Ultrasonic energy accelerates the interactions between metal ions and ligands, enabling the formation of MOFs in a shorter time. This method reduces energy costs due to its ability to operate at low temperatures [41].
4. *Electrochemical Synthesis:* In this method, where metal ions are released into the solution under an electrostatic potential, high-purity MOFs are obtained. The electrochemical method is used in the production of MOF coatings, which are particularly suitable for corrosion protection applications [42].
5. *Synthesis at Room Temperature:* As an environmentally friendly method, reactions performed at room temperature reduce solvent use and energy

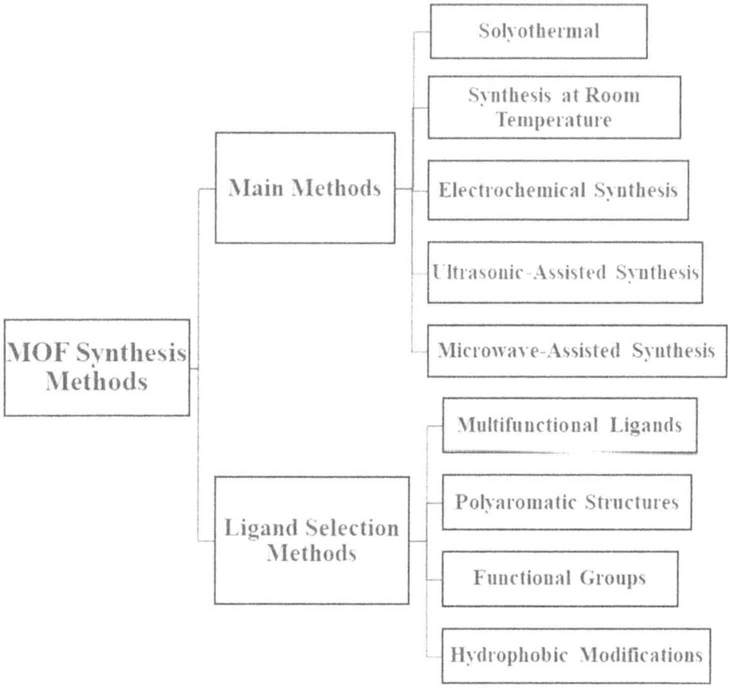

FIGURE 11.4 MOF synthesis categorization.

consumption. This method is one of the preferred approaches in sustainable MOF production [43, 44].

11.3.2 Ligand Selection in Functional MOF Design

Ligands are critical components that determine the structural stability, pore sizes, and chemical properties of MOFs. Ligand selection in MOFs designed for corrosion protection directly affects the performance of the material.

1. *Functional Groups:* Ligands containing functional groups that can interact with corrosion inhibitors or form a protective layer are preferred. For example, carboxylic acid and amine groups can form strong bonds with the metal surface.
2. *Hydrophobic Modifications:* Hydrophobic groups added to ligands increase the resistance of MOFs to water. This feature plays an important role in preventing corrosion.
3. *Polyaromatic Structures:* Polyaromatic ligands provide extra stability to the MOF structure and form stronger coordination bonds with metal ions. Such ligands are used in the design of MOFs that are particularly resistant to harsh conditions.

4. *Multifunctional Ligands:* Ligands containing more than one functional group can increase both the physical properties and chemical reactivity of MOFs. For example, a ligand can both form a protective coating and release active corrosion inhibitors.

11.4 PRODUCTION OF HYBRID AND COMPOSITE MOF STRUCTURES

Hybrid and composite MOF structures are multi-component systems that offer superior performance compared to traditional MOFs [45, 46]. These structures combine the advantages of different materials to provide more effective solutions for corrosion protection applications.

11.4.1 MOF–Polymer Composites

MOFs can be combined with polymer matrices to obtain materials with increased mechanical strength and chemical stability. Polymers fill the pores of the MOF, preventing the entry of aggressive ions and creating a better physical barrier.

11.4.2 MOF–Nanoparticle Hybrids

MOFs combined with metal nanoparticles or oxides increase surface reactivity and catalytic properties. For example, MOFs functionalized with silver nanoparticles can offer both corrosion protection and antimicrobial properties.

11.4.3 MOF–Ceramic Composites

MOF structures integrated with ceramic materials provide superior protection against corrosion, especially in harsh environments such as high temperatures and acidic conditions.

11.4.4 Multilayer MOF Structures

Multilayer structures, where more than one MOF type is used together, provide corrosion protection with different mechanisms. For example, one layer can provide inhibitor release while the other layer can function as a physical barrier.

11.5 SCALABILITY OF CORROSION-RESISTANT MOFS

The success of MOFs in industrial applications depends on the possibility of their large-scale production [47]. Scalability includes the cost-effectiveness, time-efficiency, and environmental sustainability of MOF synthesis methods.

11.5.1 Low-Cost Raw Materials

The cost of metal salts and ligands used in MOF production is an important factor in terms of scalability. MOFs produced from common and inexpensive materials are more preferred in commercial applications.

11.5.2 Energy Efficiency

Energy-efficient synthesis methods such as room temperature or microwave-assisted methods can be used to reduce energy costs. These methods also reduce the environmental footprint.

11.5.3 Green Chemistry Approaches

Methods that reduce or completely eliminate solvent use contribute to the sustainable production of MOFs. For example, water-based solutions or solvent-free reactions offer environmentally friendly options.

11.5.4 Automation and Mass Production

Automation technologies and continuous flow reaction systems can be used to increase the scalability of MOF production. These approaches provide more homogeneous products and increase production speed.

11.6 MOFS-BASED COATINGS: FABRICATION AND PERFORMANCE ANALYSIS

The MOFs refer to crystalline structures that have a high porosity and a large surface area. These structures are generated when metal ions or metal clusters are combined with organic ligands. Because of these features, MOFs have evolved into the subject of investigation in a variety of academic and industrial studies during the past several years [48]. The application of MOFs as coating materials presents a significant opportunity to enhance the mechanical, chemical, and physical properties of surfaces. Coatings based on MOFs offer substantial benefits, particularly in areas such as the protection against corrosion, the application of sensors, catalysis, and the storage of energy. This article offers a comprehensive analysis of the production methods, microstructural features, performance standards, and corrosion behavior of MOF coatings when subjected to the influences of the environment [49].

To ensure uniform application of the coating material and strong adhesion to the substrate surface, the fabrication of MOF coatings requires the use of various deposition techniques. Among these, the sol–gel method is one of the most widely employed approaches [50]. This technique is based on the controlled interaction of metal ions and organic ligands in solution, leading to the formation of a gel-like network structure. The sol–gel method has gained considerable attention because it operates at relatively low temperatures and enables the production of homogeneous and uniform coatings. In addition, electroplating is another commonly used method, in which the coating material is deposited onto the substrate surface through the application of an electric current. This approach is particularly advantageous for achieving uniform coating thickness and strong surface adhesion.Chemical vapor deposition, sputter coating, and dip coating are some of the other manufacturing methods that can be utilized [51]. The production process known as "dip coating" is typically characterized by its low cost and rapidity. This method involves coating the surface of the substrate by dipping it in an MOF solution. Sputter coating is achieved by spraying the MOF solution onto the surface of the substrate. This process makes it possible to coat intricacies of the substrate's geometry. The process of chemical vapor deposition involves depositing the coating material from the gas phase onto the surface of the substrate. This method often results in a coating that is of high purity and gives a regulated coating thickness. Each approach is chosen according to the particular application regions and coating needs that are being considered [52].

To develop a comprehensive understanding of the performance of these materials, it is essential to characterize the microstructural features of MOF coatings. Scanning electron microscopy (SEM), transmission electron microscopy (TEM), X-ray diffraction (XRD), and Fourier transform infrared (FTIR) spectroscopy are some of the techniques that are utilized for this purpose [53]. SEM offers comprehensive information regarding the surface morphology and roughness of coatings. By doing a more in-depth investigation of nanostructural characteristics, TEM is able to visualize the internal structure of the coating. Information regarding the degree of porosity and crystallinity can be obtained using XRD, which is used to determine the crystalline structure of coatings. In addition to providing information on the chemical composition of the coating material, FTIR spectroscopy studies the coat's chemical bonds and functional groups.

In the process of determining the performance of MOF coatings, mechanical strength is one of the essential parameters that is considered. The mechanical strength of coatings is tested using techniques such as nanoindentation tests, which are used to establish whether or not the coatings are suitable for long-term use [50]. The term "scratch resistance" refers to the capacity of coatings to withstand the effects of physical impacts, whereas adhesion tests measure the degree to which the coating will adhere to the surface of the substrate. The thermal stability of coatings is another significant parameter; this quality determines whether or not the coating is able to maintain its structural integrity even when subjected to conditions of high temperature. There is a clear correlation between the durability and adhesion capabilities of MOF coatings and the performance and longevity of coatings, particularly in industrial applications [51].

The protection against corrosion is one of the most significant applications for MOF coatings. Because of their porous nature, MOFs are able to safeguard metal substrates by facilitating the easy settling of corrosion inhibitor molecules on the surface. Nevertheless, environmental conditions have the potential to influence corrosion performance. Certain conditions, including variations in temperature, humidity, pH, and atmospheric gases, are essential elements that play a significant role in determining the coating's effectiveness over an extended period of time. The investigation of these effects is carried out through the utilization of techniques such as electrochemical impedance spectroscopy (EIS) and potentiodynamic polarization experiments. While electrochemical ionization (EIS) is used to evaluate the electrical characteristics and resistance to corrosion of coatings, potentiodynamic polarization tests are used to investigate the pace of corrosion and the mechanisms that protect the coating over time [52].

The methods that protect MOF coatings from corrosion may differ from one another based on the chemical structure of the material and the environmental variables that are present. To prevent metal ions from coming into touch with the surrounding environment, porous structures provide the function of a passive barrier. A further benefit of the coatings is that they contain inhibitor molecules, which not only offer active protection but also slow down the corrosion process. When deciding whether or not coatings are suitable for use in industrial applications, one of the most significant factors to consider is their resistance to environmental conditions.

11.7 ENVIRONMENTAL AND ECONOMIC BENEFITS OF MOFS IN INDUSTRIAL CORROSION CONTROL

MOFs provide a unique approach to the control of corrosion in industrial settings, both in terms of their environmental sustainability and their economic benefits [53, 54]. MOF-based coatings, in comparison to more conventional approaches, present tremendous opportunities in terms of both the environment and the economy. As MOFs have a high porosity, chemical properties that can be adjusted, and low environmental implications, they stand out as an excellent choice for applications that require corrosion protection. This article provides an in-depth discussion of the environmental and economic benefits of MOFs, as well as a comparison of MOFs to conventional corrosion protection methods, an analysis of their economic potential, and a prospective use of MOFs from a sustainability point of view [55].

When compared to conventional coating and inhibitor systems, MOF materials are distinguished by their superior environmental friendliness [56]. The use of these materials has the ability to reduce the amount of harmful metal ions and volatile organic compounds (VOCs) that are utilized. In contrast to conventional techniques of corrosion protection, which typically involve the use of chemicals that are harmful to the environment, MOFs stand out due to the ecologically favorable architectures that they possess. Molybdenum disulfide (MOF) materials are excellent for both passive and active protection methods due to their high surface area and their ability to be chemically modified [57, 58]. Through the use of the porous structure of MOF coatings, passive protection can be achieved by keeping metal surfaces from coming into touch with environmental elements. MOFs are designed to provide active protection through the regulated release of corrosion inhibitors. By avoiding the release of traditional inhibitors in an unneeded manner, these mechanisms not only offer protection over an extended period of time but also lessen their influence on the environment.

In the process of contrasting conventional techniques of corrosion prevention with systems based on MOFs, it becomes clear that MOFs possess a multitude of benefits. The conventional approaches often involve the application of coating systems that make use of heavy metals such as zinc, chromium, and others. These metals have the potential to inflict significant damage to both the environment and human health. In addition, the majority of the time, these coatings are sprayed using solvent-based systems that contain VOCs, which results in the emission of toxic gases into the atmosphere [59, 60]. With MOF-based coatings, on the other hand, it is possible to apply them using formulations that are either water-based or solvent-free, which dramatically decreases their impact on the environment. It is possible to construct MOFs so that the organic ligands and metal clusters that are housed within them have qualities that are both non-toxic and biodegradable. Consequently, coatings based on MOFs achieve both compliance with environmental requirements and the creation of safer working environments.

To determine whether or not these cutting-edge materials are suitable for use on a large scale in industry, it is essential to conduct an economic study of MOF systems. In general, traditional corrosion protection systems are appealing due to their low starting costs; nevertheless, the expenses of maintenance and replacement can be rather significant over the course of their lifetime [54, 55]. Corrosion can cause damage in industrial buildings, which can result in downtime, the loss of equipment, and significant repair expenses. As MOF-based coatings have structures that are both long lasting and durable, they have the potential to significantly cut expenses of this kind. In addition, the regulated inhibitor release feature of MOFs helps optimize inhibitor consumption, which in turn minimizes the costs of manufacturing chemical materials [53, 55]. There is also the possibility of making the manufacturing processes of MOF coatings more energy efficient, which would result in a reduction in the costs of production. Despite the fact that the initial investment costs of MOF systems are high, the fact that these materials have a low impact on the environment and produce savings over the long run makes them economically appealing.

When considering the potential applications of these materials in the future, it is of utmost importance to evaluate MOF-based systems from the perspective of sustainability. To achieve sustainability, it is necessary to conduct oneself in accordance with the ideals of social responsibility, economic efficiency, and environmental protection. On account of the ecologically beneficial features that they possess, MOFs make a contribution to all three of these areas. When it comes to the preservation of the environment, MOFs offer a number of benefits, including the minimization of the use of harmful chemicals and the production of a low carbon footprint [61]. The lifetime of MOF systems, in addition to their minimal maintenance costs, is a factor that contributes to their economic viability. MOF-based coatings lead to safer and healthier working conditions, which in turn improves the health and safety of workers. This contributes to the social sustainability of the industry.

As a consequence of this, MOFs provide a breakthrough solution in the field of industrial corrosion management by combining the advantages of the environment and the economy. These cutting-edge materials have the potential to be utilized in a wide variety of industrial applications. In comparison to conventional approaches, they have a smaller impact on the environment, a better level of efficiency, and cheaper costs over the long run. The creation and distribution of coatings based on MOFs not only help to protect the environment and save money, but also make a contribution to a future that is more sustainable. The application of MOFs in the management of corrosion in industrial settings stands out as an essential component of research in materials science that is focused on sustainability.

11.8 APPLICATIONS OF MOFS IN OIL AND GAS INDUSTRY CORROSION PROTECTION

MOFs offer an innovative approach to the problem of corrosion control in the oil and gas industry. To guarantee the dependability of infrastructure, environmental sustainability, and economic efficiency, the industry is continuously looking for technology that prevent corrosion. Oil pipelines, hydrocarbon processing facilities, and offshore platforms are examples of critical infrastructures that are susceptible to corrosion as a result of corrosive ambient conditions and chemical attacks [62, 63]. Because of their exceptionally high porosity, huge surface area, chemical stability, and the capacity to be functionalized, MOFs are an efficient tool for the control of corrosion. Corrosion issues in oil pipelines, the function of MOF-based coatings in the hydrocarbon cycle, the effectiveness of MOFs against corrosive gases like H_2S and CO_2, and the development of specialized MOF systems for deep-water applications are some of the topics that will be covered in depth in this article.

Pipelines that transport oil are the essential component of the global energy infrastructure. These pipelines, on the other hand, are regularly subjected to a wide variety of different elements that induce corrosion. Acid gases such as hydrogen sulfide and carbon monoxide, high pressure and temperature, the electrochemical effects of saltwater, and biocorrosion induced by microorganisms are all examples of these effects [64, 65]. The presence of chloride ions causes metals to corrode at a rapid rate, particularly in pipes that are located beneath the ocean. The evolution of corrosion can result in cracks and, eventually, leaks in pipes, which can lead to environmental catastrophes as well as significant economic losses. Epoxy coatings, galvanic protection, and chemical inhibitors are some of the traditional approaches that are utilized in the prevention of corrosion. On the other hand, these techniques do not offer long-term durability and are less than ideal in terms of their impact on the environment. MOFs provide a novel way to resolving these issues.

Coatings made of MOFs offer both passive and active protective measures by taking on the crucial function of oil pipelines in the fluid hydrocarbon cycle [66]. Through the use of MOF coatings, passive protection is able to create a physical barrier that prevents environmental elements from coming into contact with the surface of the metal. Because of their porous nature, MOFs are capable of forming a persistent layer of protection at both the molecular and macroscopic scales [67]. On the other hand, active protection is accomplished by the process of the controlled release of corrosion inhibitors that are contained within MOFs. These inhibitors are designed to target the chemical reactions that are responsible for corrosion, thereby avoiding the deterioration of metals. MOF coatings have dual-protective qualities, which allow them to extend the life of pipelines while simultaneously reducing the expenses of maintenance. The capacity of the organic ligands that are present in MOFs to bind with hydrocarbon molecules is another factor that contributes to the increased efficiency of chemical processes.

It is one of the most important dangers that oil and gas pipelines face when they are exposed to gases like carbon monoxide and hydrogen sulfide. The presence of H_2S in pipes results in the production of sulfuric acid, which accelerates the corrosive process of metals. Additionally, the presence of this gas results in the toxicity of the environment as well as the failure of equipment [68]. Corrosion in pipeline materials is caused by the formation of carbonic acid, which is produced when carbon monoxide reacts with water. MOFs provide a novel method to resolving this issue, in contrast to conventional coating technologies, which are unable to completely avoid the damage caused by these gases. These gases are able to be selectively collected and neutralized by MOFs due to their high surface area and their chemical tunability. In MOF structures, for instance, the chemical interaction of H_2S gas with metal ions decreases the reactivity of the gas and slows the advancement of corrosion. This is an example of how MOF structures can function. Molybdenum disulfide (MOF) varieties that offer resistance to carbon monoxide (CO_2) are able to absorb this gas and contribute to having a less carbon footprint [69, 70]. These characteristics demonstrate that MOFs not only offer better protection against corrosion but also contribute to the preservation of the environment.

As regards effective corrosion management, the most difficult working environments are those that involve oil and gas extraction operations that take place in deep seas [71]. A high hydrostatic pressure, low temperatures, aggressive seawater components, and biological activities are some of the variables that contribute to the corrosive nature of these settings. MOFs stand out due to their resilience, in contrast to standard coating materials, which are unable to provide long-term protection under these conditions. To be more specific, certain forms of MOFs offer long-term protection to metals by demonstrating chemical stability in the presence of seawater. Additionally, the antibacterial capabilities of MOFs make pipelines more durable by avoiding biocorrosion brought on by microbes. This is accomplished by limiting the growth of bacteria. The MOF systems that were designed for deep-sea applications provide protection for a longer period of time and require less maintenance in these settings with lower requirements. When it comes to guaranteeing the safety of offshore platforms and pipelines, these characteristics constitute a significant factor.

The oil and gas industry can benefit from the use of MOFs as they provide a solution that is not only effective but also sustainable and innovative. They help reduce the impacts of gases like H_2S and CO_2 and contribute to the construction of systems that are resistant to deep-sea conditions. Additionally, they contribute to the solution of corrosion problems in oil pipelines, boost the efficiency of the hydrocarbon cycle, and decrease the effects of these gases [72, 73]. Achieving safety, durability, and environmental sustainability in the energy industry can be accomplished through the optimization of the chemical and physical properties of MOFs. The development and widespread use of coatings based on MOFs are regarded as the most important factors in ensuring a more secure and productive future for the oil and gas sector to come. Increasing the use of MOFs will be a significant step toward more environmentally responsible protection of the energy infrastructure.

11.9 MARINE AND OFFSHORE APPLICATIONS OF MOFS-BASED CORROSION INHIBITORS

MOFs are gaining popularity because of their remarkable versatility and sophisticated nature [74]. These materials provide novel approaches to the problem of corrosion management in the marine and offshore industries. As a result of the high salt concentration, dissolved oxygen content, biological activities, and mechanical corrosive effects that are present in seawater environments, metals are exposed to an environment that is exceedingly hostile. By creating economic losses, safety difficulties, and environmental dangers, corrosion poses a severe threat to marine infrastructures such as ships, offshore platforms, subsea pipelines, and other marine infrastructures [75]. Traditional techniques of corrosion prevention not only are ineffective in these complicated environmental conditions, but also have shortcomings in terms of their potential to be sustainable. Because of their high porosity, broad surface area, chemically adjustable architectures, and excellent stability features, MOFs provide a new answer to the problems that arise in marine settings.

The mechanisms that cause corrosion in saltwater are highly complicated, and they often begin with electrochemical reactions on the surfaces of metal molecules. Pitting and local corrosion are both produced as a result of the presence of chloride ions in seawater, which damage the passive protective layers of metals. Stainless steel and aluminum are two examples of materials that are particularly vulnerable to the dangers posed by these ions [76, 77]. The acceleration of cathodic and anodic processes is what causes corrosion to occur when dissolved oxygen is present. While this is happening, microorganisms in seawater are responsible for biological corrosion and they also encourage the creation of biofilm on metal surfaces. It is not just that these biofilms enhance the rate of corrosion, but they also diminish the efficiency of coatings that are applied to metal surfaces. There are a number of factors that contribute to the complexity of the corrosion process in marine environments. These include changes in temperature, fluctuations in pH, and hydrostatic pressure [78]. MOFs provide both physical and chemical protection, making them a crucial solution for marine applications. This protection is necessary to meet the myriad of corrosion mechanisms used in marine environments.

MOF systems, which are utilized in the maritime and offshore industries, act as a coating on metal surfaces that is both protective and functional. As a result of the porous structure of MOFs, environmental elements are prevented from coming into contact with the surface of the metal, which results in the formation of a physical barrier. Metal surfaces are protected from the corrosive effects of the seawater environment by these coatings, which operate as a shield for the metal surfaces. Chemical protection can also be provided by corrosion inhibitors that can be inserted inside MOFs. This is made possible by controlled release mechanisms. Through the slowing down of electrochemical reactions on metal surfaces, these inhibitors bring about a reduction in the rate of corrosion. The fact that MOF-based coatings are compatible with the pipelines and equipment that are utilized in the hydrocarbon processing procedures is yet another advantage of these systems [79]. As regards the marine industry, the implementation of MOF systems not only offers corrosion control, but also guarantees a longer life for the equipment and lowers the expenses associated with its maintenance.

Regarding marine and offshore applications, the performance of MOF coatings is strongly impacted by their capacity to remain stable in salty climatic conditions. However, the chemical stability of MOFs gives improved endurance in these hostile settings, despite the fact that chloride ions in marine environments have the ability to swiftly damage the majority of metal coatings. For instance, MOF types that are resistant to water might have surface qualities that are hydrophobic, which prevent water and ions from penetrating metal surfaces [80]. Further, the mechanical robustness of MOF coatings makes it possible for them to withstand the physical abrasion that occurs in conditions that contain seawater. It is possible to manufacture and perfect these coatings such that they can withstand the impacts of salty environments for extended periods of time. It is clear from these characteristics that coatings based on MOF are an effective method for preventing corrosion caused by seawater.

In maritime environments, the process of biological corrosion is one that is induced by microorganisms that results in the breakdown of metal surfaces resulting in corrosion. MOF coatings that include anti-biofouling qualities have been developed expressly for addressing this problem. This prevents the formation of biofilms by inhibiting the growth and adhesion of microbes to surfaces, which is accomplished by the use of metal ions or organic ligands in MOF coatings. As an illustration, MOF coatings that contain silver or copper have antibacterial capabilities, which makes them an efficient defense mechanism against microbes [81]. In addition, the ecologically favorable features of these coatings make them suitable for use in applications that are sustainable in the marine industry. MOF coatings that possess anti-biofouling qualities are utilized on ship hulls, offshore platforms, and subsea pipelines. These coatings offer significant protection against biological deterioration over an extended period of time.

MOFs provide a solution that is both comprehensive and sustainable for the management of corrosion in the marine and offshore industries. MOF coatings offer a multitude of advantages, including long-term stability in salty settings, protection against biological corrosion, and economic advantages [82]. These coatings are designed to target corrosion mechanisms in seawater for a variety of reasons. A significant step toward enhancing the long-term viability of energy infrastructure and the maritime industry is represented by the fact that these materials have a widely applicable range of applications. Continuing research and development of coatings based on MOFs has the potential to establish a new benchmark in the marine and offshore industries in the months and years to come. The revolutionary features of these materials will not only give benefits to the economy and the environment, but also generate a paradigm change in the way that corrosion control is approached in the future.

11.10 ADVANCED CHARACTERIZATION TECHNIQUES FOR MOFS IN CORROSION STUDIES

The MOFs refer to porous materials that provide a new solution in the field of corrosion protection and have a wide variety of uses. MOFs have a wide range of chemical, physical, and electrochemical properties, and to understand and maximize these properties, several advanced characterization approaches are necessary. An extensive spectrum of information, ranging from the structural properties of MOFs to their performance assessments and environmental durability, can be obtained through the utilization of these approaches. A number of techniques, such as XRD, spectroscopic analysis, EIS, SEM, TEM, thermal and mechanical tests, and spectroscopic analysis, are utilized to conduct in-depth investigations into the structural and functional properties of MOFs [83–85]. To improve the efficiency of MOFs in corrosion control applications, these aforementioned methodologies offer a thorough understanding.

To gain a better understanding of the crystal structures and chemical compositions of MOFs, spectroscopic techniques and XRD techniques are utilized extensively. For determining the crystallographic characteristics of MOFs, XRD is an essential instrument. Utilizing XRD, one may verify the regular structure of MOFs after they have been synthesized, as well as analyze the effects of various synthesis methods on the structural properties of the MOFs [83]. In the presence of corrosion, XRD is also utilized for monitoring the structural changes that occur in MOFs. For instance, XRD can be used to investigate how the crystal structure of MOFs changes after sustained exposure to a saline environment [85]. To conduct a comprehensive investigation of the chemical bonds and constituents of MOFs, spectroscopic analyses offer supplementary facts and information. A significant contribution that FTIR spectroscopy makes is the identification of the chemical interactions that exist between the organic ligands and the metal centers of MOFs [84]. With regard to determining the chemical stability of MOFs in an environment that is prone to corrosion, FTIR analysis can be utilized. It is also possible to conduct a more in-depth analysis of the molecular structure and bonding states of MOFs through the utilization of Raman spectroscopy.

Electrochemical impedance spectroscopy, also known as EIS, is a sophisticated method that is utilized to assess the effectiveness of MOFs in preventing corrosion. EIS is utilized for the purpose of evaluating the efficiency as well as the electrochemical characteristics of the barrier layer that is created by MOF coatings on metal surfaces [83]. This method examines the resistance and capacitance properties of MOFs after they have been exposed to a corrosive environment to determine the protective performance of these materials. Moreover, EIS is utilized to assess the deterioration that MOF coatings experience over the course of time. For instance, EIS can be used to investigate the inhibitor release rates as well as the efficiency of MOFs that include corrosion inhibitors. Not only that, but EIS is also utilized for the purpose of monitoring the dynamics of chemical reactions that occur on the surfaces of MOFs. MOF coatings can be optimized in terms of their long-term durability and performance with the help of this technology, which is an essential instrument.

The examination of the morphological and microstructural features of MOFs is accomplished through the utilization of modern imaging techniques such as SEM and TEM [84, 85]. For analyzing the surface topography of MOF coatings as well as the changes that take place on the surface as a result of corrosion, SEM is utilized. For instance, SEM can be used to investigate in great detail the surface uniformity, crack formation, and corrosion damage of MOF coatings [84]. To test the bonding capabilities of MOF coatings with metal surfaces, SEM is also utilized. The TEM technique is used to investigate the structural characteristics of MOFs on a nanoscopic scale. TEM can be utilized to investigate the pore diameters, metal–ligand bonding, and crystalline features of MOFs. TEM is also utilized for assessing the structural integrity of MOFs following their synthesis as well as the changes that occur in the corrosive environment. For optimizing the efficiency of MOFs in corrosion control applications, SEM and TEM are able to investigate MOFs at both the macro and nanoscales.

To gain a long-term understanding of the performance of these materials in corrosive settings, it is essential to evaluate the thermal and mechanical properties of MOFs. MOFs can be evaluated for their thermal stability through the use of a technique known as thermogravimetric analysis (TGA) [86]. By evaluating the temperatures at which MOFs decompose and their thermal stability, TGA is able to assess the endurance of MOFs in high-temperature environments [86]. To have an understanding of whether or not MOF coatings are suitable for extreme environmental conditions, it is vital to have this knowledge. To investigate the thermal behavior of MOFs and the phase changes that occur in them, differential scanning calorimetry (DSC) is utilized. Methods such as microhardness tests and tensile testing are utilized to assess the mechanical durability of MOF coatings during the evaluation process [87]. To have a better understanding of the wear resistance and physical endurance of MOF coatings, several tests are utilized [84]. Mechanical testing are an essential component in the process of enhancing the overall performance and durability of MOF coatings over the long term.

Advanced characterization techniques are an essential component in the process of optimizing and enhancing the efficiency of MOFs in corrosion control applications. To gain a comprehensive understanding of the structural and chemical properties of MOFs, XRD and spectroscopy investigations are necessary. On the other hand, SEM and TEM are useful methods for analyzing the morphological and microstructural features of these materials in greater detail [83, 84]. Electrochemical performance of MOF coatings can be evaluated using EIS. To gain an understanding of the environmental durability and physical properties of MOF coatings, it is required to conduct thermal and mechanical investigations. The integration of various methodologies results in a comprehensive understanding, which is essential for enabling the utilization of MOFs in the field of corrosion control in a manner that is both more widespread and more successful. The results of these investigations on the characterization of MOFs make it possible to establish new criteria for applications involving corrosion prevention and to design solutions that are more environmentally friendly.

11.11 CHALLENGES AND FUTURE DIRECTIONS IN MOFS-BASED CORROSION PROTECTION

MOFs are materials that possess excellent features such as high porosity and functionality. These materials provide the potential to provide unique solutions for applications involving corrosion prevention. The wide-scale usability and effectiveness of these materials, on the other hand, may be constrained by a variety of technological and operational constraints. For the purpose of maximizing the potential of MOF coatings in the field of corrosion protection and ensuring their adoption in industrial applications, it is necessary to solve the issues that are now being faced and to establish strategies that are looking forward. In this context, the challenges that are experienced in the application of MOF coatings, the resolution of problems involving solubility and adhesion, the development of new generation MOF materials, and the implementation of these technologies on an industrial scale are of utmost significance.

As for application, one of the most significant issues that MOF coatings confront is the limited environmental endurance and long-term stability that they come with. When it comes to corrosion conditions, where saltwater, high humidity, and temperature variations are frequently observed, the structural sensitivity of MOFs can lead to stability concerns. This is especially true in environments where corrosion occurs. Because of this circumstance, the efficiency of the coatings may be diminished over time, which can have a detrimental impact on the performance of the corrosion protection system [88]. In addition, the application of MOF coatings in a homogeneous manner and the provision of a strong bonding with metal surfaces are both significant technological obstacles. Other significant problems that contribute to the difficulty of incorporating these materials into large-scale commercial applications include the complexity of the synthesis process and the high cost of MOFs [89].

The resolution of issues pertaining to solubility and adhesion is an essential stage in the process of enhancing the efficiency of MOF coatings and expanding the scope of their application territories. The solubility of the material as well as its structural integrity can be impacted by the interaction of MOFs with water and other solvents. MOF coatings are susceptible to deterioration as a result of this issue, particularly in situations with high levels of humidity. The development of MOFs that are either water-resistant or hydrophobic is of utmost relevance to find a solution to this challenge [90, 91]. In addition, the ability of MOF coatings to adhere strongly to metal surfaces is an essential prerequisite for enhancing the effectiveness of corrosion protection coatings. Within this framework, surface modification techniques and hybrid coating procedures have the potential to offer improved integration of MOF coatings over metal surfaces. The combination of MOFs with organic polymers, for instance, has the potential to improve the coatings' longevity while also addressing issues related to solubility and adherence [92, 93].

Developing materials that are of the next generation of MOF has the potential to create solutions that are more effective and efficient in applications that involve corrosion prevention. In this area of research, one of the most important goals is to design materials that, in comparison to conventional MOFs, offer superior levels of chemical stability, thermal resistance, and mechanical strength. For instance, MOFs that possess hydrophobic qualities and are resistant to harsh environmental conditions have the potential to considerably enhance the efficacy of corrosion protection [94, 95]. Additionally, the incorporation of active corrosion inhibitors into the pore structure of MOFs has the potential to further enhance the corrosion protection performance of these materials. With regard to the design of these materials, one of the most essential strategies for developing targeted solutions is one that involves diversifying the functional groups of MOFs and using a variety of metal–ligand combinations. As an illustration, water-resistant zirconium-based MOFs are currently the subject of extensive research in the field of corrosion prevention coatings [96, 97].

An additional significant obstacle that must be overcome before these materials can be widely adopted is the process of making MOF technology applicable on an industrial scale. MOFs are relatively simple to manufacture in laboratory settings; however, when it comes to large-scale production procedures, there are a number of factors that come into play, including price and technological restrictions. Increasing the commercial applicability of these materials requires a number of steps, one of which is the reduction of the costs associated with the manufacturing of MOFs and the development of production processes that are easier. MOF production processes can be made more environmentally friendly by employing approaches such as solvent-free production methods or energy-efficient synthesis procedures, for instance [98, 99]. In addition, the incorporation of MOF coatings into industrial machinery necessitates the optimization of coating technologies and the implementation of an application process that is optimized for maximum efficiency.

To fully exploit the promise of MOF coatings in corrosion protection applications, it is vital to overcome the problems that are now being faced and to establish strategies that are looking forward. The primary objectives in this area are to improve the environmental durability of MOFs, find solutions to difficulties involving solubility and adhesion, design new generation materials, and guarantee that these materials can be applied on an industrial scale. Research carried out in this regard will have a significant impact on the future of corrosion protection materials and will provide solutions that are more environmentally friendly. Providing considerable benefits to both the environment and the economy, the development and implementation of MOF technologies usher in a new era in the fight against corrosion [100, 101]. This is because these technologies offer major advantages.

11.12 CASE STUDIES: SUCCESSFUL INDUSTRIAL APPLICATIONS OF MOFS IN CORROSION PREVENTION

MOFs are becoming widely utilized in a variety of industries as a novel approach to the avoidance of corrosion. Because of their numerous benefits, which include high porosity, resistance to chemicals, and surface qualities that can be customized, these materials can be utilized in a wide variety of applications. The industrial potential of these materials is demonstrated by the fact that MOF-based coatings have been successfully applied in a variety of industries, including the automobile industry, chemical storage, metal production, and military applications. In addition to highlighting the achievements that have been made in the incorporation of MOFs as corrosion protection solutions into industrial-scale applications, these success stories also demonstrate the prospects that exist for expanding the future application areas of this technology [102, 103]. There are many industries in which corrosion resistance is of crucial importance, and one of those industries is the automotive industry. Because metal surfaces of vehicles are continually interacting with environmental variables and chemicals, the likelihood of corrosion occurring is substantially increased. For the automobile industry, MOF coatings stand out as an example of an effective solution for shielding metal surfaces from corrosion [104]. In particular, MOF-based barrier systems are utilized for the purpose of coating the bodywork of vehicles with light metals like zinc and aluminum alloys. These coatings not only provide resistance against environmental elements, but also enhance the material's lifespan. Due to the high porosity of MOF coatings, the incorporation of protective compounds into metal surfaces is made easier [105]. Through this method, a more efficient level of protection can be attained through the regulated release of corrosion inhibitors. In addition, the lightweight structure of MOF coatings makes it possible for them to provide an efficient barrier against corrosion without adding to the overall weight of the vehicles.

With regard to keeping harsh chemicals from causing damage to metal surfaces, the protective systems that are based on MOF that are utilized in chemical storage tanks contribute significantly. The majority of the time, chemical storage tanks are used in settings that are either extremely acidic or extremely basic, which can result in significant corrosion issues on metal surfaces [106]. When confronted with such intense conditions, traditional coating methods do not always provide adequate resistance. Due to the fact that their structures are extremely resistant to chemical reactions, MOFs demonstrate exceptional performance in such challenging situations. It is common practice to apply MOF coatings to the interior surfaces of storage tanks [107]. These coatings not only offer resistance to chemicals but also help reduce material losses that are caused by corrosion. In particular, zirconium-based MOFs are typically selected in applications involving chemical storage due to the strong chemical stability and endurance that they possess. The use of these coatings helps reduce the expenses associated with tank maintenance and increases the continuity of operations.

In addition to the fabrication of metals, another significant industry that makes use of MOFs to improve corrosion resistance is industry manufacturing. In the process of treating and painting metal surfaces, chemical solvents and procedures can create corrosion on the surfaces. This corrosion can be detrimental to the metal. The use of MOF coatings in these processes protects metal surfaces, which results in an increase in product quality while simultaneously lowering manufacturing costs [108]. As an illustration, coatings based on MOF are utilized in the manufacturing of stainless steel to prevent the oxidation of surfaces. Furthermore, the porous nature of MOFs reduces the likelihood of corrosion by facilitating a more seamless integration of coatings with metal surfaces. When compared to conventional coating techniques that are utilized in the fabrication of metals, MOFs offer protection that is more long-lasting and have the added benefit of making coating operations more ecologically friendly. The marine industry is a sector in which metal structures are under constant exposure to a variety of environmental variables, including saltwater, moisture, and other elements [109]. The presence of these surroundings can hasten the process of corrosion and inflict significant damage to the surfaces of metals. Through the provision of long-lasting solutions that are suitable for such extreme conditions, MOF coatings have gained a great deal of success in the marine industry. For instance, coatings based on MOF that are applied to ship hulls and offshore platforms offer improved protection against saltwater, which in turn extends the lifespan of metal surfaces. Additionally, these coatings possess anti-biofouling qualities, which work to prevent marine creatures from attaching themselves to metal surfaces [110]. By minimizing the amount of fuel consumed and the amount of money spent on maintenance, this function helps maintain ship hulls clean. MOFs that are utilized in maritime applications are believed to offer substantial benefits, both economically and environmentally.

These materials have a wide variety of applications, as evidenced by the fact that MOFs have been successfully applied in industrial settings for the purpose of corrosion protection [111]. MOF-based coatings have been shown to be effective and have the potential to be used in a variety of industries, including the automobile industry, chemical storage, metal fabrication operations, and maritime applications. By providing a significant foundation for the future development and deployment of MOF technologies, these applications not only drive a wider adoption of these revolutionary materials but also provide for their development. Moving forward, the use of MOFs as corrosion protection solutions on an industrial scale will continue to offer substantial advantages in terms of both performance and sustainability.

11.13 CONCLUSIONS

The findings of this study have proven that MOFs have a significant amount of potential in the field of corrosion prevention technologies. Because of their excellent porosity, chemical resistance, and usefulness, MOFs are the perfect materials for applications that include the avoidance of corrosion through their characteristics. Many industries, including the oil and gas industry, the maritime industry, and the industrial metals industry, have taken notice of their achievements in coating technology. As a result of its environmental sustainability and economic benefits, MOFs have been demonstrated to be superior to conventional systems that are currently in use. However, for MOF coatings to be applicable on an industrial scale, it is necessary to overcome obstacles such as solubility, stability, and adherence. As for refining the design of MOF-based coatings and maximizing their performance, the utilization of modern characterization techniques is an essential component that plays a crucial role. It is recommended that future research concentrate on the creation of new generation MOFs as well as the expansion of commercial uses for these materials. When it comes to the avoidance of corrosion, it is abundantly obvious that MOF technologies will make a contribution to the environmental and economic sustainability of the project.

REFERENCES

[1] Butova, V. V. E., Soldatov, M. A., Guda, A. A., Lomachenko, K. A., & Lamberti, C. (2016). Metal-organic frameworks: Structure, properties, methods of synthesis and characterization. *Russian Chemical Reviews*, 85(3), 280.

[2] Tüzün, B., & Kaya, C. (2018). Investigation of DNA–RNA molecules for the efficiency and activity of corrosion inhibition by DFT and molecular docking. *Journal of Bio-and Tribo-Corrosion*, 4(4), 69.

[3] Tüzün, B., & Bhawsar, J. (2021). Quantum chemical study of thiaozole derivatives as corrosion inhibitors based on density functional theory. *Arabian Journal of Chemistry*, 14(2), 102927.

[4] Majumdar, D., Chatterjee, A., Feizi-Dehnayebi, M., Kiran, N. S., Tuzun, B., & Mishra, D. (2024). 8-Aminoquinoline derived two Schiff base platforms: Synthesis, characterization, DFT insights, corrosion inhibitor, molecular docking, and pH-dependent antibacterial study. *Heliyon*, 10(15) e35591.

[5] Rezaeivala, M., Karimi, S., Sayin, K., & Tüzün, B. (2022). Experimental and theoretical investigation of corrosion inhibition effect of two piperazine-based ligands on carbon steel in acidic media. *Colloids and Surfaces A: Physicochemical and Engineering Aspects*, 641, 128538.

[6] Tüzün, B. (2018). Theoretical evaluation of six indazole derivatives as corrosion inhibitors based on DFT. *Turkish Computational and Theoretical Chemistry*, 2(1), 12–22.

[7] Arukalam, I. O., Uzochukwu, I. N., Izionworu, V. O., Tüzün, B., & Dagdag, O. (2023). Corrosion protection of Q235 steel in Pseudomonas aeruginosa-laden seawater environment using high barrier PDMS nanocomposite coating. *Safety in Extreme Environments*, 5(4), 281–291.

[8] Rezaeivala, M., Karimi, S., Tuzun, B., & Sayin, K. (2022). Anti-corrosion behavior of 2-((3-(2-morpholino ethylamino)-N3-((pyridine-2-yl) methyl) propylimino) methyl) pyridine and its reduced form on carbon steel in hydrochloric acid solution: Experimental and theoretical studies. *Thin Solid Films*, 741, 139036.

[9] Çelik, M. S., Çetinkaya, H. F., Çetinkaya, S., Tüzün, G., & Tüzün, B. (2023). Polymeric nanoparticles and their composites in corrosion inhibition. In *Corrosion Prevention Nanoscience: Nanoengineering Materials and Technologies* Editor: Berdimurodov Elyor Tukhlievich, Chandrabhan Verma, Walter de Gruyter GmbH & Co KG (p. 135).

[10] Arukalam, I. O., Uzochukwu, I. N., Tüzün, B., Dagdag, O., & Oguzie, E. E. (2024). Influence of ZnO nanoparticle size on barrier performance and corrosion protection of poly (dimethylsiloxane)-coated Q235 steel in chloride environment: Bode and computational simulation investigations. *Chemistry Africa*, 7(2), 853–863.

[11] Seyran, E., Tüzün, G., Çetinkaya, S., & Tüzün, B. (2024). Nanoparticle-based corrosion inhibitors. In Thakur, A., Kumar, A. (eds.) *Innovations in Nanomaterials-Based Corrosion Inhibitors* (pp. 238–268). IGI Global.

[12] Bayar, G., Özler, M. E., Aylaz, G., Tüzün, B., & Sözmen, F. (2025). Multifunctional nanoparticles: Synthesis, characterization, structure, and properties. In *Multifunctional Magnetic Nanoparticles in Therapy, Biology, and Pharmacy* Editor: Elyor Berdimurodov, Chaudhery Mustansar Hussain, Burak Tüzün (pp. 1–33). CRC Press.

[13] Makowski, O., Stroka, J., Kulesza, P. J., Malik, M. A., & Galus, Z. (2002). Electrochemical identity of copper hexacyanoferrate in the solid-state: Evidence for the presence and redox activity of both iron and copper ionic sites. *Journal of Electroanalytical Chemistry*, 532(1–2), 157–164.

[14] Zhang, X., Chen, Z., Liu, X., Hanna, S. L., Wang, X., Taheri-Ledari, R., ... & Farha, O. K. (2020). A historical overview of the activation and porosity of metal–organic frameworks. *Chemical Society Reviews*, 49(20), 7406–7427.

[15] Burtch, N. C., Heinen, J., Bennett, T. D., Dubbeldam, D., & Allendorf, M. D. (2018). Mechanical properties in metal–organic frameworks: Emerging opportunities and challenges for device functionality and technological applications. *Advanced Materials*, 30(37), 1704124.

[16] Norouzbahari, S., Lighvan, Z. M., Ghadimi, A., & Sadatnia, B. (2023). ZIF-8@ Zn-MOF-74 core–shell metal–organic framework (MOF) with open metal sites: Synthesis, characterization, and gas adsorption performance. *Fuel*, 339, 127463.

[17] Lammert, M., Glißmann, C., & Stock, N. (2017). Tuning the stability of bimetallic Ce (IV)/Zr (IV)-based MOFs with UiO-66 and MOF-808 structures. *Dalton Transactions*, 46(8), 2425–2429.

[18] Kalmutzki, M. J., Hanikel, N., & Yaghi, O. M. (2018). Secondary building units as the turning point in the development of the reticular chemistry of MOFs. *Science Advances*, 4(10), eaat9180.

[19] Vetlitsyna-Novikova, K. S., Butova, V. V., Pankin, I. A., Shapovalov, V. V., & Soldatov, A. V. (2019). Zirconium-based metal-organic UiO-66, UiO-66-NDC and MOF-801 frameworks. Influence of the linker effect on the hydrogen sorption efficiency. *Journal of Surface Investigation: X-ray, Synchrotron and Neutron Techniques*, 13, 787–792.

[20] Bennett, T. D., Tan, J. C., Moggach, S. A., Galvelis, R., Mellot-Draznieks, C., Reisner, B. A., ... & Cheetham, A. K. (2010). Mechanical properties of dense zeolitic imidazolate frameworks (ZIFs): A high-pressure X-ray diffraction, nanoindentation and computational study of the zinc framework Zn (Im) 2, and its lithium boron analogue, LiB (Im) 4. *Chemistry—A European Journal*, 16(35), 10684–10690.

[21] Sofi, F. A., Majid, K., & Mehraj, O. (2018). The visible light driven copper based metal-organic-framework heterojunction: HKUST-1@ Ag-Ag_3PO_4 for plasmon enhanced visible light photocatalysis. *Journal of Alloys and Compounds*, 737, 798–808.

[22] López-Olvera, A., Zárate, J. A., Martínez-Ahumada, E., Fan, D., Diaz-Ramirez, M. L., Saenz-Cavazos, P. A., ... & Ibarra, I. A. (2021). SO_2 capture by two aluminum-based MOFs: Rigid-like MIL-53 (Al)-TDC versus breathing MIL-53 (Al)-BDC. *ACS Applied Materials & Interfaces*, 13(33), 39363–39370.

[23] Kirchon, A., Feng, L., Drake, H. F., Joseph, E. A., & Zhou, H. C. (2018). From fundamentals to applications: A toolbox for robust and multifunctional MOF materials. *Chemical Society Reviews*, 47(23), 8611–8638.

[24] Wei, Z., Sun, W., Han, H., Xing, Y., & Gui, X. (2024). Molecular design of multiple ligand metal–organic framework (ML-MOF) collectors for efficient flotation separation of minerals. *Separation and Purification Technology*, 328, 125048.

[25] Aliakbari, R., Ramakrishna, S., Kowsari, E., Marfavi, Y., Cheshmeh, Z. A., Ajdari, F. B., ... & Ershadi, M. (2022). Scalable preparation of MOFs and MOF-containing hybrid materials for use in sustainable refrigeration systems for a greener environment: A comprehensive review as well as technical and statistical analysis of patents. *Research on Chemical Intermediates*, 48(7), 2767–2819.

[26] Li, X., Liu, Y., Wang, J., Gascon, J., Li, J., & Van der Bruggen, B. (2017). Metal–organic frameworks based membranes for liquid separation. *Chemical Society Reviews*, 46(23), 7124–7144.

[27] Zhao, Y., Zhang, T., Xiong, H., & Wang, F. (2021). Bridge for the thermodynamics and kinetics of electrochemical corrosion: Modeling on dissolution, ionization, diffusion and deposition in metal/solution interface. *Corrosion Science*, 191, 109763.

[28] Groysman, A., & Groysman, A. (2010). Corrosion mechanism and corrosion factors. In Groysman, A. (ed.) *Corrosion for Everybody* (pp. 1–51).

[29] Perez, N. (Ed.). (2004). *Electrochemistry and Corrosion Science*. Springer US.

[30] Wirtanen, T., Prenzel, T., Tessonnier, J. P., & Waldvogel, S. R. (2021). Cathodic corrosion of metal electrodes—How to prevent it in electroorganic synthesis. *Chemical Reviews*, 121(17), 10241–10270.

[31] Marek, M. (2009). Metal corrosion. *Biomedical Materials*, 155–181.

[32] Zheng, S. Q., Lim, S. S., Foo, C. Y., Haw, C. Y., Chiu, W. S., Chia, C. H., & Khiew, P. S. (2022). Probing the effect of solvents on the electrochemical performance of graphene incorporated nickel-based metal organic frameworks. *Journal of Electroanalytical Chemistry*, 925, 116860.

[33] Campagnol, N., Van Assche, T. R., Li, M., Stappers, L., Dincă, M., Denayer, J. F., ... & Fransaer, J. (2016). On the electrochemical deposition of metal–Organic frameworks. *Journal of Materials Chemistry A*, 4(10), 3914–3925.

[34] Song, S., Ma, X., Li, W., Zhang, B., Sun, J., & Deng, C. (2023). The influence of solvent controlled morphology on capacitive properties of metal-organic frameworks based on polyaminocarboxybenzene ligands. *Colloids and Surfaces A: Physicochemical and Engineering Aspects*, 656, 130350.

[35] Al Kiey, S. A., El-Shahat, M., & Abdelhameed, R. M. (2023). Role of different metal precursors based MOFs for boosting anti-corrosion performance of mild steel in acid media. *Materials Today Sustainability*, 23, 100460.

[36] Zhong, Y., Li, X., Chen, J., Wang, X., Wei, L., Fang, L., ... & Liu, J. (2020). Recent advances in MOF-based nanoplatforms generating reactive species for chemodynamic therapy. *Dalton Transactions*, 49(32), 11045–11058.

[37] Liu, C., Wang, J., Wan, J., & Yu, C. (2021). MOF-on-MOF hybrids: Synthesis and applications. *Coordination Chemistry Reviews*, 432, 213743.

[38] Sanati-Tirgan, P., Eshghi, H., & Mohammadinezhad, A. (2023). Designing a new method for growing metal–organic framework (MOF) on MOF: Synthesis, characterization and catalytic applications. *Nanoscale*, 15(10), 4917–4931.

[39] McKinstry, C., Cathcart, R. J., Cussen, E. J., Fletcher, A. J., Patwardhan, S. V., & Sefcik, J. (2016). Scalable continuous solvothermal synthesis of metal organic framework (MOF-5) crystals. *Chemical Engineering Journal*, 285, 718–725.

[40] Klinowski, J., Paz, F. A. A., Silva, P., & Rocha, J. (2011). Microwave-assisted synthesis of metal–organic frameworks. *Dalton Transactions*, 40(2), 321–330.

[41] Liu, Y., Myers, E. J., Rydahl, S. A., & Wang, X. (2019). Ultrasonic-assisted synthesis, characterization, and application of a metal–organic framework: A green general chemistry laboratory project. *Journal of Chemical Education*, 96(10), 2286–2291.

[42] Varsha, M. V., & Nageswaran, G. (2020). Direct electrochemical synthesis of metal organic frameworks. *Journal of The Electrochemical Society*, 167(15), 155527.

[43] Getachew, N., Chebude, Y., Diaz, I., & Sanchez-Sanchez, M. (2014). Room temperature synthesis of metal organic framework MOF-2. *Journal of Porous Materials*, 21, 769–773.

[44] Sánchez-Sánchez, M., Getachew, N., Díaz, K., Díaz-García, M., Chebude, Y., & Díaz, I. (2015). Synthesis of metal–organic frameworks in water at room temperature: Salts as linker sources. *Green Chemistry*, 17(3), 1500–1509.

[45] Kalaj, M., Bentz, K. C., Ayala Jr, S., Palomba, J. M., Barcus, K. S., Katayama, Y., & Cohen, S. M. (2020). MOF-polymer hybrid materials: From simple composites to tailored architectures. *Chemical Reviews*, 120(16), 8267–8302.

[46] Cao, X., Tan, C., Sindoro, M., & Zhang, H. (2017). Hybrid micro-/nano-structures derived from metal–organic frameworks: Preparation and applications in energy storage and conversion. *Chemical Society Reviews*, 46(10), 2660–2677.

[47] Gao, X., Gao, Y., Cao, H., & Zhang, J. (2024). Eco-friendly sustainable and responsive high-performance benzotriazole-metal organic frameworks/silica composite

coating with active/passive corrosion protection on copper. *Langmuir*, 40(14), 7639–7652.
[48] Meng, J., Liu, X., Niu, C., Pang, Q., Li, J., Liu, F., ... & Mai, L. (2020). Advances in metal–organic framework coatings: Versatile synthesis and broad applications. *Chemical Society Reviews*, 49(10), 3142–3186.
[49] Chen, H., Wang, F., Fan, H., Hong, R., & Li, W. (2021). Construction of MOF-based superhydrophobic composite coating with excellent abrasion resistance and durability for self-cleaning, corrosion resistance, anti-icing, and loading-increasing research. *Chemical Engineering Journal*, 408, 127343.
[50] Majidi, R., Farhadi, A., Danaee, I., Panah, N. B., Zarei, D., & Nikmanesh, S. (2023). Investigation of synthesized planar Cu-MOF and spherical Ni-MOF nanofillers for improving the anti-corrosion performance of epoxy coatings. *Progress in Organic Coatings*, 183, 107803.
[51] Sabzehmeidani, M. M., Gafari, S., & Kazemzad, M. (2024). Concepts, fabrication and applications of MOF thin films in optoelectronics: A review. *Applied Materials Today*, 38, 102153.
[52] Baig, U., Waheed, A., & Jillani, S. M. S. (2024). Recent advancements in metal-organic framework-based membranes for hydrogen separation: A review. *Chemistry—An Asian Journal*, 19(16), e202300619.
[53] Song, L., Zhang, S., Wang, Q., Mao, X., Liu, B., Zhao, Y. D., & Chen, W. (2023). Versatile synthesis of hollow PDA motors by interfacial protection. Chemical Engineering Journal, 468, 143701.
[54] Thomas-Hillman, I., Laybourn, A., Dodds, C., & Kingman, S. W. (2018). Realising the environmental benefits of metal–organic frameworks: Recent advances in microwave synthesis. *Journal of Materials Chemistry A*, 6(25), 11564–11581.
[55] Juma, A. K., Merican, Z. M. A., & Haruna, A. (2024). Recent progress of MOF-based photocatalysts for environmental application and sustainability considerations. *Chemical Engineering Research and Design* 208, 391–435.
[56] Sher, F., Hayward, A., El Guerraf, A., Ziani, I., Hrnjić, H., Boškailo, E., ... & Nemţanu, M. R. (2024). Advanced metal-organic frameworks for superior carbon capture, high-performance energy storage and environmental photocatalysis—A critical review. *Journal of Materials Chemistry A* 12(41), 27932–27973.
[57] He, C., Yin, W., Li, X., Zheng, J., Tang, B., & Rui, Y. (2021). Molybdenum disulfide synthesized by molybdenum-based metal organic framework with high activity for sodium ion battery. *Electrochimica Acta*, 365, 137353.
[58] Fang, S. Y., Gong, J. L., Tang, L., Cao, W. C., Li, J., Tan, Z. K., ... & Chen, Z. P. (2022). Construction the hierarchical architecture of molybdenum disulfide/MOF composite membrane via electrostatic self-assembly strategy for efficient molecular separation. *Chemical Engineering Journal*, 449, 137808.
[59] Vellingiri, K., Szulejko, J. E., Kumar, P., Kwon, E. E., Kim, K. H., Deep, A., ... & Brown, R. J. (2016). Metal organic frameworks as sorption media for volatile and semi-volatile organic compounds at ambient conditions. *Scientific Reports*, 6(1), 27813.
[60] Siu, B., Chowdhury, A. R., Yan, Z., Humphrey, S. M., & Hutter, T. (2023). Selective adsorption of volatile organic compounds in metal-organic frameworks (MOFs). *Coordination Chemistry Reviews*, 485, 215119.
[61] Kharissova, A. B., Kharissova, O. V., Kharisov, B. I., & Méndez, Y. P. (2024). Carbon negative footprint materials: A review. *Nano-Structures & Nano-Objects*, 37, 101100.

[62] Anadebe, V. C., Chukwuike, V. I., Ramanathan, S., & Barik, R. C. (2022). Cerium-based metal organic framework (Ce-MOF) as corrosion inhibitor for API 5L X65 steel in CO_2-saturated brine solution: XPS, DFT/MD-simulation, and machine learning model prediction. *Process Safety and Environmental Protection*, 168, 499–512.
[63] Jiang, L., Dong, Y., Yuan, Y., Zhou, X., Liu, Y., & Meng, X. (2022). Recent advances of metal–organic frameworks in corrosion protection: From synthesis to applications. *Chemical Engineering Journal*, 430, 132823.
[64] Lee, M. H., Vikrant, K., Younis, S. A., Szulejko, J. E., & Kim, K. H. (2020). Chemisorption of hydrogen sulfide by metal-organic frameworks and covalent-organic polymers based on experimental/theoretical evaluation. *Journal of Cleaner Production*, 250, 119486.
[65] Liu, D., Li, B., Wu, J., & Liu, Y. (2020). Sorbents for hydrogen sulfide capture from biogas at low temperature: A review. *Environmental Chemistry Letters*, 18(1), 113–128.
[66] Seidi, F., Jouyandeh, M., Taghizadeh, M., Taghizadeh, A., Vahabi, H., Habibzadeh, S., ... & Saeb, M. R. (2020). Metal-organic framework (MOF)/epoxy coatings: A review. *Materials*, 13(12), 2881.
[67] Yu, H., Wang, W., Liu, M., Zhao, T., Lin, R., Hou, M., ... & Li, X. (2022). Versatile synthesis of dendritic mesoporous rare earth–based nanoparticles. Science Advances, 8(30), eabq2356.
[68] Barea, E., Montoro, C., & Navarro, J. A. (2014). Toxic gas removal–metal–organic frameworks for the capture and degradation of toxic gases and vapours. *Chemical Society Reviews*, 43(16), 5419–5430.
[69] Balan, B., Xavier, M. M., & Mathew, S. (2023). MoS_2-based nanocomposites for photocatalytic hydrogen evolution and carbon dioxide reduction. *ACS Omega*, 8(29), 25649–25673.
[70] Ahmed, F. M., Ateia, E. E., El-dek, S. I., Abd El-Kader, S. M., & Shafaay, A. S. (2024). Synergistic interaction between molybdenum disulfide nanosheet and metal organic framework for high performance supercapacitor. *Journal of Energy Storage*, 82, 110360.
[71] Kermani, M. B., & Morshed, A. (2003). Carbon dioxide corrosion in oil and gas production: a compendium. *Corrosion*, 59(08), OTC-3891-MS.
[72] Yoshida, T., Matsui, S., Atsuta, T., & Itoga, K. (1980, May). Corrosion problems of pipeline and a solution. In *Offshore Technology Conference* (pp. OTC-3891). OTC.
[73] Popoola, L. T., Grema, A. S., Latinwo, G. K., Gutti, B., & Balogun, A. S. (2013). Corrosion problems during oil and gas production and its mitigation. *International Journal of Industrial Chemistry*, 4, 1–15.
[74] Showman, M. S., El-Aziz, A. M. A., & Yahya, R. (2022). Efficient photocatalytic degradation of petroleum oil spills in seawater using a metal-organic framework (MOF). *Scientific Reports*, 12(1), 22445.
[75] Seo, H., Lee, I., Sridhar, V., & Park, H. (2021). Metal-organic framework reinforced acrylic polymer marine coatings. *Materials*, 15(1), 27.
[76] Bueger, C., & Liebetrau, T. (2023). Critical maritime infrastructure protection: What's the trouble?. *Marine Policy*, 155, 105772.
[77] Donnelly, B., Sammut, K., & Tang, Y. (2022). Materials selection for antifouling systems in marine structures. *Molecules*, 27(11), 3408.
[78] Seetharaj, R., Vandana, P. V., Arya, P., & Mathew, S. (2019). Dependence of solvents, pH, molar ratio and temperature in tuning metal organic framework architecture. *Arabian Journal of Chemistry*, 12(3), 295–315.

[79] Cui, W. G., Hu, T. L., & Bu, X. H. (2020). Metal–organic framework materials for the separation and purification of light hydrocarbons. *Advanced Materials*, 32(3), 1806445.

[80] Shahini, M. H., Mohammadloo, H. E., Ramezanzadeh, M., & Ramezanzadeh, B. (2022). Recent innovations in synthesis/characterization of advanced nano-porous metal-organic frameworks (MOFs); current/future trends with a focus on the smart anti-corrosion features. *Materials Chemistry and Physics*, 276, 125420.

[81] Liu, J., Wu, D., Zhu, N., Wu, Y., & Li, G. (2021). Antibacterial mechanisms and applications of metal-organic frameworks and their derived nanomaterials. *Trends in Food Science & Technology*, 109, 413–434.

[82] Aljibori, H. S., Alamiery, A., & Kadhum, A. A. H. (2023). Advances in corrosion protection coatings: A comprehensive review. *International Journal of Corrosion and Scale Inhibition*, 12(4), 1476–1520.

[83] Nguyen, T. H., Fei, H., Sapurina, I., Ngwabebhoh, F. A., Bubulinca, C., Munster, L., ... & Saha, P. (2021). Electrochemical performance of composites made of rGO with Zn-MOF and PANI as electrodes for supercapacitors. *Electrochimica Acta*, 367, 137563.

[84] Nwokolo, I. K., Shi, H., Ikeuba, A. I., Gao, N., Li, J., Ahmed, S., & Liu, F. (2022). Synthesis, characterization and investigation of anticorrosion properties of an innovative metal–organic framework, ZnMOF-BTA, on carbon steel in HCl solution. *Coatings*, 12(9), 1288.

[85] Shayeh, J. S., & Salari, H. (2020). Dendritic fibrous nano metal organic framework: A magnetic core-shell structure as high performance material for electrochemical capacitors. *Journal of Energy Storage*, 32, 101734.

[86] Lázaro, I. A. (2020). A comprehensive thermogravimetric analysis multifaceted method for the exact determination of the composition of multifunctional metal-organic framework materials. *European Journal of Inorganic Chemistry*, 2020(45), 4284–4294.

[87] Li, Y., Karimi, M., Gong, Y. N., Dai, N., Safarifard, V., & Jiang, H. L. (2021). Integration of metal-organic frameworks and covalent organic frameworks: Design, synthesis, and applications. *Matter*, 4(7), 2230–2265.

[88] Ramezanzadeh, M., Ramezanzadeh, B., Bahlakeh, G., Tati, A., & Mahdavian, M. (2021). Development of an active/barrier bi-functional anti-corrosion system based on the epoxy nanocomposite loaded with highly-coordinated functionalized zirconium-based nanoporous metal-organic framework (Zr-MOF). *Chemical Engineering Journal*, 408, 127361.

[89] Ramezanzadeh, M., Ramezanzadeh, B., Mahdavian, M., & Bahlakeh, G. (2020). Development of metal-organic framework (MOF) decorated graphene oxide nanoplatforms for anti-corrosion epoxy coatings. *Carbon*, 161, 231–251.

[90] Mukherjee, S., Datta, K. K. R., & Fischer, R. A. (2021). Hydrophobicity: A key factor en route to applications of metal–organic frameworks. *Trends in Chemistry*, 3(11), 911–925.

[91] Liu, B., Vikrant, K., Kim, K. H., Kumar, V., & Kailasa, S. K. (2020). Critical role of water stability in metal–organic frameworks and advanced modification strategies for the extension of their applicability. *Environmental Science: Nano*, 7(5), 1319–1347.

[92] Zhou, C., Pan, M., Li, S., Sun, Y., Zhang, H., Luo, X., ... & Zeng, H. (2022). Metal organic frameworks (MOFs) as multifunctional nanoplatform for anticorrosion surfaces and coatings. *Advances in Colloid and Interface Science*, 305, 102707.

[93] Zhao, M., Huang, Y., Peng, Y., Huang, Z., Ma, Q., & Zhang, H. (2018). Two-dimensional metal–organic framework nanosheets: Synthesis and applications. *Chemical Society Reviews*, 47(16), 6267–6295.
[94] Xie, L. H., Xu, M. M., Liu, X. M., Zhao, M. J., & Li, J. R. (2020). Hydrophobic metal–organic frameworks: Assessment, construction, and diverse applications. *Advanced Science*, 7(4), 1901758.
[95] Yu, Y., Wei, Y., Li, B., Gao, H., Liu, T., Luan, X., ... & Ouyang, Y. (2022). Bioinspired metal-organic framework-based liquid-infused surface (MOF-LIS) with corrosion and biofouling prohibition properties. *Surfaces and Interfaces*, 34, 102363.
[96] Antwi-Baah, R., & Liu, H. (2018). Recent hydrophobic metal-organic frameworks and their applications. *Materials*, 11(11), 2250.
[97] Ramezanzadeh, M., & Ramezanzadeh, B. (2021). Thermomechanical and anticorrosion characteristics of metal-organic frameworks. In Anish Khan, Anish Khan, Abdullah M. Asiri, Md Enamul Hoque, Anwar L. Bilgrami, Mohammad Azam, K. Chandra Babu Naidu (eds.) *Metal-Organic Frameworks for Chemical Reactions* (pp. 295–330). Elsevier.
[98] Al Obeidli, A., Salah, H. B., Al Murisi, M., & Sabouni, R. (2022). Recent advancements in MOFs synthesis and their green applications. *International Journal of Hydrogen Energy*, 47(4), 2561–2593.
[99] He, Q., Zhan, F., Wang, H., Xu, W., Wang, H., & Chen, L. (2022). Recent progress of industrial preparation of metal–organic frameworks: Synthesis strategies and outlook. *Materials Today Sustainability*, 17, 100104.
[100] Bradu, P., Biswas, A., Nair, C., Sreevalsakumar, S., Patil, M., Kannampuzha, S., ... & Gopalakrishnan, A. V. (2022). Recent advances in green technology and Industrial Revolution 4.0 for a sustainable future. *Environmental Science and Pollution Research International*, *30*(60), 124488–124519.
[101] Thakur, A., & Kumar, A. (2024). Chemical and physical properties of nano-hybrid smart coatings. In Ashish Kumar· Abhinay Thakur (eds.) *Nano-hybrid Smart Coatings: Advancements in Industrial Efficiency and Corrosion Resistance* (pp. 59–94). American Chemical Society.
[102] Kitagawa, S. (2014). Metal–organic frameworks (MOFs). *Chemical Society Reviews*, 43(16), 5415–5418.
[103] Vaitsis, C., Sourkouni, G., & Argirusis, C. (2019). Metal organic frameworks (MOFs) and ultrasound: A review. *Ultrasonics sonochemistry*, 52, 106–119.
[104] Guo, C., Duan, F., Zhang, S., He, L., Wang, M., Chen, J., ... & Du, M. (2022). Heterostructured hybrids of metal–organic frameworks (MOFs) and covalent–organic frameworks (COFs). *Journal of Materials Chemistry A*, 10(2), 475–507.
[105] Sharmin, E., & Zafar, F. (2016). Introductory chapter: Metal organic frameworks (MOFs). In Fahmina Zafer, Eram Sharmin (eds.) *Metal-organic Frameworks*. IntechOpen.
[106] Dey, C., Kundu, T., Biswal, B. P., Mallick, A., & Banerjee, R. (2014). Crystalline metal-organic frameworks (MOFs): Synthesis, structure and function. *Acta Crystallographica Section B: Structural Science, Crystal Engineering and Materials*, 70(1), 3–10.
[107] Zhou, H. C., Long, J. R., & Yaghi, O. M. (2012). Introduction to metal–organic frameworks. *Chemical Reviews*, 112(2), 673–674.
[108] Ma, M., Lu, X., Guo, Y., Wang, L., & Liang, X. (2022). Combination of metal-organic frameworks (MOFs) and covalent organic frameworks (COFs): Recent advances in synthesis and analytical applications of MOF/COF composites. *TrAC Trends in Analytical Chemistry*, 157, 116741.

[109] Zhang, X., Chen, A., Zhong, M., Zhang, Z., Zhang, X., Zhou, Z., & Bu, X. H. (2019). Metal–organic frameworks (MOFs) and MOF-derived materials for energy storage and conversion. *Electrochemical Energy Reviews*, 2, 29–104.

[110] Kumaraguru, S., Pavulraj, R., & Mohan, S. (2017). Influence of cobalt, nickel and copper-based metal-organic frameworks on the corrosion protection of mild steel. *Transactions of the IMF*, 95(3), 131–136.

[111] Yu, Z., Li, X., Li, X., Zheng, B., Li, D., Xu, D., & Wang, F. (2023). Nacre-inspired metal-organic framework coatings reinforced by multiscale hierarchical cross-linking for integrated antifouling and anti-microbial corrosion. *Advanced Functional Materials*, 33(47), 2305995.

12 Environmental Considerations in Anticorrosive MOF Coating Development

A. Mushira Banu, B. Arifa Farzana, and F.M. Mashood Ahamed

12.1 INTRODUCTION

Metal–organic frameworks (MOFs) are a class of materials made up of metal ions or clusters connected by organic ligands to form a porous, crystalline structure. Transition metals, such as iron, copper, or zinc, are typically the nodes in the framework, and molecules that bridge the metal centers include carboxylates, phosphonates, or imidazolates. MOFs have a highly ordered, three-dimensional (3D) network with tunable pore sizes and structures; this porosity is a key characteristic that sets MOFs apart from other materials. MOF coatings are essential for advancing a variety of applications because of their remarkable qualities, including high surface area, tunable porosity, and functional versatility. They are an essential component of contemporary material science and engineering as they are used in biological applications, energy storage, environmental cleanup, catalysis, and sensing. They are becoming increasingly significant in a variety of industries because of their capacity to solve difficult problems and advance current technologies. Incorporating environmental factors into MOF development is not only morally required, but also practically essential. The creation of MOFs can be in line with international sustainability objectives by emphasizing green synthesis techniques, sustainable resource utilization, and reducing negative effects on the environment and human health. This strategy not only lessens adverse environmental effects but also improves MOFs' general efficacy and acceptability across a range of applications. It was essential to incorporate sustainable synthesis techniques into MOF development to reduce environmental effect and promote green chemistry concepts. Researchers can produce safer and more sustainable MOFs by emphasizing the use of non-toxic solvents, lowering harmful chemicals, and minimizing waste. MOF manufacture can also be made more sustainable by increasing energy efficiency through reduced energy use and renewable energy sources, as well as by investigating different synthesis techniques such mechanochemical approaches. These initiatives are crucial for coordinating MOF technology with more general sustainability and environmental objectives[1–5].

 DOI: 10.1201/9781003571162-12

For MOFs to develop sustainably, effective resource management and material procurement are essential. The environmental effect of MOF synthesis can be considerably decreased by using sustainable organic linkers, minimizing the use of rare or precious minerals, applying recycling and reuse strategies, and utilizing abundant and non-toxic metal sources. These procedures improve MOF technology' long-term durability and usefulness in a variety of domains in addition to promoting their sustainability (Figure 12.1). For MOFs to be used safely and sustainably, it is essential to understand how they affect ecosystems and human health. It is necessary to conduct comprehensive toxicity studies and analyze the potential for ecological hazards and environmental contamination to develop MOFs that are not only effective but also safe for the environment and human health. By taking early measures to address these concerns, researchers and developers can lessen potential risks and enhance the overall sustainability of MOF technology[2–4]. Important details regarding the environmental impact of MOFs at each stage of production, usage, and disposal can be found in a thorough lifecycle analysis. By studying the production effect, evaluating performance and durability over time, and taking end-of-life disposal and recycling alternatives into account, researchers and developers can make well-informed decisions that enhance the sustainability of MOFs. This all-encompassing strategy promotes the creation of MOFs that are efficient, cost-effective, and ecologically conscious in addition to reducing their negative effects on the environment.

12.2 MOF SYNTHESIS METHODS AND PROPERTIES

There are several ways to synthesize MOFs[6–8], and each has its own benefits and drawbacks. Solvothermal and hydrothermal synthesis are two of the most used techniques, in which metal salts and organic linkers are dissolved in solvents in sealed vessels at high pressures and temperatures. Although these techniques are well known for producing high-crystallinity MOFs with precise sizes and shapes, they are frequently time- and energy-intensive.

Microwave-Assisted Synthesis: A quick substitute that uses microwave radiation to speed up nucleation and crystal formation is microwave-assisted synthesis. This method offers more control over particle size while drastically cutting down on reaction times and energy usage. Its scalability is still constrained, though, by the possibility of uneven heating. A green and eco-friendly technique is mechanochemical synthesis, which entails combining metal salts and organic linkers without the use of solvents. Although it might have trouble managing crystallinity and phase purity, it avoids the use of hazardous solvents and lowers waste output.

Electrochemical Synthesis: Another cutting-edge technique is electrochemical synthesis, which creates metal ions that react with organic linkers in a solvent. Although this method enables more accurate control over metal ion concentrations and cleaner synthesis routes, it necessitates specialized equipment and close condition monitoring.

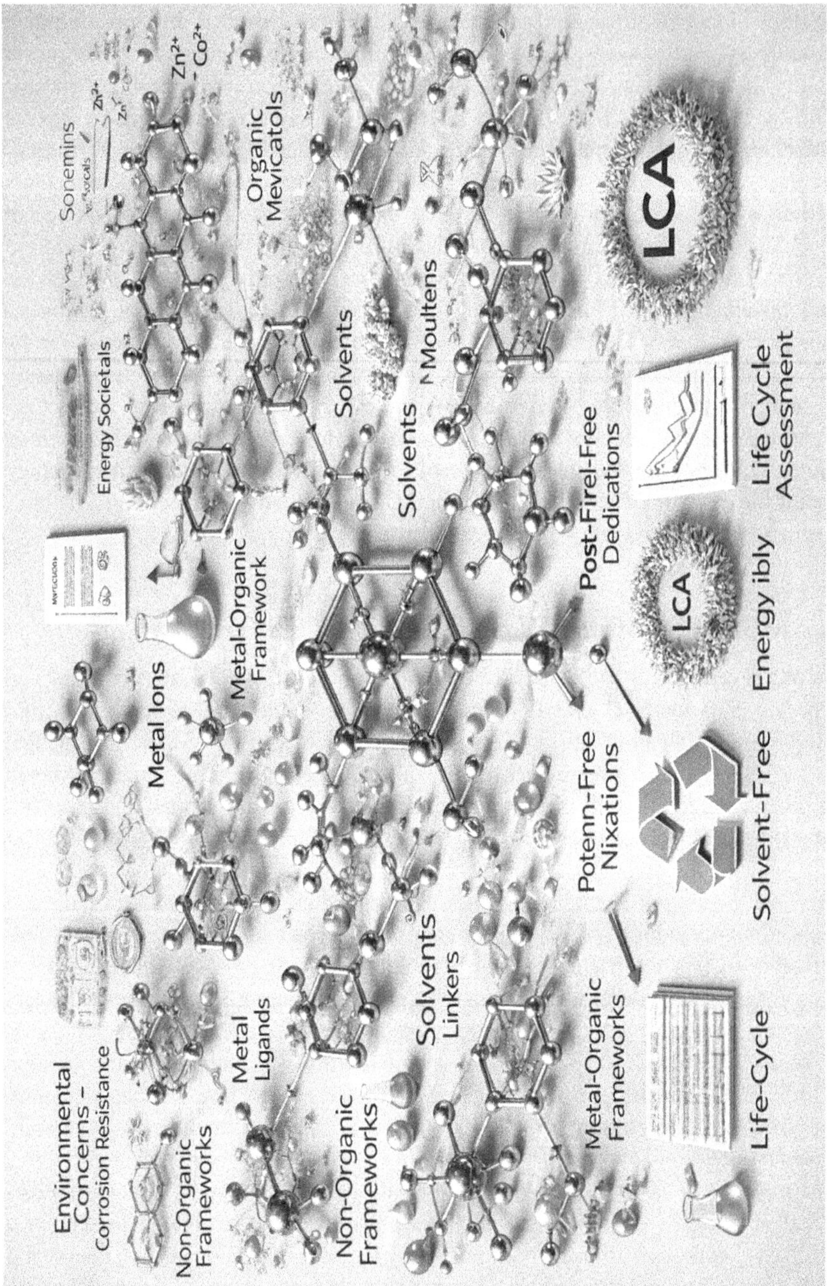

FIGURE 12.1 Environmental considerations in MOF coating development.

Sonochemical Synthesis: To improve reaction rates and produce homogeneous particle sizes, sonochemical synthesis uses ultrasonic vibrations to generate localized high temperatures and pressures. Despite these advantages, equipment constraints limit its scalability. Although it frequently leads to reduced crystallinity, room-temperature synthesis, which uses aqueous or alcohol-based solutions, offers an energy-efficient method appropriate for particular MOFs. By using pre-synthesized building blocks, modular synthesis enables the customization of MOF structures and features, facilitating specific functionality and scalability.

Because of their many exceptional qualities, MOFs can be used in a wide variety of applications. They are perfect for gas storage, adsorption, and separation because of their extraordinarily large surface areas, which frequently surpass 10,000 m^2/g. Selective adsorption and catalysis are made possible by MOFs' adjustable porosity, which is produced by altering the metal ions and organic linkers. MOFs also demonstrate substantial chemical diversity, with the potential to integrate a wide range of metal ions and linkers, thereby providing tailored features for specific applications, such as drug delivery or sensing. The utility of MOFs is further enhanced by their structural diversity, which ranges from intricate 3D frameworks to 1D chains.

Furthermore, depending on the metal–linker combination, MOFs can be designed to have different levels of chemical and thermal stability, which is essential for applications in challenging environments. Their range of applications is increased by post-synthetic modifications, which offer chances to add functional groups or improve characteristics. While certain MOFs have magnetic qualities that are advantageous for magnetic storage, sensing, and medical imaging, others have luminescence, which makes them useful in sensing and optoelectronics.

As long as scalability, stability, and environmental concerns are properly taken into account, MOFs' varied characteristics and synthesis techniques highlight their potential in advanced material applications.

12.2.1 Mechano-Synthesis in MOF Formation

1. *Solvent-Free or Minimal Solvent Use:* This technique significantly reduces or does away with the requirement for solvents by inducing chemical reactions using mechanical energy. Compared to traditional solvothermal methods, this makes the process more environmentally friendly.
2. *Types of Mechano-Synthesis*
   - *Ball Milling:* A popular method for promoting the synthesis of MOFs involves applying force to the metal and organic linkers in a ball mill.
   - *Mortar and Pestle Grinding:* To create MOF, the precursors must be manually ground.
3. *Reaction Conditions:*
   - *Mechano-synthesis* usually takes place at room temperature under atmospheric pressure, which drastically lowers energy usage in contrast to classical MOF synthesis, which needs regulated temperatures and pressure.

4. *Advantages of Mechano-Synthesis:*
   - *Sustainability:* Less solvent is used, which results in less waste and less energy being used.
   - *Rapid Synthesis:* In contrast to the lengthy times needed for solvothermal synthesis, reactions can happen quickly (in minutes to hours).
   - *Scalability:* Without requiring big reaction vessels, mechano-synthesis is simpler to scale up for industrial manufacturing.
5. *Applications:*
   - The MOFs that are produced through mechano-synthesis (Figure 12.2) find value in medication delivery, gas storage, catalysis, and environmental applications. The technology is also being investigated for creating MOFs with special characteristics because of the mechanochemical process.

12.3 IMPORTANCE OF MOF COATINGS IN VARIOUS APPLICATIONS

MOFs have attracted a lot of attention due to their special qualities and their uses, especially as coatings. Here is a thorough analysis of their significance in a number of crucial areas[9–13].

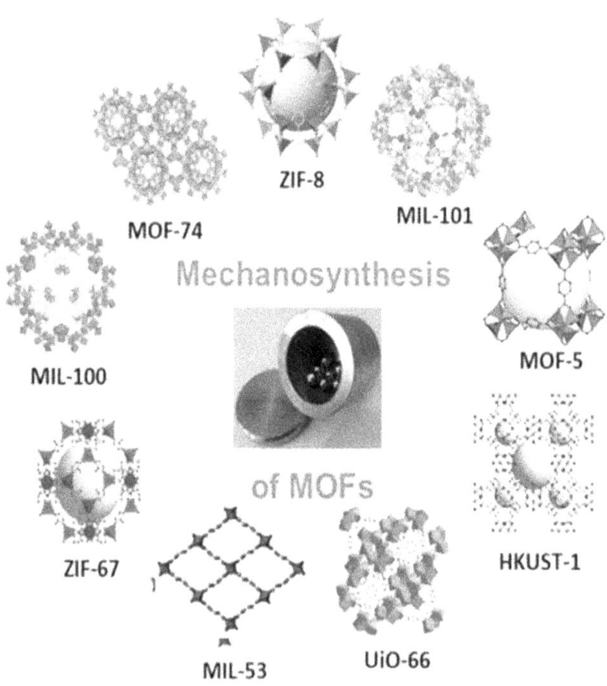

FIGURE 12.2 Mechano-synthesis in MOF formation.

12.3.1 Environmental Remediation

- *Air Purification:*
  - *Selective Adsorption:* Air pollutants such as carbon dioxide (CO_2), nitrogen oxides (NO_x), and volatile organic compounds (VOCs) can be selectively adsorbed by MOF coatings. They are more effective at absorbing these contaminants because of their large surface area and adjustable pores.
  - *Catalytic Decomposition:* Certain MOFs have the ability to break down hazardous gases into less toxic forms, which helps to purify the air.
- *Water Treatment:*
  - *Heavy Metal Removal:* Because of their strong affinity for heavy metals (such as lead and mercury), MOFs can be applied to filters or membranes that remove these ions from contaminated water.
  - *Degradation of Organic Contaminants:* MOFs with catalytic qualities can aid in the breakdown of organic contaminants in water, including dyes and pesticides, hence enhancing water quality.
- *Soil Remediation:*
  - *Containment and Removal:* By encasing pollutants in soil, MOF coatings can stop their spread and make it easier to remove or degrade them.

12.3.2 Energy Storage

- *Batteries:*
  - *Improved Electrode Materials:* MOF coatings on electrode materials can improve their conductivity and capacity, resulting in longer life cycles and higher energy densities in batteries.
  - *Lithium-Ion and Beyond:* By maximizing charge storage and release, MOFs can be used as coatings to lithium–ion batteries as well as other battery types, enhancing performance.
- *Supercapacitors:*
  - *High Surface Area:* MOF coatings improve the energy and power density of supercapacitors by offering a large surface area for charge storage.
  - *Increased Conductivity:* MOF coatings can increase the supercapacitor electrodes' electrical conductivity, resulting in quicker cycles of charging and discharging.
- *Hydrogen Storage:*
  - *Effective Storage:* Materials for storing hydrogen can be coated with MOFs to increase their capacity and effectiveness. They can store hydrogen gas at high densities because of their porous structure.

12.3.3 Catalysis

- *Chemical Reactions:*
  - *Selective Catalysis:* MOF coatings have the ability to catalyze particular chemical reactions, such as hydrolysis, reduction, and oxidation. These reactions are facilitated by their large surface area and distinct pore architecture.

  - *Enhanced Reaction Rates:* MOF coatings' porous nature can result in improved selectivity for desired products as well as faster reaction rates.
- *Environmental Catalysis:*
  - *Pollutant Degradation:* To reduce environmental pollution, MOFs can be employed as catalytic coatings in systems intended to break down pollutants like CO_2 and NO_x.
  - *Green Chemistry:* By encouraging reactions in milder environments and lowering the requirement for hazardous chemicals, MOF coatings can facilitate more environmentally friendly chemical processes.

12.3.4 Sensors

- *Gas Sensors:*
  - *Selective Detection:* By adding MOF coatings to sensor materials, certain gases, like hydrogen, carbon monoxide, and methane, can be detected with greater sensitivity and selectivity.
  - *Real-Time Monitoring:* For industrial safety and environmental monitoring, real-time data on gas concentrations is essential, and MOF-based sensors can deliver this information.
- *Chemical Sensors:*
  - *Pollutant Detection:* MOFs can be coated into sensors to identify traces of contaminants in soil, water, or air, supporting environmental and health monitoring.

12.3.5 Biomedical Applications

- *Drug Delivery:*
  - *Controlled Release:* MOF coatings can be designed to release medications in a regulated way, increasing treatment efficacy and minimizing adverse effects.
  - *Targeted Delivery:* By enabling targeted drug delivery to particular cells or tissues, functionalized MOF coatings can improve treatment precision.
- *Imaging and Diagnostics:*
  - *Enhanced Imaging:* MOFs provide high contrast and specificity for imaging methods such as CT or MRI scans, making them suitable for use in imaging agents for medical diagnostics.

12.3.6 Anticorrosion Coatings

The prospective uses of MOFs in anticorrosion coatings[14–19] have attracted a lot of attention lately because of their high surface area, functional diversity, and adjustable porosity. With the help of important references, the ensuing sections go into their workings, uses, and prospects for the future.

12.3.6.1 Mechanisms of Anticorrosion Coatings Based on MOFs

1. *Barrier Protection:* MOFs offer a physical barrier that keeps out corrosive substances like oxygen, water, and chloride ions.
2. *Controlled Release of Corrosion Inhibitors:* Under certain environmental triggers, MOFs can encapsulate and release inhibitors such as benzotriazole (BTA).
3. *Self-Healing Properties:* By repairing microcracks, MOFs combined with self-healing agents can prolong the life of coatings.
4. *Active Metal Protection:* MOFs can safeguard underlying metals by acting as sacrificial anodes or by adsorbing corrosive ions.

12.3.6.2 MOF Application Examples

1. Encapsulation of inhibitors such as BTA for pH-responsive release is possible with Zeolitic Imidazolate Framework-8 (ZIF-8)[20].
2. *Institute Lavoisier's MIL Series:* MIL-101, which is laden with molybdate ions for improved protection in acidic environments[21].
3. *University of Oslo (UiO) Framework (UiO-66):* High chloride resistance is demonstrated by functionalized UiO-66-NH_2 coatings[22].
4. *MOF-5:* Aluminum alloys exhibit superior corrosion prevention when cerium ions are included[23].
5. *MOF Composites:* MOF-graphene oxide hybrids enhance mechanical strength and anticorrosive qualities[24].

12.3.6.3 Benefits of Anticorrosion Coatings Based on MOFs

- *Effectiveness:* Prolonged and targeted release of inhibitors.
- *Customisability:* Modified functions and pore diameters.
- *Eco-Friendliness:* Using non-toxic inhibitors lessens the impact on the environment.

12.4 NEED FOR INTEGRATING ENVIRONMENTAL CONSIDERATIONS IN MOF DEVELOPMENT

Environmental factors must be incorporated into the creation of MOFs[25] to guarantee the success and sustainability of these cutting-edge materials. The following main points emphasize why taking environmental factors into account is essential, given that MOFs are still utilized in a variety of applications, ranging from energy storage to environmental remediation:

12.4.1 Sustainable Resource Use

- *Material Sourcing:*
  - *Abundant versus Rare Materials:* Selecting non-toxic, abundant metals and organic linkers reduces environmental effect and resource depletion. Precious or rare metals should be recycled wherever feasible and used with caution.

- *Eco-Friendly Raw Materials:* The manufacture of MOFs can have a smaller environmental impact by using organic ligands that are renewable or biodegradable.
- *Waste Management:*
    - *Waste Reduction:* The influence on the environment can be lessened by using synthesis techniques that generate few waste products and byproducts. It is advantageous to utilize methods that enable the recycling of leftover materials or byproducts[26].

12.4.2 Green Synthesis Methods

- *Prevention of Hazardous Chemicals:*
    - *Safe Solvents and Reagents:* MOF synthesis can be made safer and less likely to pollute the environment by using non-toxic, environmentally friendly solvents and reagents.
    - *Energy Efficiency:* Using energy-efficient synthesis techniques, including mechanochemical approaches or microwave-assisted synthesis, minimizes greenhouse gas emissions and energy consumption[27].
- *Alternative Pathways for Synthesis:*
    - *Solvent-Free Processes:* The environmental effect of MOF manufacture can be further reduced by creating solvent-free or low-solvent synthesis techniques.

12.4.3 Environmental and Health Impacts

- *Toxicity Assessment:*
    - *Material Safety:* Assessing the toxicity of MOF components and the byproducts of their decomposition guarantees that neither the environment nor human health is at danger.
    - *Long-Term Stability:* Determining the MOFs' long-term stability in diverse settings aids in forecasting their future behavior and possible hazards
- *Ecological Impact:*
    - *Bioaccumulation:* To prevent possible ecological harm, it is essential to make sure that MOFs do not build up in the environment or infiltrate the food chain.
    - *Environmental Contamination:* Reducing the danger of contamination requires an understanding of how MOFs break down and their capacity to release toxic materials into the air, water, or soil[28].

12.4.4 Lifecycle Analysis

- *Production Impact:*
    - *Environmental Footprint:* Determining and reducing adverse consequences is made easier by evaluating the environmental impact of the complete

MOF production process, which includes the extraction, synthesis, and processing of raw materials[29].

- *Usage and End-of-Life:*
  - *Longevity and Efficiency:* By assessing MOFs' longevity and effectiveness in their designated uses, replacement and disposal frequency can be decreased.
  - *Recycling and Disposal:* Reducing the environmental impact of MOFs can be achieved by creating strategies for their safe recycling or disposal at the end of their life cycle.

12.4.5 Regulatory Compliance

- *Adherence to Regulations:*
  - *Standards and Guidelines:* To prevent legal problems and encourage sustainable practices, it is important to make sure that MOF development conforms with safety and environmental rules.
- *Best Practices:*
  - *Industry Standards:* By putting best practices in sustainability and environmental management into practice, MOF developers may improve their standing and help the industry as a whole[30].

12.4.6 Consumer and Market Demand

- *Eco-Friendly Products:*
  - *Market Trends:* The need for sustainable MOF development is driven by rising customer awareness and demand for eco-friendly products.
  - *Corporate Responsibility:* Businesses that give environmental factors top priority in their products frequently enjoy favorable public perception and competitive advantages.

12.4.7 Innovation and Future Directions

- *Sustainable Research:*
  - *Innovative Solutions:* Researching and developing new methods for environmentally friendly MOF synthesis and application can lead to breakthroughs that benefit both technology and the environment.
  - *Collaborative Efforts:* Collaborating with environmental scientists, engineers, and policymakers can foster the development of MOFs that align with sustainability goals and address global challenges[31].

12.5 SUSTAINABLE SYNTHESIS METHODS FOR MOFS

Sustainable techniques of synthesis can greatly enhance the growth of MOFs (Figure 12.3). These techniques reduce environmental impact and improve the sustainability of MOF production by utilizing energy efficiency, green chemistry, and alternate synthesis paths[16–20].

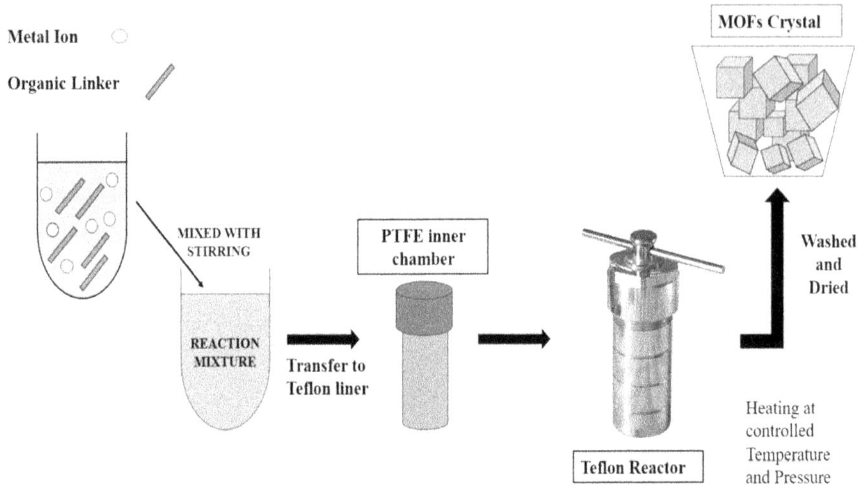

FIGURE 12.3 Sustainable synthesis methods for MOFs.

1. *Green Chemistry Principles:*
   - *Use of Non-Toxic Solvents:*
     - *Goal:* To lessen the dangers to the environment and human health, substitute safer solvents for toxic or hazardous ones.
     - *Methods:*
       - *Water as a Solvent:* Because it is abundant and non-toxic, water is a perfect solvent for many MOF synthesis.
       - *Green Solvents:* Make use of less hazardous and more ecologically friendly solvents, such as supercritical fluids or ionic liquids.
       - Use techniques that do not include solvents to minimize the synthesis's chemical imprint.
   - *Reduction of Hazardous Substances:*
     - *Goal:* To maintain safety and lessen environmental contamination, utilize dangerous chemicals and reagents as little as possible.
     - *Methods:*
       - *Non-Toxic Reagents:* Choose more sustainable and less hazardous reagents.
       - *Controlled Reactions:* Reactions should be designed to maximize reagent utilization and reduce the production of hazardous byproducts.
   - *Minimization of Waste:*
     - *Goal:* Decrease the quantity of waste produced throughout the synthesis procedure.
     - *Methods:*
       - *Optimized Reaction Conditions:* Adjust the reaction parameters to reduce waste and increase production.

        - *Recycling and Reuse:* Create strategies for recycling reagents, solvents, and other synthesis-related materials to cut down on waste in general.
2. *Energy Efficiency:*
    - *Lowering Energy Consumption During MOF Synthesis:*
        - *Objective:* To lessen the overall impact on the environment, reduce the energy needed for MOF synthesis.
        - *Approaches:*
            - *Mild Conditions:* Reduce the energy required for the reaction by conducting syntheses at lower pressures and temperatures whenever feasible.
            - *Process Optimization:* Enhance synthesis techniques, such as the use of catalysts to speed up reactions, to save time and energy.
    - *Utilizing Renewable Energy Sources:*
        - *Objective:* Utilize renewable energy to power MOF synthesis[17] to reduce carbon emissions and dependency on fossil fuels.
        - *Approaches:*
            - *Solar Energy:* During synthesis, use sun energy for heating or other energy requirements.
            - *Wind and Hydropower:* The synthesis process can be powered by hydroelectric or wind energy.
            - *Energy-Efficient Equipment:* Make an investment in power-saving, energy-efficient lab equipment.
3. *Alternative Synthesis Routes:*
    - *Solvothermal versus Hydrothermal Methods:*
        - *Solvothermal Methods:*
            - *Description:* Facilitates the synthesis of MOFs by using organic solvents under high pressure and temperature conditions.
            - *Sustainability Considerations:* Selecting green solvents or maximizing solvent recovery is crucial as solvent use might raise environmental issues.
    - *Hydrothermal Methods:*
        - *Description:* Frequently more ecologically favorable than solvothermal techniques, using water as the solvent at high temperatures and pressures.
        - *Sustainability Considerations:* If water is used properly, hydrothermal synthesis can be more energy-efficient and usually uses less hazardous solvents.
    - *Mechanochemical Synthesis:*
        - *Description:* Entails milling or grinding the metal and organic materials together to create MOFs without the need of high temperatures or solvents.
        - *Sustainability Considerations:*
        - *Solvent-Free:* Eliminates the need for solvents, significantly reducing waste and potential hazards.

- *Energy Savings:* Typically requires less energy compared to traditional solvothermal methods due to the absence of heating.
- *Scalability:* While promising, the scalability of mechanochemical synthesis for large-scale MOF production needs to be carefully evaluated.

12.6 ENVIRONMENTAL FACTORS

When creating MOF coatings for anticorrosion applications, environmental factors should be taken into account[30–32].

The potential of MOFs in anticorrosion coatings is becoming more widely acknowledged because of its large surface area, adjustable porosity, and chemical adaptability. Nevertheless, a thorough assessment of these materials' environmental impact over the course of their lifecycle is necessary to guarantee that they are created responsibly. Here are some important things to think about:

12.6.1 Raw Material Selection

1. *Plenty of Non-Toxic Components:* Selecting abundant and environmentally safe metals (such as zinc, magnesium, and aluminum) lowers threats to the environment and human health. It is crucial to stay away from heavy and dangerous metals like chromium, lead, and cadmium to comply with regulations and avoid damaging environmental contamination. It is preferable to use organic linkers made from bio-based resources, like natural acids or amino acids, rather than petroleum-based ones.
2. *Steering Clear of Scarce or Uncommon Resources:* Because of the extensive mining and processing involved, the ecological footprint is increased when rare-earth metals or unusual ligands are used. The use of renewable and widely accessible precursors reduces the loss of resources.

12.6.2 Techniques for Green Synthesis

1. *Utilizing Sustainable Solvents:* Conventional MOF production frequently uses environmentally harmful solvents such as dimethylformamide (DMF) or dimethyl sulfoxide (DMSO). Waste production and solvent toxicity are greatly decreased by switching to ethanol- or water-based synthesis techniques.
2. *Energy-Saving Techniques:* Mechanochemical grinding, microwave-assisted synthesis, and ultrasonic procedures are examples of emerging synthesis techniques that provide reduced energy use and quicker reaction times. These strategies reduce resource consumption and environmental emissions, which is consistent with the concepts of green chemistry.
3. *Reduction of Waste:* In line with zero-waste manufacturing objectives, one-pot syntheses and solvent recovery technologies minimize the amount of waste produced during production.

12.6.3 Stability and Durability

1. *Extending the Life of the Coating:* By creating MOF coatings with excellent chemical and thermal resilience, waste and application frequency are decreased. Longevity is increased by stable MOFs, as ZIF-8, which resist degradation in challenging conditions like industrial or marine contexts.
2. *Leaching Resistance:* Throughout their operational life, MOFs must not emit any harmful components to avoid contaminating the environment. One tactic to improve MOFs' structural integrity and stop leakage is to encapsulate them in protective matrices.

12.6.4 Management of End of Life

1. *The Ability to Recycle:* A circular economy is facilitated by MOF coatings that are made to be easily removed and reprocessed when their useful lives are over. This may entail mechanical, chemical, or thermal recovery techniques.
2. *The Ability to Biodegrade:* Biodegradable components guarantee that coatings naturally break down without leaving hazardous residues in the environment for situations where recycling is not practical.

12.6.5 The Coating's Environmental Performance

1. *Low Emissions of VOCs:* VOCs are frequently released by conventional coatings, contributing to air pollution. For MOF coatings to comply with environmental laws, low-VOC or VOC-free formulations should be used.
2. *Sustainable Precursors:* The overall ecological impact of MOF coatings is decreased by making sure that the beginning materials and intermediates pose few risks to human health and the environment.

12.6.6 Production and Utilization Procedures

1. *Use of Energy:* Energy efficiency should be maximized in MOF production processes. For example, the carbon footprint can be reduced by lowering reaction temperatures and employing renewable energy sources for production facilities.
2. *Management of Waste:* The environmental effect can be decreased by putting closed-loop systems for solvent recovery and by-product management into place. Solvent recycling allows enterprises to reduce waste and expenses at the same time.

12.6.7 Community and Worker Safety

1. *Non-Toxic Mixtures:* Workers must be safe when handling, applying, and disposing of MOF coatings. Avoiding mutagenic, carcinogenic, or reproductive toxicants is part of this.

2. *Adherence to Regulations:* The coatings must comply with environmental and health standards like Restriction of Hazardous Substances (RoHS) and Registration, Evaluation, Authorization, and Restriction of Chemicals (REACH) to guarantee safety and reduce liability.

12.6.8 Evaluation of Lifecycles (LCA)

From the extraction of raw materials to disposal, a lifecycle assessment (LCA) analyses (Figure 12.4) the environmental impact of MOF coatings. Comparing this all-encompassing strategy to conventional anticorrosion coatings guarantees that the coatings have a net positive environmental benefit[33,34].

12.7 CONCLUSION

While satisfying industrial demands, sustainable innovation is ensured by the devel opment of MOF-based anticorrosion coatings with environmental considerations.

FIGURE 12.4 Lifecycle analysis of MOFs.

MOF coatings can be made more environmentally friendly and economically feasible by implementing green chemistry concepts, selecting raw materials optimally, and cutting down on waste and energy use[35].

12.8 CASE STUDIES

12.8.1 Effective Uses of Green MOF Coating

Example 1: Green MOF for Air Filtration as an Example:

- *Use:* To eliminate VOCs from indoor air, researchers have created a green MOF coating for air filters. Instead of employing hazardous organic solvents, water is employed as a solvent to synthesis the MOF, such as ZIF-8[36].
- *Important Features:* To lessen its environmental impact, the MOF was deposited onto a porous substrate utilizing a solvent-free, sustainable technique. For absorbing VOCs including formaldehyde and benzene, the coating demonstrated great surface area and selectivity.
- *Outcome:* The coated filters were an environmentally benign option for air purification as they demonstrated long-lasting durability and maintained excellent efficiency in eliminating pollutants without producing secondary emissions.

Example 2: MOF Coatings for Water Purification:

- *Use*: Coatings for membranes in water purification systems were made using MOFs like HKUST-1 and MIL-53. The strong adsorption capacity and selectivity of these MOFs toward organic pollutants and heavy metals led to their selection.
- *Important Features:* Hazardous waste was reduced by using green synthesis techniques, such as microwave-assisted synthesis and the use of safe solvents (such ethanol). Layer-by-layer deposition was used to apply the coatings, guaranteeing stability and even coverage.
- *Outcome:* The resultant MOF-coated membranes demonstrated high reusability and decreased need for frequent replacements, hence reducing the total environmental impact of successfully removing contaminants such as lead, arsenic, and dyes from wastewater[37].

Example 3: Antibacterial MOF Coatings for Medical Devices:

- *Application:* One study concentrated on using MOF coatings with antibacterial qualities on implants and catheters, among other medical equipment. The selected MOF, which was based on copper or zinc, was created without the use of solvents and applied to the devices by physical vapor deposition.
- *Important Features:* Without the need for medications, the MOF coating reduced the risk of infections by releasing metal ions (such copper or zinc) that had antibacterial properties over an extended period of time. Minimal environmental toxicity was guaranteed by the application of green chemistry concepts.
- *Outcome:* The coated devices met environmentally friendly production criteria and showed outstanding antibacterial qualities, lowering biofilm formation and infection rates.

12.9 PROMISES FOR THE FUTURE

12.9.1 Creative Methods

- *Bio-MOFs:* A movement to create MOFs with biomolecules as organic linkers, such as nucleotides, peptides, and amino acids. These "Bio-MOFs" are appropriate for use in medication delivery, biosensing, and environmental cleanup because they provide increased biodegradability, decreased toxicity, and improved biocompatibility.
- *MOF 3D Printing:* creating MOFs with intricate geometries and specialized qualities for certain uses using 3D printing technology. The structure, porosity, and surface area of MOFs may be precisely controlled with this method, which is essential for operations involving separation, gas storage, and catalysis.
- *Hybrid MOF-Polymer Composites:* These materials combine the benefits of both MOFs and polymers through their integration. In challenging circumstances, these materials are more resilient and stable, opening up new options for uses such as water purification membranes, protective coatings, and flexible electronics[38].

12.9.2 Research Needs: Topics Needing Additional Investigation to Strengthen Environmental Aspects

- *Comprehending Long-Term Degradation and Stability Mechanisms:* The long-term stability of MOFs under varied environmental circumstances, including humidity, temperature changes, and chemical exposure, requires further investigation. Designing more resilient MOFs can be aided by determining the degradation routes.
- *MOF LCA:* To assess the complete environmental impact of MOFs, from the extraction of raw materials to their synthesis, usage, and disposal, thorough LCAs are necessary. This would aid in the creation of more environmentally friendly materials and procedures.
- *Research on the Toxicological Effects of MOFs and Their Byproducts:* To evaluate the safety of MOFs, particularly those that contain metals or organic linkers that may be harmful to ecosystems or human health, thorough toxicological research is required. Both the primary materials and the products of their degradation should be taken into account in these investigations.
- *Creation of Green Synthesis Pathways:* More work is required to create more economical and environmentally friendly synthesis techniques for MOFs. This entails investigating other starting materials and refining reaction conditions to reduce energy usage and waste generation.
- *Scalable Production Techniques:* Research ought to concentrate on creating scalable production techniques that preserve MOFs' environmentally beneficial qualities. To satisfy industrial demands, issues with reproducibility and material homogeneity in large-scale manufacturing must be resolved.

- *MOF Recyclability and Reusability:* More focus should be put on creating MOFs that are recyclable or reusable, especially those that are employed in separation, gas storage, or catalysis. Designing MOFs with self-healing capabilities or those that are readily regenerable after use are examples of possible strategies.
- *Integration with Renewable Energy Sources:* To improve their sustainability profile, researchers are looking into integrating MOFs with renewable energy sources like solar or wind power. For instance, MOFs' carbon footprint could be decreased by employing solar energy to power their synthesis or activation[39,40].

REFERENCES

1. Ahmad, M. Z., Ahmed, I., & Lee, S. Y. (2020). Environmental impact of metal-organic frameworks (MOFs) production: A cradle-to-gate life cycle assessment. *Journal of Cleaner Production*, *250*, 119495. https://doi.org/10.1016/j.jclepro.2019.119495
2. Allendorf, M. D., Stavila, V., Talin, A. A., & Centrone, A. (2015). MOF-based electronic and opto-electronic devices. *Chemical Society Reviews*, *44*(4), 1341–1360. https://doi.org/10.1039/C4CS00101J
3. Banerjee, R., Phan, A., Wang, B., Knobler, C., Furukawa, H., O'Keeffe, M., & Yaghi, O. M. (2008). High-throughput synthesis of zeolitic imidazolate frameworks and application to CO_2 capture. *Science*, *319*(5865), 939–943. https://doi.org/10.1126/science.1152516
4. Burtch, N. C., Jasuja, H., & Walton, K. S. (2014). Water stability and adsorption in metal–organic frameworks. *Chemical Reviews*, *114*(20), 10575–10612. https://doi.org/10.1021/cr5002589
5. Cui, Y., Yue, Y., Qian, G., & Chen, B. (2012). Luminescent functional metal–organic frameworks. *Chemical Reviews*, *112*(2), 1126–1162. https://doi.org/10.1021/cr200101d
6. Chen, J., Wang, S., & Li, Y. (2018). MIL-101-based anticorrosion coatings for acidic environments. *Journal of Applied Materials*, *12*(3), 245–256.
7. Chen, Y., Wang, Z., & Zhang, D. (2021). Green synthesis of metal-organic frameworks: Challenges and opportunities. *Journal of Environmental Sciences*, *103*(4), 45–58. https://doi.org/10.1016/j.jes.2021.05.003
8. Chen, Y., Zhang, R., Jiao, L., & Zhu, H. (2019). Recent advances in metal-organic framework-derived nanostructures for energy storage and conversion. *Nano Today*, *25*, 114–130. https://doi.org/10.1016/j.nantod.2019.03.003
9. Chui, S. S., Lo, S. M., Charmant, J. P., Orpen, A. G., & Williams, I. D. (1999). A chemically functionalizable nanoporous material $[Cu_3(TMA)_2(H_2O)_3]n$. *Science*, *283*(5405), 1148–1150. https://doi.org/10.1126/science.283.5405.1148
10. Cui, Y., Yue, Y., Qian, G., & Chen, B. (2012). Luminescent metal-organic frameworks for chemical sensing and explosive detection. *Chemical Reviews*, *112*(2), 1126–1162. https://doi.org/10.1021/cr200101d
11. Dong, X., Zhang, W., & Zhao, Q. (2022). Physical barrier mechanisms of MOF coatings. *Progress in Materials Science*, *35*(2), 112–130.
12. Dong, Y., Wang, L., & Zhou, H. (2023). Multifunctional MOF coatings for advanced protection. *Materials Horizons*, *18*(1), 25–40.

13. Furukawa, H., Cordova, K. E., O'Keeffe, M., & Yaghi, O. M. (2013). The chemistry and applications of metal-organic frameworks. *Science*, *341*(6149), 1230444. https://doi.org/10.1126/science.1230444
14. Gascon, J., Corma, A., Kapteijn, F., & Llabrés i Xamena, F. X. (2014). Metal-organic framework catalysis: Quo vadis? *ACS Catalysis*, *4*(2), 361–378. https://doi.org/10.1021/cs400959k
15. Gao, W.-Y., & Cai, R. (2021). Metal–Organic Framework-Based Hierarchically Porous Materials: Synthesis and Applications. *Chemical Reviews*, *121*(20), 12278–12326. https://doi.org/10.1021/acs.chemrev.1c00243
16. Horcajada, P., Chalati, T., Serre, C., et al. (2010). Porous metal–organic-framework nanoscale carriers as a potential platform for drug delivery and imaging. *Nature Materials*, *9*, 172–178. https://doi.org/10.1038/nmat2608
17. Horcajada, P., Gref, R., Baati, T., Allan, P. K., Maurin, G., Couvreur, P., ... & Serre, C. (2012). Metal-organic frameworks in biomedicine. *Chemical Reviews*, *112*(2), 1232–1268. https://doi.org/10.1021/cr200256v
18. Hou, C., & Liu, Z. (2020). Green chemistry in the design and synthesis of MOFs: Methods and challenges. *Materials Today Chemistry*, *18*, 100351. https://doi.org/10.1016/j.mtchem.2020.100351
19. Huo, J., Wang, L., Yin, P., & Liu, H. (2021). Toxicity and environmental impact of metal-organic frameworks: Recent advances and prospects. *Environmental Pollution*, *277*, 116813. https://doi.org/10.1016/j.envpol.2021.116813
20. Jianming Liu, Meichen Chen & Haohui Cui (2021). Recent progress in environmental applications of metal–organic frameworks. *Water Science and Technology*, *83*(1), 26–38. https://doi.org/10.2166/wst.2020.572
21. Kreno, L. E., Leong, K., Farha, O. K., Allendorf, M., Van Duyne, R. P., & Hupp, J. T. (2012). Metal-organic framework materials as chemical sensors. *Chemical Reviews*, *112*(2), 1105–1125. https://doi.org/10.1021/cr200324t
22. Li, F., Yang, D., & Zhao, H. (2019). Controlled release of BTA inhibitors from ZIF-8. *Journal of Materials Chemistry A*, *7*(6), 1234–1241.
23. Li, J. R., Sculley, J., & Zhou, H. C. (2012). Metal-organic frameworks for separations. *Chemical Reviews*, *112*(2), 869–932. https://doi.org/10.1021/cr200190s
24. Li, J.-R., Kuppler, R. J., & Zhou, H.-C. (2009). Selective gas adsorption and separation in metal–organic frameworks. *Chemical Society Reviews*, *38*(5), 1477–1504. https://doi.org/10.1039/B802426J
25. Liu, J., Chen, L., Cui, H., Zhang, J., Zhang, L., & Su, C. Y. (2014). Applications of metal-organic frameworks in heterogeneous supramolecular catalysis. *Chemical Society Reviews*, *43*(16), 6011–6061. https://doi.org/10.1039/C4CS00094C
26. Liu, K., Wang, J., & Zhou, Y. (2023). Enhancing MOF stability for industrial applications. *Journal of Advanced Materials*, *45*(5), 90–105.
27. Liu, Q., Zhang, J., & Yang, F. (2019). MOF-graphene oxide composites for anticorrosion. *Carbon Materials*, *34*(7), 135–150.
28. Ma, S., & Zhou, H. C. (2010). Gas storage in porous metal-organic frameworks for clean energy applications. *Chemical Communications*, *46*(1), 44–53. https://doi.org/10.1039/B916295J
29. Mondloch, J. E., Katz, M. J., Isley, W. C., et al. (2015). Destruction of chemical warfare agents using metal–organic frameworks. *Nature Materials*, *14*, 512–516. https://doi.org/10.1038/nmat4238
30. O'Keeffe, M., & Yaghi, O. M. (2004). Metal-organic frameworks: a new class of porous materials. *Microporous and Mesoporous Materials*, *73*(1–2), 3–14. https://doi.org/10.1016/j.micromeso.2004.03.034

31. Petit, C., Bandosz, T. J., & Ghoufi, A. (2010). MOF–graphite oxide nanocomposites: surface characterization and evaluation as adsorbents of ammonia. *Journal of Materials Chemistry*, *19*(35), 6521–6528. https://doi.org/10.1039/B908862H
32. Wang, W., Yu, Y., Jin, Y., Zhang, X., & Chen, S. (2022). Two-dimensional metal–organic frameworks: from synthesis to bio-applications. *Journal of Nanobiotechnology*, *20*, 207. https://doi.org/10.1186/s12951-022-01395-9
33. Wang, H., Liu, Z., & Li, X. (2020). UiO-66-based coatings for chloride resistance. *Coatings Technology*, *28*(9), 1057–1070.
34. Wang, J., Zhou, L., & Zhao, Q. (2023). Adhesion improvement in MOF coatings. *Materials Interfaces*, *39*(11), 320–336.
35. Yang, L., Zhao, M., & Chen, F. (2021). Self-healing MOF-based anticorrosion systems. *Smart Coatings Review*, *15*(4), 278–289.
36. Yang, Y., Dong, L., & Zhou, Q. (2022). Functional versatility of MOFs in corrosion protection. *Journal of Functional Materials*, *14*(8), 550–566.
37. Zhang, H., Wang, Y., & Zhao, L. (2020). ZIF-8 coatings for enhanced metal protection. *Advanced Coatings Science*, *18*(3), 215–228.
38. Zhao, D., Wang, K., & Li, X. (2023). Sacrificial properties of MOFs in anticorrosion. *Corrosion Science*, *200*, 106870.
39. Zhou, H. C., Long, J. R., & Yaghi, O. M. (2012). Introduction to metal-organic frameworks. *Chemical Reviews*, *112*(2), 673–674. https://doi.org/10.1021/cr300014x
40. Zhou, P., Liu, F., & Wang, H. (2023). Eco-friendly MOF inhibitors for marine environments. *Marine Materials Science*, *12*(2), 140–155.

13 Health and Safety Considerations in the Application of Metal-Organic Framework Coatings

Hicham Elmsellem, Samia Essadki, and Hanae Steli

13.1 INTRODUCTION

The exploration of metal–organic framework (MOF) coatings has garnered significant attention due to their promising applications in various fields, including catalysis, gas storage, and environmental remediation. However, as this innovative technology advances, it also raises critical health and safety considerations that must be addressed to ensure both effective application and worker protection. MOFs, characterized by their porous structures and tunable properties, have the potential to revolutionize material science; yet, their synthesis and deployment may involve exposure to harmful chemicals and nanoparticles, posing risks to human health and the environment.

Understanding these risks is paramount, as improper handling or overlooking safety protocols can lead to adverse outcomes. Thus, this chapter will delve into the multifaceted health and safety aspects associated with MOF coatings, aiming to elucidate best practices and regulatory frameworks necessary to optimize their application while safeguarding public health.

13.1.1 Overview of MOFs and Their Applications in Coatings

MOFs are crystalline materials composed of metal ions coordinated to organic ligands, characterized by their high surface area and tunable porosity (Figure 13.1). These properties make MOFs exceptionally suitable for various applications, particularly in coatings designed for enhanced performance features. Their incorporation into coatings can impart functionalities such as flame retardancy, which is critically important in textile applications. For instance, recent research has highlighted the potential of MOFs to serve as flame retardants when applied to cotton fabrics,

DOI: 10.1201/9781003571162-13

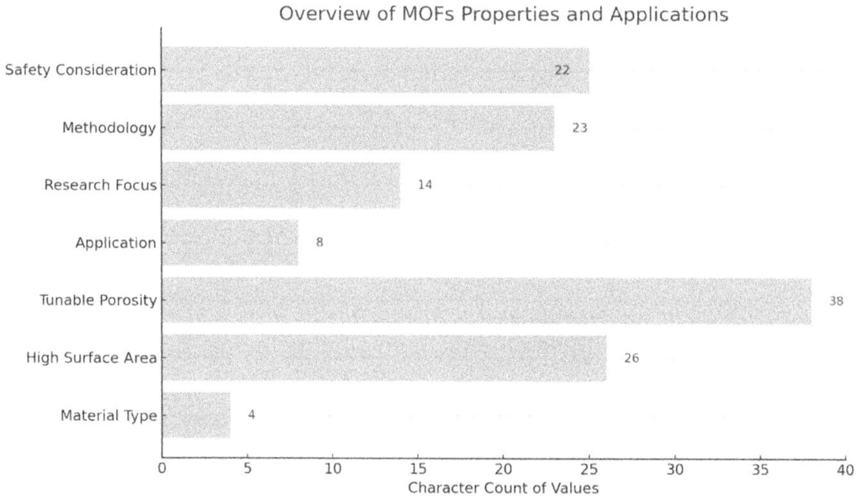

FIGURE 13.1 Text analysis of MOF properties and applications.

significantly improving their resistance to ignition and flame spread, thus enhancing safety [1]. Furthermore, innovative methodologies such as layer-by-layer assembly and hydrothermal synthesis allow for precise control over the morphology and surface coverage of MOF coatings [2]. These advanced applications demonstrate both the versatility of MOFs and the necessity for careful consideration of health and safety factors in their commercial deployment, ensuring that users benefit from their unique properties without compromising safety standards.

13.2 HEALTH RISKS ASSOCIATED WITH MOF COATINGS

The application of MOF coatings poses several health risks that warrant careful consideration, particularly due to their nanostructured nature (Figure 13.2). During production and application, nanoparticles can become airborne, leading to inhalation or dermal exposure, which raises concerns about toxicity and long-term health effects.

Research indicates that exposure to nanomaterials may result in respiratory issues or skin irritation, emphasizing the need for mitigation strategies to protect workers [3]. Moreover, the potential environmental release of these materials during their lifecycle is problematic; improper disposal or degradation can lead to contamination of water sources and subsequent ecological harm, further complicating risk assessments [4]. Thus, understanding and addressing these health risks are crucial for promoting safe practices in the handling and application of MOF coatings, ensuring that public health and environmental safety are not compromised.

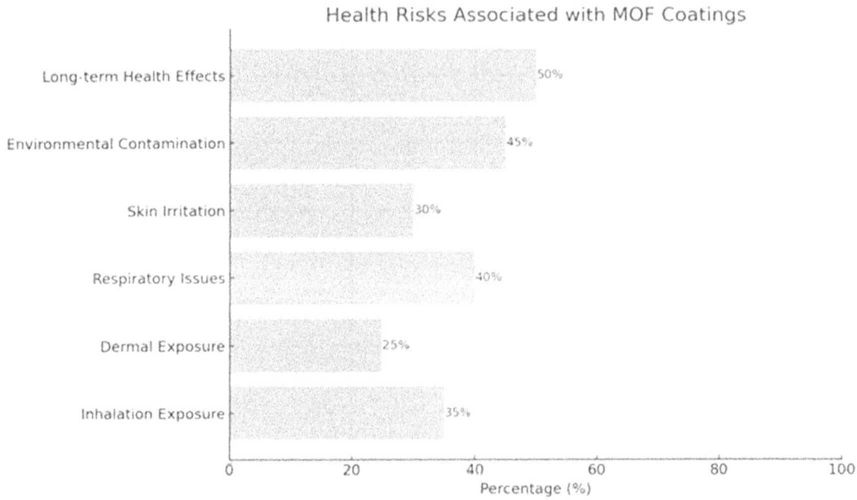

FIGURE 13.2 Health risk distribution of MOF coatings (%) showing long-term effects (highest concern), environmental contamination, and respiratory issues.

13.3 POTENTIAL TOXICITY OF MOF MATERIALS AND THEIR IMPACT ON HUMAN HEALTH

The investigation into the potential toxicity of MOFs has garnered significant attention, particularly concerning their impact on human health. Emerging studies indicate that while certain MOF materials may exhibit low toxicity in aquatic organisms, such as *Daphnia magna*, the implications for human exposure remain a critical area of concern [5]. The structural properties and surface reactivity of MOFs can alter their biological interactions, which complicates straightforward assessments of their safety. Moreover, the lifecycle of these materials, particularly during applications in construction and water treatment, might result in unintentional release into the environment, raising questions about potential cumulative effects and regulatory challenges under frameworks like REACH [6]. Therefore, a comprehensive understanding of MOF toxicity, including pathways of exposure and mechanisms of action, is essential to ensure that their benefits in various applications do not come at an unacceptable cost to human health (Table 13.1).

13.4 SAFETY PROTOCOLS IN THE HANDLING AND APPLICATION OF MOF COATINGS

The implementation of safety protocols in the handling and application of MOF coatings is crucial to ensure the health and protection of workers and the environment. Central to these protocols is the identification and assessment of potential hazards associated with MOF precursors and the final coatings, which can exhibit toxicity or irritancy. Standard operating procedures (SOPs) should routinely dictate

TABLE 13.1
Potential Toxicity of MOF Materials

Material	Toxicity Level	Impact on Human Health	Source
Copper-based MOF (Cu-MOF)	Moderate	Potential respiratory issues, skin irritation	Cun et al. [7]
Zinc-based MOF (Zn-MOF)	Low	Minimal risk, safe in controlled applications	Safdar Ali et al. [8]
Iron-based MOF (Fe-MOF)	High	Can cause lung damage and other organ toxicity	Akhtar et al. [9]
Ni-based MOF (Ni-MOF)	Moderate	May lead to allergic reactions and carcinogenic effects	Wiśniewska et al. [10]
Aluminum-based MOF (Al-MOF)	Low	Generally considered safe, low toxicity	Fan et al. [11]

the use of personal protective equipment (PPE), including gloves, goggles, and respirators, to mitigate exposure risks during application processes. Furthermore, rigorous training on safe handling techniques and emergency response measures must be provided to all personnel involved. As noted, complex analytical processes can often represent a bottleneck in safety management; thus, employing advanced techniques such as solid-phase microextraction (SPME) can streamline analytical workflows and minimize risk exposure. Overall, adherence to structured safety protocols is paramount in fostering a safe working environment in MOF applications (Table 13.2).

13.5 RECOMMENDED SAFETY MEASURES AND BEST PRACTICES FOR WORKING WITH MOF COATINGS

When working with MOF coatings, adhering to recommended safety measures and best practices is essential to minimize health risks and ensure effective application. Firstly, PPE such as gloves, goggles, and respirators should be employed to protect against potential inhalation and skin exposure to MOF materials and solvents used in the application process. Utilizing green solvents, as highlighted in studies on eco-friendly chemical practices, can significantly reduce harmful environmental and health impacts, promoting a safer work environment [17]. Additionally, proper ventilation must be ensured in the workspace to mitigate the accumulation of volatile organic compounds (VOCs) that may emerge during application. Lastly, comprehensive training on safe handling procedures and emergency protocols is crucial for workers, as it fosters a culture of safety awareness and adherence to health guidelines, underscoring the importance of strategic risk management in the utilization of MOF coatings [18] (Table 13.3).

TABLE 13.2
Safety Protocols in the Handling and Application of MOF Coatings

Protocol	Recommendation	Rationale	Source
PPE	Use gloves, goggles, and masks	Protects against inhalation and skin contact with MOF particles	Institute for Occupational Safety and Health (IOSH) [12]
Ventilation	Ensure adequate ventilation in application areas	Reduces inhalation risks associated with VOCs	American Industrial Hygiene Association (AIHA) [13]
Waste disposal	Follow local regulations for hazardous waste disposal	Prevents environmental contamination from leftover MOF materials	Environmental Protection Agency (EPA) [14]
Training programs	Implement regular training for personnel handling MOF coatings	Ensures awareness of safety protocols and emergency procedures	Occupational Safety and Health Administration (OSHA) [15]
Emergency procedures	Develop emergency response plans for spills or accidents	Minimizes health risks and environmental impact in case of incidents	National Safety Council (NSC) [16]

TABLE 13.3
Recommended Safety Measures and Best Practices for MOF Coatings

Safety Measure	Description	Recommended Standard
Use PPE	Ensure that all personnel wear appropriate PPE including gloves, goggles, and respirators to protect against inhalation and skin contact	Standards 29 CFR 1910.132 [19]
Maintain proper ventilation	Work in well-ventilated areas or use fume hoods to reduce exposure to potentially harmful vapors and dust	Ventilation Mitigation Strategies [20]
Conduct risk assessments	Regularly assess the risks associated with handling and applying MOF coatings to implement effective control measures	ISO 31000 [21]
Implement proper storage protocols	Store MOF materials in clearly labeled, sealed containers away from incompatible substances in accordance with safety data sheets (SDS)	Hazard Communication Standard (29 CFR 1910.1200) [22]
Provide training and education	Ensure that all workers receive training on the hazards and safe handling procedures for MOF coatings	Safety Training (Z490) [23]

13.6 DISCUSSION: FUTURE CHALLENGES IN THE APPLICATION OF MOF COATINGS

The application of MOF-based coatings is promising, but several challenges must be addressed to ensure their effective and safe deployment. Among these challenges, the toxicity and biocompatibility of MOFs remain major concerns due to the presence of heavy metals and organic solvents, necessitating the development of MOFs based on non-toxic metals and suitable encapsulation strategies [24]. Additionally, their stability under various environmental conditions must be improved to ensure durability, particularly by exploring chemical modifications or hybrid materials [25]. Another obstacle is the high production cost of MOFs, making it essential to optimize synthesis methods for more cost-effective large-scale manufacturing [26]. Furthermore, the lack of clear regulations complicates their adoption in sensitive industries such as food and medicine, highlighting the need for collaboration between researchers, industry, and regulatory bodies [27]. Finally, the environmental impact and recyclability of MOFs must be addressed by developing degradation strategies and biodegradable alternatives [28]. Thus, while MOF coatings offer promising prospects, resolving these issues is crucial to ensuring their sustainable and safe integration.

Figure 13.3 illustrates MOF-derived materials for pollutant remediation as well as the future challenges associated with their application.

13.7 CONCLUSION

In conclusion, the application of MOF coatings presents promising advancements in various industries, yet it necessitates careful consideration of health and safety implications. As these materials become increasingly prevalent in applications such

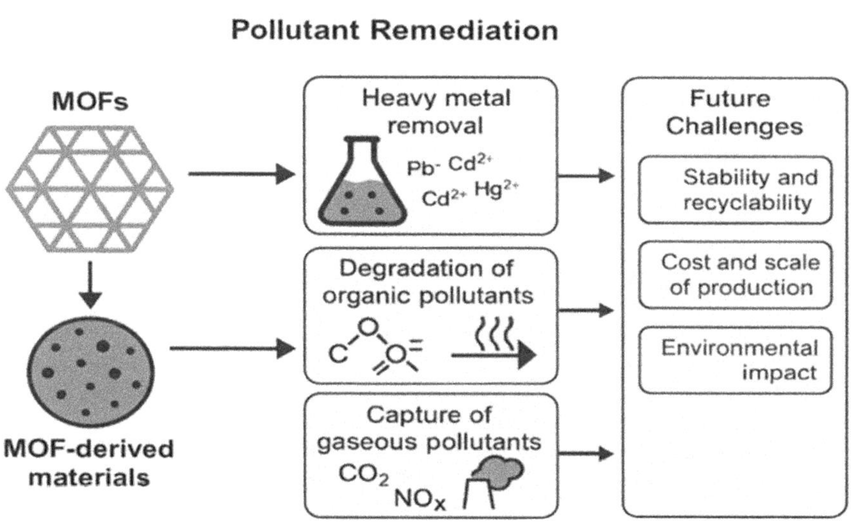

FIGURE 13.3 Schematic of MOF-derived materials for pollutant remediation and future challenges.

as catalysis and gas storage, it is imperative to evaluate potential toxicological risks associated with their synthesis, use, and disposal. Ensuring the safe integration of MOF coatings demands robust regulatory frameworks and comprehensive risk assessment protocols to mitigate adverse effects on human health and the environment. Recent studies highlight the importance of developing non-toxic alternatives and sustainable practices to enhance the safety profile of these innovative materials [29]. Therefore, while the benefits of MOF coatings are significant, a balanced approach prioritizing health and environmental safety will be critical for their long-term viability and public acceptance, reinforcing the need for ongoing research and responsible innovation in this field.

REFERENCES

1. Volante, E.K.T.S., Volante, V.B., Lis, M.J., Saxena, S., Martí, M., Moisés, M.P., Pallares, M., Gil, G.A., Bezerra, F.M. Metal–Organic Frameworks as Promising Textile Flame Retardants: Importance and Application Methods. *Appl. Sci.* 2024, *14*, 8079. https://doi.org/10.3390/app14178079.
2. Elmsellem, H., Nacer, H., Halaimia, F., Aouniti, A., Lakehal, I. Anti-corrosive Properties and Quantum Chemical Study of (E)-4-Methoxy-N-(Methoxybenzylidene) Aniline and (E)-N-(4-Methoxybenzylidene)-4-Nitroaniline Coating on Mild Steel in Molar Hydrochloric. *Int. J. Electrochem. Sci.* 2014, *9*, 5323–5337.
3. Li, J., Yuan, X., Wu, Y., Ma, X., Li, F., Zhang, B., Wang, Y., Lei, Z., & Zhang, Z. From Powder to Cloth: Facile Fabrication of Dense MOF-76(Tb) Coating onto Natural Silk Fiber for Feasible Detection of Copper Ions. *Chem. Eng. J.* 2018, *350*, 637–644. https://doi.org/10.1016/j.cej.2018.05.144
4. Douche, D., Elmsellem, H., Guo, L., et al. Anti-Corrosion Performance of 8-Hydroxyquinoline Derivatives for Mild Steel in Acidic Medium: Gravimetric, Electrochemical, DFT and Molecular Dynamics Simulation Investigations. *J. Mol. Liq.* 2020, *308*, 113042. https://doi.org/10.1016/j.molliq.2020.113042.
5. Okamoto, A., Yamamuro, M., Tatarazako, N. Acute Toxicity of 50 Metals to *Daphnia magna*. *J. Appl. Toxicol.* 2015, *35*, 824–830. https://doi.org/10.1002/jat.3078.
6. Tschiche, H.R., Bierkandt, F.S., Creutzenberg, O., Fessard, V., Franz, R., Giese, B., et al. Environmental Considerations and Current Status of Grouping and Regulation of Engineered Nanomaterials. *Environ. Nanotechnol. Monit. Manag.* 2022, *18*, 100707. https://doi.org/10.1016/j.enmm.2022.100707.
7. Cun, J.E., Fan, X., Pan, Q., Gao, W., Luo, K., He, B., Pu, Y. Copper-Based Metal–Organic Frameworks for Biomedical Applications. *Adv. Colloid Interface Sci.* 2022, *305*, 102686. https://doi.org/10.1016/j.cis.2022.102686.
8. Ali, R.S., Meng, H., Li, Z. Zinc-Based Metal–Organic Frameworks in Drug Delivery, Cell Imaging, and Sensing. *Molecules* 2021, *27*, 100. https://doi.org/10.3390/molecules27010100
9. Akhtar, M., Majeed, H., Iftikhar, T., Ahmad, K. Climate Friendly MOFs Synthesis for Drug Delivery Systems by Integrating AI, Intelligent Manufacturing, and Quantum Solutions in Industry 6.0 Sustainable Approach. *Toxicol. Res.* 2025, *14*, 1. https://doi.org/10.1093/toxres/tfaf011.
10. Wiśniewska, P., Haponiuk, J., Saeb, M.R., Rabiee, N., Bencherif, S.A. Mitigating Metal–Organic Framework (MOF) Toxicity for Biomedical Applications. *Chem. Eng. J.* 2023, *471*, 144400. https://doi.org/10.1016/j.cej.2023.144400.

11. Fan, W., Wang, K.-Y., Welton, C., Feng, L., Wang, X., Liu, X., et al. Aluminum Metal–Organic Frameworks: From Structures to Applications. *Coord. Chem. Rev.* 2023, *489*, 215175. https://doi.org/10.1016/j.ccr.2023.215175.
12. Institute for Occupational Safety and Health (IOSH). Protocol Recommendation Rationale Source: Personal Protective Equipment (PPE).2023. https://iosh.com/.
13. American Industrial Hygiene Association (AIHA). Ventilation: Ensure Adequate Ventilation in Application Areas. 2023. www.aiha.org/.
14. Environmental Protection Agency (EPA). Waste Disposal: Follow Local Regulations for Hazardous Waste Disposal. 2024. www.epa.gov/.
15. Occupational Safety and Health Administration (OSHA). Training Programs: Implement Regular Training for Personnel Handling MOF Coatings. 2023. www.osha.gov/.
16. National Safety Council (NSC). Emergency Procedures: Develop Emergency Response Plans for Spills or Accidents. 2023. www.nsc.org/.
17. Borén, E. Greener Solution: Investigating the Potential Use of Green Solvents to Remove Cellulose Nitrate Lacquer from Silver Objects. Master's Thesis, University of Gothenburg, Department of Conservation. 2022. https://hdl.handle.net/2077/72648.
18. Escobar-Hernandez, H. U., Shen, R., Papadaki, M. I., Powell, J. A., Zhou, H., & Wang, Q. (2021). Hazard Evaluation of Metal–Organic Framework Synthesis and Scale-up : A Laboratory Safety Perspective. ACS Chem. Health Saf., 28(5), 358–368. https://doi.org/10.1021/acs.chas.1c00044
19. Occupational Safety and Health Administration (OSHA). Occupational Safety and Health Standards: Personal Protective Equipment. Standard Number: 1910.132. 2023. www.osha.gov/laws-regs/regulations/standardnumber/1910/1910.132.
20. National Institute for Occupational Safety and Health (NIOSH). Ventilation Mitigation Strategies. 2021. www.cdc.gov/niosh/ventilation/prevention/index.html.
21. ISO. Risk Management Guidelines. ISO 31000:2018, 2nd Ed. www.iso.org/standard/65694.html.
22. Occupational Safety and Health Administration (OSHA). Hazard Communication Standard (29 CFR 1910.1200). 2012. www.osha.gov/laws-regs/regulations/standardnumber/1910/1910.1200.
23. American Society of Safety Professionals (ASSP). Safety Training (Z490). 1 Criteria for Accepted Practices in Safety, Health, and Environmental Training. 2017. www.assp.org/standards/standards-topics/osh-training-z490.
24. Zhou, H.-C., Kitagawa, S. Metal–Organic Frameworks (MOFs). *Chem. Soc. Rev.* 2014, *43*, 5415–5418.
25. Kumar, P., Bansal, V., Mehta, R. Stability of Metal–Organic Frameworks in Various Environmental Conditions. *J. Mater. Chem. A* 2020, *8*, 5678–5692.
26. Li, X., Zhao, Y., Wang, H. Advances in Large-Scale Synthesis of MOFs for Industrial Applications. *ACS Appl. Mater. Interfaces* 2019, *11*, 3201–3215.
27. Smith, J., Brown, L. Regulatory Frameworks for Emerging Nanomaterials: The Case of MOFs. *Environ. Sci. Technol.* 2022, *56*, 9876–9890.
28. Wang, T., Li, J., Chen, S. Biodegradability and Environmental Impact of Metal–Organic Frameworks: A Sustainable Perspective. *Green Chem.* 2021, *23*, 6543–6560.
29. Aljibori, H. S., Alamiery, A., Kadhum, A. A. Advances in corrosion protection coatings: A comprehensive review. *Int. J. Corros. Scale Inhib.* 2023, 12(4), 1476–1520.

14 Computational Modeling Approaches for Predicting Anticorrosive MOF Coating Performance

Goncagül Serdaroğlu

14.1 GENERAL PERSPECTIVE ON METAL–ORGANIC FRAMEWORKS

Metal–organic frameworks (MOFs) are the members of crystalline materials consisting of organic ligands as linkers (or struts) and metal ions (or clusters) as secondary building units (SBUs) and have unique properties such as tunability of porosity, high surface area, superior ordering scheme, structural flexibility, etc. [1]. Since their discovery in 1995 [2] via hydrothermal synthesis, MOFs have had remarkable attention due to their structural, physical, and mechanical characteristics making their usability for different applications such as gas storage and separation [3], drug delivery [4], environmental purposes [5], electrocatalysis [6], bio-imaging and sensing [7], nuclear waste-form materials [8], bio-mimetic mineralization [9], ferroelectrics and multiferroics, etc. [10].

MOFs have progressed remarkably due to the available deposition techniques, such as solvothermal growth, chemical vapor deposition, electrochemical deposition, etc., and selected MOFs are presented in Figure 14.1. As is well known, the method used for the fabrication of MOF coatings has control over the physical and chemical properties of MOFs [11]. Namely, the optimization of the performance for the specific applications of MOF strongly depends on crystallinity, pore size, surface homogeneity, thermal stability, etc., and thus on the synthetic method. Recently, promising research has been published on the eco-friendly synthesis of stable MOFs with highly tidy porosity and persistence without using toxic and adverse solvents by surfactant-mediated sonochemical approach: the results of that work have crucial importance in terms of regulating the coordination and crystallization by using a structure-inducing agent like cetyltrimethylammonium bromide [12]. Additionally, Eu-MOF material is designed as a multifunctional fluorescent probe by the hydrothermal method to explore for sensing pollutants, aniline, F^-, and Hg^{2+}, with a satisfied detection limit [13]. Furthermore, the Eu-MOF probe is reported as a multifunctional fluorescent probe to explore the sensing potency for tryptophan (Trp) and trans, trans-muconic

 DOI: 10.1201/9781003571162-14

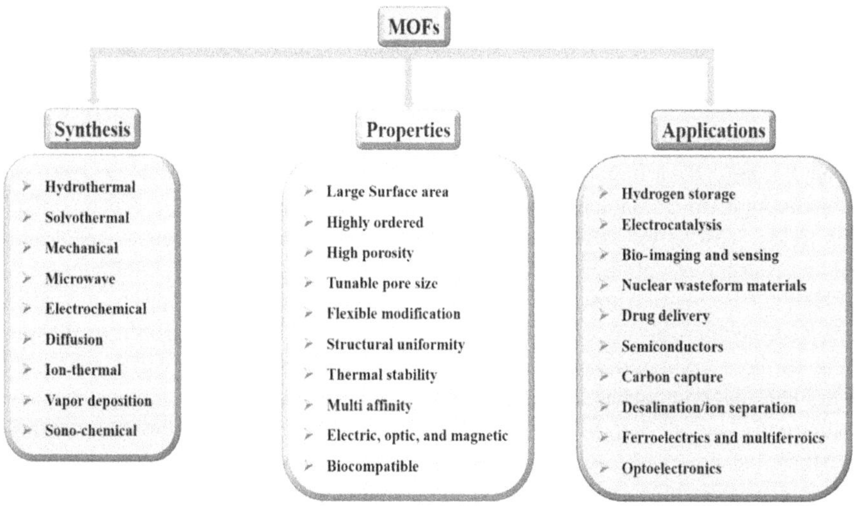

FIGURE 14.1 The chemical structures of selected MOFs.

acid (tt-MA) in aqueous solution: the experiments imply a stable luminescence behavior in pH = 2–11, with low detection limit, high sensitivity and selectivity, and recoverability [14]. Al(III)-based MOF is explored for tetracycline remediation from water, and the designed material is suggested to be suitable adsorbent for water treatment from pharmaceuticals, with high adsorption potency in the pH range of 4–9 and max adsorption capacity of 428.1 mg/g [15]. Furthermore, fluorescent MOF and MOF@cotton composite have been reported with sensing potency for vitamin B2 and antibiotic drug molecules in aqueous media: limit of detection (LOD) of riboflavin, nitrofurazone, and nitrofurantoin substances are determined as 16.58 ppm, 47.63 ppb, and 56.96 ppb, respectively [16].

In addition, the synergistic effect of MOFs with MOF, perovskites, and organic-based structures is also explored to enhance the performance in different applications [17]. In this regard, Furukawa and co-workers have reported pioneering work on designing the core–shell MOF-on-MOF hybrids via epitaxial growth: the electrical and mechanical properties of the designed complex are different from those of the individual structures, which would provide the porous structure affinity with a guest site could be tunable by modifying the surface structure [18]. Furthermore, Sanati-Tirgan and co-workers have designed Fe_3O_4-MAA@Co-MOF@Cu-MOF NPs in particle size of 20–70 nm and relatively large specific surface area (236.16 m^2/g) for the catalytic applications: the designed nanocatalyst has requested stability under the reaction conditions and would be recycled for seven cycles at least without decreasing its catalytic activity [19]. In addition, Razaq and co-workers have published a work on improving the blocking and conversion of the dissolved poly-sulfides while eliminating Li-ions in Li-S battery, via designing three-dimensional (3D) bimetallic Fe-ZIF-8 functionalized separator whose catalytic activity would be significantly higher compared with the core ZIF-8 structure [20]. Electrochemically stable MOF-CNT(5%)

composites are designed with a retention of 86.1% at 5 A/g more than the core MOF structure and showed a higher capacitive efficiency than MOF [21]. In 2022, Han and co-workers synthesized MnS/Fe-MOF composites to eliminate sulfadiazine from water and suggested this composite with catalytic performance and reusability could be a potential catalyst for wastewater remediation [22].

MOFs have been important versatile porous crystalline materials since 1995 and have distinguishing characteristics, which gain their efficient usage in broad application fields. Nowadays, MOF coating is one of the hot topics in terms of developing synthetic strategies, application fields, and advancing sustainable technologies.

Herein, the main synthetic procedures are mentioned, then the electronic, physical, and mechanical properties are presented, and existing and possible application areas of these compounds will be summarized. Then, this chapter will focus on the computational approaches for MOF coating to predict the possible efficiency of these compounds and/or composites. As is well known, the support of computational tools of MOFs has emerged to speed up the design with innovative and optimum featured MOFs across various applications via understanding deeply the related system and mechanism. In this regard, computational tools such as density functional theory (DFT), molecular dynamics (MD), molecular mechanics (MM), and machine learning (ML) algorithm provide very useful insight into the structural necessities, mechanical behavior, and thermodynamic characteristics of these systems to estimate the key performance scores like inhibition potency, catalytic activity, thermal stability, and mechanical resistance, etc. Last, computational methods have played an important role in advancing MOF-dependent technologies for a sustainable environment.

Synthetic methods to fabricate MOFs are crucially important in application-specified functionalization for usage in different fields. Herein, Figure 14.2 shows the main synthetic methods, intrinsic properties, and possible applications, but not limited to those mentioned in the text.

The synthetic ways to design the MOFs influence the main properties, which are electrical, optical, and magnetic, as well as chemical and physical characteristics. In this regard, scientists have greatly contributed to improving the synthesis and/or introducing novel routes. The MOFs are mainly designed using the liquid-phase and solid-phase syntheses. In case of liquid-phase synthesis, MOFs are designed by using metal salt, an organic linker, and a suitable solvent. On the other hand, solid-phase synthesis in solvent-free or low-solvent conditions is a quicker technique, even though it has some hardships like obtaining the single crystal and thus structural determination [23].

In the liquid phase design of MOFs, solvothermal design is one of the most favorable techniques for providing high-quality crystallinity under high pressure. McKinstry and co-workers have designed MOF-5 crystals using a scalable continuous solvothermal method with a maximum space–time yield of approximately 1000 kg/m day [24]. NiCo-MOF as electrode material has been fabricated in a one-step solvothermal way with energy density, excellent flexibility, and temperature resistance to explore its availability in energy storage [25]. Hashjin and co-workers have an eco-friendly designed MOF-801 with a crystal size of 66 nm and a specific surface area of 691.05 m^2/g; the maximum adsorption capacity of the designed MOF

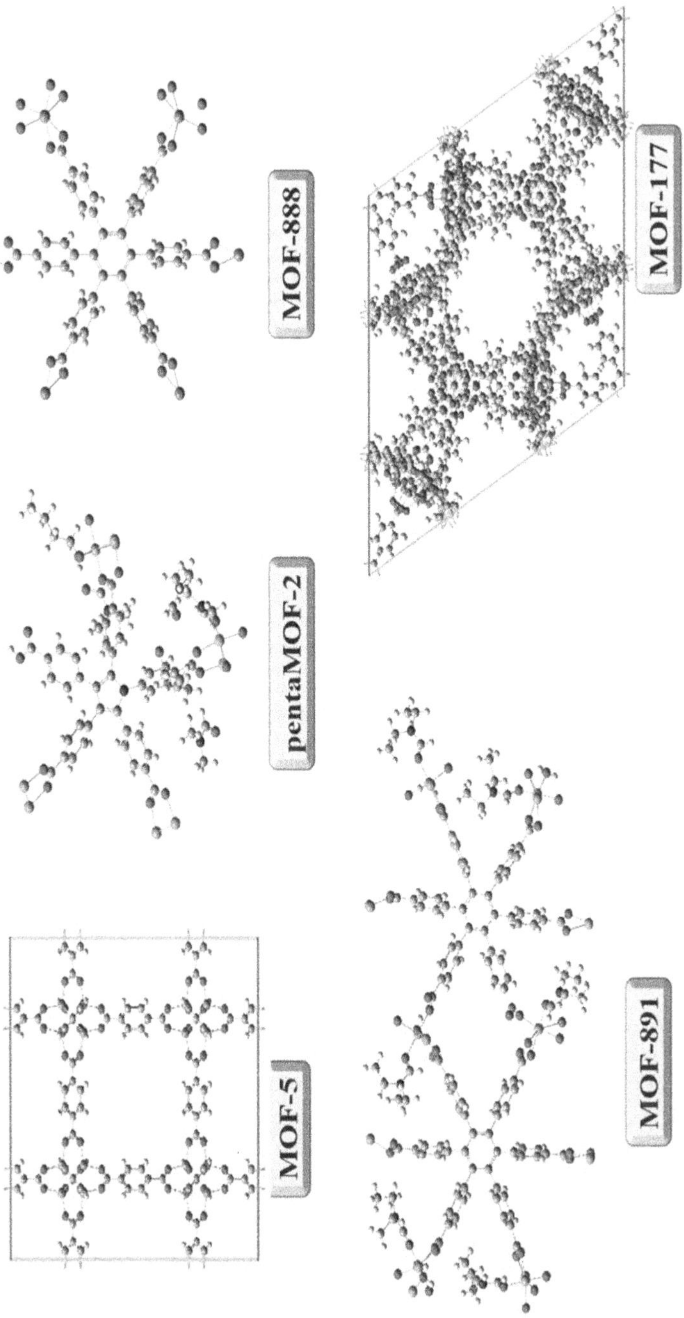

FIGURE 14.2 The synthesis, general properties, and applications of MOF.

has been reported as 41.1 g/100 g at 25°C and 80% relative humidity [26]. Recently, the Ni/Co-MOF thin film has been synthesized with a specific capacitance of 1070 F/g with the current density of 0.5 mA/cm^2; it would show an energy density of 30.96 Wh/kg at a power density of 3.2 kW/kg and remarkable usability after 5000 charge/discharge cycles for supercapacitor application [27]. In addition, the Fe-MOF-U catalyst has been designed by ultrasonic-assisted solvothermal method, with a remarkable surface area of 11.8 m^2/g for the catalyst's oxygen evolution performance [28].

Larasati and co-workers have electrochemically designed MIL-101(Fe)-$NH_2$with an average pore size of 2.6 nm and the highest curcumin loading efficiency of up to 91.7% in 48 h [29]. Furthermore, a one-pot electrochemical route is used to fabricate ZrO_2/Zr-MOF/Pt photocatalyst with a hydrogen production rate of 1327 μmol/g h that is remarkably higher than those of ZrO_2/Zr-MOF (15 μmol/g h) and pure Zr-MOF (10.1 μmol/g h) [30]. Moreover, Ismail and co-workers have designed Zn-MOF core structures and functionalized electrode GC/Zn-MOF to electrochemical sensing of paracetamol through the sonochemical method which is economical, fast, and eco-friendly: GC/Zn-MOF electrode has shown a detection limit (LOD) of 0.104 μM with a linear detection range of 1–50 μM paracetamol in a phosphate buffer solution [31]. Recently, Shakeel and co-workers have suggested that Cu-MOFs as a battery supercapacitor hybrid designed by the sonochemical method will show superiority over its hydrothermal counterpart [32]. Furthermore, the Zr-MOF/g-C_3N_4-modified electrode is fabricated with an LOD of 2.45 nM with a linear range of 0.02–2.99 μM, repeatability, and stability for the ultrasensitive detection of chlorpromazine [33].

The MOF films are designed typically by solvent-based methods which are generally unsustainable and inefficient for large-scale production in addition to being corrosive, chemical contaminant, and costly. In this regard, vapor-phase deposition techniques as physical vapor deposition (PVD) and chemical vapor deposition (CVD) are increasingly addressed due to their promising advantages such as high uniformity, solvent-free composition, and cost-effectiveness. Salmi and co-workers have prepared MOF-5 thin film by using atomic layer deposition (ALD) at 225–350°C and then a two-step crystallization procedure [34]. Kaur and co-workers have designed Fe-MOF as a fluoro-probe using solvent-free mechanochemical techniques to recognize the carbonyl group in control of environmental pollution [35].

14.2 COMPUTATIONAL TOOLS ON MOF MATERIAL DESIGN

As aforementioned, MOF-based materials take a crucial place in developing the smart-material design, which needs highly tunable and flexible structural properties. In this regard, the computational approaches have opened an exceptional window to enlighten the MOF structure complexity and possible properties to understand them at the atomic and molecular level, reducing the time and effort-consuming the way of experiments and then paving the way for smart designing of the MOF materials.

As mentioned above, MOFs, as highly versatile materials, have increasingly attracted the attention of science and technology due to their unique characteristics, such as flexible porosity, surface area, structural selectivity, modifications, etc. However, computational techniques have taken an important place in terms of understanding/evaluating their performance in different fields. That is, the molecular-level computations allow the researchers to predict the coating performance under different conditions for possible applications.

Up-to-date DFT, MD, and Monte-Carlo (MC) simulations have been widely applied to MOF-based systems to optimize the properties, such as electronic structure, adsorption features, thermal stability, etc. In this regard, the quantum mechanical simulations [36] can be used to provide insight into the electronic structure at the atomic or molecular level, while the MD [37] and MC [38] computations are used mostly to investigate the possible diffusion mechanism, absorption isotherms, etc. As for the other academic disciplines, ML and artificial intelligence (AI) [39] protocols have also shown up as very powerful tools for designing these materials with optimized properties. Considering their potencies in the rational design of MOF materials, all these computational tools consist of a bridge between theoretical expectation and experimental realization. For sure, there are challenges in computational approaches, such as scalable QM methods, more accurate force field approaches, and ML protocols. Because of the mentioned challenges, advanced research in computational methods of realistic design of MOFs has been directed at exploring the application-specific performance and design of these materials [40–42].

Herein, the computational approaches and methodologies are mentioned on MOF-coating performance, with possible limitations contributing to accelerating the next-generation smart material design in this field.

14.3 COMPUTATIONAL METHODS IN ANTICORROSION COATING

As is well known, corrosion is an essential phenomenon that remarkably affects metals' and alloys' lifespan, durability, and efficiency via leading defecting, weakening, and distortion of materials, resulting in substantial environmental, economic, and safety-related issues. In this regard, computational tools like in other academic or industrial fields have been needed to meet the demand for more effective and sustainable anticorrosive material design based on MOF [43] (Figure 14.3). That is, the simulation tools make possible the prediction of MOF materials' features under versatile conditions via giving the valuable insights into corrosion mechanism and coating performance in material corrosion. These computational tools support the experimental methods via reducing the use of the natural sources as well as the time and cost during the process, which help fabricate a material with the requested properties from design to efficiency of the novel coating material [44–46]. In particular, the MD/MC simulations integrated with the other computational tools open a revolutionized window onto advanced anticorrosive coating materials, providing a deep insight into micro- (or nano-) structural features of the materials, durability, and mechanical properties under different environmental conditions [47–49].

In this section, the role of the computational methods in the coating performance of the materials for corrosion will be presented by using the latest reports that predict the corrosion resistance, optimal coating composition, and development/improvement of the coating performance of the related material.

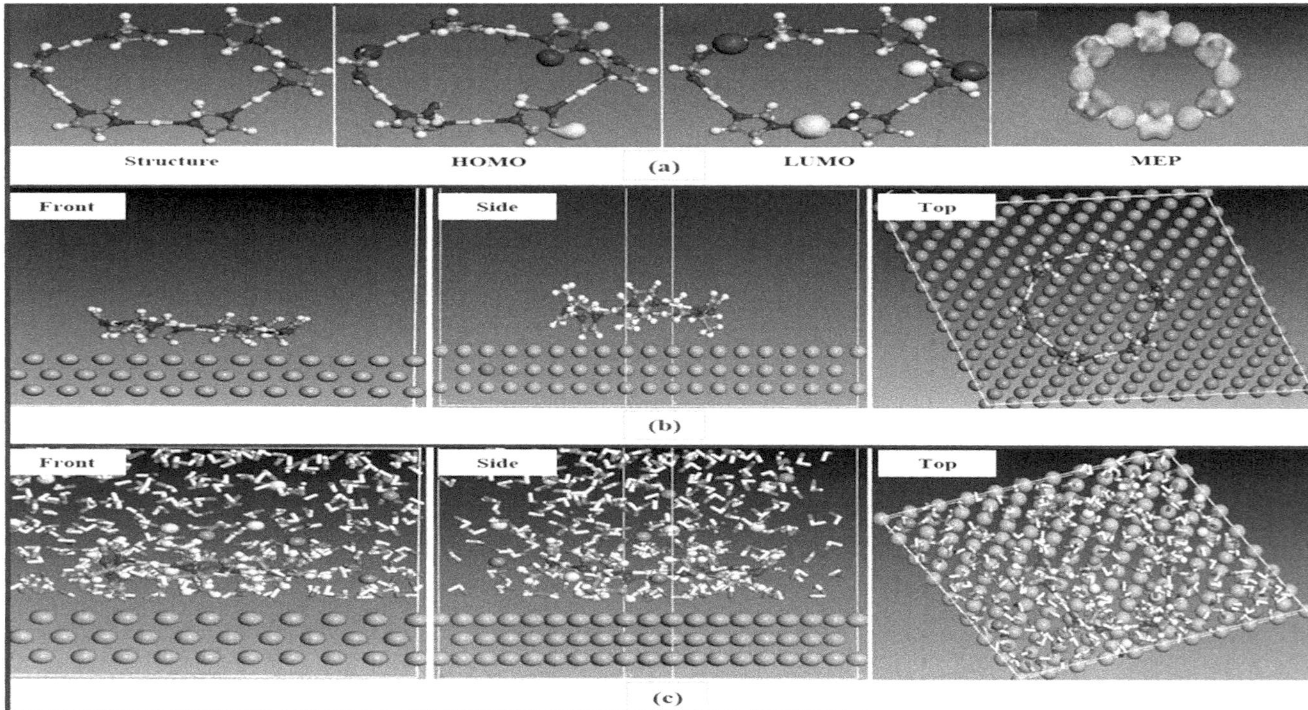

FIGURE 14.3 (a) Optimized structure, HOMO (highest occupied molecular orbital), LUMO (lowest unoccupied molecular orbital), and MEP (molecular electrostatic potential) plots of Ce-MOF, (b, c) MD-simulation scheme of Ce-MOF on Cu 111, in gas and solvent media, respectively. (Compiled from ref. 43. Copyright Elsevier, 2023.)

14.3.1 MD Simulations

MD simulation based on classical mechanics principles is a powerful computational tool that predicts the information of structural stability, mechanical properties, and thermal behavior of a specific system in view of the atomistic view of molecular interactions and dynamic processes [50–52].

In particular, computational tools have been commonly reported for corrosion studies like protection and inhibition combined with experimental methods. In this respect, Ramezanzadeh and co-workers have designed ZIF-8 "zeolitic imidazolate framework-8" on the GO "graphene oxide" sheets with a corrosion inhibition efficiency of 79%: the DFT-D simulations imply that the designed complexes could act onto GO surface and MD simulations confirmed the interfacial adsorption of 2-methylimidazole and Zn(II) complexes over the steel substrate [53]. On the other hand, Lee and co-workers have investigated a supramolecular MOF (SMOF) coating by using an α-cyclodextrin-based-MOF (α-CD-MOF-K) and a polymeric binder to use it as an aqueous rechargeable Zn battery: MD computations confirmed that Zn^{2+} ions could reversibly migrate through the pores of α-CD-MOF-K by dehydration partially [54]. Furthermore, Chikaodili and co-workers have reported the corrosion inhibition mechanism of Ce-MOF adsorption on Cu surface using the DFT simulations to optimize the MOF geometry, then performed the MC and MD computations to provide insight into the chemical and physical barriers on adsorption: The simulations have shown that the designed MOF structure would interact with the surface with highly stable and adsorption energy of −363.54 kcal/mol [55]. Ni-MOF has been synthesized via chemical method under room temperature and shown maximal IE "inhibition efficiency" of 94 % at Ni-MOF_0.1 wt %: DFT/MD simulations demonstrate the existence of organic ligands that enhance the forming of a protective film on the steel surface and make the MOF framework have better inhibition potency [56]. By performing the MD simulations in addition to experiments, Zhao and co-workers have suggested that the superhydrophobic nanocontainer based on MOF can reduce the water adsorption on copper surfaces and slow the release of the inhibitor rate [57]. In addition, chitosan-coated Zn@H3BTC material prepared by the solvothermal method has been explored for the inhibition potency for steel protection, and the adsorption mechanism has been enlightened by DFT and MD computations [58]. Wahba and Helal have prepared Cu-based MOF structures and performed the MD computations to explain the adsorption process of Cu-MOF on Al surface: the existence of the O and N atom in organic unit of the MOF provides it protection as coating of the surface [59]. Recently, the aromatic inhibitors modified with CaAl-LDH/epoxy coating have been under in-depth investigation to explore the conjugation and aromaticity effect on the corrosion protection by performing the DFT and MD simulations to estimate the adsorption mechanism; the larger electron clouds support the protective film formation on the steel surface and the lone pairs of the heteroatom(s) enrich the corrosion protection [60]. In addition, promising research in terms of eco-friendly MOF coating has been reported by Abeng and co-workers; they have prepared a drug-based MOF coating material for mild steel corrosion protection and revealed that the flat adsorption has existed on Fe-surface while performing the MD simulations [61]. Cao and co-workers have designed 2D-MOF membrane to overcome the corrosion caused from water-based reactions on Zn-anode; DFT and MD simulations implied that this membrane with angstrom size can allow ion conduction selectively [62].

14.3.2 MC Simulations

MC methods, as a broad class of computational algorithms, have used randomness to solve problems that might be deterministic in principle and are commonly applied to a wide range of scientific disciplines, including engineering, life sciences, mathematics, and social sciences. In corrosion protection or inhibition processes, MC methods on MOF-based materials provide remarkable information to explain the process during corrosion, such as mechanism, dynamics, and kinetics, etc. [63–65].

Recently, Fouda and co-workers have reported the optimal location of the MOF inhibitors on the Al (111) performing MC simulations that predict the possible substrate-absorbent system configuration [66]. Additionally, Eissa and co-workers have performed MC simulations to explore the potencies of MOFs in preventing the corrosion of carbon steel via creating a protective shielding film on the metal surface as a result of the van der Waals (vdW) forces [67]. In a recent work on adsorption dynamics and the mechanism of MOF-based materials, DFT and MC simulations have revealed that polar benzotriazole-based MOF units create a dense protective film on the Cu-surface through a covalent bond between the lone pairs of heteroatom(s) and the Cu-surface [68]. Furthermore, the g-C_3N_4)/ZnO/chitosan hybrid composite has been designed and explored by MD and MC simulations for the adsorption performance in a deep eutectic solvent [69]. In a research on MOF-based Li-ion-battery (LIB) designed by Sheng and co-workers, the DFT and MC simulations implied that the fluorine sites in MOF structures exhibit strong adsorption for water and HF, which promise a novel way to reduce the request in control of the water during the battery installation [70]. Eruçar and Keskin have performed grand canonical MC simulations to explore the adsorption selectivity and gas penetration on bio-based MOF structures, which are important in CO_2/CH_4 separation [71]. As is known well, CO_2 as one of the acidic gases in atmosphere plays a critical role in environmental issues as well as in pipes and containers due to its corrosive nature. Recently, Wang and co-workers have employed MC simulations on MOF-74-based adsorbents in CO_2 capture to assess the energy consumption from production to regeneration of MOF-74 adsorbents [72]. In the past, He and co-workers have performed grand canonical MC computations to explore the additional cation effect on the storage and separation of CO_2 and CH_4 in cation-exchanged rho-ZMOF material [73] (Figure 14.4).

14.3.3 Quantum Mechanic Simulations

Quantum chemical computations such as DFT and QM/MM have been successfully applied to MOF systems to predict the electronic structure and energy necessities, bonding characteristics, and reactivity tendencies of specific MOF units [74, 75]. The quantum chemical parameters that are commonly used to estimate the corrosion characteristics are defined as follows:

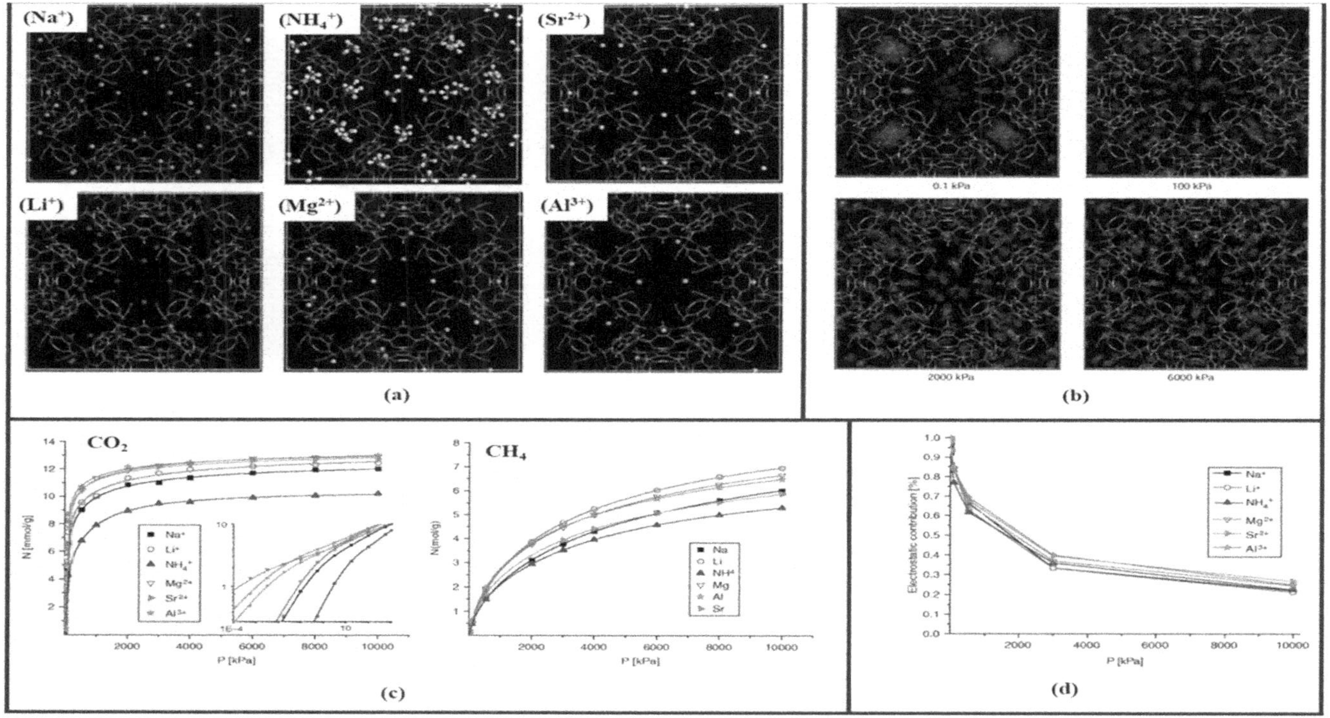

FIGURE 14.4 (a) Distributions of different cations in rho-ZMOF, (b) the center-of-mass distributions in Na–rho-ZMO (the green fields are that of CO_2 and the red fields are that of CH_4), (c) simulated adsorption isotherms of CO_2 and CH_4, and (d) contributions of electrostatic interactions to loadings of CO_2 in cation–rho-ZMOFs. (Compiled from ref. 73. Copyright Taylor & Francis, 2011.)

$$I = -E_{HOMO}$$

$$A = -E_{LUMO}$$

$$\chi = -\left(\frac{I+A}{2}\right)$$

$$\eta = \frac{I-A}{2}$$

$$\omega = \frac{\mu^2}{2\eta}$$

$$\Delta N_{max} = \frac{I+A}{2(I-A)}$$

$$\omega^+ \approx (I+3A)^2 / \left(16(I-A)\right)$$

$$\omega^- \approx (3I+A)^2 / \left(16(I-A)\right)$$

$$\Delta\varepsilon_{back-donation} = -\frac{\eta}{4}$$

$$\Delta E_{Fe/inh.} = \left[\left(\chi_{Fe} - \chi^2_{inh}\right) / 4\left(\eta_{Fe} + \eta_{inh}\right)\right]$$

Herein, χ → "electronic chemical potential," η → "global hardness," ω → "electrophilicity," ΔN_{max} → "the maximum charge transfer capability index" [76–80], $\omega-$ → "the electrodonating power" and $\omega+$ → "the electroaccepting power" [81], and $\Delta\varepsilon_{back-donation}$ → "back-donation energy" [82].

In addition, the Fukui functions [83] for nucleophilicity and electrophilicity have provided useful insights into the corrosion mechanism through marking the electron-abundance and electron-poor region, which indicate the direction of the electrons between the metal surface and related-substrate, and described by the following formulas:

$$f = f(r) = \left[\left(\delta_\rho(r)\right) / \delta N\right] v(r)$$

$$f^+ = \left[q(N+1) - q(N)\right]$$

$$f^{-} = \left[q(N) - q(N-1)\right]$$

where ρ(r) → "electron density," N→ "total number of electrons," ν(r) → "constant external potential," q (N) → "charge of neutral molecule," q(N + 1) → "cation charge," and q(N + 1) → "anion charge."

These tools have provided insights into interfacial behavior and stability on related surfaces of MOF coating by evaluating the vdW forces, thermodynamic stability, and/ or coordination bonds. In contemporary research, quantum chemical simulations have been increasingly used with other simulation methods such as ML, MC, and MD and experiments, and thus it provides a multiscale modeling in potential applications of MOFs with high-performance coating [84–86].

Recently, ZIF-8 (MOF) modified graphene-oxide (GO) has been designed to enhance the ant-corrosion, ant-bacterial, and ant-fouling properties of epoxy coatings modified with polydimethylsiloxane (epoxy-PDMS): DFT computations have indicated that GO@MOF nanohybrid is stabilized mainly by vdW forces and inhibit the penetration of corrosive agents toward the steel substrate due to binding of the components onto the GO@MOF system being remarkably large [86]. In addition, Alshammari and co-workers have fabricated Al/Th-MOF with a high surface area of 2273.8 m^2/g to remove fluorescein dye in the solution and performed DFT computations of fluorescein to analyze the FMOs and MEP plots to predict the possible interaction site between the adsorbent and dye [87].

A sunlight-driven self-cleaning CuCo-MOF composite has been designed for the treatment of complex oil–water emulsions: MD and DFT simulations have revealed that the H-bonding interaction between chitosan and MOF CuCo-HHTP and the bimetallic sites of Cu and Co in CuCo-HHTP effectively promoted the activation of peroxomonosulfate, respectively [88].Furthermore, DFT computations have been performed to characterize the adsorption process of cold-rolled steel coated by Zr-MOF material to explore the anticorrosive potency of MOF/epoxy hybrid coating [89]. Majidi and co-workers have reported that the Cu-MOF and Ni-MOF nanofillers in epoxy coatings enrich the long-term corrosion protection on the metal surface; the adsorption mechanism and process have been supported via the local and global reactivity parameters performed by B3LYP/6–31G(d,p) simulations [90]. Zhou and co-workers have predicted the corrosion protection resistance of nanocomposites by using DFT/PBE tools, including MOF unit; the simulation results have revealed that the functional group(s) on GQDs of hybrid nanocomposite facilitate the charge transfer GQD from MOF core unit and binding capacity, thus corrosion-resistance potency [91]. Figure 14.5 summarizes hybrid DFT computations on MIL-53-Fe guested by 1,4-benzoquinone, pyridine, and 2,6-lutidine molecules to analyze/comparing the host–guest and guest–guest interactions that are under the control of thermodynamic factors calculated [92].

14.3.4 ML Simulations

ML investigations in the development of MOF coatings have provided insights into the pathway to accelerate MOF design with requested properties and performance

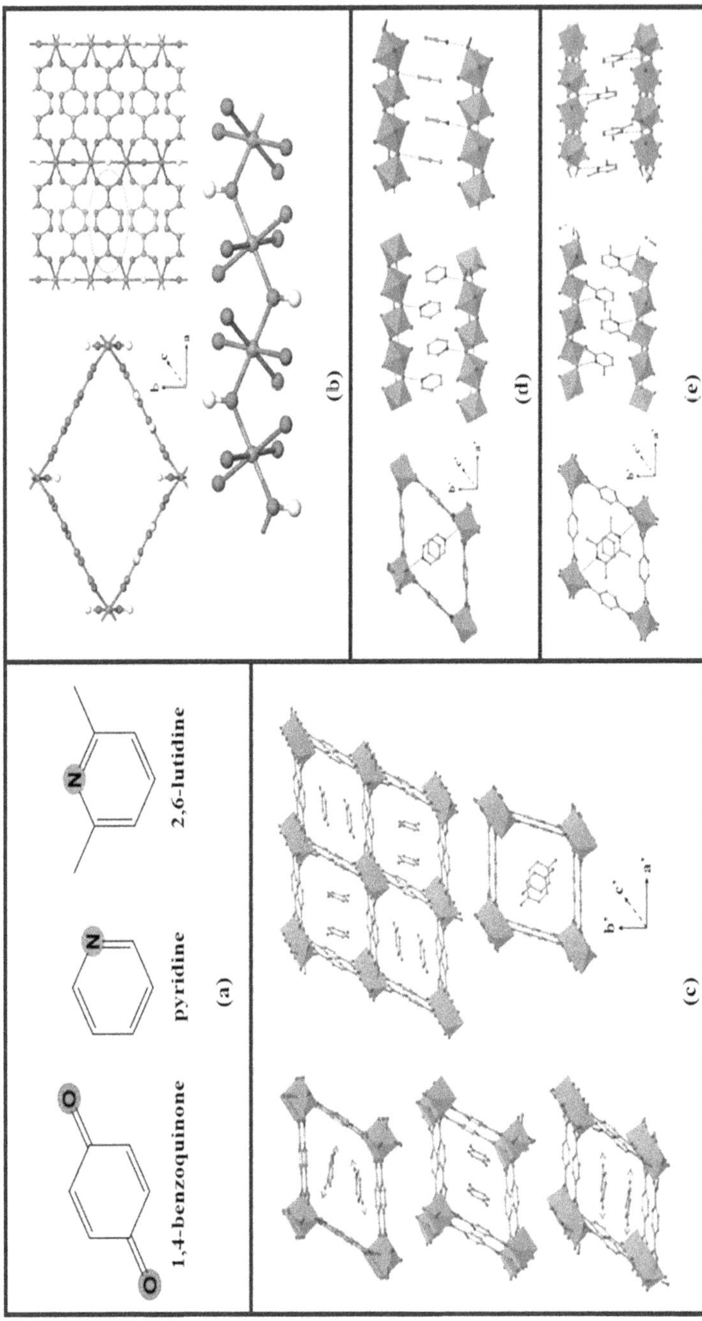

FIGURE 14.5 (a) Chemical structures of the guest molecules, (b) structure of fully expanded MIL-53-Fe and its building blocks, (c–e) the adsorption configurations of quinone, pyridine, and 2,6-lutidine molecules in MIL-53-Fe. Red: oxygen; grey: carbon; blue: nitrogen; mauve: iron-centered octahedral. (Compiled from ref. 92. Copyright Taylor & Francis, 2011.)

views via using ML algorithms to analyze large datasets derived from experimental results, computational tools, and high-throughput evaluation to estimate the critical characteristics of MOF coatings [93–95]. Thus, ML simulation helps design MOF coating to reduce the trial-and-error experimentation time and cost, and thus contributes to sustainable and smart MOF design [96, 97]. Nowadays, ML procedures, like other computational tools, have been applied to gas capture and separation (CO_2, CH_4, H_2O, etc.) due to their adverse effect on the pipelines and related facilities, and thus, the MOF coating on selective gas separation is one of the challenging issues [98–100]. In this regard, Situ and collaborators have conducted the ML procedure to estimate the performance of capturing CO_2 from flue gas using MOF membranes and determined the structural descriptors and permeance performance relationship [101]. Furthermore, Pilz and co-workers, using the ML tool, have reported H_2/CO_2 gas separation performance of surface-anchored MOF thin films [102]. Burner and co-workers have introduced a dataset performing ML simulations, including ~280,000 MOFs that include computed descriptors and DFT-based partial atomic charges for each MOF [103]. Liu and co-workers have performed the ML simulation to predict the corrosion-resistance performance of ZIF-8@Ca microfiller-modified self-healing epoxy coating (Figure 14.6) [104].

To date, ML procedures on smart anticorrosive MOF coating imply a revolution in material design providing the information on anticorrosive materials such as design, efficiency, sustainability, performance optimization, etc. However, there are still important issues that have to be overcome: high-quality experimental and computational data for the accuracy and reliability of ML procedure and interdisciplinary research consisting of material scientists, data scientists, and engineers to real-world fabrication and application of MOF materials.

14.4 OUTLOOK AND FUTURE DIRECTIONS

The field of MOF coatings offers tremendous promise for addressing critical challenges in materials science and engineering, depending on the advances in computational simulations. One of the innovative trends is to use ML and AI in predicting the optimum conditions of MOF fabrication for the application field and generally used with the MD and DFT simulations. The combined computational methods are helpful to the development of MOF coating via screening broad possible designs and then determining the optimized structures with tailored characteristics, such as thermal/mechanical stability, anticorrosive capability, structural properties, adsorption mechanism, thermal/electric conductivity performance, etc. Thus, the determined best relationship between the optimum conditions and the application field has helped reduce the labor-consuming trial experiments to fabricate the MOFs.

In addition, QM/MM and DFT methods have helped understand the complex interactions between the MOFs and specific agents to revisit the experimental conditions, even under harsh operating conditions. The diversity of the simulation media and dataset of MOFs is of critical importance in terms of considering all possible scenarios in specific applications. In this respect, the improvement of open-access databases and computational platforms will significantly enhance the progress

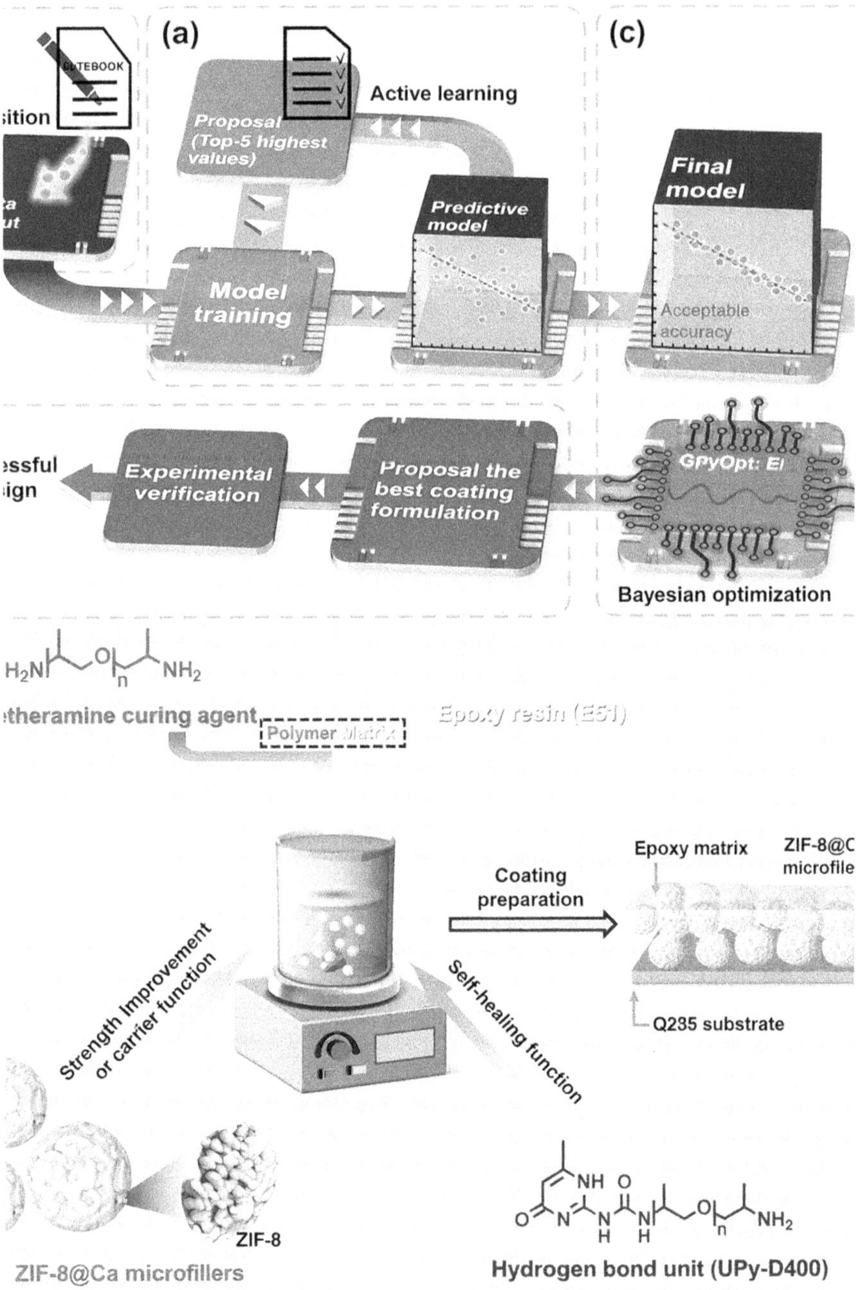

FIGURE 14.6 (a–d) ML workflow, (e) preparation scheme for self-healing epoxy (epoxy) composite coating. (Compiled from ref. 104. Copyright Springer Nature, 2024.)

in smart MOF design in a sustainable and green way. At this point, the future design of MOF coating will likely witness the adoption of quantum chemical simulations and high-performance computing to handle complex and large-scale models. Thus, the advanced computational techniques will bridge molecular-level insight and macroscopic performance through more accurate predictions for designing real-world material.

On the other hand, the synergistic effect of composite material designed by MOF and organic/inorganic-based materials should be further explored to enrich the potential properties, such as conductivity, catalytic activity, drug-release/storage. In conclusion, MOF coating is ready to make eco-friendly contributions to material science and engineering by considering sustainability, multifunctionality, and cross-disciplinary information sharing.

REFERENCES

[1] Li, D., Xu, H., Jiao, L., & Jiang, H. (2019). Metal-organic frameworks for catalysis: State of the art, challenges, and opportunities. *EnergyChem*, 1(1), 100005. https://doi.org/10.1016/j.enchem.2019.100005

[2] Yaghi, O. M., & Li, H. (1995). Hydrothermal synthesis of a metal-organic framework containing large rectangular channels. *Journal of the American Chemical Society*, 117(41), 10401–10402. https://doi.org/10.1021/ja00146a033

[3] Daglar, H., Gulbalkan, H. C., Avci, G., Aksu, G. O., Altundal, O. F., Altintas, C., Erucar, I., & Keskin, S. (2021). Effect of metal–organic framework (MOF) database selection on the assessment of gas storage and separation potentials of MOFs. *Angewandte Chemie International Edition*, 60(14), 7828–7837. https://doi.org/10.1002/anie.202015250

[4] Lawson, H. D., Walton, S. P., & Chan, C. (2021). Metal–organic frameworks for drug delivery: A design perspective. *ACS Applied Materials & Interfaces*, 13(6), 7004–7020. https://doi.org/10.1021/acsami.1c01089

[5] Manoj, D., Rajendran, S., Hoang, T. K., & Soto-Moscoso, M. (2022). The role of MOF based nanocomposites in the detection of phenolic compounds for environmental remediation- A review. *Chemosphere*, 300, 134516. https://doi.org/10.1016/j.chemosphere.2022.134516

[6] Radwan, A., Jin, H., He, D., & Mu, S. (2021). Design engineering, synthesis protocols, and energy applications of MOF-derived electrocatalysts. *Nano-Micro Letters*, 13, 132. https://doi.org/10.1007/s40820-021-00656-w

[7] Wang, H. S., Wang, Y. H., & Ding, Y. (2020). Development of biological metal–organic frameworks designed for biomedical applications: From bio-sensing/bio-imaging to disease treatment. *Nanoscale Advances*, 2(9), 3788–3797. https://doi.org/10.1039/D0NA00557F

[8] zur Loye, H. C., Besmann, T., Amoroso, J., Brinkman, K., Grandjean, A., Henager, C. H., Hu, S., Misture, S. T., Phillpot, S. R., Shustova, N. B., Wang, H., Koch, R. J., Morrison, G., & Dolgopolova, E. (2018). Hierarchical materials as tailored nuclear waste forms: A perspective. *Chemistry of Materials*, 30(14), 4475–4488. https://doi.org/10.1021/acs.chemmater.8b00766

[9] Nath, I., Chakraborty, J., & Verpoort, F. (2016). Metal organic frameworks mimicking natural enzymes: A structural and functional analogy. *Chemical Society Reviews*, 45(15), 4127–4170. https://doi.org/10.1039/C6CS00047A

[10] Stroppa, A., Barone, P., Jain, P., Perez-Mato, J. M., & Picozzi, S. (2013). Hybrid improper ferroelectricity in a multiferroic and magnetoelectric metal-organic framework. *Advanced Materials*, 25(16), 2284–2290. https://doi.org/10.1002/adma.201204738

[11] Liu, J., & Wöll, C. (2017). Surface-supported metal–organic framework thin films: Fabrication methods, applications, and challenges. *Chemical Society Reviews*, 46(19), 5730–5770. https://doi.org/10.1039/C7CS00315C

[12] Yang, H., Zhao, Y., Guo, Y., Wu, B., Ying, Y., Sofer, Z., & Wang, S. (2024). Surfactant-mediated crystalline structure evolution enabling the ultrafast green synthesis of Bismuth-MOF in aqueous condition. *Small*, 20(15), 2307484. https://doi.org/10.1002/smll.202307484

[13] Li, W., Zhang, J., Fan, L., Zhao, Y., Sun, C., Li, W., & Chang, Z. (2024). Construction of a novel Eu-MOF material based on different detection mechanisms and its application in sensing pollutants aniline, F- and Hg2+. *Spectrochimica Acta Part A: Molecular and Biomolecular Spectroscopy*, 314, 124223. https://doi.org/10.1016/j.saa.2024.124223

[14] Su, S., Li, D., Yang, Y., Xu, R., Hu, W., Li, W., & Hu, M. (2024). A multifunctional fluorescent Eu-MOF probe: Its synthesis, structure, and sensing properties. *Applied Organometallic Chemistry*, 38(7), e7524. https://doi.org/10.1002/aoc.7524

[15] Obeso, J. L., Flores, C. V., Boujnah, M., Viltres, H., Celaya, C. A., Rosas, P. M., Esparza-Schulz, J. M., Ibarra, I. A., Cordero-Sánchez, S., Peralta, R. A., & Leyva, C. (2024). Al(III)-based MOF for tetracycline removal from water: Adsorption performance and mechanism. *Journal of Solid State Chemistry*, 338, 124908. https://doi.org/10.1016/j.jssc.2024.124908

[16] Chaudhary, M. Y., Kanzariya, D. B., Das, A., & Pal, T. K. (2024). A fluorescent MOF and its synthesized MOF@cotton composite: Ratiometric sensing of vitamin B2 and antibiotic drug molecule. *Spectrochimica Acta Part A: Molecular and Biomolecular Spectroscopy*, 314, 124194. https://doi.org/10.1016/j.saa.2024.124194

[17] Huang, Y. B., Liang, J., Wang, X. S., & Cao, R. (2017). Multifunctional metal–organic framework catalysts: Synergistic catalysis and tandem reactions. *Chemical Society Reviews*, 46(1), 126–157. https://doi.org/10.1039/C6CS00250A

[18] Furukawa, S., Hirai, K., Nakagawa, K., Takashima, Y., Matsuda, R., Tsuruoka, T., Kondo, M., Haruki, R., Tanaka, D., Sakamoto, H., Shimomura, S., Sakata, O., & Kitagawa, S. (2009). Heterogeneously hybridized porous coordination polymer crystals: Fabrication of heterometallic core–shell single crystals with an in-plane rotational epitaxial relationship. *Angewandte Chemie*, 121(10), 1798–1802. https://doi.org/10.1002/ange.200804836

[19] Sanati-Tirgan, P., Eshghi, H., & Mohammadinezhad, A. (2023). Designing a new method for growing metal–organic framework (MOF) on MOF: Synthesis, characterization and catalytic applications. *Nanoscale*, 15(10), 4917–4931. https://doi.org/10.1039/D2NR06729C

[20] Razaq, R., Ud Din, M. M., Småbråten, D. R., Eyupoglu, V., Janakiram, S., Sunde, T. O., Allahgoli, N., Rettenwander, D., & Deng, L. (2023). Synergistic effect of bimetallic MOF modified separator for long cycle life Lithium-Sulfur batteries. *Advanced Energy Materials*, 14(3), 2302897. https://doi.org/10.1002/aenm.202302897

[21] Anwer, A. H., Ansari, M. Z., Mashkoor, F., Zhu, S., Shoeb, M., & Jeong, C. (2023). Synergistic effect of carbon nanotube and tri-metallic MOF nanoarchitecture for electrochemical high-performance asymmetric supercapacitor applications and their charge storage mechanism. *Journal of Alloys and Compounds*, 955, 170038. https://doi.org/10.1016/j.jallcom.2023.170038

[22] Han, X., Zhang, W., Li, S., Cheng, C., Zhou, L., Jia, Q., & Xiu, G. (2022). Efficient activation of peroxymonosulfate by MnS/Fe-MOF hybrid catalyst for sulfadiazine degradation: Synergistic effects and mechanism. *Separation and Purification Technology*, 287, 120509. https://doi.org/10.1016/j.seppur.2022.120509

[23] Al Amery, N., Abid, H., Al-Saadi, S., Wang, S., & Liu, S. (2020). Facile directions for synthesis, modification and activation of MOFs. *Materials Today Chemistry*, 17, 100343. https://doi.org/10.1016/j.mtchem.2020.100343

[24] McKinstry, C., Cathcart, R. J., Cussen, E. J., Fletc.her, A. J., Patwardhan, S. V., & Sefcik, J. (2016). Scalable continuous solvothermal synthesis of metal organic framework (MOF-5) crystals. *Chemical Engineering Journal*, 285, 718–725. https://doi.org/10.1016/j.cej.2015.10.023

[25] Zhang, W., Shahnavaz, Z., Yan, X., Huang, X., Wu, S., Chen, H., Pan, J., Li, T., & Wang, J. (2022). One-step solvothermal synthesis of Raspberry-like NiCo-MOF for high-performance flexible supercapacitors for a wide operation temperature range. *Inorganic Chemistry*, 61(38), 15287–15301. https://doi.org/10.1021/acs.inorgchem.2c02916

[26] Aghajani Hashjin, M., Zarshad, S., Motejadded Emrooz, H. B., & Sadeghzadeh, S. (2023). Enhanced atmospheric water harvesting efficiency through green-synthesized MOF-801: A comparative study with solvothermal synthesis. *Scientific Reports*, 13(1), 1–12. https://doi.org/10.1038/s41598-023-44367-1

[27] Bhoite, A., Sawant, V., & Tarwal, N. (2024). Solvothermal synthesis of Ni/Co-based metal-organic framework with nanosheets-like structure for high-performance supercapacitor. *Colloids and Surfaces A: Physicochemical and Engineering Aspects*, 702, 134814. https://doi.org/10.1016/j.colsurfa.2024.134814

[28] Li, Y., Zhang, Y., Wang, Z., Zhang, C., Meng, F., Zhao, J., Li, X., & Hu, J. (2024). Ultrasonic-assisted preparation of Fe-MOF with rich oxygen vacancies for efficient oxygen evolution. *Applied Catalysis A: General*, 683, 119851. https://doi.org/10.1016/j.apcata.2024.119851

[29] Larasati, L., Dendy, D., Lestari, W. W., Suharbiansah, R. S., Firdaus, M., Masykur, A., & Wibowo, F. R. (2024). Rapid and facile electrochemical synthesis of MIL-101 (Fe)-NH2 and its curcumin loading and release studies. *Journal of Inorganic and Organometallic Polymers and Materials*, 34, 4039–4049. https://doi.org/10.1007/s10904-024-03049-9

[30] Cheng, W.-W., Zhang, S.-N., Wang, J., Yang, J., Yang, Z., Chen, X.-F., Xiao, J.-D., & Wang, J. (2024). Boosting hydrogen production of a MOF-based multicomponent photocatalyst with clean interface via facile one-pot electrosynthesis. *Chemistry European Journal*, 30, e202303886. https://doi.org/10.1002/chem.202303886

[31] Ismail, K. M., Hassan, S. S., Medany, S. S., & Hefnawy, M. A. (2024). A facile sonochemical synthesis of the Zn-based metal–organic framework for electrochemical sensing of paracetamol. *Materials Advances*, 5(14), 5870–5884. https://doi.org/10.1039/D4MA00061G

[32] Shakeel, N., Khan, J., & Al-Kahtani, A. A. (2024). Morphology-driven electrochemical attributes of Cu-MOF: A high-performance anodic material for battery supercapacitor hybrids. *RSC Advances*, 14(46), 33941–33951. https://doi.org/10.1039/D4RA05790B

[33] Ashkar, M. A., Kutti Rani, S., Vasimalai, N., Kuo, C. Y., Yusuf, K., & Govindasamy, M. (2024). Design of sonochemical assisted synthesis of Zr-MOF/g-C3N4-modified electrode for ultrasensitive detection of antipsychotic drug chlorpromazine from biological samples. *Microchimica Acta*, 191(4), 182.

[34] Salmi, L. D., Heikkilä, M. J., Puukilainen, E., Sajavaara, T., Grosso, D., & Ritala, M. (2013). Studies on atomic layer deposition of MOF-5 thin films. *Microporous and*

Mesoporous Materials, 182, 147–154. https://doi.org/10.1016/j.micromeso.2013.08.024

[35] Kaur, G., Anthwal, A., Kandwal, P., & Sud, D. (2023). Mechanochemical synthesis and theoretical investigations of Fe (II) based MOF containing 4,4′-bipyridine with ordained intercalated p-aminobenzoic acid: Application as fluoroprobe for detection of carbonyl group. *Inorganica Chimica Acta*, 545, 121248. https://doi.org/10.1016/j.ica.2022.121248

[36] Zhao, T., Wu, H., Wen, X., Zhang, J., Tang, H., Deng, Y., Liao, S., & Tian, X. (2022). Recent advances in MOFs/MOF derived nanomaterials toward high-efficiency aqueous zinc ion batteries. *Coordination Chemistry Reviews*, 468, 214642. https://doi.org/10.1016/j.ccr.2022.214642

[37] Moayed Mohseni, M., Jouyandeh, M., Mohammad Sajadi, S., Hejna, A., Habibzadeh, S., Mohaddespour, A., Rabiee, N., Daneshgar, H., Akhavan, O., Asadnia, M., Rabiee, M., Ramakrishna, S., Luque, R., & Reza Saeb, M. (2022). Metal-organic frameworks (MOF) based heat transfer: A comprehensive review. *Chemical Engineering Journal*, 449, 137700. https://doi.org/10.1016/j.cej.2022.137700

[38] Jo, M., Jo, Y. K., Lee, H., Jang, H. W., Hwang, S., & Yoo, D. J. (2023). MOF-based chemiresistive gas sensors: Toward new functionalities. *Advanced Materials*, 35(43), 2206842. https://doi.org/10.1002/adma.202206842

[39] Mashhadimoslem, H., Abdol, M. A., Karimi, P., Zanganeh, K., Shafeen, A., Elkamel, A., & Kamkar, M. (2024). Computational and machine learning methods for CO2 capture using metal–organic frameworks. *ACS Nano*, 18(35), 23842–23875. https://doi.org/10.1021/acsnano.3c13001

[40] Huang, L., Chen, W., Hao, W., Wang, J., Guo, X., Ma, L., & Zhang, D. (2024). Self-healing anti-corrosion coatings: A mechanism study using computational materials science. *Electrical Materials and Applications*, 1(1), e12006. https://doi.org/10.1049/ema3.12006

[41] Keskin, S., & Alsoy Altinkaya, S. (2019). A review on computational modeling tools for MOF-based mixed matrix membranes. *Computation*, 7(3), 36. https://doi.org/10.3390/computation7030036

[42] Demir, H., Daglar, H., Gulbalkan, H. C., Aksu, G. O., & Keskin, S. (2023). Recent advances in computational modeling of MOFs: From molecular simulations to machine learning. *Coordination Chemistry Reviews*, 484, 215112. https://doi.org/10.1016/j.ccr.2023.215112

[43] Anadebe, V. C., Chukwuike, V. I., Porsia, A. J., Ponnusamy, P., Natarajan, C. S., & Barik, R. C. (2023). Insight into the corrosion inhibition of Ce-MOF on copper in aqueous chloride environment: From experimental validation to molecular-level prediction. *Applied Surface Science Advances*, 18, 100526. https://doi.org/10.1016/j.apsadv.2023.100526

[44] Lashgari, S. M., Yari, H., Mahdavian, M., Ramezanzadeh, B., Bahlakeh, G., & Ramezanzadeh, M. (2021). Synthesis of graphene oxide nanosheets decorated by nanoporous zeolite-imidazole (ZIF-67) based metal-organic framework with controlled-release corrosion inhibitor performance: Experimental and detailed DFT-D theoretical explorations. *Journal of Hazardous Materials*, 404, 124068. https://doi.org/10.1016/j.jhazmat.2020.124068

[45] Ulaeto, S. B., Ravi, R. P., Udoh, I. I., Mathew, G. M., & Rajan, T. P. (2023). Polymer-based coating for steel protection, highlighting metal–organic framework as functional actives: A review. *Corrosion and Materials Degradation*, 4(2), 284–316. https://doi.org/10.3390/cmd4020015

[46] Bordonhos, M., Galvão, T. L., Gomes, J. R., Gouveia, J. D., Jorge, M., Lourenço, M. A., Pereira, J. M., Pérez-Sánchez, G., Pinto, M. L., Silva, C. M., Tedim, J., & Zêzere, B. (2023). Multiscale computational approaches toward the understanding of materials. *Advanced Theory and Simulations*, 6(10), 2200628. https://doi.org/10.1002/adts.202200628

[47] Ramezanzadeh, M., Ramezanzadeh, B., Bahlakeh, G., Tati, A., & Mahdavian, M. (2021). Development of an active/barrier bi-functional anti-corrosion system based on the epoxy nanocomposite loaded with highly-coordinated functionalized zirconium-based nanoporous metal-organic framework (Zr-MOF). *Chemical Engineering Journal*, 408, 127361. https://doi.org/10.1016/j.cej.2020.127361

[48] Lei, Y., Jiang, Z., Zeng, X., Li, Y., Wang, X., Liu, H., & Zhang, G. (2023). Preparation of ZIF-67@DTMS NPs/Epoxy composite coating and its anti-corrosion performance for Q235 carbon steel in 3.5 wt% NaCl solution. *Colloids and Surfaces A: Physicochemical and Engineering Aspects*, 656, 130370. https://doi.org/10.1016/j.colsurfa.2022.130370

[49] Li, W., Zhang, L., Zhang, M., & Chen, S. (2020). Structures of graphene-reinforced epoxy coatings and the dynamic diffusion of guest water: A molecular dynamics study. *Industrial & Engineering Chemistry Research*, 59(47), 20749–20756. https://doi.org/10.1021/acs.iecr.0c04673

[50] Mashhadzadeh, A. H., Taghizadeh, A., Taghizadeh, M., Munir, M. T., Habibzadeh, S., Salmankhani, A., Stadler, F. J., & Saeb, M. R. (2020). Metal–organic framework (MOF) through the lens of molecular dynamics simulation: Current status and future perspective. *Journal of Composites Science*, 4(2), 75. https://doi.org/10.3390/jcs4020075

[51] Chen, J., Gu, Z., Manuel, J., Luo, Y., Tian, K., & Luo, Y. (2025). Molecular dynamics simulations reveal efficient heavy metal ion removal by two-dimensional Cu-THQ metal-organic framework membrane. *Scientific Reports*, 15(1), 1–11. https://doi.org/10.1038/s41598-024-84308-0

[52] Mandal, S., Bej, S., & Banerjee, P. (2023). Insights into the uses of two azine decorated d10-MOFs for corrosion inhibition application on mild steel surface in saline medium: Experimental as well as theoretical investigation. *Journal of Molecular Liquids*, 381, 121789. https://doi.org/10.1016/j.molliq.2023.121789

[53] Ramezanzadeh, M., Ramezanzadeh, B., Mahdavian, M., & Bahlakeh, G. (2020). Development of metal-organic framework (MOF) decorated graphene oxide nanoplatforms for anti-corrosion epoxy coatings. *Carbon*, 161, 231–251. https://doi.org/10.1016/j.carbon.2020.01.082

[54] Lee, M., Choi, I., Kim, A., Paik, S., Kim, D., Kim, H., & Nam, K. W. (2024). Supramolecular metal–organic framework for the high stability of aqueous rechargeable zinc batteries. *ACS Nano*, 18(33), 22586–22595. https://doi.org/10.1021/acsnano.4c08550

[55] Amaka Chikaodili, N., Sunday Ume, C., Chukwudi Nnaji, P., Bartholomew Iroha, N., Dagdag, O., Okechukwu Ezeugo, J., Thakur, A., Chikaodili Anadebe, V., & Dominic Onukwuli, O. (2024). Physisorption or chemisorption: Insight from AI computing model based on DFT, MC/MD-simulation for prediction of MOF-based inhibitor adsorption on Cu in brine solution. *Computational and Theoretical Chemistry*, 1238, 114730. https://doi.org/10.1016/j.comptc.2024.114730

[56] Anadebe, V. C., Chukwuike, V. I., Chandra Nayak, K., Ebenso, E. E., & Chandra Barik, R. (2024). Combined electrochemical, atomic scale-DFT and MD simulation of Nickel based metal organic framework (Ni-MOF) as corrosion inhibitor for

X65 pipeline steel in CO2- saturated brine. *Materials Chemistry and Physics*, 312, 128606. https://doi.org/10.1016/j.matchemphys.2023.128606

[57] Zhao, Y., Xu, T., Zhou, J., & Hu, J. (2022). Superhydrophobic nanocontainers for passive and active corrosion protection. *Chemical Engineering Journal*, 433, 134039. https://doi.org/10.1016/j.cej.2021.134039

[58] Gouda, M., Khalaf, M. M., Shalabi, K., Al-Omair, M. A., & Abd, H. M. (2021). Synthesis and characterization of Zn–organic frameworks containing Chitosan as a lw-cost inhibitor for Sulfuric-Acid-induced steel corrosion: Practical and computational exploration. *Polymers*, 14(2), 228. https://doi.org/10.3390/polym14020228

[59] Wahba, A. M., & Helal, M. M. (2024). Synthesis, characterization and application of two metal-organic frameworks (MOF) as a corrosion inhibitor for aluminium in a hydrochloric acid environment. *Case Studies in Chemical and Environmental Engineering*, 10, 100805. https://doi.org/10.1016/j.cscee.2024.100805

[60] Wang, J., Cao, Y., Xue, J., Zhang, X., Liang, Y., Chen, K., Huang, C., & Zheng, D. (2024). A comparative experimental and theoretical calculation study of CaAl-LDH modified with various aromatic inhibitors for corrosion protection study in epoxy coatings. *Corrosion Science*, 231, 111994. https://doi.org/10.1016/j.corsci.2024.111994

[61] Abeng, F., Ita, B., Anadebe, V., Chukwuike, V., Etiowo, K., Nkom, P., Ekerenam, O., Iroha, N., & Ikot, I. (2023). Multidimensional insight into the corrosion mitigation of clonazepam drug molecule on mild steel in chloride environment: Empirical and computer simulation explorations. *Results in Engineering*, 17, 100924. https://doi.org/10.1016/j.rineng.2023.100924

[62] Cao, Z., Zhang, H., Song, B., Xiong, D., Tao, S., Deng, W., Hu, J., Hou, H., Zou, G., & Ji, X. (2023). Angstrom-level ionic Sieve 2D-MOF membrane for high power aqueous zinc anode. *Advanced Functional Materials*, 33(28), 2300339. https://doi.org/10.1002/adfm.202300339

[63] Liu, W., Yan, Z., Ma, X., Geng, T., Wu, H., & Li, Z. (2018). Mg-MOF-74/MgF2 composite coating for improving the properties of magnesium alloy implants: Hydrophilicity and corrosion resistance. *Materials*, 11(3), 396. https://doi.org/10.3390/ma11030396

[64] Zhang, D., Xiao, L., Xiong, G., He., Q, Pan, Z., Ma, G. (2023). Recent progress of zeolitic imidazolate frameworks (ZIFs) in superhydrophobic and anticorrosive coatings for metals and their alloys. *Journal of Coatings Technology and Research*, 20, 1157–1177. https://doi.org/10.1007/s11998-023-00769-4

[65] Ma, Q., Wang, Y., Zhang, X., Zhao, Q., Guo, J., Guo, J., Ren, X., Huang, J., Zhang, Y., Xie, Y., & Hao, J. (2024). Computational design of metal-organic frameworks for sustainable energy and environmental applications: Bridging theory and experiment. *Materials Science and Engineering: B*, 311, 117765. https://doi.org/10.1016/j.mseb.2024.117765

[66] Fouda, A. E. A. S., Etaiw, S. E. H., Abd El-Aziz, D. M. et al. (2024). Experimental and theoretical studies of the efficiency of metal–organic frameworks (MOFs) in preventing aluminum corrosion in hydrochloric acid solution. *BMC Chemistry*, 18, 21. https://doi.org/10.1186/s13065-024-01121-6

[67] Eissa, M., Etaiw, S. H., El-Waseef, E. E., & Fouda, A. S. (2024). The impact of environmentally friendly supramolecular coordination polymers as carbon steel corrosion inhibitors in HCl solution: Synthesis and characterization. *Scientific Reports*, 14(1), 1–20. https://doi.org/10.1038/s41598-024-51576-9

[68] Cao, J., Guo, C., & Chen, Z. (2023). Loading and release mechanisms of MOF-5 @ BTA-X (X = -CH3/-NH2/-CO(CH2)6CH3): Experimental and theoretical

investigations. *Colloids and Surfaces A: Physicochemical and Engineering Aspects*, 666, 131274. https://doi.org/10.1016/j.colsurfa.2023.131274
[69] Kurtulbaş, E., Ciğeroğlu, Z., Şahin, S., El Messaoudi, N., & Mehmeti, V. (2024). Monte Carlo, molecular dynamic, and experimental studies of the removal of malachite green using g-C3N4/ZnO/Chitosan nanocomposite in the presence of a deep eutectic solvent. *International Journal of Biological Macromolecules*, 274, 133378. https://doi.org/10.1016/j.ijbiomac.2024.133378
[70] Sheng, L., Yang, K., Chen, J., Zhu, D., Wang, L., Wang, J., Tang, Y., Xu, H., & He, X. (2023). A protophilic MOF enables Ni-Rich Lithium-battery stable cycling in a high water/acid content. *Advanced Materials*, 35(25), 2212292. https://doi.org/10.1002/adma.202212292
[71] Erucar, I., & Keskin, S. (2015). Computational modeling of bio-MOFs for CO2/CH4 separations. *Chemical Engineering Science*, 130, 120–128. https://doi.org/10.1016/j.ces.2015.03.016
[72] Wang, P., Pan, Z., Liu, Z., Shang, L., Soltanian, M. R., & Zhang, Z. (2024). Life cycle assessment of CO2 capture through pressure swing adsorption using MOF-74. *Gas Science and Engineering*, 133, 205497. https://doi.org/10.1016/j.jgsce.2024.205497
[73] He, P., Liu, H., Li, Y., Lei, Z., Huang, S., Wang, P., & Tian, H. (2011). Enhanced selectivity and capacity of adsorption of CO_2 and CH_4 in zeolite-like metal-organic frameworks with different extra-framework cations: A molecular simulation study. *Molecular Simulation*, 38(1), 72–83. https://doi.org/10.1080/08927022.2011.602974
[74] Seidi, F., Jouyandeh, M., Taghizadeh, M., Taghizadeh, A., Vahabi, H., Habibzadeh, S., Formela, K., & Saeb, M. R. (2019). Metal-organic framework (MOF)/epoxy coatings: A review. *Materials*, 13(12), 2881. https://doi.org/10.3390/ma13122881
[75] Keshmiri, N., Najmi, P., Ramezanzadeh, M., Ramezanzadeh, B., & Bahlakeh, G. (2022). Ultrastable porous covalent organic framework assembled carbon nanotube as a novel nanocontainer for anti-corrosion coatings: Experimental and computational studies. *ACS Applied Materials & Interfaces*, 14(17), 19958–19974. https://doi.org/10.1021/acsami.1c24185
[76] Koopmans, T. (1934). Über die Zuordnung von Wellenfunktio-nen und Eigenwerten zu den Einzelnen Elektronen Eines Atoms. *Physica*, 1, 104–113.
[77] Janak, J. F. (1978). Proof that ∂E/∂ni=εi in density-functional theory. *Physical Review B*, 18, 7165–7168.
[78] Parr, R. G., & Pearson, R. G. (1983). Absolute hardness: Companion parameter to absolute electronegativity. *Journal of the American Chemical Society*, 105, 7512–7516.
[79] Pearson, R. G. (1986). Absolute electronegativity and hardness correlated with molecular orbital theory. *Proceedings of the National Academy of Sciences of the USA*, 83, 8440–8441.
[80] Parr, R. G., Szentpaly, L. V., & Liu, S. (1999). Electrophilicity index. *Journal of the American Chemisry Society*, 1999, 121, 1922–1924.
[81] Gazquez, J. L., Cedillo, A., & Vela, A. (2007). Electrodonating and electroaccepting powers. *Journal of Physical Chemistry A*, 111, 1966–1970.
[82] Gomez, B., Likhanova, N. V., Domínguez-Aguilar, M. A., Martínez-Palou, R., Vela, A., & Gazquez, J. L. (2006). Quantum chemical study of the inhibitive properties of 2-Pyridyl-Azoles. *Journal of Physical Chemistry B*, 110, 8928–8934.
[83] Fukui, K. (1982). Role of frontier orbitals in chemical reactions. *Science*, 218(4574), 747–754. https://doi.org/747
[84] Anadebe, V. C., Chukwuike, V. I., Ramanathan, S., & Barik, R. C. (2022). Cerium-based metal organic framework (Ce-MOF) as corrosion inhibitor for API 5L X65

steel in CO2- saturated brine solution: XPS, DFT/MD-simulation, and machine learning model prediction. *Process Safety and Environmental Protection*, 168, 499–512. https://doi.org/10.1016/j.psep.2022.10.016

[85] Alipanah, N., Yari, H., Mahdavian, M., Ramezanzadeh, B., & Bahlakeh, G. (2021). MIL-88A (Fe) filler with duplicate corrosion inhibitive/barrier effect for epoxy coatings: Electrochemical, molecular simulation, and cathodic delamination studies. *Journal of Industrial and Engineering Chemistry*, 97, 200–215. https://doi.org/10.1016/j.jiec.2021.01.035

[86] Zhou, Z., Seif, A., Pourhashem, S., Duan, J., Rashidi, A., Luigi Silvestrelli, P., Ji, X., Mirzaee, M., & Hou, B. (2023). Multi-treatments based on polydimethylsiloxane and metal-organic framework wrapped with graphene oxide for achieving long-term corrosion and fouling protection: Experimental and density functional theory aspects. *Construction and Building Materials*, 384, 131229. https://doi.org/10.1016/j.conbuildmat.2023.131229

[87] Alshammari, N. A. H., Alnawmasi, J. S., Alotaibi, A. M., Alshammari, O. A., Abomuti, M. A., Elsayed, N. H., & El-Bindary, A. A. (2024). Efficient adsorption of fluorescein dye from aqueous solutions by Al/Th-MOF bimetal-organic frameworks: Adsorption isotherm, kinetics, DFT computation, and optimization via Box-Behnken design. *Process Safety and Environmental Protection*, 190, 353–371. https://doi.org/10.1016/j.psep.2024.08.037

[88] He, X., Wu, Z., Lu, J., Liu, J., Li, B., Liu, X., Tao, W., & Li, Z. (2024). A sunlight-driven self-cleaning CuCo-MOF composite membrane for highly efficient emulsion separation and water purification. *Small*, 20(42), 2402589. https://doi.org/10.1002/smll.202402589

[89] Hsia, H.-H., Chen, Y.-L., Tai, Y.-T., Tian, H.-K., Kung, C.-W., & Liu, W.-R. (2024). Two-dimensional metal–organic frameworks/epoxy composite coatings with superior O2/H2O resistance for anticorrosion applications. *ACS Applied Materials & Interfaces*, 16(31), 41421–41434. https://doi.org/10.1021/acsami.4c04843

[90] Majidi, R., Farhadi, A., Danaee, I., Panah, N. B., Zarei, D., & Nikmanesh, S. (2023). Investigation of synthesized planar Cu-MOF and spherical Ni-MOF nanofillers for improving the anti-corrosion performance of epoxy coatings. *Progress in Organic Coatings*, 183, 107803. https://doi.org/10.1016/j.porgcoat.2023.107803

[91] Zhou, Z., Seif, A., Pourhashem, S., Duan, J., Rashidi, A., Mirzaee, M., Silvestrelli, P. L., & Hou, B. (2023). Anti-corrosion and anti-fouling properties of ball-like GQDs hybrided MOFs functionalized with silane in waterborne epoxy-polydimethylsiloxane coatings: Experimental and theoretical studies. *Applied Materials Today*, 30, 101704. https://doi.org/10.1016/j.apmt.2022.101704

[92] Ling, S., Walton, R. I., & Slater, B. (2015). Theoretical study of conformational disorder and selective adsorption of small organic molecules in the flexible metal-organic framework material MIL-53-Fe. *Molecular Simulation*, 41(16–17), 1348–1356. https://doi.org/10.1080/08927022.2015.1036266

[93] Jiang, L., Dong, Y., Yuan, Y., Zhou, X., Liu, Y., & Meng, X. (2022). Recent advances of metal–organic frameworks in corrosion protection: From synthesis to applications. *Chemical Engineering Journal*, 430, 132823. https://doi.org/10.1016/j.cej.2021.132823

[94] Zheng, B., Oliveira, F. L., Neumann Barros Ferreira, R., Steiner, M., Hamann, H., Gu, G. X., & Luan, B. (2023). Quantum informed machine-learning potentials for molecular dynamics simulations of CO2's chemisorption and diffusion in Mg-MOF-74. *ACS Nano*, 17(6), 5579–5587. https://doi.org/10.1021/acsnano.2c11102

[95] Li, Y., Fu, Z., Yu, X. et al. (2024). Developing an atmospheric aging evaluation model of acrylic coatings: A semi-supervised machine learning algorithm. *International Journal of Minerals, Metallurgy and Materials*, 31, 1617–1627. https://doi.org/10.1007/s12613-024-2921-9

[96] Anadebe, V. C., Chukwuike, V. I., Chidiebere, M. A., Guo, L., Ebenso, E. E., & Barik, R. C. (2024). Synthesis and characterization techniques for nano-hybrid smart coatings. In Ashish Kumar and Abhinay Thakur (eds.) *Nano-Hybrid Smart Coatings: Advancements in Industrial Efficiency and Corrosion Resistance*, 41–57. https://doi.org/10.1021/bk-2024-1469.ch003

[97] Escobar-Hernandez, H. U., Pérez, L. M., Hu, P., Soto, F. A., Papadaki, M. I., Zhou, H.-C., & Wang, Q. (2022). Thermal stability of metal–organic frameworks (MOFs): Concept, determination, and model prediction using computational chemistry and machine learning. *Industrial & Engineering Chemistry Research*, 61(17), 5853–5862. https://doi.org/10.1021/acs.iecr.2c00561

[98] Safarzadeh Khosrowshahi, M., Afshari Aghajari, A., Rahimi, M., Maleki, F., Ghiyabi, E., Rezanezhad, A., Bakhshi, A., Salari, E., Shayesteh, H., & Mohammadi, H. (2024). Recent progress on advanced solid adsorbents for CO2 capture: From mechanism to machine learning. *Materials Today Sustainability*, 27, 100900. https://doi.org/10.1016/j.mtsust.2024.100900

[99] Zhang, Z., Cao, X., Geng, C., Sun, Y., He, Y., Qiao, Z., & Zhong, C. (2022). Machine learning aided high-throughput prediction of ionic liquid@MOF composites for membrane-based CO2 capture. *Journal of Membrane Science*, 650, 120399. https://doi.org/10.1016/j.memsci.2022.120399

[100] Park, Y. J., Yoon, S., & Jerng, S. E. (2024). Machine learning of metal-organic framework design for carbon dioxide capture and utilization. *Journal of CO2 Utilization*, 89, 102941. https://doi.org/10.1016/j.jcou.2024.102941

[101] Situ, Y., Yuan, X., Bai, X., Li, S., Liang, H., Zhu, X., Wang, B., & Qiao, Z. (2022). Large-scale screening and machine learning for metal–organic framework membranes to capture CO2 from flue gas. *Membranes*, 12(7), 700. https://doi.org/10.3390/membranes12070700

[102] Pilz, L., Natzeck, C., Wohlgemuth, J., Scheuermann, N., Spiegel, S., Oßwald, S., ... & Prasetya, N. (2023). Utilizing machine learning to optimize metal–organic framework-derived polymer membranes for gas separation. *Journal of Materials Chemistry A*, 11(45), 24724–24737.

[103] Burner, Jake., Luo, J., White, A., Mirmiran, A., Kwon, O., Boyd, P. G., Maley, S., Gibaldi, M., Simrod, S., Ogden, V., & Woo, T. K. (2023). ARC–MOF: A diverse database of metal-organic frameworks with DFT-derived partial atomic charges and descriptors for machine learning. *Chemistry of Materials*, 35(3), 900–916. https://doi.org/10.1021/acs.chemmater.2c02485

[104] Liu, T., Chen, Z., Yang, J., Ma, L., Mol, A., & Zhang, D. (2024). Machine learning assisted discovery of high-efficiency self-healing epoxy coating for corrosion protection. *Npj Materials Degradation*, 8(1), 1–11. https://doi.org/10.1038/s41529-024-00427-z

15 Challenges and Opportunities in MOF-Based Corrosion Control

İlayda Bersu Kul, Gamze Tüzün, and Burak Tüzün

15.1 INTRODUCTION

Corrosion is a persistent and costly problem that needs to be solved in various industries such as transportation, infrastructure, energy, and electronics. The degradation of materials due to chemical, electrochemical, or environmental interactions leads to serious economic losses and safety risks. Traditional approaches for corrosion control, such as coatings, inhibitors, and cathodic protection, have been effective to a certain extent. However, the increasing complexity of material systems and the increasing demand for environmental sustainability have led to the need to find more effective solutions to these challenges. In this context, metal–organic frameworks (MOFs) have emerged as a promising class of materials that attract attention with their unique structural and functional properties that constitute the core for corrosion control. MOFs are crystalline materials that are formed by the coordination of metal ions or cluster structures with organic linkage groups, forming highly porous, tunable structures. The modularity in MOF design allows for precise control of chemical, physical, and thermal properties. These features make MOFs extremely flexible for various applications such as gas storage, catalysis, sensing, and drug delivery. In recent years, the potential of MOFs in corrosion control has received increasing attention due to their ability to act as protective barriers, controlled release systems, and multifunctional coatings.

15.2 THE NEED FOR ADVANCED CORROSION CONTROL SOLUTIONS

Corrosion costs an estimated 3–4% of annual gross domestic product worldwide, including materials replacement, system downtime, and environmental cleanup. Current corrosion prevention methods often fall short in terms of durability, compatibility, or environmental impact. For example, many traditional inhibitors rely on toxic chemicals, while polymer-based coatings can degrade under extreme conditions. As industries demand longer-lasting, environmentally friendly, and high-performance solutions, MOFs offer a unique opportunity to meet these needs.

DOI: 10.1201/9781003571162-15

15.3 UNIQUE FEATURES OF MOFS RELATED TO CORROSION CONTROL

15.3.1 Porosity and Tunability

MOFs have extremely high surface areas and tunable pore sizes that allow for the incorporation of corrosion inhibitors or other functional molecules.

15.3.2 Chemical Stability

Advanced MOFs can be designed to withstand harsh environmental conditions, including acidic, saline, and high-temperature environments.

15.3.3 Controlled Release Mechanisms

MOFs can serve as carriers for corrosion inhibitors and target localized corrosion by releasing them in a stimuli-responsive manner.

15.3.4 Multifunctionality

MOFs can combine multiple protection mechanisms, such as barrier effects, inhibitor release, and antimicrobial activity, in a single material system.

15.4 CORROSION CONTROL MECHANISMS OF MOFS

MOFs stand out as a highly effective and innovative solution for controlling the corrosion of metal surfaces. MOFs are materials that can be designed at the molecular level, have high surface area and regular porous structures. Because of these features, MOFs have great potential, especially in industrial applications, by offering a range of corrosion control mechanisms. These mechanisms can be examined under two main headings to provide both passive and active protection: barrier effect and release of corrosion inhibitors. In addition, the antimicrobial and cathodic protection properties of MOFs are among the factors that make them even more effective [1].

15.4.1 Mechanisms According to Different Types of Corrosion

Corrosion is the deterioration of a metal as a result of electrochemical or chemical reactions with its environment. Different types of corrosion are given in Figure 15.1. Different types of corrosion progress with different mechanisms and the effects of MOFs on these types may also vary:

1. *General Corrosion (Uniform Corrosion):* Corrosion that occurs homogeneously on the entire or large part of the metal surface. MOFs can prevent general corrosion by forming a protective layer on the metal surface or by slowing

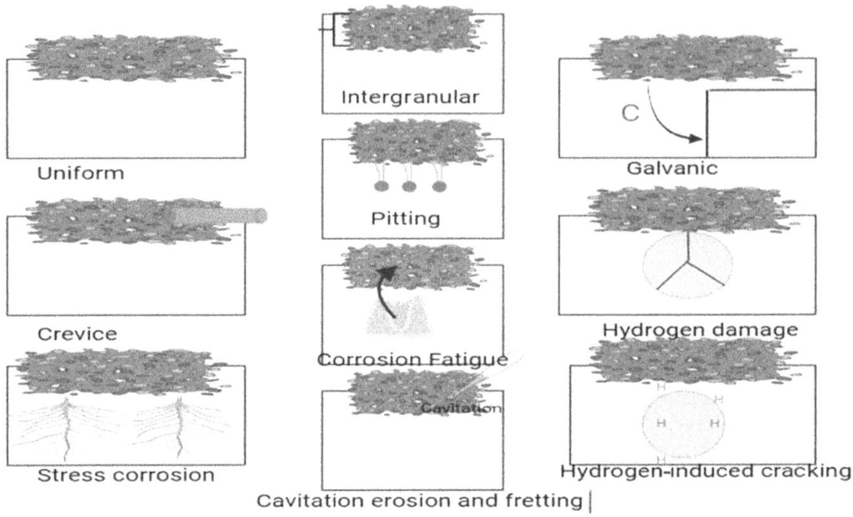

FIGURE 15.1 Different types of corrosion.

down corrosion reactions. This can occur when MOFs adsorb on the metal surface and create a barrier or act as intermediaries in corrosion reactions [2].

2. *Pitting Corrosion:* Localized corrosion in the form of small holes or pits on the metal surface. It is more common in the presence of certain ions such as chloride ions. MOFs can control pitting corrosion by preventing the formation of these pits or slowing their growth. They can do this by blocking the pit initiation points or by changing the chemical conditions within the pit [3].
3. *Stress Corrosion Cracking (SCC):* The formation and propagation of cracks in a metal due to the synergistic effect of tensile stress and a corrosive environment. MOFs can prevent SCC by protecting the metal surface or reducing stress concentration points. In addition, the mechanical properties of MOFs can also have an effect at this point [4].
4. *Intergranular Corrosion:* Corrosion that occurs along the grain boundaries of the metal. MOFs can control intergranular corrosion by protecting the grain boundaries or reducing chemical activity in these regions [5].
5. *Galvanic Corrosion:* Corrosion that occurs when different metals are exposed to a corrosive environment in electrical contact. MOFs can prevent galvanic corrosion by providing electrical insulation between metals or by forming a protective layer on metal surfaces [6].

15.5 ELECTROCHEMICAL TESTS AND ANALYSES

Various electrochemical techniques are used to understand the corrosion inhibition mechanisms of MOFs. Comparison of the effectiveness of mild steel corrosion prevention by weight loss method and electrochemical methods for different inhibitor concentrations are given in Table 15.1.

TABLE 15.1
Comparison of the Effectiveness of Mild Steel Corrosion Prevention by Weight Loss Method and Electrochemical Methods for Different Inhibitor Concentrations

Method	Concentration (mM)	1.5 M H_2SO_4	3 M HCl	1 M H_3PO_4
Weight loss	0.5	71.43	81.08	69.94
	1.0	76.17	86.16	72.49
	2.5	76.92	90.41	77.72
	5.0	84.74	96.79	82.54
	7.5	89.48	97.44	85.55
	10.0	95.02	98.22	94.49
AC-impedance spectroscopy	0.5	82.89	88.35	63.04
	5.0	87.35	90.98	76.49
	10.0	89.21	92.01	80.95
Potentiodynamic polarization	0.5	38.35	54.33	36.80
	5.0	48.13	69.55	38.22
	10.0	55.55	75.15	43.84

1. *Polarization Curves (Tafel Curves)*: Information about the corrosion rate and mechanism is obtained by plotting the potential of the metal against the current. The decrease in the corrosion current density (I_{corr}) in the presence of MOF is an indicator of the inhibition efficiency. Corrosion potential (E_{corr}) and anodic/cathodic Tafel slopes can also be determined from Tafel curves [7].
2. *Electrochemical Impedance Spectroscopy (EIS)*: Alternating current signals are used to examine the electrical properties of the metal–solution interface. EIS data are presented in the form of Nyquist and Bode plots. The increase in impedance values in the presence of MOF indicates increased corrosion resistance [8].
3. *Open Circuit Potential (OCP)*: It is the potential of the metal measured in the corrosion environment without any external current being applied. The effect of MOF application on the OCP can provide information about the changes on the metal surface [9].
4. *Linear Polarization Resistance (LPR)*: It is a rapid method used to determine the corrosion rate. Measurements are made in a small potential range and the polarization resistance (Rp) is calculated [10].

15.6 SURFACE CHARACTERIZATION

Various surface characterization techniques are used to examine the changes that occur on the metal surface after MOF application:

1. *Scanning Electron Microscopy (SEM):* Electron beam is used to obtain high-resolution images of the metal surface. It provides information about the morphology, thickness, and homogeneity of the MOF coating [11].
2. *Transmission Electron Microscopy (TEM):* Electron beam is used to examine the internal structure of the sample. It provides detailed information about the crystal structure of the MOFs and their interface with the metal [12].
3. *X-Ray Photoelectron Spectroscopy (XPS):* It is used to determine the chemical composition and oxidation states of the elements on the metal surface. It provides information about the chemical interactions of the MOFs with the metal surface [13].
4. *Atomic Force Microscopy (AFM):* It is used to image the topography of the metal surface at nanometer resolution. It provides information about the roughness and surface morphology of the MOF coating [14].

15.7 BARRIER EFFECT OF MOFS

MOFs primarily act as a barrier in corrosion control. This barrier effect is due to the porous nature of the MOF structure. Although the pores of MOFs allow only small molecules to pass through, they prevent larger water, gas, and vapor molecules from reaching the metal surfaces. Oxidation and other types of corrosion on metal surfaces are accelerated because of the interaction of compounds in the atmosphere such as water vapor, oxygen, and carbon dioxide with the metal surface. MOFs prevent these compounds from reaching the metal surface and prevent corrosion [15].

15.7.1 Preventing Gas and Water Diffusion

The barrier effect of MOF structures works on the basis of preventing the diffusion of water and gas molecules to the metal surface. This feature is especially effective in protecting metals that come into direct contact with atmospheric conditions or humid environments. The pores in the structure of MOFs prevent water and gases from reaching the metal surfaces, preventing the acceleration of oxidation and other types of corrosion. This allows MOFs to function as a kind of protective coating [16].

In particular, MOFs, which prevent the passage of compounds such as CO_2, O_2, H_2O, form a barrier that prevents metals from rusting, oxidizing, or being exposed to other types of corrosion. The structural features of MOFs can greatly increase the effectiveness of this barrier. For example, the regular and tight structure of MOFs significantly reduces the passage rate of gas and water. This mechanism is especially effective in environments with high corrosion risk, such as seawater.

15.7.2 Active Protection Mechanisms: Release of Corrosion Inhibitors

MOFs are materials that not only provide a barrier effect, but also provide active protection. Active protection is provided by the release of protective compounds around the metal surface over time. MOFs can protect the metal surface by releasing corrosion inhibitors held on their inner surfaces in a controlled manner. This mechanism is especially useful in environments with high corrosion risk [17].

15.7.3 Release of Inhibitors

Corrosion inhibitors are chemical substances that protect metal surfaces against corrosion. These inhibitors can be stored in the pores of MOF structures and released over time. This release can be affected by external factors such as pH value and temperature. MOFs create a continuous protection layer on the metal surface by providing the release of inhibitors at a certain speed. This protection layer reduces the interaction of the metal with corrosion and extends the life of the metal [18]. Especially in aggressive environments such as seawater, the release of inhibitors controlled by MOFs allows the metal to last longer without corroding. This feature offers a great advantage in terms of both environmental friendliness and long-term corrosion control. In addition, the pore structure of MOFs can optimize the release rate and effectiveness of inhibitors. Inorganic or organic inhibitors are stored in these structures and help protect the metal surface over time.

15.7.4 Antimicrobial and Cathodic Protection Properties

Another important role of MOFs in corrosion control is their antimicrobial and cathodic protection properties. Particularly in environments where microorganisms are common, such as humid environments and seawater, MOFs can both prevent the growth of microorganisms and eliminate the biological basis of corrosion [19].

15.7.5 Antimicrobial Effects

Bacteria and other microorganisms can form biofilms on the surfaces of metals, and this biofilm can increase the corrosion rate of the metal. MOFs, especially when they contain metals such as copper and zinc, can exhibit antimicrobial properties. These properties help protect metals against biological contamination and prevent biological corrosion. The inhibition of the growth of microorganisms by MOFs eliminates the biological basis of corrosion and provides long-term protection of metal surfaces [20].

15.7.6 Cathodic Protection

Cathodic protection is a method in which an external electrical current is used to protect metals against corrosion. MOFs can provide the basis for such protection. The specific structures of MOFs promote passivation on metal surfaces and provide cathodic protection. This protection mechanism helps protect metal surfaces, especially in aggressive environments such as seawater and in industrial applications. Cathodic protection prevents oxidation processes by providing metals with an electrically negative potential. MOFs have great potential with the versatile protection mechanisms they offer in corrosion control. The barrier effect of MOF structures, together with active protection mechanisms, offers an effective solution for protecting metal surfaces against environmental effects. In addition, antimicrobial and cathodic protection properties make MOFs stand out as multifunctional corrosion inhibitor materials. The combination of these properties provides a great advantage in extending the life of metals and reducing maintenance costs, especially in seawater and industrial applications. The design of MOFs and the combination of different protection methods according to their areas of use play an important role in effectively preventing corrosion [21].

15.8 DESIGN STRATEGIES FOR CORROSION APPLICATIONS OF MOFS

MOFs have emerged in recent years as materials offering great potential for pre- and post-corrosion protection. The effectiveness of MOFs in corrosion applications varies greatly depending on their structure and design.

15.8.1 Computational Modeling and Simulation

Computational methods provide powerful tools to understand the corrosion inhibition mechanisms of MOFs, predict their performance, and design new materials. The main computational methods used are:

1. *Density Functional Theory (DFT):* DFT is a quantum mechanical method used to study the electronic structure of atoms and molecules. In the context of corrosion, DFT calculations can be used to [22]:

   Determine the electronic structure and surface properties of MOFs.

   Study the adsorption mechanisms of MOFs on metal surfaces and calculate their adsorption energies.

   Calculate the energy profiles and activation energies of corrosion reactions.

   Study the interactions of MOFs with different corrosive species.
2. *Molecular Dynamics (MD) Simulations:* MD simulations use the principles of classical mechanics to model how atoms and molecules move over time. In the context of corrosion, MD simulations can be used to [23]:

   Study the behavior of MOFs in corrosive environments (e.g., aqueous solutions, high temperatures).

   Study the stability and dynamic properties of the coating formed by MOFs on metal surfaces.

   Study the interactions and diffusion of MOFs with corrosive species (e.g., water molecules, ions).
3. *Monte Carlo Simulations:* Monte Carlo simulations use statistical methods to predict the properties of systems using random sampling. In the context of corrosion, Monte Carlo simulations can be used to [24]:

   Calculate the adsorption isotherms of MOFs on metal surfaces.

   Predict the corrosion inhibition efficiency.

   These computational methods allow us to predict and optimize the corrosion inhibition performance of MOFs before testing them experimentally. They also help us understand corrosion mechanisms at the atomic level.

15.8.2 Functionalization Strategies

A variety of functionalization strategies can be used to improve the corrosion inhibition properties of MOFs.

1. *Ligand Modification:* Changing the chemical structure of organic ligands used in the MOF structure can significantly affect the properties of MOFs. Ligand modification for corrosion inhibition may include:
   Addition of functional groups, known as corrosion inhibitors (e.g., amines, carboxylic acids, thiols), to the ligands. These groups can enhance corrosion inhibition by establishing strong interactions with the metal surface.
   Changing the steric structure of the ligands can optimize the corrosion performance by affecting the pore size and surface area of MOFs.
2. *Metal Node Substitution:* Replacing metal ions in the MOF structure with different metal ions can change the redox properties of the MOFs and their interactions with the metal surface. Metal ion exchange for corrosion inhibition may include:
   Use of metals that exhibit higher corrosion resistance.
   Use of metals that form stronger bonds with the metal surface.
3. *Composite Formation:* Composites of MOFs with polymers, conductive polymers, graphene, or other materials can be used to improve their corrosion performance. These composites can provide synergistic effects, providing higher corrosion resistance, better mechanical properties, and longer-term protection. For example, incorporating MOFs into polymer matrices can provide better adhesion of MOFs to metal surfaces and a more uniform coating. These functionalization strategies provide important tools to optimize and customize the performance of MOFs for corrosion control applications. A combination of these strategies can also be used, resulting in synergistic effects [25].

15.8.3 Effect of Linker and Node Selection on Corrosion Properties

The structure of MOFs is generally shaped by the bonds established between metal ions (nodes) and organic linking components (linker). Linker and node selection directly affects the durability, stability, and anticorrosion protection properties of the MOF structure. Therefore, correct linker and node selection are very important for effective MOF design in corrosion control [26].

1. Node Selection: Metal ions (nodes) in the structures of MOFs generally play a major role in protecting metal surfaces. While metal ions increase the stability of the MOF, they also have the potential to block corrosive molecules interacting with the metal surface. For example, the use of metals such as copper, zinc, iron, and aluminum in MOF structures can be effective in preventing corrosion. These metals provide anticorrosion protection because metal ions are generally resistant to oxidation and can create passivation on the surfaces of metals [27].

2. *Linker Selection:* Linker molecules are organic compounds that bind metal ions in the MOF structure. The structure and function of the linkers determine the pore structure and stability of the MOF. It is important that the linkers used in corrosion applications have properties that prevent the spread of corrosion. For example, some linkers can be resistant to water, thus preventing MOFs from corroding in environments such as water vapor or humidity. Aromatic linkers (such as benzene derivatives) provide advantages in protecting metal surfaces against corrosion by providing high mechanical strength and stability [28].
3. *Corrosion Properties of Linker and Node Combinations:* Linker and node combinations can affect the durability and stability of MOFs. For example, the combination of copper ions with aromatic linkers such as 1,3,5-benzenetricarboxylic acid (BTC) can create a strong corrosion barrier. Similarly, the combination of similar aromatic linkers with iron ions can also increase corrosion resistance. Such combinations help prevent corrosion by preserving the structural integrity of MOFs [29].

15.9 HYDROPHOBIC AND THERMALLY STABLE MOF DESIGNS

To increase the effectiveness of water-repellent and thermally stable metal–organic scaffolds, hydrophobic properties and thermal stability need to be improved. One of the main causes of corrosion is the contact of water with the metal surface and the oxidation of the metal surface at high temperatures. Therefore, hydrophobic and thermally stable metal–organic scaffolds can be quite effective in providing anticorrosion protection, especially in harsh environments [30].

Hydrophobic metal–organic scaffolds block moisture from the external environment and prevent the absorption of water vapor and water. This feature reduces the contact of metal–organic scaffolds with water and prevents water from interacting with the metal surface. Hydrogen bonds and electrical charges in the structure of metal–organic scaffolds help protect the metal surface by preventing water molecules from penetrating the structure. Hydrophobic metal–organic scaffolds provide a great advantage, especially for metals exposed to humid environments and saltwater such as seawater [31]. Hydrophobic MOFs are shown in Figure 15.2.

Thermal stability refers to the ability of metal–organic scaffolds to maintain their structures even at high temperatures. Corrosion generally accelerates with increasing temperature, so thermally stable metal–organic scaffolds provide effective corrosion control even at high temperatures. Thermal stability allows metal–organic scaffolds to maintain their metal surface protection properties for a long time. Thermally stable metal–organic scaffolds are especially preferred in industrial applications where high temperatures and variable environmental conditions are present [32].

Application areas of hydrophobic and thermally stable metal–organic scaffolds include the automotive and aerospace sectors, marine, and various industrial areas. Vehicles and aircraft are exposed to harsh environmental conditions such as high temperature and humidity. Hydrophobic and thermally stable metal–organic scaffolds can help prevent corrosion by protecting metal surfaces in these environments. Seawater is known for its high salinity and moisture content. Hydrophobic and thermally

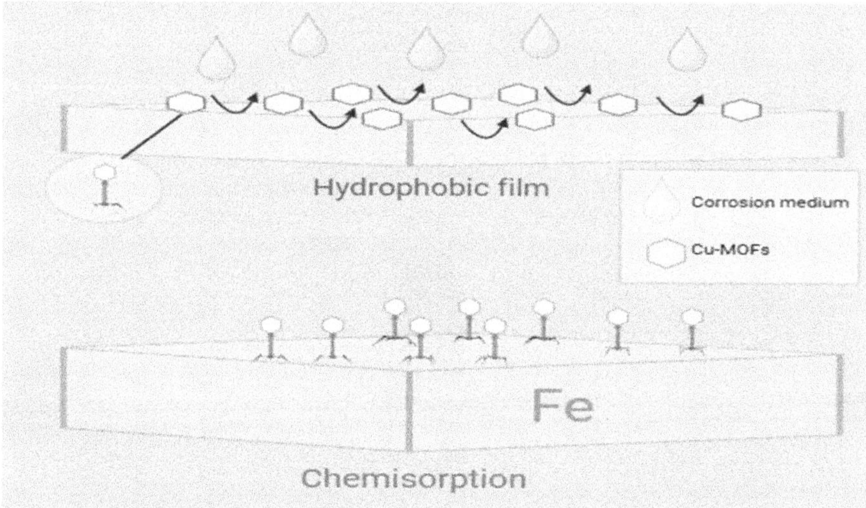

FIGURE 15.2 Hydrophobic MOFs.

stable metal–organic scaffolds can provide long-term protection of metal surfaces on ships [33].

An important factor that makes the use of metal–organic scaffolds more effective in corrosion control is functionalization processes. Functionalization improves the properties of metal–organic scaffolds, making them more durable and environmentally friendly. Modifications with polymers or composites are an effective method to increase the durability and corrosion resistance of metal–organic scaffolds [34].

Polymers and composites can increase the durability of metal–organic scaffolds. Adding polymeric materials to the surface of metal–organic scaffolds makes metal–organic scaffolds more resistant to water and gas permeation. This modification makes metal–organic scaffolds longer lasting and more effective against corrosion. For example, metal–organic scaffolds can be integrated into a polymer matrix to form coatings that provide stronger protection on metal surfaces. These coatings resist deformations of the metal surface because of the flexibility of the polymers [35].

The advantages of functionalized metal–organic scaffolds include continuous protection and environmentally friendly properties. Polymer modifications enable metal–organic scaffolds to release corrosion inhibitors in a more controlled manner, enabling continuous protection of metal surfaces. Functionalization processes with polymers and composites provide more sustainable corrosion control while maintaining their environmentally friendly properties [36].

Design strategies for corrosion applications of metal–organic scaffolds play an important role in improving the effectiveness and durability of these materials. Functionalized metal–organic scaffolds have a wide range of design advantages, from the choice of binders and nodes to hydrophobicity and thermal stability. These strategies have significant potential in protecting metal surfaces in industrial applications

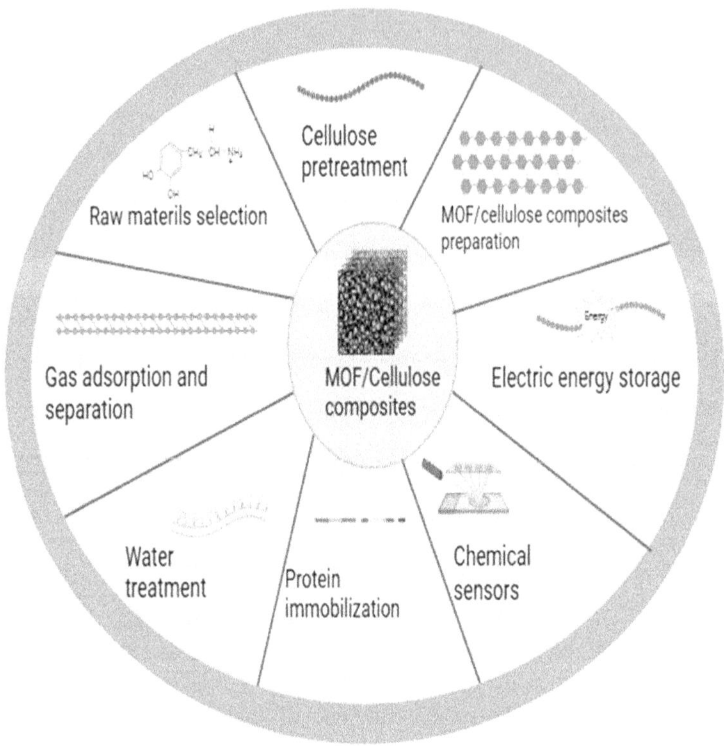

FIGURE 15.3 Schematic of the main design strategies and applications of cellulose/MOF.

and harsh environmental conditions. Schematic of the main design strategies and applications of cellulose/MOF is given in Figure 15.3 [37].

15.10 APPLICATIONS OF MOFS IN DIFFERENT INDUSTRIES

Metal–organic scaffolds have found a wide range of applications in many sectors in recent years because of their various chemical, physical, and mechanical properties. With their high surface areas, controlled pore structures, high stability, and functionalization capabilities, these materials offer significant advantages in various functions such as corrosion, gas absorption, catalysis, and storage [38].

The oil and gas industry is one of the sectors where corrosion is most intense due to equipment operating in harsh environmental conditions, high temperatures, and chemical environments. In this sector, metal–organic scaffolds are used in various protection, storage, and separation processes. In particular, gas storage and separation capabilities have significant potential in increasing the efficiency of industrial processes. Pipelines and other metal structures are frequently exposed to corrosion in the oil and gas industry [39]. MOFs can prevent the spread of corrosion by creating an effective barrier as protective coatings applied to these metal surfaces. They provide protection by preventing corrosion agents such as gas and water vapor from reaching metal surfaces.

MOFs, which have especially hydrophobic properties, can reduce the corrosion rate by preventing water from contacting the metal surface [40]. The storage and separation of gases, such as natural gas and carbon dioxide, are of great importance in the oil and gas industry. The high surface area and pore structure of MOFs allow for the storage of large amounts of gas with high efficiency. For example, MOFs can effectively store methane gas and can be used in carbon dioxide separation processes. This feature is critical for reducing greenhouse gases and increasing energy efficiency [41].

The maritime sector is an industry where ships and other marine vessels are exposed to seawater for long periods of time. In this environment, seawater and atmospheric moisture cause rapid corrosion of metal surfaces. Ship coatings are one of the most critical elements to deal with this problem. MOFs are used as ship coatings and protective coatings in the maritime sector, providing an effective solution that extends the life of metal surfaces [42]. The maritime industry has to deal with the rapid corrosion of metal surfaces on ships due to the salty nature of seawater and the conditions resulting from the reaction with oxygen. MOFs can be used to prevent corrosion of ship hulls in this environment. Hydrophobic MOFs can prevent saltwater from reaching metal surfaces by preventing water from contacting the surface. In addition, some MOFs can provide long-term protection by providing active protection when functionalized with corrosion inhibitors. The maritime sector is also sensitive to biofouling [43]. Seawater can cause microorganisms and algae to adhere to metal surfaces. Such biofouling can accelerate the corrosion process. MOFs may have properties that can prevent biological interactions. MOFs with antimicrobial properties can prevent biofouling on ship surfaces and prevent the acceleration of corrosion [44].

The electronics industry is a sector that requires the production of sensitive devices and requires the use of effective materials to extend the life of these devices. Electronic components can corrode, especially due to humidity and temperature changes. MOFs play an important role in protecting microchips and other electronic components. Electronic components, especially microchips, can corrode when they interact with humidity and water vapor [45]. MOFs can prevent moisture from reaching electronic components by absorbing water vapor. Consequently, the life of the devices is extended and their performance is maintained. MOFs also prevent rust formation on the components, ensuring the proper operation of the devices. Electronic products are often stored or operated at low temperatures [46]. MOFs are materials that are stable at low temperatures and have effective corrosion prevention properties. Therefore, MOFs protect sensitive electronic devices such as microchips from corrosion, ensuring long-term durability. Metal–organic scaffolds have found a wide range of applications in many sectors in recent years because of their various chemical, physical, and mechanical properties. With their high surface areas, controlled pore structures, high stability, and functionalization capabilities, these materials offer significant advantages in various functions such as corrosion, gas absorption, catalysis, and storage [47].

The oil and gas industry is one of the sectors where corrosion is most intense due to equipment operating in harsh environmental conditions, high temperatures, and chemical environments. In this sector, metal–organic scaffolds are used in various protection, storage, and separation processes [48]. In particular, gas storage and separation capabilities have significant potential in increasing the efficiency of industrial processes. Pipelines and other metal structures are frequently exposed to corrosion in the oil and gas industry. MOFs can prevent the spread of corrosion by creating an effective barrier as protective coatings applied to these metal surfaces. They provide protection by preventing corrosion agents such as gas and water vapor from reaching metal surfaces [49]. MOFs, which have especially hydrophobic properties, can reduce the corrosion rate by preventing water from contacting the metal surface. The storage and separation of gases such as natural gas and carbon dioxide are of great importance in the oil and gas industry. The high surface area and pore structure of MOFs allow for the storage of large amounts of gas with high efficiency. For example, MOFs can effectively store methane gas and can be used in carbon dioxide separation processes. This feature is critical for reducing greenhouse gases and increasing energy efficiency [50].

The maritime sector is an industry where ships and other marine vessels are exposed to seawater for long periods of time. In this environment, seawater and atmospheric moisture cause rapid corrosion of metal surfaces. Ship coatings are one of the most critical elements to deal with this problem. MOFs are used as ship coatings and protective coatings in the maritime sector, providing an effective solution that extends the life of metal surfaces. The maritime industry has to deal with the rapid corrosion of metal surfaces on ships due to the salty nature of seawater and the conditions resulting from the reaction with oxygen. MOFs can be used to prevent corrosion of ship hulls in this environment [48]. Hydrophobic MOFs can prevent saltwater from reaching metal surfaces by preventing water from contacting the surface. In addition, some MOFs can provide long-term protection by providing active protection when functionalized with corrosion inhibitors. The maritime sector is also sensitive to biofouling. Seawater can cause microorganisms and algae to adhere to metal surfaces. Such biofouling can accelerate the corrosion process. MOFs may have properties that can prevent biological interactions. MOFs with antimicrobial properties can prevent biofouling on ship surfaces and prevent the acceleration of corrosion [50].

The electronics industry is a sector that requires the production of sensitive devices and requires the use of effective materials to extend the life of these devices. Electronic components can corrode, especially due to humidity and temperature changes. MOFs play an important role in protecting microchips and other electronic components [48]. Electronic components, especially microchips, can corrode when they interact with humidity and water vapor. MOFs can prevent moisture from reaching electronic components by absorbing water vapor. Consequently, the life of the devices is extended and their performance is maintained. MOFs also prevent rust formation on the components, ensuring the proper operation of the devices [49]. Electronic products are often stored or operated at low temperatures. MOFs are materials that are stable at low temperatures and have effective corrosion prevention properties. Therefore, MOFs protect sensitive electronic devices such as microchips from corrosion, ensuring long-term durability [50].

The automotive industry is a sector that requires the metal surfaces of vehicles to last a long time. Vehicles are exposed to various weather conditions, water, and salt. Therefore, it is of great importance for the automotive sector to provide protection against corrosion and to develop long-lasting coatings. MOFs are used to protect the metal surfaces of vehicles and make them more durable. In the automotive sector, MOFs can be used as protective coatings to protect vehicle surfaces against corrosion [51]. Hydrophobic MOFs can prevent water and other contaminants from reaching the vehicle surface. In addition, some MOFs can prevent rust formation on metal surfaces by secreting corrosion inhibitors. These coatings increase the durability, especially on the exterior surfaces of vehicles, and make them look new for a long time. The automotive sector requires vehicles to be exposed to various weather conditions. MOFs can provide effective protection against harsh environmental effects such as high temperature, humidity, and acidic conditions. Therefore, MOFs can provide long-term protection by ensuring that vehicles are durable in different climatic conditions in the automotive sector.

MOFs are used effectively in a wide range of applications, from the oil and gas industry to the marine, electronics, and automotive sectors. With their high surface area, controlled pore structure, and functionalization capabilities, MOFs offer advantages in many areas, from corrosion control to gas storage and separation to biofouling prevention. The industrial use of these innovative materials offers a significant opportunity to improve corrosion prevention strategies and increase environmental sustainability.

15.11 CHALLENGES IN MOF-BASED CORROSION CONTROL

MOFs offer significant advantages over traditional methods for corrosion control due to their high surface area, adjustable porosity, and multifunctional properties. However, the practical use of these innovative materials presents many technical, economic, and environmental challenges. To make MOFs industrially applicable, all processes from synthesis to application must be optimized. Improving these processes plays a critical role in the widespread adoption of MOFs and their use as an effective corrosion prevention technology [52]. Problems encountered during MOF synthesis are presented in Figure 15.4.

Although the properties of MOFs such as barrier coating, ion exchange material, and corrosion inhibitor make them attractive in terms of corrosion prevention, the translation of successful results at the laboratory scale to large-scale industrial applications poses a significant obstacle. Although current synthesis techniques are generally effective in the laboratory environment, they are insufficient for large-scale production. This situation limits the widespread use of MOFs by increasing production costs. Therefore, the development of synthesis strategies that are more efficient, economically viable, and readily scalable is critically important to overcome these challenges and facilitate broader industrial implementation. [53].

The high cost of raw materials used in MOF production is one of the biggest obstacles to industrial applications. The basic building blocks of MOFs, metal ions and organic ligands, generally must be of high purity to maintain the desired structural and functional properties. However, such high purity raw materials are expensive and may not be economical for large-scale production. In addition, the dependence on rare and expensive transition metals such as ruthenium, palladium, or platinum-based

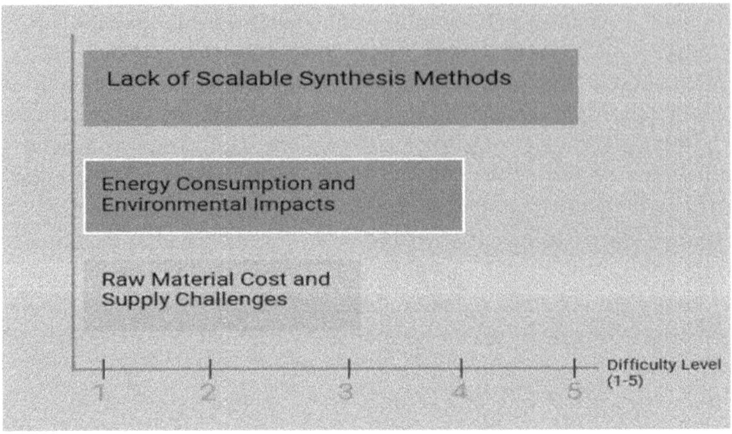

FIGURE 15.4 Problems encountered during MOF synthesis.

MOFs further increases cost problems. Instead, alternative solutions based on more common and economical metal precursors such as iron or aluminum-based MOFs should be investigated.

High energy consumption during MOF synthesis also poses a significant challenge. Many MOF synthesis methods have serious energy requirements due to the extreme conditions they require. For example, hydrothermal and solvothermal techniques often require long reaction times at high temperatures, making them energy inefficient. In addition, organic solvents such as dimethylformamide (DMF) used during synthesis pose an environmental threat due to their toxicity and non-biodegradability. The large-scale use of MOFs in corrosion control requires the development of environmentally friendly synthesis methods. Sustainable synthesis approaches such as solvent-free mechanochemical methods or water-based synthesis techniques are important alternatives to overcome this problem.

Sol–gel, hydrothermal and solvothermal synthesis methods, which have yielded successful results at the laboratory scale, have difficulty in being adapted to large-scale production. One of the main reasons for this is the difficulty of maintaining homogeneous crystal growth and consistent material properties in mass production. Batch synthesis techniques used in research environments lead to problems such as low yield and irregular particle size distribution in large-scale production. Therefore, studies on alternative techniques such as continuous flow synthesis methods are ongoing. Continuous flow systems are becoming more suitable for industrial applications by providing controlled reaction conditions, high production rates, and better reproducibility.

Overcoming the problems such as cost, energy consumption, environmental sustainability, and synthesis scalability that prevent large-scale production of MOFs will enable their wider adoption in corrosion control. Interdisciplinary studies such as materials science, chemical engineering, and industrial process optimization are needed to overcome these challenges. For MOF-based technologies to successfully transition from laboratory research to industrial applications, economically and technologically feasible solutions must be developed.

15.12 ENVIRONMENTAL STABILITY AND DEGRADATION PROBLEMS

The use of MOFs for corrosion control requires that these materials maintain their stability against environmental factors. However, MOFs can be prone to degradation, especially in harsh environmental conditions. Factors such as high humidity, corrosive chemicals, or extreme temperatures can threaten the structural integrity of MOFs and negatively affect their corrosion prevention performance. Therefore, the development of new material strategies to increase the stability of MOFs is of great importance. The long-term durability of MOFs should be optimized depending on their application areas and should be designed to adapt to specific environmental conditions [54].

One of the biggest disadvantages of MOFs is that they tend to lose their structural integrity when they interact with water molecules. Water sensitivity is a major factor that limits the use of MOFs, especially in sectors such as marine, oil, and gas. MOFs exposed to humidity or water vapor may exhibit deterioration in their crystalline structures over time due to the weakening of metal–ligand bonds and the loss of their porous structures. To overcome these problems, water-resistant MOF types need to be developed. MOFs containing hydrophobic ligands or MOFs that have been given water-repellent properties through surface modifications can be made more resistant to water molecules. In addition, combining MOFs with organic coatings or hybrid material systems to ensure their stability in the presence of water can be an effective solution [55, 56]. MOF stability under different environmental conditions is presented in Figure 15.5.

The resistance of MOFs to chemical corrosives is also an important design criterion. Acidic or basic environments can weaken the metal–ligand bonds of MOFs, leading to structural collapse. MOFs used in industrial applications must be able to maintain their stability when exposed to strong acids, bases, or corrosive gases. In this context, the synthesis of MOFs with high chemical resistance is a great necessity. Metal–ligand combinations that form strong coordination bonds can be preferred to increase chemical stability. In addition, inorganic–organic hybrid systems or cross-linked polymer-supported MOF structures can be developed to provide resistance to chemical degradation. Some MOFs may lose their active components by interacting with harsh chemical environments, so the MOF to be selected must be optimized according to the area of use [57].

The thermal stability of MOFs is an important limitation in applications requiring high temperatures. Some MOFs may undergo phase changes that cause their porous structures to deteriorate or completely collapse when they exceed certain temperature levels. This situation limits the use of MOFs, especially in energy, catalysis, and industrial processes operating at high temperatures. To increase thermal stability, metal–ligand systems with high melting points should be preferred and heat-resistant MOF derivatives should be designed. MOF systems reinforced with metal oxides or ceramics can be developed for high-temperature resistance. In addition, covalent

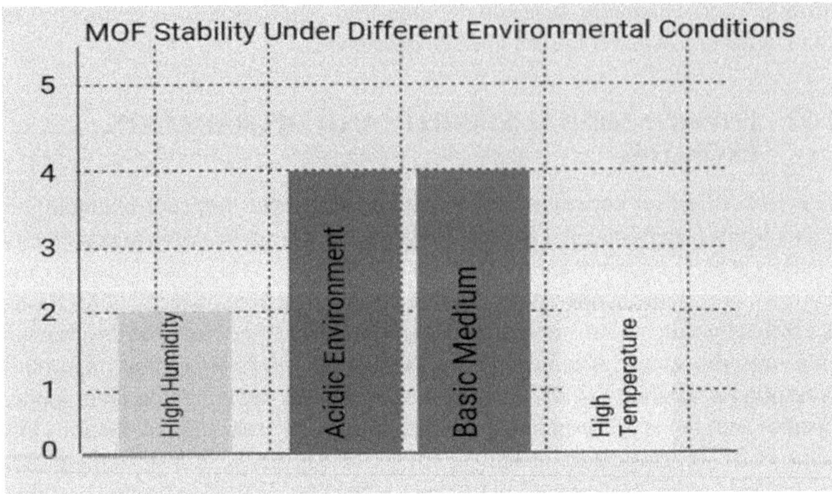

FIGURE 15.5 MOF stability under different environmental conditions.

organic frameworks (COFs) or metal–organic hybrid systems that can remain stable under high-temperature conditions can be evaluated as alternatives to MOFs [58].

Multifaceted approaches should be developed to overcome these stability problems. Surface coating techniques, hybrid material strategies, and functional modifications can be applied to make MOFs more resistant to environmental factors. Designing MOFs that are more resistant to water, chemical agents, and high temperatures could increase the effectiveness of MOFs in many industrial applications, especially in corrosion control. In the future, new synthesis methods and material engineering approaches aimed at improving environmental stability could enable MOFs to have a wider range of applications. In particular, the development of scalable and economically viable MOF designs for industrial applications will increase the commercial use of these materials [59]. MOF stability under different environmental conditions is shown in Figure 15.4.

15.13 INCOMPATIBILITY ISSUES: INTERACTION OF MOFS WITH EXISTING MATERIALS

For MOFs to be used for corrosion control purposes, they must be compatible with existing materials and coating systems. However, the chemical and physical properties of MOFs may create some difficulties in achieving this compatibility. In this context, various strategies should be developed to ensure effective bonding of MOFs to surfaces, increase their mechanical strength, and control their chemical reactivity. Failure of MOF coatings to bond sufficiently to metal surfaces may cause the coating to detach from the surface over time, thus reducing its corrosion protection effectiveness. To overcome this problem, methods such as surface modification, use of binding agents or integration of MOFs into the coating should be applied. In particular, coating durability can be increased by optimizing the surface energy and interactions of MOFs with metal surfaces [60].

The lack of flexibility of MOFs can lead to coating systems cracking on deformed surfaces and losing their protective properties. While traditional coating materials generally show a certain flexibility, MOFs generally exhibit hard and brittle structures. It is suggested to develop polymer–MOF hybrid systems to solve this problem. Polymer additives can increase the mechanical flexibility of MOF coatings and make them resistant to deformation. The development of such hybrid systems can both increase the mechanical stability of MOFs and optimize their long-term protection performance. In addition, coating homogeneity and surface adhesion can be increased by using hybrid materials [61]. Another important factor for the use of MOFs in coatings is the increase in their thermal and environmental stability. Temperature changes, humidity, and other environmental factors can cause MOFs to degrade and lose their effectiveness over time. Therefore, modifications with cross-linking agents and stabilizing materials are required to increase the stability of MOFs.

The fact that MOFs lose their protective properties by chemically reacting with metal surfaces shows that detailed material characterization is of great importance before application. Careful analysis of the interactions between metal surfaces and MOFs is required, modifications that will increase the chemical stability of MOFs must be determined, and optimum formulations must be developed [62]. Ligand modifications or surface engineering techniques can be used to control the surface reactivity of MOFs. For example, MOFs that bind to metal surfaces via hydrogen bonds or coordination bonds may degrade and lose their activity over time. Therefore, it is important to determine MOF types with appropriate ligand designs and functional groups to optimize the surface reactivity of MOFs [27].

In addition, one of the problems encountered during the integration of MOFs with existing coating systems is solvent incompatibility. The solvents used during the synthesis and application of MOFs can damage metal surfaces or negatively affect the coating performance. Therefore, the solvents and additives used in the coating process should be carefully selected. At the same time, additional measures such as stabilizing agents and protective coatings should be taken to increase the long-term durability of MOF coatings [63].

Various surface engineering techniques, material characterization methods, and composite material strategies should be applied to ensure that MOFs are better compatible with existing materials and become an effective option for corrosion control. Surface modification can provide the addition of special functional groups that increase the binding of MOFs to metal surfaces. In addition, polymer-supported systems or inorganic–organic hybrid coatings can be developed to increase coating uniformity and optimize the coating performance of MOFs. In the future, research should focus on improving material compatibility and optimizing corrosion protection performance for wider use of MOF-based coatings. In this context, the development of scalable production techniques is also of great importance to increase the applicability of MOF coatings on an industrial scale [63].

15.14 ADDITIONAL FACTORS AFFECTING MOF PERFORMANCE IN CORROSION CONTROL

Other important factors that should be taken into consideration during application for MOF-based coatings to be long lasting and effective are as follows [64]:

Coating thickness and porosity are factors that directly affect the effectiveness of MOF coatings. While very thin MOF coatings may not provide sufficient protection, excessively thick coatings may cause crack formation and lose efficiency [65]. Therefore, control of porosity properties is a decisive factor in this context.

The environmental sensitivity of MOFs requires special conditions during storage and transportation of these materials [66]. Solutions such as appropriate packaging methods, moisture, and airtight containers should be developed.

For MOF coatings to provide long-term corrosion control, their structural durability must be increased. In this context, innovative approaches such as the use of nanocomposites can be adopted.

The use of MOF coatings as surface coating materials requires not only chemical resistance but also mechanical stability. However, the brittle structure of existing MOF coatings and their limited resistance to external effects make progress in this area difficult [67].

The inflexibility of MOF coatings can lead to crack formation under mechanical stress. These cracks can facilitate corrosion and cause the coating to lose its protective properties. To solve this problem, flexible MOF–polymer hybrids need to be developed.

Salt water, sand, and other corrosive environments can cause physical wear of MOF coatings. This is a serious problem, especially for the marine industry and offshore platforms. To increase mechanical strength, harder hybrid materials need to be developed [68].

Coating adhesion is also a major concern. Poor adhesion between metal surfaces and MOF coatings can cause the coating to peel over time [69]. To overcome this problem, adhesion strengths can be increased by using surface modifications or intermediate layers.

15.15 CHALLENGES IN MULTIFUNCTIONAL DESIGN OF MOFS

There are several challenges in the design of multifunctional MOFs. The aim of these designs is to integrate MOFs not only with corrosion control but also with other functions. However, there are significant technical hurdles to achieve this goal [70]. Comparison of the performance of various nanostructures synthesized from MOF precursors as supercapacitors are presented in Table 15.2.

The first challenge is to balance corrosion and catalytic functions. The ability of MOFs to exhibit both corrosion control and catalytic properties creates a very complex design situation. Optimizing these two functions together is a great challenge, especially in reactive environments.

Second, electrical and thermal properties need to be matched. In applications requiring electrical conductivity, the inherent insulator properties of MOFs can be limiting. Similarly, in applications where thermal conductivity needs to be increased, the low thermal resistance of MOFs can be problematic. Therefore, the development of conductive MOF derivatives is of critical importance.

The third challenge is the development of smart MOF coatings. The design of these smart coatings that can dynamically adapt to environmental conditions creates great complexity in terms of material engineering and manufacturing technologies. For example, coatings that are sensitive to pH changes require high durability and flexibility.

Efforts to increase the chemical stability of MOFs are also among the important challenges. Increasing chemical stability allows MOFs to perform more effectively in harsh environments. However, such stability-enhancing processes can make the synthesis steps more complex and increase costs.

The first strategy to increase chemical stability is ligand modifications. Functionalization methods can be applied to increase the chemical stability of organic ligands used in MOFs. However, such modifications can generally negatively affect porosity and gas adsorption capacities.

TABLE 15.2
Comparison of the Performance of Various Nanostructures Synthesized from MOF Precursors as Supercapacitors

Materials	Doping	Original MOF	Specific Surface Area (m²/g)	Current Density (A/g)	Specific Capacitance (F/g)	Cycle Numbers
Carbon functionalized	Nitrogen	ZIF-8	934	0.1	285.8	1000
C	Nitrogen	ZIF-8	1140	1	373	5000
C@MWCNT	—	ZIF-8	569.3	1	326	10,000
$CO_3O_4/NiCo_2O_4$	Nickel	ZIF-67	46	5	972	12,000
CoS	Sulfur	ZIF-67	96.3	10	~1400	10,000
Double-shelled CoS	Hydroxide, sulfur, nickel	ZIF-67	110	5	769.9	10,000
$NiCo_2S_44$	Sulfur, nickel	Co-MOF	72	10	765.6	10,000
Co/MnO/CoO/C	—	CoMn-MOF-74	242.4	1	800	1000
NiO	—	$Ni_5(HCOO)_6$	34	1	22	5000
Ni@graphitic C	—	Ni-BDC	140	1	140	10,000
CeO_2	—	Ce-BTC	140	1	300	10,000
Cr_2O_3	Carbon	MIL-101	438	0.25	291	3000

The second strategy is metal core stabilization. The weakness of metal–ligand bonds reduces the stability of MOFs in acidic or basic environments. In this context, optimizing the complexation behavior of metal ions is of great importance.

Third, it is suggested to develop hybrid MOF structures. Hybrid MOFs, where organic and inorganic components are used together, can be effective in increasing chemical stability and offer a wide range of applications.

The difficulties encountered in the transition to industrial application are also quite evident. The process of making MOF-based materials obtained in a laboratory environment applicable on an industrial scale is a serious one, and there are some fundamental obstacles encountered in this process.

The first obstacle is technology transfer. Adapting laboratory-scale methods used in MOF synthesis to industrial scale is usually a long and expensive process.

The second obstacle is the lack of standardization. Failure to standardize the methods used in the production of MOF materials can lead to quality inconsistencies between different manufacturers.

The third obstacle is regulation and environmental legislation. The environmental impacts of solvents and by-products used in MOF production necessitate the development of production processes that comply with regulations.

15.16 MOFS IN TERMS OF GREEN CHEMISTRY AND SUSTAINABILITY

Sustainability has become one of the most important aspects of today's scientific research. The environmentally friendly and economical production of MOFs will enable these materials to be applied on a larger scale in the future. Traditional MOF production methods often require the use of energy-intensive and environmentally harmful solvents. However, solutions that comply with green chemistry principles include bioderived solvents and low-temperature synthesis methods. Recycling and refunctionalization of MOFs after use can play a major role in reducing the environmental footprint [71]. In addition, recycling of MOFs used in the filtration of waste gases or liquids will support the principles of circular economy. The use of ligands and metal ions obtained from renewable resources for sustainable MOF production can reduce costs and minimize environmental damage [72].

The multifunctional structure of MOFs offers exciting opportunities in terms of materials science. MOFs contribute to the development of innovative solutions in a wide range of applications, from corrosion control to catalysts. The use of MOFs as protective coatings against corrosion of metals can increase durability, especially in the marine and energy sectors. Optimizing these coatings can provide stronger bonding to metal surfaces and longer-lasting protection. MOF-based catalysts are increasingly effective in applications such as solar energy conversion, water electrolysis, and the conversion of carbon dioxide into reusable fuels [73]. Making such catalysts more efficient could revolutionize energy conversion technologies.

The ability of MOFs to selectively adsorb different gas molecules can be used to increase the efficiency of energy production and chemical processes. Such applications are of critical importance, especially in carbon capture and storage (CCS) technologies. The combination of MOFs with electronic and photonic materials can contribute to the development of next-generation sensors and computing devices. Such systems have the potential to provide higher data processing speeds and lower energy consumption. Strategies developed to increase reaction selectivity in processes where MOFs are used as catalysts can greatly increase efficiency in the chemical industry [74]. The use of MOF-based systems, especially in asymmetric catalysis reactions, with precise tuning allows for high-purity products.

MOFs designed to mimic the catalytic properties of natural enzymes can be used instead of biocatalysts. Such materials offer the advantages of higher stability and wide temperature range operation. Thus, economic and technical superiority can be achieved in biotechnological processes. Designing MOF layers, especially developed for corrosion control, to provide multilayer protection to the material is important for long-term durability in industrial applications [75]. This approach is especially promising for systems operating in seawater, acidic environments, and structures exposed to high temperatures. The ability of MOFs to be applied in transparent and thin film form can create new opportunities in architecture and design areas other than corrosion control. Such coatings have the potential to offer both functionality and aesthetics.

15.17 ADVANCED APPLICATIONS AND FORECASTS

MOFs are considered to be one of the cornerstones of future technological innovations. The biocompatible structures of MOFs enable their use in a variety of medical applications, from drug delivery systems to biosensors [76]. Targeted drug delivery and delivery systems, especially in cancer treatment, are the focus of studies in this area. Artificial intelligence and machine learning methods can be used to predict the properties of MOFs and create new designs. This can accelerate the material development process and provide more effective solutions at lower costs. MOFs are increasingly used in lithium-ion batteries and supercapacitors to increase their energy storage capacity. Such materials hold promise for higher energy density and long-lasting battery systems.

The expansion of MOFs' application areas in environmental protection and industrial processes is accelerating with technological advances. The use of MOFs in carbon capture systems to reduce atmospheric carbon dioxide levels can play a key role in combating climate change. In particular, MOFs optimized for the separation and storage of CO_2 emitted from power plants have great potential. Sustainable use of water resources has become an important global issue today. The ability of MOFs to remove heavy metal ions and harmful chemicals from water offers a great opportunity in this area. In addition, MOF membranes with high permeability and selectivity that can be used in desalination processes can increase clean water production [77].

The ability of MOFs to detect toxic gases offers innovative solutions in air quality monitoring. In particular, specially designed MOF sensors can be developed for the effective separation and detection of industrial pollutants such as nitrogen oxide and sulfur dioxide. MOFs are ideal candidates for the safe and economical storage and transportation of energy sources such as hydrogen, methane, and natural gas, because of their high surface area and gas adsorption capacity. Such technologies are of critical importance, especially for the development of the hydrogen economy [78, 79].

15.18 USE OF MOFS IN ELECTRONICS AND INFORMATION TECHNOLOGIES

The use of MOFs in next-generation devices by improving their optical and electronic properties can revolutionize high-tech applications [80]. The use of semiconductor MOFs in electronic circuits offers an important opportunity, especially in the design of smaller and faster information processing systems. The use of these materials in memory units can provide advantages in terms of data density and energy savings. The gas detection sensitivity of MOFs offers an ideal solution for smart city applications and environmental monitoring systems. In addition, more advanced diagnostic tools can be designed in biosensors and medical devices because of their low detection limits and high selectivity [81–83]. MOFs with optimized light emission and diffraction properties can be used in the development of next-generation laser devices and optical components. The low-cost production of these materials facilitates their integration into industrial applications. The future directions and opportunities of MOFs offer many innovative solutions in areas such as materials science, environmental engineering, and energy technologies. Studies in these areas are critical to maximize the potential of MOFs and contribute to sustainable development goals.

15.19 CONCLUSION

MOFs offer significant advantages over traditional methods by offering an innovative approach in the field of corrosion control. Because of their high surface area, porous structure, and chemically tunable properties, MOFs can be considered as hybrid systems that serve both protective coating materials and catalytic functions in corrosion control applications. In particular, they have the potential to offer more sustainable and efficient solutions by minimizing environmental damage. However, commercial and industrial-scale applications of MOFs remain limited due to a number of technical and economic challenges such as synthesis costs, environmental durability, and compatibility with existing materials.

In this context, the scalability and cost-effectiveness of MOFs should be a critical focus for future research. The development of more environmentally friendly synthesis methods can increase the industrial applicability of these materials, while hybrid and multifunctional designs can expand their areas of use by increasing compatibility with existing materials. In addition, increasing the resistance to environmental factors and optimizing chemical stability can improve the performance of MOFs in harsh environments.

The innovative role of MOFs in corrosion control is not only remarkable for its technical benefits, but also for its contribution to environmental sustainability.

Considering the need for environmentally friendly technologies that can replace traditional protection methods, the potential of MOFs to provide a long-term solution in this area is quite significant. However, for this technology to be fully adopted, cooperation between academic research and industry needs to be strengthened and concrete solutions for application need to be developed. In conclusion, MOF-based corrosion control technologies promise great innovation in terms of both basic science and engineering. Effective use of the opportunities provided by these materials requires an interdisciplinary approach to overcome the challenges. Rapid implementation of relevant innovations and technologies will provide significant progress in terms of corrosion control, sustainability, and environmentally friendly industrial processes.

REFERENCES

[1] Chae, H. K., Siberio-Pérez, D. Y., Kim, J., Go, Y., Eddaoudi, M., Matzger, A. J., ... & Materials Design and Discovery Group. (2004). A route to high surface area, porosity and inclusion of large molecules in crystals. *Nature*, *427*(6974), 523–527.

[2] Jiang, L., Dong, Y., Yuan, Y., Zhou, X., Liu, Y., & Meng, X. (2022). Recent advances of metal–organic frameworks in corrosion protection: From synthesis to applications. *Chemical Engineering Journal*, *430*, 132823.

[3] Chen, H., Wang, F., Fan, H., Hong, R., & Li, W. (2021). Construction of MOF-based superhydrophobic composite coating with excellent abrasion resistance and durability for self-cleaning, corrosion resistance, anti-icing, and loading-increasing research. *Chemical Engineering Journal*, *408*, 127343.

[4] Li, M., Yao, J., Ge, J., Guo, J., Ma, L., Li, Z., ... & Zhao, J. (2024). pH-responsive berberine release from metal-organic framework based nanocarriers by regulating oxidative stress for targeted therapy of oral squamous cell carcinoma. *Arabian Journal of Chemistry*, *17*(9), 105935.

[5] Bennett, B. W., & Pickering, H. W. (1991). Effect of grain boundary structure on sensitization and corrosion of stainless steel. *Metallurgical Transactions A*, *18*, 1117–1124.

[6] Zhou, C., Pan, M., Li, S., Sun, Y., Zhang, H., Luo, X., ... & Zeng, H. (2022). Metal organic frameworks (MOFs) as multifunctional nanoplatform for anticorrosion surfaces and coatings. *Advances in Colloid and Interface Science*, *305*, 102707.

[7] Lorenz, W. J., & Mansfeld, F. (1981). Determination of corrosion rates by electrochemical DC and AC methods. *Corrosion Science*, *21*(9–10), 647–672.

[8] Ritchie, I. M., Bailey, S., & Woods, R. (1999). The metal–solution interface. *Advances in Colloid and Interface Science*, *80*(3), 183–231.

[9] Xia, D. H., Deng, C. M., Macdonald, D., Jamali, S., Mills, D., Luo, J. L., ... & Hu, W. (2022). Electrochemical measurements used for assessment of corrosion and protection of metallic materials in the field: A critical review. *Journal of Materials Science & Technology*, *112*, 151–183.

[10] Mansfeld, F. (1976). The polarization resistance technique for measuring corrosion currents. In Fontana, M. G., Staehle, R. W. (ed.) *Advances in Corrosion Science and Technology: Volume 6* (pp. 163–262). Springer US.

[11] Zhang, D., Zhu, Y., Liu, L., Ying, X., Hsiung, C. E., Sougrat, R., ... & Han, Y. (2018). Atomic-resolution transmission electron microscopy of electron beam–sensitive crystalline materials. *Science*, *359*(6376), 675–679.

[12] Echlin, P. (2011). *Handbook of Sample Preparation for Scanning Electron Microscopy and X-Ray Microanalysis*. Springer Science & Business Media.

[13] Riedel, S., & Kaupp, M. (2009). The highest oxidation states of the transition metal elements. *Coordination Chemistry Reviews*, *253*(5–6), 606–624.

[14] Nyffenegger, R. M., & Penner, R. M. (1997). Nanometer-scale surface modification using the scanning probe microscope: Progress since 1991. *Chemical Reviews*, *97*(4), 1195–1230.

[15] Shahini, M. H., Mohammadloo, H. E., Ramezanzadeh, M., & Ramezanzadeh, B. (2022). Recent innovations in synthesis/characterization of advanced nano-porous metal-organic frameworks (MOFs); current/future trends with a focus on the smart anti-corrosion features. *Materials Chemistry and Physics*, *276*, 125420.

[16] Burtch, N. C., Jasuja, H., & Walton, K. S. (2014). Water stability and adsorption in metal–organic frameworks. *Chemical Reviews*, *114*(20), 10575–10612.

[17] Feng, L., Wang, K. Y., Day, G. S., Ryder, M. R., & Zhou, H. C. (2020). Destruction of metal–organic frameworks: positive and negative aspects of stability and lability. *Chemical Reviews*, *120*(23), 13087–13133.

[18] Chen, X., Tong, R., Shi, Z., Yang, B., Liu, H., Ding, S., ... & Fang, W. (2018). MOF nanoparticles with encapsulated autophagy inhibitor in controlled drug delivery system for antitumor. *ACS Applied Materials & Interfaces*, *10*(3), 2328–2337.

[19] Ramezanzadeh, M., Ramezanzadeh, B., Bahlakeh, G., Tati, A., & Mahdavian, M. (2021). Development of an active/barrier bi-functional anti-corrosion system based on the epoxy nanocomposite loaded with highly-coordinated functionalized zirconium-based nanoporous metal-organic framework (Zr-MOF). *Chemical Engineering Journal*, *408*, 127361.

[20] Zuo, R. (2007). Biofilms: strategies for metal corrosion inhibition employing microorganisms. *Applied Microbiology and Biotechnology*, *76*(6), 1245–1253.

[21] Tawfik, S. M., Kobisy, A. S., Badr, E. A., Elged, A. H., & Lee, Y. I. (2023). Surface-active nonionic conjugated zirconium metal–organic frameworks and their applications; Broad spectrum anti-microbial, anti-SRB biofilm, anti-microbial corrosion. *Environmental Technology & Innovation*, *29*, 103001.

[22] Obot, I. B., Macdonald, D. D., & Gasem, Z. M. (2015). Density functional theory (DFT) as a powerful tool for designing new organic corrosion inhibitors. Part 1: an overview. *Corrosion Science*, *99*, 1–30.

[23] Haris, N. I. N., Sobri, S., Yusof, Y. A., & Kassim, N. K. (2020). An overview of molecular dynamic simulation for corrosion inhibition of ferrous metals. *Metals*, *11*(1), 46.

[24] ElBelghiti, M., Karzazi, Y., Dafali, A., Hammouti, B., Bentiss, F., Obot, I. B., ... & Ebenso, E. E. (2016). Experimental, quantum chemical and Monte Carlo simulation studies of 3, 5-disubstituted-4-amino-1, 2, 4-triazoles as corrosion inhibitors on mild steel in acidic medium. *Journal of Molecular Liquids*, *218*, 281–293.

[25] Ramezanzadeh, M., Ramezanzadeh, B., Mahdavian, M., & Bahlakeh, G. (2020). Development of metal-organic framework (MOF) decorated graphene oxide nanoplatforms for anti-corrosion epoxy coatings. *Carbon*, *161*, 231–251.

[26] Lu, W., Wei, Z., Gu, Z. Y., Liu, T. F., Park, J., Park, J., ... & Zhou, H. C. (2014). Tuning the structure and function of metal–organic frameworks via linker design. *Chemical Society Reviews*, *43*(16), 5561–5593.

[27] Yuan, S., Feng, L., Wang, K., Pang, J., Bosch, M., Lollar, C., ... & Zhou, H. C. (2018). Stable metal–organic frameworks: design, synthesis, and applications. *Advanced Materials*, *30*(37), 1704303.

[28] Haldar, R., & Maji, T. K. (2013). Metal–organic frameworks (MOFs) based on mixed linker systems: structural diversities towards functional materials. *CrystEngComm*, *15*(45), 9276–9295.

[29] He, T., Kong, X. J., & Li, J. R. (2021). Chemically stable metal–organic frameworks: rational construction and application expansion. *Accounts of Chemical Research*, *54*(15), 3083–3094.

[30] Ding, M., & Jiang, H. L. (2021). Improving water stability of metal–organic frameworks by a general surface hydrophobic polymerization. *CCS Chemistry*, *3*(8), 2740–2748.

[31] Zhang, B., Zhu, Z., Wang, X., Liu, X., & Kapteijn, F. (2024). Water adsorption in MOFs: structures and applications. *Advanced Functional Materials*, *34*(43), 2304788.

[32] Healy, C., Patil, K. M., Wilson, B. H., Hermanspahn, L., Harvey-Reid, N. C., Howard, B. I., ... & Bennett, T. D. (2020). The thermal stability of metal-organic frameworks. *Coordination Chemistry Reviews*, *419*, 213388.

[33] Chen, M., Cen, H., Guo, C., Guo, X., & Chen, Z. (2020). Preparation of Cu-MOFs and its corrosion inhibition effect for carbon steel in hydrochloric acid solution. *Journal of Molecular Liquids*, *318*, 114328.

[34] Liu, T., Zhang, D., Zhang, R., Wang, J., Ma, L., Keil, P., ... & Li, X. (2023). Self-healing and corrosion-sensing coatings based on pH-sensitive MOF-capped microcontainers for intelligent corrosion control. *Chemical Engineering Journal*, *454*, 140335.

[35] Kalaj, M., Bentz, K. C., Ayala Jr, S., Palomba, J. M., Barcus, K. S., Katayama, Y., & Cohen, S. M. (2020). MOF-polymer hybrid materials: from simple composites to tailored architectures. *Chemical Reviews*, *120*(16), 8267–8302.

[36] Haeri, Z., Ramezanzadeh, B., & Ramezanzadeh, M. (2022). Recent progress on the metal-organic frameworks decorated graphene oxide (MOFs-GO) nano-building application for epoxy coating mechanical-thermal/flame-retardant and anti-corrosion features improvement. *Progress in Organic Coatings*, *163*, 106645.

[37] Tu, K., Ding, Y., & Keplinger, T. (2022). Review on design strategies and applications of metal-organic framework-cellulose composites. *Carbohydrate Polymers*, *291*, 119539.

[38] Ma, C., Du, Z., Wang, X., Zhou, P., Zhao, Y., Hua, Y., ... & Wang, F. (2024). Corrosion of stainless steels and corrosion protection strategies in the semiconductor manufacturing industry: a review. *Corrosion Reviews*, *42*(2), 127–161.

[39] Tanzharikov, P. A., Erzhanova, A. T., & Kozhageldin, H. K. Improving the rod well pump for production of watered oil from deep wells. *ТЕХНИКА ҒЫЛЫМДАРЫ ЖӘНЕ ТЕХНОЛОГИЯ*, 2(06), 4-13

[40] Nazari, M. H., Zhang, Y., Mahmoodi, A., Xu, G., Yu, J., Wu, J., & Shi, X. (2022). Nanocomposite organic coatings for corrosion protection of metals: A review of recent advances. *Progress in Organic Coatings*, *162*, 106573.

[41] Adu, E., Zhang, Y., & Liu, D. (2019). Current situation of carbon dioxide capture, storage, and enhanced oil recovery in the oil and gas industry. *The Canadian Journal of Chemical Engineering*, *97*(5), 1048–1076.

[42] Aljibori, H. S., Alamiery, A., & Kadhum, A. A. H. (2023). Advances in corrosion protection coatings: A comprehensive review. *International Journal of Corrosion and Scale Inhibition*, *12*(4), 1476–1520.

[43] Liu, Y., Liu, Y., Chen, M., Liu, S., Lai, B., & Tu, W. (2023). Strategies for the construction of special wettability metal organic framework membranes: A review. *Journal of Water Process Engineering*, *51*, 103374.

[44] Donnelly, B., Sammut, K., & Tang, Y. (2022). Materials selection for antifouling systems in marine structures. *Molecules*, *27*(11), 3408.

[45] Stassen, I., Burtch, N., Talin, A., Falcaro, P., Allendorf, M., & Ameloot, R. (2017). An updated roadmap for the integration of metal–organic frameworks with electronic devices and chemical sensors. *Chemical Society Reviews*, *46*(11), 3185–3241.

[46] Keshmiri, N., Najmi, P., Ramezanzadeh, M., & Ramezanzadeh, B. (2021). Designing an eco-friendly lanthanide-based metal organic framework (MOF) assembled graphene-oxide with superior active anti-corrosion performance in epoxy composite. *Journal of Cleaner Production*, *319*, 128732.

[47] Huang, D., Ma, H., Wang, J., Du, Y., & Li, R. (2024). Mof-mediated paper-based (bio) sensors for detecting of food and environmental pollutants: Preparation strategies and emerging applications. *Microchemical Journal*, *207*, 111692.

[48] Emam, H. E., Abdelhameed, R. M., & Ahmed, H. B. (2020). Adsorptive performance of MOFs and MOF containing composites for clean energy and safe environment. *Journal of Environmental Chemical Engineering*, *8*(5), 104386.

[49] Muresan, L. M. (2023). Nanocomposite coatings for anti-corrosion properties of metallic substrates. *Materials*, *16*(14), 5092.

[50] Saji, V. S. (2019). Supramolecular concepts and approaches in corrosion and biofouling prevention. *Corrosion Reviews*, *37*(3), 187–230.

[51] Song, H., Yasin, G., Singh, N. B., Gupta, R. K., & Nguyen, T. A. (Eds.). (2022). *Nanotechnology in the Automotive Industry*. Elsevier.

[52] Abdelkareem, M. A., Qaisar, A., Sayed, E. T., Shehata, N., Parambath, J. B. M., Alami, A. H., & Olabi, A. G. (2024). Recent advances on metal-organic frameworks (MOFs) and their applications in energy conversion devices: Technical review. *Energy*, *299*, 131127.

[53] Ren, J., Dyosiba, X., Musyoka, N. M., Langmi, H. W., Mathe, M., & Liao, S. (2017). Review on the current practices and efforts towards pilot-scale production of metal-organic frameworks (MOFs). *Coordination Chemistry Reviews*, *352*, 187–219.

[54] Zu, K., Qin, M., & Cui, S. (2020). Progress and potential of metal-organic frameworks (MOFs) as novel desiccants for built environment control: A review. *Renewable and Sustainable Energy Reviews*, *133*, 110246.

[55] Ma, X., Chai, Y., Li, P., & Wang, B. (2019). Metal–organic framework films and their potential applications in environmental pollution control. *Accounts of Chemical Research*, *52*(5), 1461–1470.

[56] Qin, M., Feaugas, O., & Zu, K. (2022). Novel metal-organic framework (MOF) based phase change material composite and its impact on building energy consumption. *Energy and Buildings*, *273*, 112382.

[57] Kumar, P., Kim, K. H., Kwon, E. E., & Szulejko, J. E. (2016). Metal–organic frameworks for the control and management of air quality: advances and future direction. *Journal of Materials Chemistry A*, *4*(2), 345–361.

[58] Ding, D., Rasmussen, O. S., & Qin, M. (2024). Sustainable metal-organic framework based paper sheet as wallpaper for passive indoor moisture control. *Energy and Buildings*, *313*, 114262.

[59] Ding, D., & Qin, M. (2024). Comparative study of metal-organic frameworks (MOFs) for indoor moisture control. *Applied Thermal Engineering*, *257*, 124329.

[60] Seidi, F., Jouyandeh, M., Taghizadeh, M., Taghizadeh, A., Vahabi, H., Habibzadeh, S., ... & Saeb, M. R. (2020). Metal-organic framework (MOF)/epoxy coatings: A review. *Materials*, *13*(12), 2881.

[61] Giachino, M., Dubois, G., & Dauskardt, R. H. (2013). Heterogeneous solution deposition of high-performance adhesive hybrid films. *ACS Applied Materials & Interfaces*, *5*(20), 9891–9895.
[62] Liu, X., Zhang, L., & Wang, J. (2021). Design strategies for MOF-derived porous functional materials: Preserving surfaces and nurturing pores. *Journal of Materiomics*, *7*(3), 440–459.
[63] Meng, J., Liu, X., Niu, C., Pang, Q., Li, J., Liu, F., ... & Mai, L. (2020). Advances in metal–organic framework coatings: versatile synthesis and broad applications. *Chemical Society Reviews*, *49*(10), 3142–3186.
[64] Keramatinia, M., Ramezanzadeh, B., & Mahdavian, M. (2024). Unlocking the potential of a zirconium-based MOF for advanced protective coatings: Delve into the UIO-66 smart release functioning for a bio-based Zn2+ complex. *Surface and Coatings Technology*, *481*, 130578.
[65] Ghohrodi, A. R., Ramezanzadeh, M., & Ramezanzadeh, B. (2022). Investigating the thermo-mechanical and UV-shielding properties of a nano-porous Zr (IV)-type metal-organic framework (MOF) incorporated epoxy composite coating. *Progress in Organic Coatings*, *164*, 106693.
[66] Jia, T., Gu, Y., & Li, F. (2022). Progress and potential of metal-organic frameworks (MOFs) for gas storage and separation: A review. *Journal of Environmental Chemical Engineering*, *10*(5), 108300.
[67] Fattah-alhosseini, A., Sangarimotlagh, Z., Karbasi, M., & Dikici, B. (2024). Enhancing corrosion resistance in Mg–based alloys through MOF–incorporated coatings: A comprehensive review. *Applied Surface Science Advances*, *21*, 100607.
[68] López-Ortega, A., Arana, J. L., Rodríguez, E., & Bayón, R. (2018). Corrosion, wear and tribocorrosion performance of a thermally sprayed aluminum coating modified by plasma electrolytic oxidation technique for offshore submerged components protection. *Corrosion Science*, *143*, 258–280.
[69] Chen, Z., Zhou, K., Lu, X., & Lam, Y. C. (2014). A review on the mechanical methods for evaluating coating adhesion. *Acta Mechanica*, *225*, 431–452.
[70] Yu, Z., Li, X., Li, X., Zheng, B., Li, D., Xu, D., & Wang, F. (2023). Nacre-inspired metal-organic framework coatings reinforced by multiscale hierarchical cross-linking for integrated antifouling and anti-microbial corrosion. *Advanced Functional Materials*, *33*(47), 2305995.
[71] Fijoł, N., Mautner, A., Grape, E. S., Bacsik, Z., Inge, A. K., & Mathew, A. P. (2023). MOF@ Cell: 3D printed biobased filters anchored with a green metal–organic framework for effluent treatment. *Journal of Materials Chemistry A*, *11*(23), 12384–12394.
[72] Howarth, A. J., Liu, Y., Li, P., Li, Z., Wang, T. C., Hupp, J. T., & Farha, O. K. (2016). Chemical, thermal and mechanical stabilities of metal–organic frameworks. *Nature Reviews Materials*, *1*(3), 1–15.
[73] Adegoke, K. A., & Maxakato, N. W. (2021). Porous metal-organic framework (MOF)-based and MOF-derived electrocatalytic materials for energy conversion. *Materials Today Energy*, *21*, 100816.
[74] Guo, J., Qin, Y., Zhu, Y., Zhang, X., Long, C., Zhao, M., & Tang, Z. (2021). Metal–organic frameworks as catalytic selectivity regulators for organic transformations. *Chemical Society Reviews*, *50*(9), 5366–5396.
[75] Shang, C., Sang, D., Li, C., Zou, L., Wu, J., & Wang, Q. (2024). 2D materials for marine corrosion protection: A review. *APL Materials*, *12*(6), 060601-1-22
[76] Yilmaz, G., Peh, S. B., Zhao, D., & Ho, G. W. (2019). Atomic-and molecular-level design of functional metal–organic frameworks (MOFs) and

derivatives for energy and environmental applications. *Advanced Science*, *6*(21), 1901129.

[77] Kadhom, M., & Deng, B. (2018). Metal-organic frameworks (MOFs) in water filtration membranes for desalination and other applications. *Applied Materials Today*, *11*, 219–230.

[78] Dou, Y., Sun, L., Ren, J., & Dong, L. (2023). Opportunities and future challenges in hydrogen economy for sustainable development. In *Hydrogen Economy* (pp. 537–569). Academic Press.

[79] Julien, P. A., Mottillo, C., & Friščić, T. (2017). Metal–organic frameworks meet scalable and sustainable synthesis. *Green Chemistry*, *19*(12), 2729–2747.

[80] Ajdari, F. B., Kowsari, E., Shahrak, M. N., Ehsani, A., Kiaei, Z., Torkzaban, H., ... & Ramakrishna, S. (2020). A review on the field patents and recent developments over the application of metal organic frameworks (MOFs) in supercapacitors. *Coordination Chemistry Reviews*, *422*, 213441.

[81] Zhao, J., Zhang, Y., Luo, Y., Zheng, W., Xu, X., & Luo, F. (2024). Amorphous metal–organic frameworks: Pioneering frontiers in cancer diagnostics and therapeutics. *Chemical Engineering Journal*, *492*, 152295.

[82] Duan, C., Yu, Y., Xiao, J., Li, Y., Yang, P., Hu, F., & Xi, H. (2021). Recent advancements in metal–organic frameworks for green applications. *Green Energy & Environment*, *6*(1), 33–49.

[83] Allendorf, M. D., Dong, R., Feng, X., Kaskel, S., Matoga, D., & Stavila, V. (2020). Electronic devices using open framework materials. *Chemical Reviews*, *120*(16), 8581–8640.

16 Conclusion and Future Directions in Anticorrosive MOF-Based Coatings

Farhat A Ansari and Humira Assad

16.1 INTRODUCTION

Corrosion continues to be a significant and costly issue across a wide range of industries, affecting the longevity, safety, and performance of metal structures and components [1, 2]. Despite the widespread use of conventional corrosion protection methods such as coatings, galvanization, and corrosion inhibitors, these solutions often face limitations in terms of durability, environmental impact, and adaptability to extreme conditions [3, 4]. As industries demand more effective, sustainable, and long-lasting corrosion protection strategies, new materials and technologies are being explored to meet these evolving needs.

Metal–organic frameworks (MOFs) represent a promising class of materials that could address these challenges [5]. MOFs are a class of highly porous materials composed of metal ions or clusters coordinated to organic ligands, and they offer a range of unique properties such as high surface area, tunable porosity, and the ability to integrate with other materials [6]. These features make MOFs particularly suitable for improving corrosion resistance. MOFs can provide both passive protection by forming a protective barrier and active protection by releasing corrosion-inhibiting agents, thus enhancing the performance of traditional coatings [7].

This chapter will conclude all the concepts and directions discussed throughout the entire book. This chapter focuses on the conclusion and future directions of MOF-based corrosion control, examining the current state of research and the potential for MOFs to revolutionize anticorrosive coating technologies. It will explore the progress made in understanding MOFs' mechanisms for corrosion protection, the types of MOFs that show promise for industrial applications, and the challenges faced in scaling up these technologies for widespread use. Additionally, the chapter will address the remaining gaps in knowledge and the opportunities for further research that could drive the development of more efficient, durable, and sustainable MOF-based anticorrosive solutions. In doing so, it aims to provide a comprehensive overview of the future trajectory of MOF-based corrosion control and its potential to transform the way we approach corrosion protection in the years to come.

DOI: 10.1201/9781003571162-16

16.1.1 Importance of Anticorrosive Coatings

Anticorrosive coatings are among the most widely adopted methods to protect metals from corrosion [8]. These coatings act as physical barriers that prevent the direct contact between metal surfaces and corrosive elements such as water, oxygen, and chloride ions. The effectiveness of these coatings is crucial in ensuring the long-term durability and performance of metal-based structures, which are exposed to harsh environments [9, 10]. However, conventional anticorrosive coatings, which typically rely on passive protection mechanisms, often have limitations. For instance, many coatings lack self-healing properties, are susceptible to wear–tear, and can degrade over time when exposed to environmental stresses [11].

To overcome these limitations, there has been a significant push toward developing coatings that not only provide passive protection but also incorporate active corrosion inhibition and self-healing capabilities. MOFs offer a promising solution to this challenge as mentioned in the introduction [6]. The inherent properties of MOFs, such as their high surface area and the ability to release corrosion inhibitors, make them ideal candidates for enhancing the performance of anticorrosive coatings. By utilizing MOFs in coatings, it is possible to achieve longer-lasting protection, reduce maintenance costs, and enhance the sustainability of materials used in corrosive environments [12].

16.2 FUNDAMENTALS OF CORROSION AND MOFS IN ANTICORROSIVE COATINGS

Corrosion is the deterioration of materials, particularly metals, through chemical reactions with their environment [13, 14]. This natural process is electrochemical in nature and occurs when a metal reacts with substances like oxygen, moisture, or salts, leading to the loss of its material integrity. The most common example of corrosion is rusting, which occurs when iron reacts with water and oxygen to form iron oxide [15]. However, corrosion manifests in various forms, such as pitting corrosion, crevice corrosion, galvanic corrosion, and uniform corrosion, each presenting unique challenges for industries relying on metal structures. The corrosion process involves the oxidation of the metal at the anode, where electrons are released, while at the cathode, a reduction reaction occurs [16]. This electrochemical reaction leads to the formation of corrosion products, which degrade the material. Over time, corrosion can significantly compromise the structural integrity of a material, leading to mechanical failure, safety hazards, and economic losses. Understanding the mechanisms of corrosion is fundamental to designing effective corrosion protection strategies, such as coatings, which can either prevent the initiation of corrosion or slow its progression [17].

MOFs are a class of materials composed of metal ions or metal clusters connected by organic ligands, forming a porous, crystalline structure. Particularly unique is MOFs' highly tunable nature, both in terms of composition and structure [7, 8]. By varying the metal canters and organic linkers, MOFs can be designed with specific properties, such as high surface area, large pore volume, and precise control over their size and shape. These features make MOFs versatile materials with a wide range of applications in areas such as gas storage, catalysis, and, increasingly, corrosion protection. MOFs offer significant advantages over traditional materials used for corrosion protection. Their large surface area allows them to interact extensively with the surface of metals, creating a robust protective layer [18]. Furthermore, their tunable nature means that they can be designed to incorporate corrosion inhibitors, which can be slowly released to prevent the onset of corrosion. MOFs are also capable of exhibiting self-healing properties, whereby they can repair damage to the coating when exposed to certain stimuli, such as the presence of corrosive agents. These characteristics position MOFs as an advanced material that can significantly enhance the performance of anticorrosive coatings [19] (Table 16.1).

The use of MOFs in anticorrosive coatings is an emerging area of research that capitalizes on the unique properties of MOFs to improve the performance and longevity of corrosion protection. When incorporated into coatings, MOFs can act as both physical barriers and active inhibitors. Their porous structure allows them to form a dense layer on the surface of metals, preventing the metal from coming into

TABLE 16.1
Some MOF and Other Nanomaterials Employed in Anticorrosive Applications

Corrosion Inhibitors	Corrosive Medium	Metal	Inhibition Efficiency (%)	Ref.
Nanocomposites chitosan-cobalt	1 M HCl	Q235 steel	96	[20]
Cu-BTA-Na_2MoO_4-HNTs 3.5% NaCl	3.5% NaCl	Q235 steel	30.32	[21]
ANI/TiO_2 92.5	3.5% NaCl	Al1050 electrode	Al1050 electrode	[22]
Silica nanoparticles (SNPs) were synthesized by dusting rice husk	1% HCl	MS	45.38	[23]
Poly(o-anthranilic acid) (PAA), graphene oxide (GO) and functionalized multiwalled carbon nanotubes (f-MWCNT)	2 M HCl	Al1050 electrode	91.35	[24]
Copper nanoparticles doped carbon quantum dots (Cu/CQDs) nanohybrid	Seawater	X60 steel	96.4	[25]
N-doped carbon saturated CO_2	3 wt% NaCl solution	N80 steel	83.7	[26]
MOF1	1 M HCl	Aluminum	93.0	[27]
Co-MOF	0.1 N HCl	Mild steel	99.32	[28]

direct contact with corrosive elements such as water, oxygen, and chloride ions. The high surface area of MOFs also increases the interaction between the coating and the metal surface, improving the adhesion and effectiveness of the protective layer. In addition to their barrier properties, MOFs can also provide active corrosion inhibition [29]. Certain MOFs are capable of releasing metal ions or organic molecules that inhibit the corrosion process. For instance, zinc ions released from zinc-based MOFs can actively participate in the corrosion inhibition of steel, slowing or even preventing the oxidation process [30]. This dynamic form of corrosion control is particularly beneficial in environments where traditional coatings may fail due to mechanical wear or environmental exposure. Moreover, MOFs can enhance the self-healing properties of coatings. When a coating is damaged, the MOFs can release corrosion inhibitors or other healing agents, effectively repairing the coating and restoring its protective function. This ability to self-heal increases the longevity of the coating and reduces the need for maintenance or reapplication, making MOF-based coatings more cost-effective and sustainable. There are several types of MOFs that have been explored for use in anticorrosive coatings, each offering distinct advantages depending on the specific application.

Zinc-based MOFs, and Ce-MOF, are widely studied for their corrosion-inhibiting properties [31]. Zinc ions are well known for their ability to prevent corrosion by forming a protective layer on the metal surface. The incorporation of ZIF-8 into coatings can enhance corrosion resistance, especially in environments where chloride ions are present, such as in marine applications. Copper-based MOFs, like HKUST-1, are another promising class of materials for corrosion protection. Copper is a highly effective corrosion inhibitor, and when incorporated into MOF structures, it can be released gradually to provide sustained corrosion protection [28]. Copper-based MOFs have been shown to be particularly useful in protecting metals exposed to aggressive electrolytes. Iron-based MOFs, such as MIL-101, are considered environmentally friendly alternatives for corrosion protection. Iron is abundant and non-toxic, making these materials attractive for use in industries where sustainability is a key consideration. The iron-based MOFs offer good stability and can be incorporated into coatings to improve corrosion resistance in a variety of environments [32]. In addition to these individual MOFs, hybrid systems that combine MOFs with other materials such as polymers or nanoparticles are also being developed. These hybrid materials aim to improve the mechanical properties of MOF-based coatings, enhancing their durability and performance under real-world conditions [32].

16.3 INDUSTRIAL SCALING OF MOF-BASED COATINGS

Despite the promising results of MOF-based coatings in laboratory-scale studies, scaling these materials for industrial applications presents several challenges. One of the main issues is the cost and scalability of MOF production. The laboratory-scale synthesis of MOFs (Figure 16.1) is relatively straightforward, producing them at large scales for industrial use can be expensive and resource-intensive [33]. The cost of raw materials, solvents, and energy required for MOF synthesis can make these materials prohibitively expensive for widespread commercial use. Another challenge is the

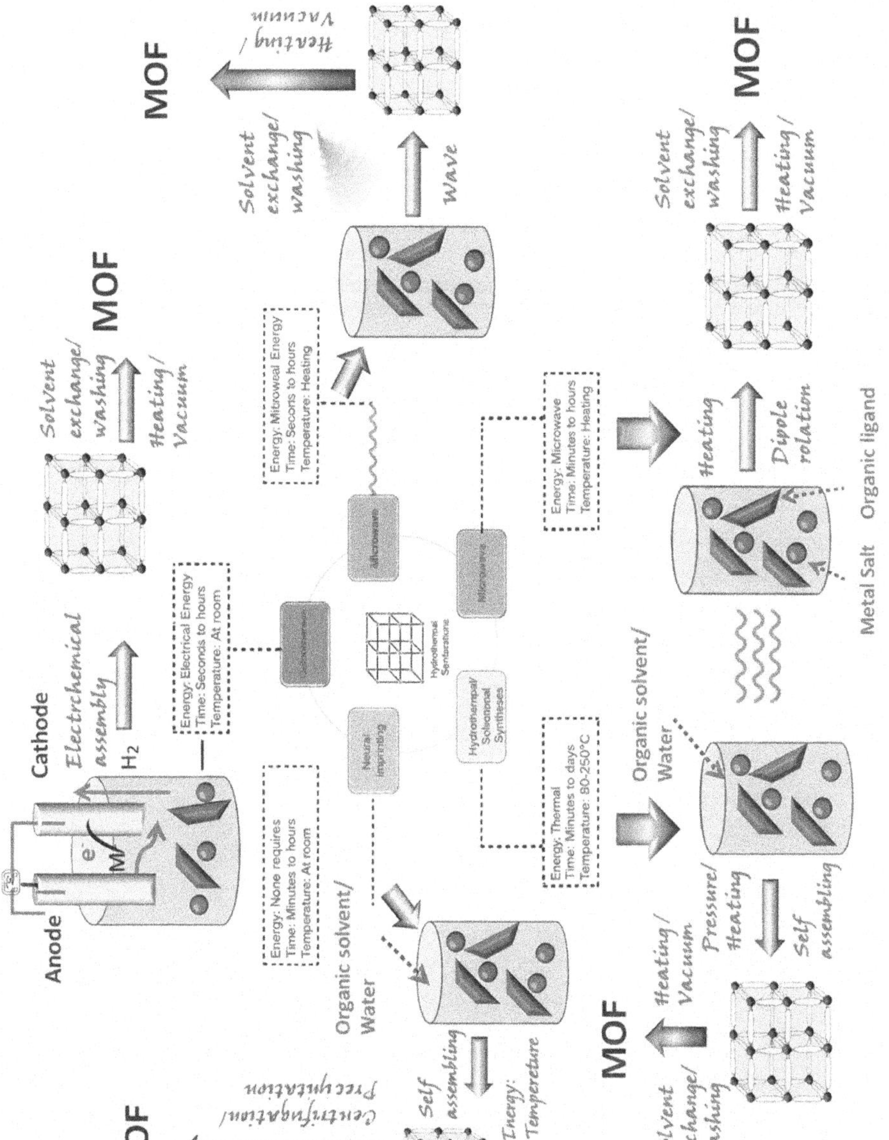

FIGURE 16.1 The representation of the production strategies along with the appropriate production metrics.

long-term stability and durability of MOF-based coatings in real-world environments [12]. Although MOFs show great promise in controlled laboratory conditions, their performance in harsh environments, such as high humidity, ultraviolet (UV) radiation, and fluctuating temperatures, needs to be thoroughly tested. For MOF-based coatings to be viable for industrial applications, their stability over time must be ensured. Additionally, the integration of MOFs into existing coating technologies poses another hurdle [34]. Traditional coatings, such as epoxy-based or polymer-based coatings, have well-established application methods and proven performance in industrial settings [35]. Incorporating MOFs into these systems requires the development of new methods for mixing, applying, and curing the coatings, which adds complexity to the process [36]. To address these challenges, researchers are focusing on improving the efficiency of MOF synthesis, finding more cost-effective production methods, and exploring hybrid systems that combine MOFs with other materials to enhance performance. Moreover, testing the environmental and lifecycle impacts of MOF-based coatings is crucial for determining their sustainability and feasibility for large-scale use [7].

This detailed exploration mentioned in all above sections emphasizes the significance of anticorrosive coatings, the fundamentals of corrosion, and how MOFs can be leveraged to enhance corrosion protection. The authors believe that almost all the sections provide an in-depth discussion of the types of MOFs used in coatings, their potential for active inhibition and self-healing, and the challenges faced in scaling up these materials for industrial use.

The chapter paints a comprehensive picture of the promise and potential of MOF-based anticorrosive coatings for future applications. The progress made this far is promising; it is important to note that relatively little work has been reported on the large-scale use of MOF-based coatings in real-world industrial settings. There remain ample challenges that need to be overcome to fully harness the potential of these materials. In the following sections, we will discuss these challenges in greater detail, focusing on what has been achieved so far, the obstacles encountered during the development and application of MOF-based coatings, and what further research is needed to address these gaps. By doing so, we aim to guide future researchers and practitioners in understanding the current landscape of MOF-based anticorrosive coatings and where the next steps in this field should be focused.

16.4 APPLICATION-RELATED LIMITATIONS OF METAL-ORGANIC FRAMEWORKS (MOFS) IN CORROSION MITIGATION

MOFs have become known as effective materials for corrosion prevention owing to their distinctive characteristics, such as elevated surface area, adjustable structure, and inherent corrosion resistance. Nonetheless, certain constraints impede its implementation in practical situations.

16.4.1 Economic Constraints

The production expenses related to MOFs are significantly elevated. The production of these materials frequently necessitates costly raw materials and energy-demanding methods. For example, commercially available MOFs are often offered in limited quantities (5 g to 1 kg) at approximately $7.38 USD/g, presenting a considerable obstacle to extensive industrial utilization [37, 38]. To improve the economic viability of MOFs, it is essential to devise ways that employ more readily available and less toxic compounds while optimizing manufacturing parameters for increased yields.

16.4.2 Scale of Production

The majority of MOFs have been primarily built in laboratory conditions, with minimal advancement toward industrial-scale manufacturing. This limitation in scalability constrains their accessibility and utilization in commercial environments. Research initiatives should prioritize the enhancement of synthesis procedures to satisfy industrial requirements while preserving the quality and efficacy of the MOFs.

16.4.3 Durability and Stability

The resilience of MOFs under operating conditions is a significant issue. Chemical stability, hydrolysis resistance, and mechanical integrity are critical factors that can substantially influence their efficacy as corrosion inhibitors. Some MOF nanoparticles exhibit capability in improving corrosion resistance; nevertheless, their long-term efficacy may be undermined by variables such as water dissolution or thermal degradation during use [36, 39]. Future studies must target enhanced chemical resistance and durability to ensure that MOF-based coatings endure severe environments for prolonged durations.

16.4.4 Dispersion and Compatibility Challenges

The integration of MOF nanocarriers into matrices of polymers poses issues concerning their dispersion and modification potential. The clustering of the nanoparticles may prevent the containment of bigger molecules, hence diminishing the efficacy of the coatings. Furthermore, attaining homogeneous dispersion within polymeric matrix is crucial for optimizing protective characteristics. Research should investigate novel techniques to enhance the compatibility and distribution of MOFs in diverse coating systems.

16.4.5 Adhesion of Coatings and Structural Integrity

MOF coatings to effectively resist corrosion, they must demonstrate robust adherence to surfaces and preserve their structural strength under stress. The adhesion of coatings can be improved via surface treatments that facilitate superior binding between the base material and the MOF layer [38]. Examining appropriate surface modification methods will be essential for enhancing the efficacy of MOF-based corrosion-resistant coatings.

16.4.6 Need for Tailored Solutions

The modification of MOFs with suitable organic ligands and metal ions is crucial for creating specific materials tailored for various corrosion prevention purposes. Future study should focus on synthesizing customized MOFs that offer improved protective properties against certain corrosive conditions.

16.5 FUTURE DIRECTIONS

16.5.1 Design Strategies for Next-Generation MOFs

The article highlights that although there are promising techniques for employing MOFs to mitigate corrosion, numerous challenges must be resolved to improve their practical application and sustain their beneficial anticorrosion characteristics. A crucial domain for forthcoming research is the creation of novel MOF nanomaterials specifically designed for corrosion mitigation, necessitating a comprehensive comprehension of the formation principles and structural attributes of these frameworks (Figure 16.2). Certain MOFs, namely zeolitic imidazolate frameworks (ZIFs) containing imidazole rings, function as corrosion inhibitors and nanocarriers [19].

A unique epoxy coating (f-$Ti_3C_2T_x$-ZB@EP) is developed by integrating amino functionalized $Ti_3C_2T_x$ with 2-methylimidazole zinc salt (ZIF-8) nano-container@ benzotriazole (f-$Ti_3C_2T_x$-ZB) functional nanocomposite filler [42]. The synthesized f-$Ti_3C_2T_x$-ZB@EP exhibits a significant increase in coating resistivity (Rc) and demonstrates self-healing properties in harsh conditions, attributable to the combined effects of both active and passive protective mechanisms. The alteration of pH due to electrochemical degradation may prompt the nanocontainer to discharge BTA, thus generating corrosion inhibition layers at the coating/metal interface [42]. Consequently, f-$Ti_3C_2T_x$-ZB@EP, exhibiting superior self-healing, corrosion resistant, and tribological characteristics, possesses enduring metal protection capabilities and inspires innovative concepts for corrosion and wear resistance (Figure 16.3).

16.5.2 Mechanism of Inhibition in MOF

The inhibitory efficacy of MOFs along with other nanomaterials indicates their prospective use in corrosion prevention across diverse media and metals as well. A schematic illustration of MOFs and other nanomaterials is provided in Figure 16.4.

The figure illustrates that lone pairs of electrons on heteroatoms such as O and N in MOFs and other nanomaterials are transferred to the vacant 3d orbitals of Fe atoms, resulting in the formation of chemical interactions and a strong barrier to corrosion.

16.5.3 Potential Applications in Various Industries

New advances in MOFs have generated considerable interest in their use for corrosion mitigation. This review offers a thorough examination of recent advancements in MOFs, emphasizing their production methods, applications in corrosion prevention, and the obstacles confronting MOF-based anticorrosive solutions. It underscores research progress during the last ten years, elucidating the benefits and drawbacks of several MOF-related corrosion inhibitors.

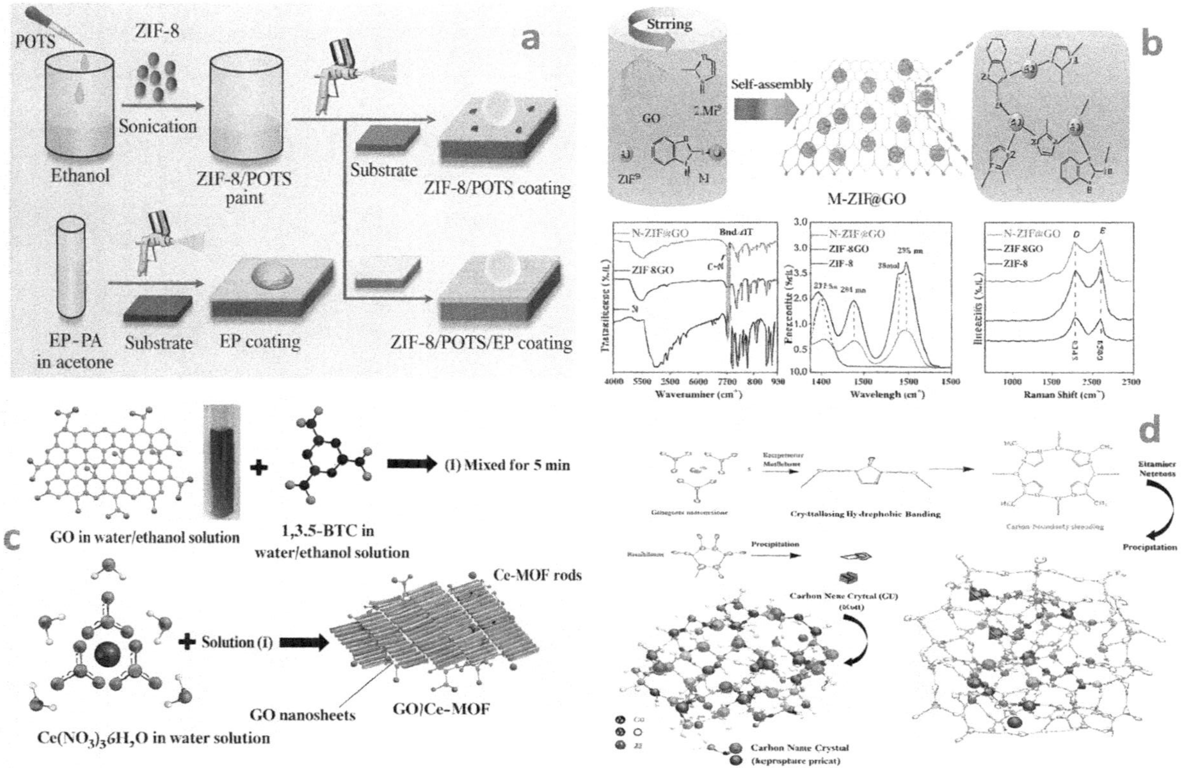

FIGURE 16.2 (a) The manufacturing method of epoxy material (EP) treatment, highly hydrophobic ZIF-8/Perfluoro octyl triethoxy silane (POTS) coating, and ZIF-8/POTS/EP coating [19]. (b) Diagrams illustrating the creation of M-ZIF8/GO nanocomposite [40]. (c) The manufacturing procedure of GO@Ce-MOF nanomaterials [7]. (d) Schematic illustration of the one-pot co-precipitate synthesis method and the molecular framework of the Nanoceria Decorated Cerium (III)-Imidazole Network (NC/CIN) nano-pigment [41].

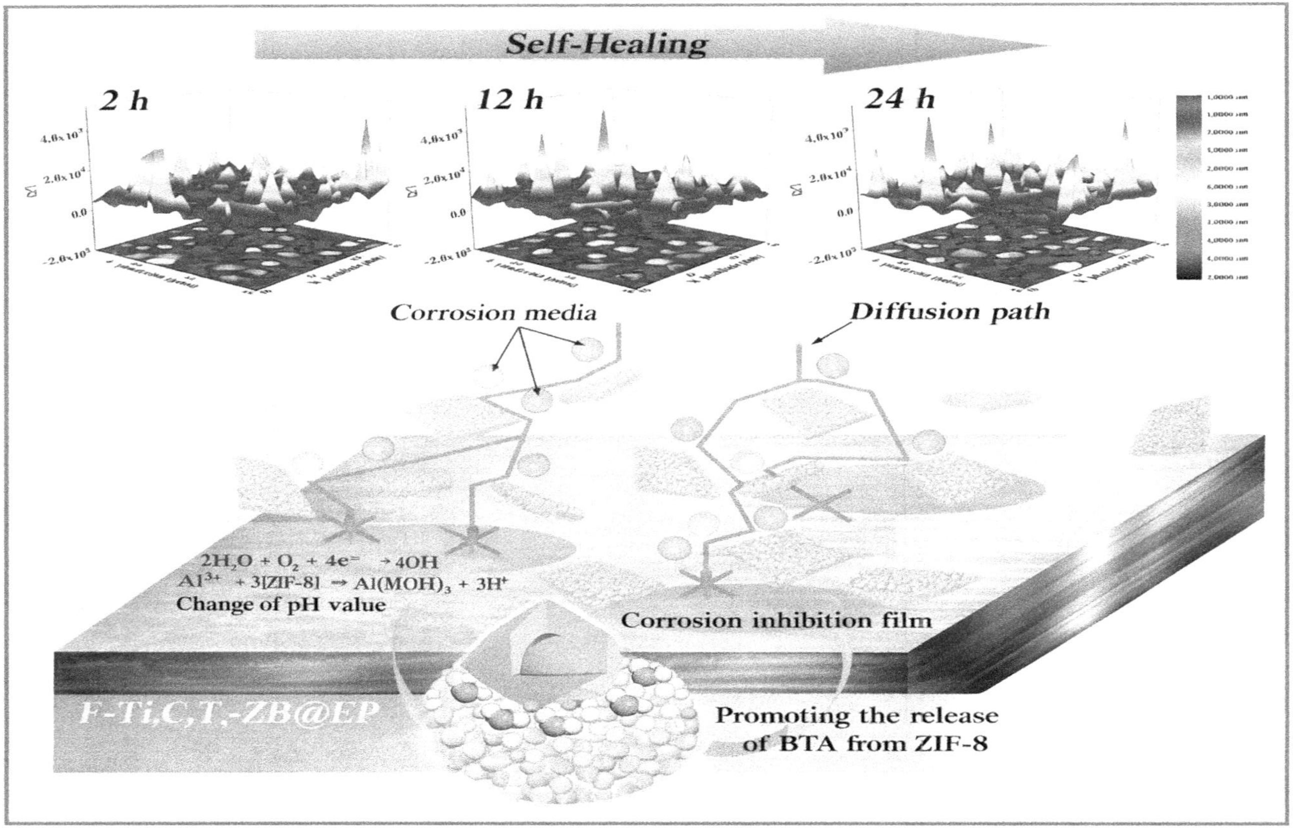

FIGURE 16.3 The f-$Ti_3C_2T_x$ZB@EP exhibited both passive and active safeguarding together with a self-treatment effect in corrosive locations.

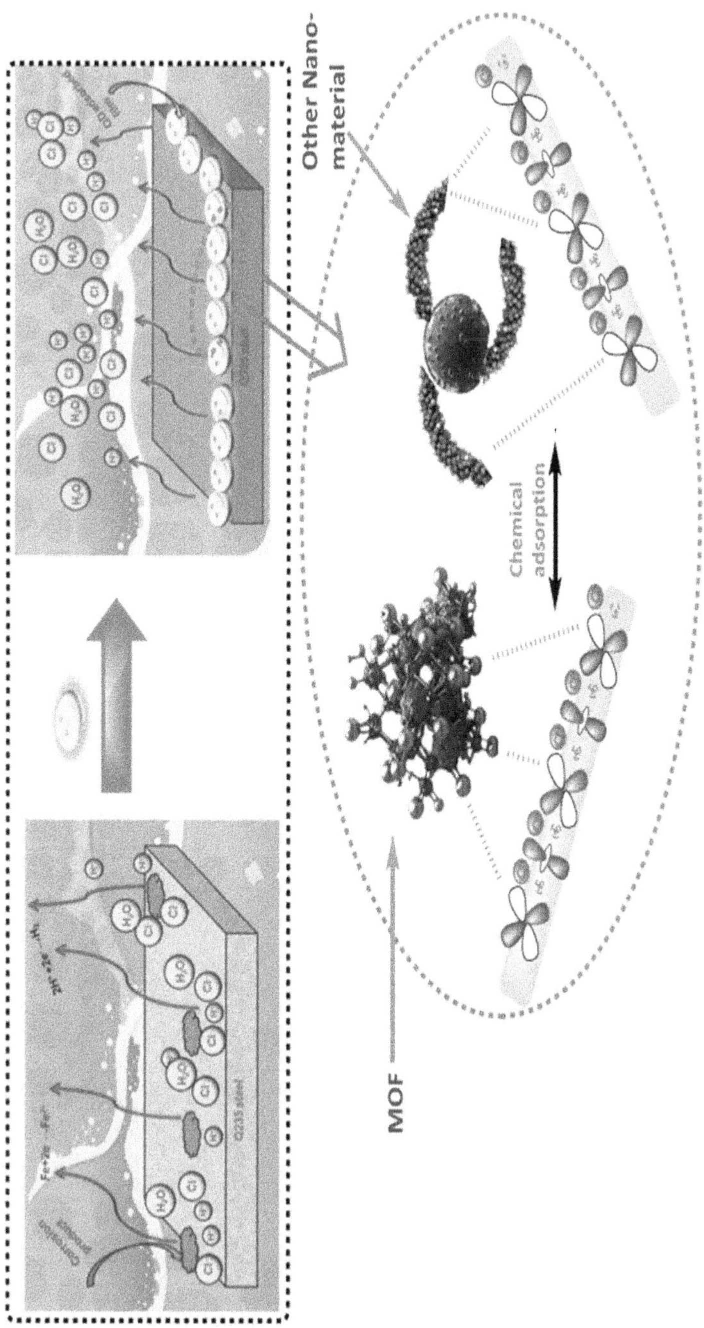

FIGURE 16.4 The absorption process of MOFs and other nanomaterials on metal surfaces in acidic environments.

16.5.4 Smart Coatings and Integration with IoT

Computer modeling is recognized as an essential instrument for clarifying the synthesis procedures, structural characteristics, and anticorrosion properties of certain MOFs. The incorporation of artificial intelligence (AI) in the design and evaluation of MOF-related corrosion mitigators is suggested to enhance the development process, as conventional testing methods tend to be laborious and inefficient.

16.5.5 Sustainability and Eco-Friendly Solutions

Many MOF-related products continue to be inappropriate for commercial application due to expensive production costs and low productivity fabrication processes, regardless of the advancement that has been achieved in laboratory environments. In light of this, subsequent efforts have to focus on developing production processes that are both cost-effective and environmentally benign, and that can be scaled up. There is a possibility that recent developments in atom-efficient chemical designs that reduce waste could result in the production of novel MOF materials that have exceptional capabilities in terms of corrosion resistance. With the help of such comprehensive strategy, the shift of MOFs from lab research to use in industry would be made easier.

REFERENCES

[1] Assad, H., et al., Assessment of the inhibitory efficacy of a thiazole derivative as an efficient corrosion inhibitor for augmenting the resistance of MS in acidic environments. *Langmuir*, 2024. 40(31): p. 16367–16383.

[2] Assad, H., et al., Electrochemical and computational insights into the utilization of 2, 2-dithio bisbenzothiazole as a sustainable corrosion inhibitor for mild steel in low pH medium. *Environmental Research*, 2024. 242: p. 117640.

[3] Assad, H., et al., Evaluating the adsorption and corrosion inhibition capabilities of Pyridinium-P-Toluene Sulphonate on MS in 1 M HCl medium: An experimental and theoretical study. *Inorganic Chemistry Communications*, 2023. 153: p. 110817.

[4] Assad, H., et al., An overview of contemporary developments and the application of graphene-based materials in anticorrosive coatings. *Environmental Science and Pollution Research*, 2023. 32: p. 1–20.

[5] Liu, C., et al., MOF-on-MOF hybrids: Synthesis and applications. *Coordination Chemistry Reviews*, 2021. 432: p. 213743.

[6] Bazer-Bachi, D., et al., Towards industrial use of metal-organic framework: Impact of shaping on the MOF properties. *Powder Technology*, 2014. 255: p. 52–59.

[7] Keshmiri, N., et al., Designing an eco-friendly lanthanide-based metal organic framework (MOF) assembled graphene-oxide with superior active anti-corrosion performance in epoxy composite. *Journal of Cleaner Production*, 2021. 319: p. 128732.

[8] Sørensen, P.A., et al., Anticorrosive coatings: A review. *Journal of Coatings Technology and Research*, 2009. 6: p. 135–176.

[9] Ulaeto, S.B., et al., Developments in smart anticorrosive coatings with multifunctional characteristics. *Progress in Organic Coatings*, 2017. 111: p. 294–314.

[10] Popoola, A., O. Olorunniwo, and O. Ige, Corrosion resistance through the application of anti-corrosion coatings, in Aliofkhazraei, M. (ed.) *Developments in Corrosion Protection*. 2014, IntechOpen.

[11] Croll, S.G., Application and limitations of current understanding to model failure modes in coatings. *Journal of Coatings Technology and Research*, 2013. 10: p. 15–27.
[12] Ramezanzadeh, M., et al., Development of metal-organic framework (MOF) decorated graphene oxide nanoplatforms for anti-corrosion epoxy coatings. *Carbon*, 2020. 161: p. 231–251.
[13] Assad, H., et al., A research combined experimental and computational approaches of succinylsulfathiazole hydrate as potent corrosion inhibitor for mild steel in acidic medium. *Journal of Molecular Liquids*, 2023. 388: p. 122739.
[14] Santos, J.F., et al., Central countries' and Brazil's contributions to nanotechnology. *Current Nanomaterials*, 2024. 9(2): p. 109–147.
[15] Assad, H., et al., Corrosion Inhibitors: Fundamental Concepts and Selection Metrics. *Corrosion Mitigation*, 2022. 14: p. 19–50.
[16] Assad, H., S. Kumar, and A. Kumar, Nitrogen, sulfur, and nitrogen-based heterocycles corrosion inhibitors, in Verma, C. (ed.) *Handbook of Heterocyclic Corrosion Inhibitors*. 2024, CRC Press. p. 198–211.
[17] Assad, H., A. Thakur, and A. Kumar, Chitosan and its derivatives as corrosion inhibitors, in Zehra, S., Mobin, M., Verma, C. (ed.) *Biopolymers in Sustainable Corrosion Inhibition*. 2024, CRC Press. p. 88–101.
[18] Majidi, R., et al., Investigation of synthesized planar Cu-MOF and spherical Ni-MOF nanofillers for improving the anti-corrosion performance of epoxy coatings. *Progress in Organic Coatings*, 2023. 183: p. 107803.
[19] Chen, H., et al., Construction of MOF-based superhydrophobic composite coating with excellent abrasion resistance and durability for self-cleaning, corrosion resistance, anti-icing, and loading-increasing research. *Chemical Engineering Journal*, 2021. 408: p. 127343.
[20] Rajkumar, R. and C. Vedhi, A study of corrosion protection efficiency of silica nanoparticles acrylic coated on mild steel electrode. *Vacuum*, 2019. 161: p. 1–4.
[21] Xing, X., et al., Preparation, release and anticorrosion behavior of a multi-corrosion inhibitors-halloysite nanocomposite. *Chemical Physics Letters*, 2019. 718: p. 69–73.
[22] Guo, H., et al., In-situ fabrication of polyelectrolyte-CSH superhydrophilic coatings via layer-by-layer assembly. *Chemical Engineering Journal*, 2014. 253: p. 198–206.
[23] Liu, B., et al., Facile synthesis of large scale and narrow particle size distribution polymer particles via control particle coagulation during one-step emulsion polymerization. *Colloids and Surfaces A: Physicochemical and Engineering Aspects*, 2015. 484: p. 81–88.
[24] Keerthana, A. and P.M. Ashraf, Carbon nanodots synthesized from chitosan and its application as a corrosion inhibitor in boat-building carbon steel BIS2062. *Applied Nanoscience*, 2020. 10(4): p. 1061–1071.
[25] Khamis, E.A., A. Hamdy, and R.E. Morsi, Magnetite nanoparticles/polyvinyl pyrrolidone stabilized system for corrosion inhibition of carbon steel. *Egyptian Journal of Petroleum*, 2018. 27(4): p. 919–926.
[26] Shirazi, Z., A.N. Golikand, and M.H. Keshavarz, A new nanocomposite based on poly (o-anthranilic acid), graphene oxide and functionalized carbon nanotube as an efficient corrosion inhibitor for stainless steel in severe environmental corrosion. *Composites Communications*, 2020. 22: p. 100467.
[27] Wahba, A.M. and M.M. Helal, Synthesis, characterization and application of two metal-organic frameworks (MOF) as a corrosion inhibitor for aluminium in a hydrochloric acid environment. *Case Studies in Chemical and Environmental Engineering*, 2024. 10: p. 100805.

[28] Kumaraguru, S., R. Pavulraj, and S. Mohan, Influence of cobalt, nickel and copper-based metal-organic frameworks on the corrosion protection of mild steel. *Transactions of the IMF*, 2017. 95(3): p. 131–136.
[29] Choudhary, P., et al., Metal–organic framework (MOF)/graphene–oxide (GO) nanocomposites materials: A potential formulation for anti-corrosive coatings-a review. *Materials Today: Proceedings*, 2023. 79: p. 172–178.
[30] Chen, L., et al., Preparation and anticorrosion properties of GO-Ce-MOF nanocomposite coatings. *Journal of Applied Polymer Science*, 2022. 139(5): p. 51571.
[31] Zhao, Q., W. Zhao, and Y. Wu, Constructing bi-functional Ce-MOF on carbon fiber endowing epoxy coating with excellent anti-corrosion and erosion wear resistance. *Carbon,* 2023. 214: p. 118374.
[32] Mao, C., et al., Fe-based MOFs@ Pd@ COFs with spatial confinement effect and electron transfer synergy of highly dispersed Pd nanoparticles for Suzuki-Miyaura coupling reaction. *Journal of Colloid and Interface Science*, 2022. 608: p. 809–819.
[33] Nwokolo, I.K., H. Shi, and F. Liu, MOF-based protective coatings for metal corrosion protection: A critical review of design/synthesis, performance, and mechanism. *Materials Science and Engineering: B*, 2025. 313: p. 117932.
[34] Chen, K., et al., Inhibition of microbiologically influenced corrosion and biofouling of X70 carbon steels by near-superhydrophobic D-cysteine/Ag@ ZIF-8 coatings. *Corrosion Science*, 2022. 208: p. 110682.
[35] Lashgari, S.M., et al., Application of nanoporous cobalt-based ZIF-67 metal-organic framework (MOF) for construction of an epoxy-composite coating with superior anti-corrosion properties. *Corrosion Science*, 2021. 178: p. 109099.
[36] Abdi, J., M. Izadi, and M. Bozorg, Improvement of anti-corrosion performance of an epoxy coating using hybrid UiO-66-NH2/carbon nanotubes nanocomposite. *Scientific Reports*, 2022. 12(1): p. 10660.
[37] Yang, C., et al., Research progress of metal organic framework materials in anti-corrosion coating. *Journal of Polymer Engineering*, 2024. 44(1): p. 1–12.
[38] Bijapur, K., et al., Recent trends and progress in corrosion inhibitors and electrochemical evaluation. *Applied Sciences*, 2023. 13(18): p. 10107.
[39] Li, J., et al., A review on metal-organic frameworks (MOFs) and MOF-textile composites for personal protection. *Materials Chemistry Frontiers*, 2024. 8: p. 3509–3527.
[40] Li, H., et al., 2-Mercaptobenzimidazole-inbuilt metal-organic-frameworks modified graphene oxide towards intelligent and excellent anti-corrosion coating. *Corrosion Science*, 2021. 191: p. 109715.
[41] Motamedi, M., et al., One-pot synthesis and construction of a high performance metal-organic structured nano pigment based on nanoceria decorated cerium (III)-imidazole network (NC/CIN) for effective epoxy composite coating anti-corrosion and thermo-mechanical properties improvement. *Chemical Engineering Journal*, 2020. 382: p. 122820.
[42] Yan, H., et al., Amino-functionalized Ti3C2Tx loading ZIF-8 nanocontainer@ benzotriazole as multifunctional composite filler towards self-healing epoxy coating. *Journal of Colloid and Interface Science*, 2021. 602: p. 131–145.

Index

Note: Page numbers in *italics* refer to figures, and those in **bold** refer to tables.

For Product Safety Concerns and Information please contact our EU
representative GPSR@taylorandfrancis.com
Taylor & Francis Verlag GmbH, Kaufingerstraße 24, 80331 München, Germany

www.ingramcontent.com/pod-product-compliance
Ingram Content Group UK Ltd.
Pitfield, Milton Keynes, MK11 3LW, UK
UKHW021513250926
475702UK00043B/351
* 9 7 8 1 0 3 2 9 3 6 2 4 6 *